NONLINEAR ILL-POSED PROBLEMS OF MONOTONE TYPE

Nonlinear Ill-posed Problems of Monotone Type

by

YAKOV ALBER

and

IRINA RYAZANTSEVA

 Springer

A C.I.P. Catalogue record for this book is available from the Library of Congress.

ISBN-10 1-4020-4395-3 (HB)
ISBN-13 978-1-4020-4395-6 (HB)
ISBN-10 1-4020-4396-1 (e-book)
ISBN-13 978-1-4020-4396-3 (e-book)

Published by Springer,
P.O. Box 17, 3300 AA Dordrecht, The Netherlands.

www.springer.com

Printed on acid-free paper

Printed in the Netherlands.

Contents

PREFACE

Many applied problems can be reduced to an operator equation of the first kind

$$Ax = f, \quad x \in X, \quad f \in Y, \tag{1}$$

where an operator A maps from a metric space X into a metric space Y. We are interested in those problems (1) which belong to the class of ill-posed problems. The concept of a well-posedness was introduced by J. Hadamard at the beginning of the 20th century [91].

Definition 1 *Let X and Y be metric spaces. The problem (1) is called well-posed if the following conditions are satisfied:*
(i) the problem is solvable in X for all $f \in Y$;
(ii) its solution $x \in X$ is unique;
(iii) $x \in X$ continuously depends on perturbations of the element $f \in Y$.

Problems that do not satisfy even one of the above requirements (i)-(iii) collectively form the class of ill-posed problems. Emphasize that solutions of ill-posed problems (if they exist) are unstable to small changes in the initial data. In connection with this, a common belief of many mathematicians in the past was that well-posedness is a necessary condition for the problems (1) to be mathematically or physically meaningful. This raised a debate about whether or not there is any need for methods of solving ill-posed problems. The tremendous development of science and technology of the last decades led, more often than not, to practical problems which are ill-posed by their nature. Solving such problems became a necessity and, thus, inventing methods for that purpose became a field of research in the intersection of theoretical mathematics with the applied sciences.

The fact is that, in practical computations, the date A and f of the problem (1), as a rule, are not precisely known. Therefore, it is important to study the continuous dependence of the approximate solutions on the intrinsic errors involved in the problem date. Without such knowledge, direct numerical resolution of (1) is impossible. At the same time, establishing the continuous dependence mentioned above is not an easy task. Attempts to avoid this difficulty led investigators to the new theory and conceptually new methods for stable solution of ill-posed problems. The expansion of computational technology contributed in large extent to the acceleration of this process.

Building the so-called regularization methods for solving ill-posed problems was initiated by A.N. Tikhonov [216]. He proved there the following important result:

Theorem 2 *A continuous one-to-one operator A which maps a compact subset $\mathcal{M}$ of the metric space X into metric space Y has the continuous inverse operator A^{-1} on the set $\mathcal{N} = A\mathcal{M} \subset Y$.*

On the basis of this theorem, M.M. Lavrent'ev introduced the concept of conditionally well-posed problems [125].

Definition 3 *The problem (1) is said to be conditionally well-posed if there exists nonempty set $X_1 \subseteq X$ such that*
(i) the equation (1) has a solution in X_1;
(ii) this solution is unique for each $f \in Y_1 = AX_1 \subset Y$;
(iii) the operator A^{-1} is continuous on Y_1.

The subset X_1 is called the well-posedness set of problem (1). According to Theorem 2 and Definition 3, we should impose conditions on the operator A in equation (1) such that they will define the compact set X_1. Even if those conditions are given, there are still difficulties that arise when we try to establish solvability of (1) with $f \in Y_1$. These difficulties can be overcome if X_1 is understood as the quasi-solution set of equation (1) with $f \in Y_1$ [97].

Definition 4 *An element $x \in X_1$ is called a quasi-solution of equation (1) if it minimizes the residual $\rho_Y(Ax, f)$ on the set X_1, where ρ_Y is a metric in the space Y.*

Problems for which it is not possible to construct the compact set X_1 of admissible solutions are said to be essentially ill-posed. In order to solve such problems, A.N. Tikhonov proposed the so-called regularization method [217]. To describe this method we establish the following definition assuming, for simplicity, that only the right-hand side f in equation (1) is given with some error δ.

Definition 5 *An operator $R(\alpha, f^\delta) : Y \to X$ is called a regularizing operator for equation (1) if it satisfies two requirements:*
(i) $R(\alpha, f^\delta)$ is defined for all $\alpha > 0$ and all $f^\delta \in Y$ such that $\rho_Y(f, f^\delta) \leq \delta$;
(ii) There exists a function $\alpha = \alpha(\delta)$ such that $R(\alpha(\delta), f^\delta) = x_\alpha^\delta \to x$ as $\delta \to 0$, where x is a solution of (1).

Operators $R(\alpha, f)$ lead to a variety of regularization methods. An element x_α^δ is called the regularized solution and α is called the regularization parameter. Thus, by Definition 5, any regularization method has to solve two main problems:
A) to show how to construct a regularizing operator $R(\alpha, f^\delta)$,
B) to show how the regularization parameter $\alpha = \alpha(\delta)$ should be chosen in order to ensure convergence of x_α^δ to some x as $\delta \to 0$.
The regularization method proposed in [217] is given in variational form (see also [43, 99, 130, 167, 218]). It produces a solution x_α^δ as a minimum point of the following smoothing functional:

$$\Phi_\delta^\alpha(x) = \rho_Y^2(Ax, f^\delta) + \alpha\phi(x), \tag{2}$$

where $\phi(x) : X \to R^1$ is a stabilizing functional which has the property that $\phi(z) \geq 0$ for all $z \in D(\phi) \subset X$, where $D(\phi)$ is a domain of functional ϕ. Moreover, a solution x of equation (1) needs to be an inclusion $x \in D(\phi)$.

Besides the regularization method described by the functional (2), two more variational methods for solving ill-posed problems are known [99]: the quasi-solution method given by the minimization problem

$$\|Ax^\delta - f^\delta\| = min\{\|Ax - f^\delta\| \mid x \in X_1\},$$

and the residual method in which approximate solutions x^δ for equation (1) are found by solving the minimization problem:

$$\|x^\delta\| = min \{\|x\| \mid x \in G^\delta\},$$

where

$$G^\delta = \{x \in D(A) \mid \|Ax - f^\delta\| \leq \sigma(\delta)\}.$$

Here $D(A)$ denotes a domain of operator A and the positive scalar function $\sigma(\delta)$ has the property that $\sigma(\delta) \to 0$ as $\delta \to 0$.

The increasing variety of variational regularization methods and growing interest in them are due to the following facts: (i) a priori information about (1) can be used in the process of their construction and (ii) there exists quite a wide choice of different means of solving optimization problems.

If the element x_α^δ is found by solving a parametric equation of the type

$$Ax + \alpha Bx = f^\delta, \tag{3}$$

where B is some stabilizing operator, then this procedure is an operator regularization method [54, 124, 125], and the equation (3) is a regularized operator equation.

The methods presented above were intensively studied in the case of linear equations. This is connected with the fact that for linear operators there are very powerful investigation tools contained in the spectral theory and optimization theory for quadratic functionals. In this direction many rather delicate characteristics and properties of the mentioned methods have been already obtained [42, 99, 125, 153, 217].

In spite of the fact that some nonlinear problems were solved by regularization methods a long time ago, the progress in this sense was slow. This seems to be due to researchers' wish to cover as wide classes of nonlinear problems as possible. It is clear that the spectral theory can not be applied directly to nonlinear operators. Moreover, if A is a nonlinear operator, the functionals $\rho_Y(Ax, f^\delta)$ and $\Phi_\delta^\alpha(x)$ are nonconvex, in general. In addition, the question of convexity for the set G^δ remains still open, that produces enormous difficulties in solving variational problems.

At the same time, the separate researches into nonlinear problems of the monotone type, like operator equations and variational inequalities with monotone, maximal monotone, semimonotone, pseudomonotone and accretive operators, turned out to be much more successful. From the applied point of view these classes are extremely wide (it is sufficient to recall that gradients and subgradients of convex functionals are monotone and maximal monotone operators, respectively, and nonexpansive mappings are accretive.)

A considerable part of this book is devoted to study of nonlinear ill-posed problems with operators in these classes. We use the fundamental results in the theory of monotone and accretive operators discovered during the latter half of the 20th century by R.I. Kachurovskii [102, 103], M.M. Vainberg [220, 221], F.E. Browder [52] - [58], G. Minty [149], J.-L. Lions [128], H. Brezis [49], R. Rockafellar [175] - [181], T. Kato [106, 109] and many others.

From the very beginning, the concept of ill-posedness was included in the construction and theoretical foundation of approximative methods for solving nonlinear problems with properly monotone and properly accretive operators, and according to this concept, it is necessary:
1) to assume that a solution of the original problem exists;
2) to construct a sequence of regularized solutions;
3) to determine whether this sequence converges to a solution of problem (1) and the convergence is stable when the initial data are perturbed.

Let us briefly describe the contents of the book. It consists of seven chapters that comprise fifty-seven sections. The numeration of sections is in the form X.Y, where X indicates the number of a chapter and Y denotes the current number of the section within the given chapter. The numeration of definitions, theorems, lemmas, corollaries, remarks and propositions is in the form X.Y.Z, where X and Y are as above and Z indicates a number of the statement in Section X.Y. Formulas are numbered analogously.

Chapter 1 contains the basic knowledge from convex and functional analysis, optimization, Banach space geometry and theory of monotone and accretive operators which allows the study of the proposed material without turning, generally, to other sources. Equations and variational inequalities with maximal monotone operators are main subjects of our investigations due to their most important applications. Recall that maximal monotone operators are set-valued, in general. In spite of this, the properties of maximal monotone operators, solvability criteria for equations and variational inequalities with such operators and also structures of their solution sets have been studied fully enough. However, it is not always easy to establish the maximal monotonicity of a map. Besides that, the numerical realization of maximal monotone operator means its change by single-valued sections when the property of maximal monotonicity is lost. Finally, the monotone operator of the original problem is not always maximal. Therefore, in this chapter we devote significant time to equations and variational inequalities with semimonotone, pseudomonotone and general monotone mappings. We describe in detail the known properties of the most important canonical monotone operators, so-called, normalized duality mapping J and duality mapping with gauge function J^μ. We also present new properties of these mappings expressed by means of geometric characteristics of a Banach space, its modulus of convexity and modulus of smoothness. We introduce the new Lyapunov functionals and show that they have similar properties. In the theory of monotone operators this approach is not traditional and it is first stated in monographic literature. The two last sections of the chapter are dedicated to equations with multiple-valued accretive and $d-$accretive operators.

Chapter 2 is devoted to the operator regularization method for monotone equations (that is, for equations with monotone operators). In order to preserve the monotonicity property in a regularized problem, we add to the original operator some one-parameter family of monotone operators. Thus, the resulting operators are monotone again. As a stabilizing operator, we use a duality mapping which has many remarkable properties. In the case of a Hilbert space, where the duality mapping is the identical operator, this regularization method for linear equations was first studied by M.M. Lavrent'ev [125]. In this chapter, we explore the operator regularization methods for monotone, semimonotone and accretive nonlinear equations in Hilbert and Banach spaces. The equations with single-valued and multiple-valued maximal monotone mappings are studied. Non-monotone approximations of operators and their domain perturbations is also discussed.

Observe that the establishment of boundedness of a sequence of regularized solutions of operator equations is one of the central points of the proof of convergence of any regularization method. It allows us to use the weak compactness principle for bounded sets in reflexive Banach spaces and to construct a subsequence weakly convergent to a solution of a given problem. Further progress appears only when we are able to prove that any weak accumulation point is unique and the weak convergence is in reality a strong convergence (possibly, under some additional conditions). In general, we do not redenote the said subsequence unless not doing so would cause confusion.

In Chapter 3 the problem of choosing regularization parameters is solved. We consider the residual principles − the ways of choosing the regularization parameter α by means of the equations that connect the residual of the equation (1) with error levels of the right-hand side f and operator A. In the case of an arbitrary monotone operator, the residual is understood in the generalized sense. The smoothing functional principle deals with minimization problems like (2) involving potential monotone operators. The regularization parameter α is found by the equation which connects minimal values $m_\delta(\alpha)$ of the functional $\Phi_\delta^\alpha(x)$ on X with the errors of the problem data. In this chapter we also investigate the minimal and modified residual principles.

In Chapter 4, we present regularization methods for solving monotone variational inequalities (that is, for solving variational inequalities with monotone operators) on convex closed subsets of Hilbert and Banach spaces. Variational inequalities with bounded and unbounded operators are considered. In contrast to operator equations, the problems become more complicated due to the fact that additional errors of constraint sets are possible here. At that, the proximity between original and perturbed sets is described both by the Hausdorff and Mosco criteria. We also study variational inequalities with pseudomonotone operators. Convergence of the regularization methods for variational inequalities involving non-monotone and hypomonotone approximations of operators is each investigated separately. In the last section of the chapter we discuss the possibility of stable solutions to mixed variational inequalities.

In Chapter 5 the operator regularization method is applied to solve the classical problem of calculating values of unbounded operators. Monotone, semimonotone and accretive mappings are considered. Observe that the problem of solving the equation (1) is equivalent to the problem of calculating the inverse operator on a given element f. Nevertheless, each such problem has its specific features, so that their separate investigation is expedient and useful. If the equation is unsolvable, then those elements which minimize its residual (they are called pseudo-solutions) acquire practical importance. We present in this chapter stable algorithms for finding pseudo-solutions of equation (1) with nonlinear monotone mappings based on the operator regularization method. As applications we also consider Hammerstein type equations, minimization problems, optimal control problems and fixed point problems.

The first two sections of Chapter 6 are dedicated to the variational methods of quasi-solutions and the residual for monotone problems. We establish the equivalence of these methods and also their connection with the operator regularization method. Note that the quasi-solution method and the residual method in this work are constructed in a non-traditional manner as compared with the linear case [99], though the same information about the initial problem is used. The monotonicity properties of mappings play an essential role in this approach. We proposed and studied the regularized penalty method for monotone variational inequalities. In this chapter, sufficient convergence conditions of the iterative methods have been obtained. We investigated the proximal point method, iterative regularization method, iterative-projection regularization method and Newton−Kantorovich regularization method in uniformly convex and uniformly smooth Banach spaces. Finally, we established the sufficient conditions for convergence of the continuous regularization method which is reduced to the Cauchy problem for a differential equation of the first order. All this variety of methods gives very wide possibilities to obtain strong approximations to a solution of nonlinear ill-posed problem.

In Chapter 7 we consider recurrent numerical and differential inequalities.

ACKNOWLEDGMENTS

We wish to express our deepest appreciation to Professors Dan Butnariu, Sylvie Guerre-Delabriere, Michael Drakhlin, Rafael Espínola, Alfredo Iusem, Athanassios Kartsatos, Zuhair Nashed, Mikhail Lavrent'ev, Simeon Reich, Terry Rockafellar and Jen-Chih Yao for repeatedly giving us the opportunity to discuss various aspects of this book and for their personal support.

We would like to mention the following institutions whose support made this work possible: The Technion - Israel Institute of Technology, the University of Haifa (Israel), the Nizhnii Novgorod State Technical University (Russia), the Institute of Pure and Applied Mathematics, Rio de Janeiro (Brazil), the Abdus Salam International Centre for Theoretical Physics, Trieste (Italy), the University Paris VI (France), the University of Sevilia (Spain), and the National Sun Yat-Sen University of Kaohsiung (Taiwan).

The authors are very grateful to Dr. Elisabeth Mol and Mrs. Marlies Vlot from Springer Verlag Publishers who helped us with thoughtful arrangements and patience.

Yakov Alber and Irina Ryazantseva

Chapter 1

INTRODUCTION INTO THE THEORY OF MONOTONE AND ACCRETIVE OPERATORS

1.1 Elements of Nonlinear Functional Analysis

Let X be a real linear normed space, $\|x\|$ be a norm of an element x in X, θ_X be an origin of X. Strong convergence $x_n \to x$, $n = 0, 1, ...$, of the sequence $\{x_n\} \subset X$ to $x \in X$ means that $\|x_n - x\| \to 0$ as $n \to \infty$. In this case, x is a (strong) limit point of the sequence $\{x_n\}$. If $\{x_n\}$ converges strongly to $x \in X$ then 1) any subsequence $\{x_{n_k}\} \subset \{x_n\}$ also converges to the same point, 2) the sequence $\{\|x_n - \xi\|\}$ is bounded for any $\xi \in X$.

A sequence $\{x_n\} \subset X$ is called a fundamental or Cauchy sequence, if for every $\epsilon > 0$ there is $n_0(\epsilon)$ such that $\|x_m - x_n\| < \epsilon$ for any $m \geq n_0(\epsilon)$ and $n \geq n_0(\epsilon)$. If a sequence $\{x_n\} \subset X$ converges to a limit then it is fundamental. The inverse assertion is not always true: there are examples of normed spaces which have non-convergent Cauchy sequences. We say that X is complete if every Cauchy sequence of the normed space X has a limit $x \in X$.

Let X and Y be arbitrary spaces. The expression $A : X \to Y$ means that operator A is single-valued and maps X into Y. If A is multiple-valued then we write $A : X \to 2^Y$. We further denote by

$$D(A) = \ dom \ A = \{x \in X \mid Ax \neq \emptyset\} \qquad (1.1.1)$$

the domain of the operator A, and by

$$R(A) = \{\phi \in Y \mid \phi \in Ax, \ x \in D(A)\} \qquad (1.1.2)$$

the range of A. For more details, an operator $A : X \to Y$ with $D(A) \subset X$ and $R(A) \subset Y$ is a one-to-one correspondence between sets of X and Y, which carries point $x \in D(A)$ to the point $Ax \in R(A)$. A map $A : X \to 2^Y$, which is not necessarily a one-to-one correspondence, carries some elements $x \in D(A)$ to sets $Ax \subset R(A)$.

1

Let R^1 be a set of all real numbers. An operator $\varphi : X \to R^1$ is called the functional on linear space X. By analogy with (1.1.1), a domain of the functional φ is defined as

$$dom\ \varphi = \{x \in X \mid \varphi(x) \neq \emptyset\}.$$

A functional φ is linear if 1) $\varphi(x_1 + x_2) = \varphi(x_1) + \varphi(x_2)$ and 2) $\varphi(x_n) \to \varphi(x)$ as $x_n \to x$. It is bounded if there exists a constant $M > 0$ such that $|\varphi(x)| \leq M\|x\|$. The smallest constant M satisfying this inequality is said to be the norm of the linear functional φ and it is denoted by $|\varphi|$.

Many important facts of Functional Analysis are deduced from the following theorem and its corollary.

Theorem 1.1.1 (Hahn−Banach) *Any linear functional φ defined on a subspace $\mathcal{M}$ of the normed linear space X can be extended onto the whole space with preservation of the norm; that is, there exists a linear functional F, $x \in X$, such that the equalities $F(x) = \varphi(x)$ for all $x \in \mathcal{M}$ and $|\varphi|_{\mathcal{M}} = |F|_X$ hold.*

Corollary 1.1.2 *Let X be a normed linear space and $x_0 \in X$ ($x_0 \neq \theta_X$) be an arbitrary fixed element. Then there exists a linear functional $\varphi : X \to R^1$ such that: $|\varphi| = 1$ and $\varphi(x_0) = \|x_0\|$.*

In what follows, X will denote a Banach space, i.e., a complete linear normed space. The dual space X^* of X is the set of all linear continuous functionals on X. It is known that X^* is also a Banach space. We denote by $\|\phi\|_*$ the norm of an element $\phi \in X^*$ and by $\langle \phi, x \rangle$ the dual product (dual pairing) of elements $\phi \in X^*$ and $x \in X$, that is, $\langle \phi, x \rangle$ is the value of the linear functional ϕ on the element x. One can show that $\|\phi\|_* = \sup\{\langle \phi, x \rangle \mid \|x\| = 1\}$. The Cauchy−Schwarz inequality states that $|\langle \phi, x \rangle| \leq \|\phi\|_*\|x\|$ for all $x \in X$ and for all $\phi \in X^*$.

We say that sequence $\{x_n\} \subset X$ converges weakly to $x \in X$ (and write $x_n \rightharpoonup x$) if $\langle \phi, x_n \rangle \to \langle \phi, x \rangle$ for every $\phi \in X^*$. A weak limit point of any subsequence of the sequence $\{x_n\}$ is called a weak accumulation point. If all weak accumulation points of the sequence $\{x_n\}$ coincide, i.e., if all weakly convergent subsequences have the same limit x, then x is the weak limit of $\{x_n\}$. If the sequence $\{x_n\}$ weakly converges to $x \in X$ then $\{\|x_n\|\}$ is bounded. The weak convergence of the sequence $\{x_n\}$ to x always follows from its strong convergence to the same point x. The converse is not true in general. However, in finite-dimensional spaces strong convergence and weak convergence are equivalent.

It is also possible to construct the dual space X^{**} for X^*. In general, $X \subset X^{**}$. If $X^{**} = X$ then space X is said to be reflexive. Note that, in this case, X^* is also a reflexive space. In reflexive spaces, the weak convergence of $\phi_n \in X^*$ to $\phi \in X^*$ means that for every $x \in X$ the limit $\langle \phi_n, x \rangle \to \langle \phi, x \rangle$ holds as $n \to \infty$.

We present the following important statements.

Theorem 1.1.3 (Banach−Steinhaus) *Let X be a Banach space, $\phi_n \in X^*$, and suppose that the sequence $\{\langle \phi_n, x \rangle\}$ is bounded for every $x \in X$. Then the sequence $\{\phi_n\}$ is bounded in X^*.*

Theorem 1.1.4 *Suppose that either $\{\phi_n\} \subset X^*$ strongly converges to $\phi \in X^*$ and $\{x_n\} \subset X$ weakly converges to $x \in X$ or $\{\phi_n\} \subset X^*$ weakly converges to $\phi \in X^*$ and $\{x_n\} \subset X$ strongly converges to $x \in X$. Then $\lim_{n\to\infty}\langle\phi_n, x_n\rangle = \langle\phi, x\rangle$.*

Definition 1.1.5 [161] *A Banach space X satisfies Opial's condition if for each sequence $\{x_n\}$ in X, the limit relation $x_n \rightharpoonup x$ implies*

$$\liminf_{n\to\infty} \|x_n - x\| < \liminf_{n\to\infty} \|x_n - y\|$$

for all $y \in X$ with $x \neq y$.

A set $B(a, r) \subset X$ such that $\|x - a\| \leq r$ is called the ball (or closed ball) with center a and radius r. A set $B_0(a, r) \subset X$ such that $\|x - a\| < r$ is called the open ball with center a and radius r. A set $S(a, r) \subset X$ such that $\|x - a\| = r$ is called the sphere with center a and radius r. The sphere and ball are called a unit if $r = 1$. We shall also use the denotations $B^*(\psi, r)$, $B_0^*(a, r)$ and $S^*(a, r)$, respectively, for closed ball, open ball and sphere in a dual space X^*.

A set Ω is bounded if it wholly lies inside of some ball. A set Ω is closed (weakly closed) if the conditions $x_n \to x$ ($x_n \rightharpoonup x$), where $x_n \in \Omega$ for all $n \geq 0$, imply the inclusion $x \in \Omega$.

For any $\psi \in X^*$, $\psi \neq \theta_{X^*}$, and for any $c \in R^1$, the set $H_c = \{x \in X \mid \langle\psi, x\rangle = c\}$ is called a closed hyperplane.

Definition 1.1.6 *Let $\Omega \subset X$ be a bounded set. Then the value*

$$diam\,\Omega = sup\,\{\|x - y\| \mid x, y \in \Omega\}$$

is a diameter of Ω.

We denote by $\partial\Omega$ the boundary of the set Ω, by $int\,\Omega$ the totality of its interior points and by $\overline{\Omega}$ the closure of the set Ω, that is, minimal closed set containing Ω.

Definition 1.1.7 *A set $\Omega \subseteq X$ is called*
1) convex if together with the points $x, y \in \Omega$, the whole segment $[x, y] = \lambda x + (1 - \lambda)y$, $0 \leq \lambda \leq 1$, also belongs to Ω;
2) compact if any infinite sequence of this set contains a convergent subsequence;
3) weakly compact if any infinite sequence of this set contains a weakly convergent subsequence;
4) dense in a set $M \subseteq X$ if $M \subseteq \overline{\Omega}$;
5) everywhere dense in X if $\overline{\Omega} = X$.

A closed bounded set Ω of the reflexive space is weakly compact. Hence, from any bounded sequence belonging to Ω, one can choose a subsequence which weakly converges to some element of this space.

Theorem 1.1.8 (Mazur) *Any closed convex set of a Banach space is weakly closed.*

Theorem 1.1.9 (Riesz) *A Banach space is finite-dimensional if and only if each of its bounded closed subsets is compact.*

The next assertion follows from the Hahn−Banach theorem and is known as the strong separation theorem.

Theorem 1.1.10 *Let $\Omega_1 \subset X$ and $\Omega_2 \subset X$ be two convex sets, nonempty and disjoint. Suppose that Ω_1 is closed and Ω_2 is compact. Then there exists a closed hyperplane that strongly separates Ω_1 and Ω_2.*

Definition 1.1.11 *A functional φ is called convex in dom φ if the inequality*

$$\varphi(tx + (1-t)y) \leq t\varphi(x) + (1-t)\varphi(y) \tag{1.1.3}$$

is satisfied for all $x, y \in dom\ \varphi$ and all $t \in [0,1]$. If the equality in (1.1.3) occurs only under the condition that $x = y$, then the functional φ is called strictly convex. If there exists a continuous increasing function $\gamma : [0,\infty) \rightarrow R^1$, $\gamma(0) = 0$, such that

$$\varphi(tx + (1-t)y) \leq t\varphi(x) + (1-t)\varphi(y) - t(1-t)\gamma(\|x - y\|)$$

for all $x, y \in dom\ \varphi$, then φ is called uniformly convex. The function $\gamma(t)$ is called a modulus of convexity of φ. If $\gamma(t) = ct^2$, $c > 0$, then the functional φ is strongly convex.

Definition 1.1.12 *A functional φ is called lower semicontinuous at the point $x_0 \in dom\ \varphi$ if for any sequence $x_n \in dom\ \varphi$ such that $x_n \rightarrow x_0$ there holds the inequality*

$$\varphi(x_0) \leq \liminf_{n\to\infty} \varphi(x_n). \tag{1.1.4}$$

If the inequality (1.1.4) occurs with the condition that the convergence of $\{x_n\}$ to x_0 is weak, then the functional φ is called weakly lower semicontinuous at x_0.

Theorem 1.1.13 *Let $\varphi : X \rightarrow R^1$ be a convex and lower semicontinuous functional. Then it is weakly lower semicontinuous.*

We present the generalized Weierstrass theorem.

Theorem 1.1.14 *Assume that a weakly lower semicontinuous functional φ is given on a bounded weakly closed set Ω of a reflexive Banach space X. Then it is bounded from below and reaches its greatest lower bound on Ω.*

In any normed linear space, the norm can be presented as an example of a weakly lower semicontinuous functional. Indeed, let $x_n \rightharpoonup \bar{x}$, i.e., $\langle \psi, x_n \rangle \rightarrow \langle \psi, \bar{x} \rangle$ for any $\psi \in X^*$. Choose $\psi = \bar{\psi}$ such that $\|\bar{\psi}\|_* = 1$ and $\langle \bar{\psi}, \bar{x} \rangle = \|\bar{x}\|$. This element exists by Corollary 1.1.2. Then $\langle \bar{\psi}, x_n \rangle \rightarrow \|\bar{x}\|$, that is,

$$\|\bar{x}\| = \lim_{n\to\infty} \langle \bar{\psi}, x_n \rangle \leq \liminf_{n\to\infty} \|x_n\|.$$

Geometric characteristics of Banach spaces such that weak and strong differentiability of the norm, convexity and smoothness of spaces, duality mappings and projection operators will play very important roles in this book.

Definition 1.1.15 *A functional $\varphi : X \to R^1$ is called*
1) trivial if $\varphi(x) = +\infty$ for all $x \in X$;
2) proper if $\mathcal{M}(\varphi) = \{x \mid \varphi(x) \neq +\infty\} \neq \emptyset$;
3) finite if $|\varphi(x)| < \infty$ for all $x \in X$.

In this definition, $\mathcal{M}(\varphi)$ is called the effective set of the functional φ. If φ is convex then $\mathcal{M}(\varphi)$ is also convex.

Theorem 1.1.16 *Any finite convex functional $\varphi : X \to R^1$ given on an open set of X is weakly lower semicontinuous.*

Definition 1.1.17 *Let a functional $\varphi : X \to R^1$. We say that φ is directionally differentiable at a point $x \in X$ if the limit*

$$V'(x, h) = \lim_{t \to 0} \frac{\varphi(x + th) - \varphi(x)}{t} \qquad (1.1.5)$$

exists for all $h \in X$.

Similarly, the directional differentiability can be defined when a functional φ is given on an open set U in X. In this case, in (1.1.5) $x \in U$ and $x + th \in U$.

Definition 1.1.18 *If the limit in (1.1.5) is linear continuous (with respect to h) operator, i.e., $V'(x, h) = \varphi'(x)h$, then φ is called Gâteaux differentiable (or weakly differentiable) at a point $x \in X$ and $V'(x, h)$ and $\varphi'(x)$ are called, respectively, Gâteaux differential (or weak differential) and Gâteaux derivative (or weak derivative) of the functional φ at a point x.*

By Definition 1.1.18, there exists operator $A : X \to X^*$ such that

$$V'(x, h) = \langle Ax, h \rangle$$

for any $x, h \in X$. The operator A is called the gradient of a functional $\varphi(x)$ and is denoted by $grad\ \varphi$ or φ'.

Operator A is called a potential if there exists a functional φ such that $A = grad\ \varphi$. In this case, φ is the potential of A.

Theorem 1.1.19 *If $\varphi(x)$ is a convex functional in dom φ, then its gradient $\varphi'(x)$ satisfies the inequality*

$$\langle \varphi'(x) - \varphi'(y), x - y \rangle \geq 0 \qquad (1.1.6)$$

for all $x, y \in dom\ \varphi$. If $\varphi(x)$ is uniformly convex then

$$\langle \varphi'(x) - \varphi'(y), x - y \rangle \geq 2\gamma(\|x - y\|). \qquad (1.1.7)$$

In particular, if $\varphi(x)$ is strongly convex then

$$\langle \varphi'(x) - \varphi'(y), x - y \rangle \geq 2c\|x - y\|^2. \qquad (1.1.8)$$

Theorem 1.1.20 *If $\varphi(x)$ is a convex functional in dom φ, then its gradient $\varphi'(x)$ satisfies the inequality*

$$\langle \varphi'(x), x - y \rangle \geq \varphi(x) - \varphi(y) \tag{1.1.9}$$

for all $x, y \in$ dom φ. If $\varphi(x)$ is uniformly convex then

$$\langle \varphi'(x), x - y \rangle \geq \varphi(x) - \varphi(y) + \gamma(\|x - y\|). \tag{1.1.10}$$

In particular, if $\varphi(x)$ is strongly convex then

$$\langle \varphi'(x), x - y \rangle \geq \varphi(x) - \varphi(y) + c\|x - y\|^2. \tag{1.1.11}$$

Theorem 1.1.21 *Let a functional $\varphi : X \to R^1$ be convex and differentiable at a point x_0. Then x_0 is the extremum point of φ if and only if grad $\varphi(x_0) = 0$.*

Remark 1.1.22 *By tradition, we shall denote "0" as a null on the right-hand side of any operator equation.*

The sufficient existence conditions of the functional minimum point is given by the following theorem.

Theorem 1.1.23 *Let X be a reflexive Banach space, $\varphi : X \to R^1$ be a convex lower semicontinuous proper functional and $\varphi(x) \to \infty$ for $\|x\| \to \infty$. Then there exists a point $x_0 \in$ dom φ such that $\varphi(x_0) = min\{\varphi(x) \mid x \in$ dom $\varphi\}$. The minimum point x_0 is unique if the functional φ is strictly convex.*

Theorem 1.1.24 *Let X be a reflexive Banach space, $\varphi : X \to R^1$ be a uniformly convex lower semicontinuous functional on a convex closed set $\Omega \subseteq X$. Then there exists a unique minimum point x^* of φ at Ω and inequality*

$$\gamma(\|x - x^*\|) \leq \varphi(x) - \varphi(x^*) \quad \forall x \in \Omega$$

holds. Here $\gamma(t)$ is the modulus of convexity of φ at Ω.

Definition 1.1.25 *A proper functional $\varphi : X \to R^1$ is said to be Fréchet differentiable (or strongly differentiable) at a point $x \in D$, where D is an open set in X, if there is a linear operator $F : X \to X^*$ such that, for any $x + h \in D$,*

$$\varphi(x + h) = \varphi(x) + \langle F(x), h \rangle + \omega(x, h)$$

and

$$\lim_{\|h\| \to 0} \frac{\omega(x, h)}{\|h\|} = 0.$$

The quantity $\langle F(x), h \rangle$ is called the Fréchet differential (or strong differential) and $F(x) = \varphi'(x)$ is called the Fréchet derivative (or strong derivative) of the functional φ at a point x.

In other word, if the limit in (1.1.5) exists uniformly for h on the unit sphere of X, then φ is Fréchet differentiable and $\varphi'(x)$ is the Fréchet derivative of φ at x. If a functional φ is Fréchet differentiable at a point $x_0 \in X$, then it is Gâteaux differentiable at that point. The inverse assertion is not true in general. However, if Gâteaux derivative φ' is continuous at a neighborhood of a point $x_0 \in X$, then it is, in fact, the Fréchet derivative at x_0.

Definition 1.1.26 *The space X is smooth if its norm is Gâteaux differentiable. The space X is strongly smooth if its norm is Fréchet differentiable over θ_X.*

Next we introduce one of the most important canonical operators in a Banach space.

Definition 1.1.27 *Let X be an arbitrary Banach space. The operator $J : X \to 2^{X^*}$ is called a normalized duality mapping in X if the following equalities are satisfied:*

$$\langle \zeta, x \rangle = \|\zeta\|_* \|x\| = \|x\|^2 \quad \forall \zeta \in Jx, \quad \forall x \in X.$$

It immediately follows from Definition 1.1.27 that J is a homogeneous and odd operator. This means that, respectively, $J(\lambda x) = \lambda Jx$ for $\lambda \geq 0$ and $J(-x) = -Jx$.

In general, the normalized duality mapping is a multiple-valued operator. However, it is single-valued in a smooth Banach space and then

$$Jx = 2^{-1} grad \, \|x\|^2. \tag{1.1.12}$$

Definition 1.1.28 *A space X is called strictly convex if the unit sphere in X is strictly convex, that is, the inequality $\|x + y\| < 2$ holds for all $x, y \in X$ such that $\|x\| = \|y\| = 1$, $x \neq y$.*

Definition 1.1.29 *Fix a point $x \in X$ and a number $\epsilon > 0$ and then define the function*

$$\delta_X(x, \epsilon) = inf \left\{ 1 - \frac{\|x - y\|}{2} \mid \|x\| = \|y\| = 1, \|x - y\| = \epsilon \right\}.$$

It is said to be the modulus of the local convexity of space X at a point x. If for any x with $\|x\| = 1$, $\delta_X(x, \epsilon) > 0$ for $\epsilon > 0$, then the space X is called locally uniformly convex.

Any locally uniformly convex space is strictly convex. In any reflexive Banach space, it is possible to introduce an equivalent norm such that the space will be locally uniformly convex with respect to this norm.

If the function $\delta_X(x, \epsilon)$ does not depend on x such that in Definition 1.1.29 $\delta_X(x, \epsilon) = \delta_X(\epsilon)$, then $\delta_X(\epsilon)$ is the modulus of convexity of X.

Definition 1.1.30 *A Banach space X is called uniformly convex if for any given $\epsilon > 0$ there exists $\delta > 0$ such that for all x, $y \in X$ with $\|x\| \leq 1$, $\|y\| \leq 1$, $\|x - y\| = \epsilon$ the inequality*

$$\|x + y\| \leq 2(1 - \delta)$$

holds.

The function $\delta_X(\epsilon)$ is defined on the interval $[0, 2]$, continuous and increasing on this interval, $\delta_X(0) = 0$ and $\delta_X(\epsilon) \leq 1$. A Banach space X is uniformly convex if and only if $\delta_X(\epsilon) > 0$ for $\epsilon > 0$. Observe that the function

$$g_X(\epsilon) = \frac{\delta(\epsilon)}{\epsilon} \tag{1.1.13}$$

will play a very important role in our researches. It is known that $g_X(\epsilon)$ is a continuous and non-decreasing function on the interval $[0, 2]$, and $g_X(0) = 0$.

Definition 1.1.31 *The function $\rho_X(\tau)$ defined by the formula*

$$\rho_X(\tau) = sup\left\{ \frac{\|x + y\|}{2} + \frac{\|x - y\|}{2} - 1 \;\Big|\; \|x\| = 1,\ \|y\| = \tau \right\}$$

is the modulus of smoothness of the space X.

Definition 1.1.32 *A Banach space X is called uniformly smooth if for any given $\epsilon > 0$ there exists $\delta > 0$ such that for all x, $y \in X$ with $\|x\| = 1$, $\|y\| \leq \delta$ the inequality*

$$2^{-1}(\|x + y\| + \|x - y\|) - 1 \leq \epsilon\|y\|$$

holds.

Note that $\rho_X(\tau)$ is defined on the interval $[0, \infty)$, convex, continuous and increasing on this interval, $\rho_X(0) = 0$. In addition, for any X the function $\rho_X(\tau) \leq \tau$ for all $\tau \geq 0$. A Banach space X is uniformly smooth if and only if

$$\lim_{\tau \to 0} \frac{\rho_X(\tau)}{\tau} = 0.$$

Any uniformly convex and any uniformly smooth Banach space is reflexive. A space X is uniformly smooth if and only if X^* is uniformly convex. A space X is uniformly convex if and only if X^* is uniformly smooth. A reflexive Banach space X is smooth if and only if X^* is strictly convex. A reflexive Banach space X is strictly convex if and only if X^* is smooth.

Definition 1.1.33 *The norm in X is uniformly Fréchet differentiable if*

$$\lim_{t \to 0} \frac{\|x + th\| - \|x\|}{t}$$

exixts uniformly for x and h in the unit sphere of X.

Theorem 1.1.34 (Klee–Šmulian) *A Banach space X is uniformly smooth if and only if the norm in X is uniformly Fréchet differentiable.*

Next we present several examples of uniformly convex and uniformly smooth Banach spaces.

1. A complete linear space X is called a real Hilbert space H if to every pair of elements $x, y \in X$ there is associated a real number called their scalar (inner) product and denoted by (x, y), in such a way that the following rules are satisfied:
1) $(x, y) = (y, x)$;
2) $(x + z, y) = (x, y) + (z, y)$;
3) $(\lambda x, y) = \lambda(x, y)$ for all $\lambda \in R^1$;
4) $(x, x) > 0$ for $x \neq \theta_H$ and $(x, x) = 0$ for $x = \theta_H$.
The norm of an element $x \in H$ is then defined as $\|x\| = \sqrt{(x, x)}$.

The main characterization of Hilbert spaces is the parallelogram equality

$$2\|x\|^2 + 2\|y\|^2 = \|x + y\|^2 + \|x - y\|^2, \tag{1.1.14}$$

which also enables us to define its modulus of convexity and modulus of smoothness. It can be proven that H is a uniformly convex and uniformly smooth space. Among all uniformly convex and uniformly smooth Banach spaces, a Hilbert space has the greatest modulus of convexity and smallest modulus of smoothness, that is,

$$\delta_H(\epsilon) \geq \delta_X(\epsilon) \quad \text{and} \quad \rho_H(\tau) \leq \rho_X(\tau).$$

Observe that the parallelogram equality (1.1.14) is the necessary and sufficient condition to be a Hilbert space. The formula (1.1.12) shows that the normalized duality mapping J in H is the identity operator I.

2. The Lebesgue space $L^p(G)$, $\infty > p > 1$, of the measurable functions $f(x)$ such that

$$\int_G |f(x)|^p dx < \infty, \quad x \in G,$$

and G is a measurable set in R^n, is a uniformly convex and uniformly smooth Banach space with respect to the norm

$$\|f\|_{L^p} = \left(\int_G |f(x)|^p dx \right)^{1/p}.$$

Dual to Lebesgue space $L^p(G)$ with $p > 1$ is the Lebesgue space $L^q(G)$ with $q > 1$ such that $p^{-1} + q^{-1} = 1$.

Recall that in spaces $L^p(G)$ the Hölder integral inequality

$$\int_G |f(x)g(x)| dx \leq \left(\int_G |f(x)|^p dx \right)^{1/p} \left(\int_G |g(x)|^q dx \right)^{1/q}$$

and the Minkovsky integral inequality

$$\left(\int_G |f(x) + g(x)|^p dx \right)^{1/p} \leq \left(\int_G |f(x)|^p dx \right)^{1/p} + \left(\int_G |g(x)|^p dx \right)^{1/p}$$

hold for all $f(x) \in L^p(G)$ and for all $g(x) \in L^q(G)$.

3. The space l^p, $\infty > p > 1$, of number sequences $x = \{\xi_1, \xi_2, ..., \xi_j, ...\}$, such that

$$\sum_{j=1}^{\infty} |\xi_j|^p < \infty$$

is a uniformly convex and uniformly smooth Banach space with respect to the norm

$$\|x\|_{l^p} = \Big(\sum_{j=1}^{\infty} |\xi_j|^p\Big)^{1/p}.$$

Dual to the space l^p, $\infty > p > 1$, is the space l^q, $\infty > q > 1$, such that $p^{-1} + q^{-1} = 1$.

We note the following very useful properties of the functional $\phi(x) = \|x\|^s$ on the spaces L^p and l^p, $\infty > p > 1$ (see [228]): $\phi(x)$ is uniformly convex on the whole L^p and on the whole l^p if $s \geq p \geq 2$. It is not true if $s \in (1, 2)$. In addition, the functional $\phi(x) = \|x\|_X^s$ with any $\infty > s > 1$ is uniformly convex on each convex bounded set of the uniformly convex space X.

4. Introduce the Sobolev spaces $W_m^p(G)$. For simplicity, let G be a bounded set on a plane with sufficiently smooth boundary, $C_0^\infty(G)$ be a set of functions continuous on G and equal to zero on ∂G, together with their derivatives of all orders. Consider functions $v(x, y)$ of the space $L^p(G)$, $\infty > p > 1$. If there exists the function $\chi(x, y) \in L^p(G)$ such that for all $u(x, y) \in C_0^\infty(G)$ the equality

$$\int\int_G \frac{\partial^l u}{\partial x^{l_1} \partial y^{l_2}}\, v(x, y)\, dx\, dy = (-1)^l \int\int_G u(x, y)\, \chi(x, y)\, dx\, dy, \quad l = l_1 + l_2,$$

holds, then the function $\chi(x, y)$ is said to be an l-order generalized derivative of the function $v(x, y)$. We denote it as the usual derivative:

$$\chi(x, y) = \frac{\partial^l v}{\partial x^{l_1} \partial y^{l_2}}.$$

The functions $v(x, y)$ such that

$$\int\int_G |v(x, y)|^p\, dx\, dy + \sum_{1 \leq l \leq m} \int\int_G \Big|\frac{\partial^l v}{\partial x^{l_1} \partial y^{l_2}}\Big|^p\, dx\, dy < \infty, \quad l = l_1 + l_2,$$

form a Sobolev space $W_m^p(G)$ with the norm

$$\|v\|_{W_m^p} = \Big(\int\int_G |v(x, y)|^p\, dx\, dy + \sum_{1 \leq l \leq m} \int\int_G \Big|\frac{\partial^l v}{\partial x^{l_1} \partial y^{l_2}}\Big|^p\, dx\, dy\Big)^{1/p}. \qquad (1.1.15)$$

A closure of $C_0^\infty(G)$ in the metric of $W_m^p(G)$ is denoted as $\overset{\circ}{W}_m^p(G)$. It is known that $\overset{\circ}{W}_m^p(G) \subset W_m^p(G)$, and

$$\frac{\partial^l u}{\partial^{l_1} x \partial^{l_2} y}\Big|_{\partial G} = 0 \ \text{ for } \ u \in \overset{\circ}{W}_m^p(G), \ \ l = l_1 + l_2, \ \ 0 \leq l \leq m - 1.$$

The norm

$$\|v\|_{\overset{\circ}{W}{}_m^p} = \Big(\sum_{m=m_1+m_2} \int \int_G \Big| \frac{\partial^m v}{\partial x^{m_1} \partial y^{m_2}} \Big|^p \, dx \, dy \Big)^{1/p}$$

is equivalent to (1.1.15).

The Sobolev spaces $W_m^p(G)$ and $\overset{\circ}{W}{}_m^p(G)$ are also uniformly convex and uniformly smooth for $\infty > p > 1$, and they can be defined for any bounded measurable domain G in R^n. The dual space of $W_m^p(G)$ is the Banach space $W_{-m}^q(G)$ with $\infty > q > 1$, such that $p^{-1} + q^{-1} = 1$.

Next we introduce the Friedrichs' inequality which is often used in applications. Let G be a bounded domain of points $x = x(x_1, ..., x_n)$ of R^n, the boundary ∂G be Lipschitz-continuous, $f \in W_1^2$. Then

$$\int_G f^2 dG \le k \Big[\int_G \sum_{i=1}^n \Big(\frac{\partial f}{\partial x_i} \Big)^2 dG + \int_{\partial G} f^2 d(\partial G) \Big],$$

where k is a constant independent of n and is completely determined by the domain G. A similar inequality is valid in more general spaces.

Consider the imbedding operator $\mathcal{E}$ which is defined for any function $v(x, y) \in W_l^p(G)$ and carries $v(x, y)$ into the same function considered now as an element of the space $W_m^p(G)$.

Theorem 1.1.35 (Sobolev) *The space $W_m^p(G)$ for any $l < m$ is imbedded into the space $W_l^p(G)$.*

This theorem means that every function $v(x, y)$ having all m-order generalized derivatives has also all l-order generalized derivatives for $l < m$. Moreover, there exists a constant $C > 0$ such that

$$\|v\|_{W_l^p(G)} \le C \|v\|_{W_m^p(G)}.$$

It is clear that the imbedding operator $\mathcal{E}$ is bounded and linear, consequently, continuous.

In addition, the following inclusions take place:

$$W_m^r(G) \subset W_m^p(G) \subset L^p(G), \quad 0 < p \le r < \infty.$$

More generally, let X, Y be Banach spaces and $X \subseteq Y$. We say that the operator $\mathcal{E} : X \to Y$ with $D(\mathcal{E}) = X$ is an imbedding operator of X into Y if it carries each element $x \in X$ into itself, that is, $\mathcal{E}x = x$.

Definition 1.1.36 *The space X has the Kadeč–Klee property if, for any sequence $\{x_n\}$, the weak convergence $x_n \rightharpoonup x$ and convergence of the norms $\|x_n\| \to \|x\|$ imply strong convergence $x_n \to x$.*

Definition 1.1.37 *A reflexive Banach space X is said to be E-space if it is strictly convex and has the Kadeč–Klee property.*

Hilbert spaces as well as reflexive locally uniformly convex spaces are E-spaces. Therefore, L^p, l^p, W_m^p, $\infty > p > 1$, are also E-spaces.

Theorem 1.1.38 *X is an E-space if and only if X^* is strongly smooth.*

Definition 1.1.39 *A space X is called separable if in this space there exists a countable everywhere dense set. In other words, a space X is separable if for any element $x \in X$, there exists a sequence $\{x_n\} \subset X$ such that there is a subsequence $\{x_{n_k}\}$ of $\{x_n\}$ which converges to x.*

Next we define diverse properties of an operator.

Definition 1.1.40 *An operator $A : X \to 2^Y$ is bounded if it carries any bounded set of X to a bounded set of Y. If $R(A) \subset Y$ is a bounded set, then A is called a uniformly bounded operator.*

Definition 1.1.41 *The operator $A^{-1} : Y \to 2^X$ is called inverse to an operator $A : X \to 2^Y$ if the set of values of A^{-1} at any point $y \in R(A)$ is the set $\{x \mid y \in Ax\}$.*

Definition 1.1.42 *An operator $A : X \to 2^{X^*}$ is said to be coercive if there exists a function $c(t)$ defined for $t \geq 0$ such that $c(t) \to \infty$ as $t \to \infty$, and the inequality*

$$\langle y, x \rangle \geq c(\|x\|)\|x\|$$

holds for all $x \in D(A)$ and for all $y \in Ax$.

Definition 1.1.43 *An operator $A : X \to 2^{X^*}$ is said to be coercive relative to a point $x_0 \in X$ if there exists a function $c(t)$ defined for $t \geq 0$ such that $c(t) \to \infty$ as $t \to \infty$, and the inequality*

$$\langle y, x - x_0 \rangle \geq c(\|x\|)\|x\|$$

holds for all $x \in D(A)$ and for all $y \in Ax$.

Theorem 1.1.44 *If the functional $\varphi(x)$ is uniformly convex (in particular, strongly convex) then $Ax = grad\ \varphi(x)$ is coercive.*

Definition 1.1.45 *An operator $A : X \to 2^{X^*}$ is called weakly coercive if $\|y\| \to \infty$ as $\|x\| \to \infty$ for all $y \in Ax$.*

Emphasize that a coercive operator has a bounded inverse. This fact can be easily obtained by a contradiction.

Definition 1.1.46 *An operator $A : X \to Y$ is called compact on the set Ω if it carries any bounded subset of Ω into a compact set of Y.*

Theorem 1.1.47 (Schauder principle) *Let X be a Banach space. Assume that $\Omega \subset X$ is a convex closed and bounded set. If the map A is compact on Ω and $A(\Omega) \subseteq \Omega$, then the equation $Ax = x$ has at least one solution in Ω.*

Definition 1.1.48 *An operator $A : X \to Y$ is called*

1) continuous at a point $x_0 \in D(A)$ if $Ax_n \to Ax_0$ as $x_n \to x_0$;

2) hemicontinuous at a point $x_0 \in D(A)$ if $A(x_0 + t_n x) \rightharpoonup Ax_0$ as $t_n \to 0$ for any vector x such that $x_0 + t_n x \in D(A)$ and $0 \leq t_n \leq t(x_0)$;

3) demicontinuous at a point $x_0 \in D(A)$ if for any sequence $\{x_n\} \subset D(A)$ such that $x_n \to x_0$, the convergence $Ax_n \rightharpoonup Ax_0$ holds (it is evident that hemicontinuity of A follows from its demicontinuity);

4) Lipschitz-continuous if there exists a constant $l > 0$ such that $\|Ax_1 - Ax_2\|_Y \leq l\|x_1 - x_2\|_X$ for all $x_1, x_2 \in X$;

5) strongly continuous if $x_n \rightharpoonup x$ implies $Ax_n \to Ax$;

6) completely continuous on the set Ω if it is continuous and compact on Ω;

7) weak-to-weak continuous (or sequentially weakly continuous) at a point $x_0 \in D(A)$ if for any sequence $\{x_n\} \subset D(A)$ such that $x_n \rightharpoonup x_0$, the convergence $Ax_n \rightharpoonup Ax_0$ holds.

The corresponding properties are fulfilled in $D(A)$ if they are valid at each point of $D(A)$.

Definition 1.1.49 *Let an operator $A : X \to 2^Y$. We say that*

1) A is closed at $D(A)$ if the relations $x_n \to x$, $y_n \to y$, $x_n \in D(A)$, $y_n \in Ax_n$ for all $n \geq 0$, imply the inclusion $y \in Ax$;

2) A is weakly closed at $D(A)$ if the relations $x_n \rightharpoonup x$, $y_n \rightharpoonup y$, $x_n \in D(A)$, $y_n \in Ax_n$ for all $n \geq 0$, imply the inclusion $y \in Ax$.

Definition 1.1.50 *An operator $A_1 : X \to 2^Y$ is said to be the extension of an operator $A : X \to 2^Y$ if $D(A) \subseteq D(A_1)$ and $Ax = A_1 x$ for all $x \in D(A)$.*

Definition 1.1.51 *An operator $A : X \to Y$ is called linear if it is additive, i.e.,*

$$A(x + y) = Ax + Ay \quad \forall x, y \in X,$$

and homogeneous, i.e.,

$$A(\lambda x) = \lambda Ax \quad \forall x \in X, \quad \forall \lambda \in R^1.$$

Definition 1.1.51 also yields the definition of a linear functional φ in X when $Y = R^1$, that is, $A = \varphi : X \to R^1$.

Definition 1.1.52 *Assume that X and Y are linear normed spaces and $A : X \to Y$ is a linear bounded operator. The smallest constant M satisfying the inequality $\|Ax\|_Y \leq M\|x\|_X$ is said to be the norm of the operator A, and it is denoted by $|A|$.*

Definition 1.1.53 *Let X and Y be linear normed spaces, $A : X \to Y$ be a nonlinear operator, $D(A)$ be an open set. A is said to be Fréchet differentiable (or strongly differentiable) at a point $x \in D(A)$ if there exists a linear continuous operator $A'(x) : X \to Y$ such that for all $h \in X$,*

$$A(x + h) - Ax = A'(x)h + \omega(x, h),$$

where $x + h \in D(A)$ *and*

$$\frac{\omega(x, h)}{\|h\|} \to 0 \quad \text{as} \quad \|h\| \to 0.$$

Respectively, $A'(x)h$ and $A'(x)$ are called the Fréchet differential (or strong differential) and the Fréchet derivative (or strong derivative) of the operator A at a point x.

Definition 1.1.54 *Let X and Y be linear normed spaces, $A : X \to Y$ be a nonlinear operator, $D(A)$ be an open set. If there exists a linear operator $A'(x) : X \to Y$ such that*

$$\lim_{t \to 0} \frac{A(x + h) - Ax}{t} = A'(x)h,$$

then A is called Gâteaux differentiable (or weakly differentiable). Respectively, $A'(x)h$ and $A'(x)$ are called the Gâteaux differential (or weak differential) and the Gâteaux derivative (or weak derivative) of the operator A at a point x.

Theorem 1.1.55 *Let X be a linear normed space, Y be a Banach space and A be a nonlinear operator acting from a linear dense set of X to Y which has a bounded Gâteaux differential at each point of the set $D = U \cap X$, where U is some neighborhood of $x \in X$. If the Gâteaux derivative is continuous at a point $x_0 \in D$, then it is, in fact, the Fréchet derivative there.*

Let X be a Banach space, X^* be its dual space, $A : X \to X^*$ be a nonlinear operator having the Fréchet derivatives on $D(A)$ up to order $n + 1$, and $[x, x + h] \subset D(A)$. Then

$$\langle A(x + h), y \rangle = \left\langle Ax + A'(x)h + \frac{A''(x)h^2}{2!} + \circ \circ \circ + \frac{A^{(n)}(x)h^n}{n!}, y \right\rangle \qquad (1.1.16)$$

$$+ \left\langle \frac{A^{(n+1)}(x + \theta h)h^{n+1}}{(n+1)!}, y \right\rangle \quad \forall y \in X,$$

where $\theta = \theta(y)$ satisfies the inequality $0 < \theta < 1$. We call (1.1.16) the Taylor formula and its particular case

$$\langle A(x + h) - Ax, y \rangle = \langle A'(x + \theta h)h, y \rangle \quad \forall y \in X^*$$

the Lagrange formula.

Let Ω be a convex closed subset of X. The operator P_Ω is called a metric projection operator if it assigns to each $x \in X$ its nearest point $y \in \Omega$ such that

$$\|x - y\| = min\{\|x - z\| \mid z \in \Omega\}. \qquad (1.1.17)$$

It is known that the metric projection operator P_Ω is continuous in a uniformly convex Banach space X and uniformly continuous on each bounded set of X if, in addition, X is uniformly smooth.

An element y is called the metric projection of x onto Ω and denoted by $P_\Omega x$. It exists and is unique at any point of the reflexive strictly convex space. Observe that y is often called the best approximation of x because the quantity $d(x, \Omega) = \|x - y\|$ is the distance from $x \in X$ to the subset $\Omega \subset X$.

Definition 1.1.56 *A linear bounded operator P is called a projector of the space X onto Y if $R(P) = Y$ and $P^2 = P$.*

Definition 1.1.57 *A Banach space X possesses the approximation if there exists a directed family of finite-dimensional subspaces $\{X_n\}$ ordered by inclusion, and a corresponding family of projectors $P_n : X \to X_n$ such that $|P_n| = 1$ for all $n = 0, 1, , \ldots$, and $\bigcup_n X_n$ is dense in X.*

Let us consider two sets Ω_1 and Ω_2 in X. Let

$$\beta(\Omega_1, \Omega_2) = sup\{d(x, \Omega_2) \mid x \in \Omega_1\}$$

be a semideviation of the set Ω_1 from Ω_2. Then the Hausdorff distance between the sets Ω_1 and Ω_2 is defined by the formula

$$\mathcal{H}_X(\Omega_1, \Omega_2) = max\{\beta(\Omega_1, \Omega_2), \beta(\Omega_2, \Omega_1)\}.$$

The distance $\mathcal{H}_{X^*}(\Omega_1, \Omega_2)$ between the sets Ω_1 and Ω_2 in a space X^* is introduced analogously.

Definition 1.1.58 [154] *The set sequence $\{\Omega_n\}$, $\Omega_n \subset X$, is Mosco-convergent to the set Ω (we write "M-convergent" in short), if*
(i) $\Omega = s - \liminf_{n \to \infty} \Omega_n$, that is, for any element $x \in \Omega$ it is possible to construct the sequence $\{x_n\}$ such that $x_n \in \Omega_n$ and $x_n \to x$;
(ii) $\Omega = w - \limsup_{n \to \infty} \Omega_n$, that is, if $x_n \in \Omega_n$ and $x_n \rightharpoonup x \in X$ then $x \in \Omega$.

Definition 1.1.59 *Let Ω be a set of elements $x, y, z, \ldots$. Suppose there is a binary relation defined between certain pairs (x, y) of elements of Ω, expressed by $x \prec y$, with the properties:*

$$\begin{cases} x \prec x; \\ \text{if } x \prec y \quad \text{and} \quad y \prec x \quad \text{then} \quad x = y; \\ \text{if } x \prec y \quad \text{and} \quad y \prec z \quad \text{then} \quad x \prec z. \end{cases}$$

Then Ω is said to be semi-ordered by the relation $\prec$.

Definition 1.1.60 *A semi-ordered set Ω is said to be linearly ordered if for every pair (x, y) in Ω, either $x \prec y$ or $y \prec x$.*

Lemma 1.1.61 (Zorn) *Let Ω be a nonempty semi-ordered set with the property that every linearly ordered subset of Ω has an upper bound in Ω. Then Ω contains at least one maximal element.*

Finally, we provide the following important theorem in a space R^n. Let $A : R^n \to 2^{R^n}$ be a certain operator. Denote $R(Ax)$ the closed convex hull of all limit points of sequence $\{Ax_k\}$, where $x_k \to x$, taking also into account infinite limit points in all directions.

Theorem 1.1.62 *Let $A : R^n \to 2^{R^n}$ be an operator defined on a closed convex set Ω with the boundary $\partial\Omega$, $\theta_{R^n} \in int\ \Omega$ and $(y, x) \geq 0$ if $x \in \partial\Omega$, $y \in Ax$. Then there exists a point x_0 in Ω such that $\theta_{R^n} \in R(Ax_0)$.*

1.2   Subdifferentials

Let X be a reflexive Banach space, $\varphi : X \to R^1$ be a proper functional with domain $D(\varphi) = X$.

Definition 1.2.1 *An element $w \in X^*$ satisfying the inequality*

$$\varphi(y) \geq \varphi(x) + \langle w, y - x \rangle \quad \forall y \in X, \tag{1.2.1}$$

is said to be a subgradient of the functional φ at a point x. The subgradient is often called a supporting functional.

Definition 1.2.2 *An operator $\partial\varphi : X \to 2^{X^*}$ is called a subdifferential of the functional φ if and only if (1.2.1) is fulfilled for $w \in \partial\varphi(x)$.*

Thus, the set $\{w \mid w \in \partial\varphi(x)\}$ is the totality of all subgradients of the functional φ at a point x. Establish the connection between the subdifferential and the gradient of φ.

Lemma 1.2.3 *Let φ be a proper convex functional on X. If φ is Gâteaux differentiable at a point $x \in X$, then there exists only one subgradient of the functional φ at this point and $\partial\varphi(x) = \varphi'(x)$.*

Proof. Since φ is convex, we can write

$$\varphi(x + t(y - x)) \leq \varphi(x) + t\Big(\varphi(y) - \varphi(x)\Big) \quad \forall x, y \in X, \ 0 < t < 1,$$

or

$$\frac{\varphi(x + t(y - x)) - \varphi(x)}{t} \leq \varphi(y) - \varphi(x).$$

Letting $t \to 0$ one gets

$$\langle \varphi'(x), y - x \rangle \leq \varphi(y) - \varphi(x) \quad \forall y \in X.$$

This means that $\varphi'(x) \in \partial\varphi(x)$.

Suppose that $w \neq \varphi'(x)$ and $w \in \partial\varphi(x)$. Then (1.2.1) with $x + ty$ in place of y implies

$$\frac{\varphi(x + ty) - \varphi(x)}{t} \geq \langle w, y \rangle \quad \forall y \in X,$$

and if $t \to 0$ then we obtain

$$\langle \varphi'(x) - w, y \rangle \geq 0 \quad \forall y \in X.$$

This is possible only if $\varphi'(x) = w$. ∎

Lemma 1.2.4 *Assume that $\partial\varphi$ is a single-valued hemicontinuous subdifferential of φ on X. Then φ has the Gâteaux derivative φ' and $\partial\varphi(x) = \varphi'(x)$ for all $x \in X$.*

Proof. By (1.2.1), for all $y \in X$ and $t > 0$,

$$\varphi(x + ty) - \varphi(x) \geq \langle \partial\varphi(x), ty \rangle.$$

Therefore,

$$\liminf_{t \to 0} \frac{\varphi(x + ty) - \varphi(x)}{t} \geq \langle \partial\varphi(x), y \rangle.$$

Thus, it follows from (1.2.1) that

$$\varphi(x) - \varphi(x + ty) \geq -\langle \partial\varphi(x + ty), ty \rangle.$$

By virtue of the hemicontinuity of $\partial\varphi$,

$$\limsup_{t \to 0} \frac{\varphi(x + ty) - \varphi(x)}{t} \leq \langle \partial\varphi(x), y \rangle,$$

i.e., $\varphi'(x) = \partial\varphi(x)$. ∎

Note that $D(\partial\varphi) \subseteq dom\ \varphi$. In addition, it is known by (1.2.1) that the set of values of the operator $\partial\varphi$ at each point $x \in D(\partial\varphi)$ is convex and weakly closed.

The following lemmas emphasizes the exceptional importance of the subdifferential concept for applications.

Lemma 1.2.5 *A functional $\varphi : X \to R^1$ has the minimum at a point $x \in D(\partial\varphi)$ if and only if $\theta_{X^*} \in \partial\varphi(x)$.*

Proof. By (1.2.1), if $\theta_{X^*} \in \partial\varphi(x)$ then we have $\varphi(y) \geq \varphi(x)$, that is,

$$\varphi(x) = min\{\varphi(y) \mid y \in X\}. \tag{1.2.2}$$

Let (1.2.2) hold. Then $\varphi(y) - \varphi(x) \geq 0$ for all $y \in X$. By the definition of subdifferential, it follows from this that $\theta_{X^*} \in \partial\varphi(x)$. ∎

Lemma 1.2.6 *If a functional φ on the open convex set $M \subset dom\ \varphi$ has a subdifferential, then φ is convex and lower semicontinuous on this set.*

Proof. Since M is a convex set, then $z = (1 - t)x + ty \in M$ for all $x, y \in M$ and $t \in [0, 1]$. By (1.2.1), we conclude for $w \in \partial\varphi(z)$ that

$$\varphi(x) \geq \varphi(z) + \langle w, x - z \rangle$$

and

$$\varphi(y) \geq \varphi(z) + \langle w, y - z \rangle.$$

Multiplying these inequalities by $1 - t$ and t, respectively, and adding them together, we obtain

$$(1 - t)\varphi(x) + t\varphi(y) \geq \varphi((1 - t)x + ty).$$

That means that the functional φ is convex, as claimed.

Let $x_0 \in M$, $x_n \in M$ for all $n \geq 0$, $x_n \to x_0$, $w \in \partial\varphi(x_0)$. Then by the definition of $\partial\varphi(x_0) \in X^*$, we have $\varphi(x_n) \geq \varphi(x_0) + \langle w, x_n - x_0 \rangle$. Hence,

$$\liminf_{x_n \to x_0} \varphi(x_n) \geq \varphi(x_0),$$

which means that φ is lower semicontinuous on M. ∎

Lemma 1.2.7 *If the operator $A : X \to X^*$ is hemicontinuous and $\varphi'(x) = Ax$, then*

$$\varphi(x) - \varphi(\theta_X) = \int_0^1 \langle A(tx), x \rangle dt \quad \forall x \in X.$$

Proof. Consider the function $\psi(t) = \varphi(tx)$. Then

$$\psi'(t) = \lim_{\tau \to 0} \frac{\varphi(tx + \tau x) - \varphi(tx)}{\tau} = \langle A(tx), x \rangle.$$

Taking into account the hemicontinuity of A we have

$$\varphi(x) - \varphi(\theta_X) = \psi(1) - \psi(0) = \int_0^1 \psi'(t) dt = \int_0^1 \langle A(tx), x \rangle dt. \quad \blacksquare$$

Let us give a sufficient condition for the subdifferential existence.

Theorem 1.2.8 *Let $\varphi : X \to R^1$ be a proper convex lower semicontinuous functional. Then φ has a subdifferential at int dom φ.*

From the definition of $\partial \varphi$ it follows that $\partial(\lambda \varphi) = \lambda \partial \varphi$ for all $\lambda > 0$. The additivity property of the subdifferential is established by the following theorem.

Theorem 1.2.9 *Let φ_1 and φ_2 be convex functionals on X and let there exist at least one point $z \in dom\ \varphi_1 \cap dom\ \varphi_2$ such that one of these functionals is continuous at z. Then $\partial(\varphi_1 + \varphi_2) = \partial\varphi_1 + \partial\varphi_2$.*

Definition 1.2.10 *If a functional φ on X is non-trivial, then a functional φ^* on X^*, defined by the formula*

$$\varphi^*(x^*) = sup\{\langle x^*, x \rangle - \varphi(x) \mid x \in X\}, \tag{1.2.3}$$

is called conjugate to φ.

This definition implies the Young–Fenchel inequality:

$$\varphi^*(x^*) + \varphi(x) \geq \langle x^*, x \rangle. \tag{1.2.4}$$

Theorem 1.2.11 *Assume that φ is a weakly lower semicontinuous, convex and finite functional and $\partial\varphi$ is its subdifferential. Then $x^* \in \partial\varphi(x)$ if and only if*

$$\langle x^*, x \rangle = \varphi(x) + \varphi^*(x^*). \tag{1.2.5}$$

Proof. Let $x^* \in \partial\varphi(x)$ be given. We have

$$\langle x^*, y \rangle - \varphi(y) \leq \langle x^*, x \rangle - \varphi(x) \quad \forall y \in X.$$

Hence,

$$\begin{aligned} \varphi(x) \quad + \quad & \varphi^*(x^*) = \varphi(x) + sup\ \{\langle x^*, y \rangle - \varphi(y) \mid y \in X\} \\ \leq\ & \varphi(x) + \langle x^*, x \rangle - \varphi(x) = \langle x^*, x \rangle. \end{aligned}$$

Then the Young–Fenchel inequality implies (1.2.5).

Let now (1.2.5) be valid. Then, by definition of φ^*, one gets

$$\langle x^*, x \rangle \geq \varphi(x) + \langle x^*, y \rangle - \varphi(y) \quad \forall y \in X,$$

that is, $\varphi(y) - \varphi(x) \geq \langle x^*, y - x \rangle$. Thus, $x^* \in \partial\varphi(x)$. ∎

Theorem 1.2.12 *Suppose that $\varphi : X \to R^1$ is a proper convex lower semicontinuous functional. Then $R(\partial\varphi) = X^*$ if and only if for all $w \in X^*$,*

$$\varphi(x) - \langle w, x \rangle \to +\infty \quad \text{as} \quad \|x\| \to \infty. \tag{1.2.6}$$

Proof. (i) Consider the functional $\Phi(x) = \varphi(x) - \langle w, x \rangle$ for an arbitrary $w \in X^*$. It is proper convex lower semicontinuous and

$$\lim_{\|x\| \to \infty} \Phi(x) = +\infty.$$

Then, by Theorem 1.1.23, there exists $x_0 \in X$ such that $\Phi(x)$ reaches its minimum at this point. Therefore,

$$\varphi(x) - \langle w, x \rangle \geq \varphi(x_0) - \langle w, x_0 \rangle \quad \forall x \in D(\varphi),$$

i.e., $w \in \partial\varphi(x_0)$ and $R(\partial\varphi) = X^*$.

(ii) Let $R(\partial\varphi) = X^*$ be given. Prove (1.2.6) by the contradiction. Suppose that $\{x_n\} \subset X$, $\|x_n\| \to \infty$ and the sequence $\{\Phi(x_n)\}$ is bounded. Take an element $g \in X^*$ such that $\langle g, x_n \rangle \to \infty$ as $\|x_n\| \to \infty$. By the condition, there exists an element $\bar{x} \in D(\partial\varphi)$ such that $w + g \in \partial\varphi(\bar{x})$. Then the inequality

$$\langle g, x_n \rangle \leq \varphi(x_n) - \langle w, x_n \rangle - \varphi(\bar{x}) + \langle w + g, \bar{x} \rangle$$

follows from the subdifferential definition. Thus, $\{\langle g, x_n \rangle\}$ is bounded which contradicts the choice of the element g. ∎

1.3 Monotone Operators

Let X be a reflexive Banach space, X^* its dual space and $A : X \to 2^{X^*}$.

Definition 1.3.1 *The set of pairs $(x, f) \in X \times X^*$ such that $f \in Ax$ is called the graph of an operator A and it is denoted by $\mathrm{gr}\,A$.*

Definition 1.3.2 *A set $G \subseteq X \times X^*$ is called monotone if the inequality*

$$\langle f - g, x - y \rangle \geq 0$$

holds for all pairs (x, f) and (y, g) from G.

Definition 1.3.3 *An operator $A : X \to 2^{X^*}$ is monotone if its graph is a monotone set, i.e., if for all $x, y \in D(A)$,*

$$\langle f - g, x - y \rangle \geq 0 \quad \forall f \in Ax, \ \forall g \in Ay. \tag{1.3.1}$$

It is obvious that if the operator A is monotone then the operators $A(x + x_0)$ and $Ax + w_0$ are also monotone, where $x_0 \in X$ and $w_0 \in X^*$ are fixed elements. It is not difficult to verify that if A and B are monotone operators then the sum $A + B$, the product λA, $\lambda > 0$, and the inverse operator A^{-1} are monotone operators as well. For a linear operator $A : X \to 2^{X^*}$ the monotonicity condition is equivalent to its non-negativity:

$$\langle g, x \rangle \geq 0 \quad \forall g \in Ax, \quad \forall x \in D(A).$$

Definition 1.3.4 *An operator $A : X \to 2^{X^*}$ is strictly monotone if the equality in (1.3.1) holds only for $x = y$.*

Proposition 1.3.5 *If among two monotone operators A and B at least one is strictly monotone, then the sum $A + B$ is strictly monotone.*

It is possible to give another definition of the monotone operator in a Hilbert space H.

Definition 1.3.6 *An operator $A : H \to 2^H$ is said to be monotone if*

$$\|x - y\| \leq \|x - y + \lambda(f - g)\| \tag{1.3.2}$$

for all $x, y \in D(A)$, $f \in Ax$, $g \in Ay$ and $\lambda \geq 0$.

Theorem 1.3.7 *Definitions 1.3.3 and 1.3.6 are equivalent.*

Proof. Suppose that (1.3.1) is satisfied. Then (1.3.2) follows from the equality

$$\|x - y + \lambda(f - g)\|^2 = \|x - y\|^2 + 2\lambda(f - g, x - y) + \lambda^2 \|f - g\|^2. \tag{1.3.3}$$

If (1.3.2) holds then we deduce from (1.3.3) that

$$2(f - g, x - y) + \lambda \|f - g\|^2 \geq 0.$$

Setting $\lambda \to 0$ we obtain (1.3.1). ∎

If operator A is weakly differentiable then the following definition of the monotonicity is given.

Definition 1.3.8 *A Gâteaux differentiable operator $A : X \to X^*$ with $D(A) = X$ is called monotone if*

$$\langle A'(x)h, h \rangle \geq 0 \quad \forall x, \ x + h \in X.$$

This definition is motivated by the following theorem.

Theorem 1.3.9 *Let an operator A be defined on a convex set $\Omega \in X$. If the directional derivative*

$$\left[\frac{d}{dt}\Big\langle x_2 - x_1, A\Big(x_1 + t(x_2 - x_1)\Big)\Big\rangle\right]_{t=0} \quad \forall x_1, x_2 \in \Omega$$

exists and is non-negative, then A is monotone on Ω.

Let us present some examples of monotone operators.

1. Let $\varphi : R^1 \to R^1$ be a non-decreasing function. Then an operator $A : R^1 \to 2^{R^1}$, defined by the equality

$$Ax = [\varphi(x - 0), \ \varphi(x + 0)] \quad \forall x \in dom \ \varphi,$$

is monotone.

2. Assume that $\varphi : X \to R^1$ is a proper convex functional and there exists a subdifferential $\partial \varphi : X \to 2^{X^*}$. Then the operator $\partial \varphi$ is monotone. Indeed, by Definition 1.2.2, we can write for all $x, y \in dom \ \varphi$:

$$\varphi(y) - \varphi(x) \geq \langle f, y - x \rangle, \ \ f \in \partial \varphi(x),$$

$$\varphi(x) - \varphi(y) \geq \langle g, x - y \rangle, \ \ g \in \partial \varphi(y).$$

Summing these inequalities one gets

$$\langle f - g, x - y \rangle \geq 0 \quad \forall x, y \in dom \ \varphi, \ \ f \in \partial \varphi(x), \ \ g \in \partial \varphi(y).$$

Furthermore, it follows from Lemma 1.2.3 for the Gâteaux differentiable functional that the gradient of a proper convex functional is the single-valued monotone operator.

3. Assume that H is a Hilbert space, $A : H \to H$ is a nonexpansive operator, i.e.,

$$\|Ax - Ay\| \leq \|x - y\| \quad \forall x, y \in D(A).$$

Then the operator $I - A$, where $I : H \to H$ is an identity operator, is monotone.

Indeed, the claim follows from the relations

$$(x - Ax - y + Ay, x - y) = \|x - y\|^2 - (Ax - Ay, x - y)$$

$$\geq \|x - y\|^2 - \|Ax - Ay\|\|x - y\| \geq \|x - y\|^2 - \|x - y\|^2 = 0.$$

4. Let Ω be a convex closed set in a Hilbert space H, $x \in H$ and $P_\Omega x$ be a projection of x on Ω defined by (1.1.17):

$$\|x - P_\Omega x\| = min\{\|x - z\| \mid z \in \Omega\}. \tag{1.3.4}$$

The element $u = P_\Omega x$ is unique for every $x \in H$. Prove that the operator P_Ω is monotone. First of all, we show that (1.3.4) is equivalent to the inequality

$$(P_\Omega x - x, z - P_\Omega x) \geq 0 \quad \forall z \in \Omega. \tag{1.3.5}$$

In fact, the expression

$$\|x - z\|^2 \geq \|x - P_\Omega x\|^2 + 2(x - P_\Omega x, P_\Omega x - z) \quad \forall z \in \Omega,$$

holds (see Theorem 1.1.20 and Lemma 1.2.3). Then, obviously, (1.3.4) results from (1.3.5). Let now (1.3.4) be valid and $t \in (0, 1)$ be given. Since Ω is a convex set, we have that the element $(1 - t)P_\Omega x + tz \in \Omega$ and then, by the inequality

$$\|x - P_\Omega x\|^2 - \|x - (1 - t)P_\Omega x - tz\|^2 \geq 2(x - P_\Omega x - t(z - P_\Omega x), t(z - P_\Omega x)),$$

one gets

$$(x - P_\Omega x - t(z - P_\Omega x), z - P_\Omega x) \leq 0.$$

That leads to (1.3.5) as $t \to 0$.

Similarly to (1.3.5) we can write

$$(P_\Omega y - y, z - P_\Omega y) \geq 0 \quad \forall z \in \Omega. \tag{1.3.6}$$

Presume $z = P_\Omega y$ and $z = P_\Omega x$, respectively, in (1.3.5) and in (1.3.6). Summing thus obtained inequalities we have

$$(P_\Omega x - P_\Omega y, x - y) - \|P_\Omega x - P_\Omega y\|^2 \geq 0.$$

Hence, P_Ω is monotone. Note that it also follows from the last inequality that the projection operator is nonexpansive in H. If Ω is a subspace of Hilbert space then P_Ω is a linear and orthogonal operator.

In a Banach space, a metric projection operator is not monotone and not nonexpansive, in general. However, there exists nonexpansive projections from a Banach space even into a nonconvex subset Ω [61].

5. Let $G \subset R^n$ be a bounded measurable domain. Define the operator $A : L^p(G) \to L^q(G)$, $p^{-1} + q^{-1} = 1$, $p > 1$, by the formula

$$Ay(x) = \varphi(x, |y(x)|^{p-1})|y(x)|^{p-2}y(x), \ x \in G,$$

where the function $\varphi(x, s)$ is measurable as a function on x for every $s \in [0, \infty)$ and continuous for almost all $x \in G$ as a function on s, $|\varphi(x, s)| \leq M$ for all $s \in [0, \infty)$ and for almost all $x \in G$. Note that the operator A really maps $L^p(G)$ to $L^q(G)$ because of the inequality $|Ay| \leq M|y|^{p-1}$.

Show that the operator A is monotone provided that function $s\varphi(x, s)$ is non-decreasing with respect to s. We can write the following estimates:

$$
\begin{aligned}
\langle Ay - Az, y - z \rangle &= \int_G \left(\varphi(x, |y|^{p-1})|y|^{p-2}y - \varphi(x, |z|^{p-1})|z|^{p-2}z \right)(y - z)dx \\
&\geq \int_G \varphi(x, |y|^{p-1})|y|^{p-2}(|y|^2 - |y||z|)dx \\
&\quad - \int_G \varphi(x, |z|^{p-1})|z|^{p-2}(|y||z| - |z|^2)dx \\
&= \int_G \left(\varphi(x, s_1)s_1 - \varphi(x, s_2)s_2 \right)(|y| - |z|)dx \geq 0,
\end{aligned}
$$

where $s_1 = |y|^{p-1}$, $s_2 = |z|^{p-1}$. Hence, the property of monotonicity is proved.

6. Let the operator

$$Au = -\sum_{i=1}^{n} \frac{\partial}{\partial x_i} \left[a_i\left(x, \left| \frac{\partial u}{\partial x_i} \right|^{p-1} \right) \left| \frac{\partial u}{\partial x_i} \right|^{p-2} \frac{\partial u}{\partial x_i} \right] + a_0(x, |u|^{p-1})|u|^{p-2}u,$$

be given, where the functions $a_i(x, s)$, $i = 0, 1, 2, ..., n$, have the same properties as the function $\varphi(x, s)$ in Example 5 and G is a bounded measurable set in R^n. Then the operator $A : \overset{\circ}{W}{}_1^p(G) \to (\overset{\circ}{W}{}_1^p(G))^*$ defined by the formula

$$\langle Au, v \rangle = \sum_{i=1}^{n} \int_G a_i\left(x, \left| \frac{\partial u}{\partial x_i} \right|^{p-1} \right) \left| \frac{\partial u}{\partial x_i} \right|^{p-2} \frac{\partial u}{\partial x_i} \frac{\partial v}{\partial x_i} dx$$

$$+ \int_G a_0(x, |u|^{p-1})|u|^{p-2}uv dx \quad \forall u, v \in \overset{\circ}{W}{}_1^p(G)$$

is monotone. This fact is verified by the same arguments as in Example 5 (see [83]).

7. Next we give the example from quantum mechanics [230, 231]. Consider the operator

$$Au = -a^2 \triangle u + (g(x) + b)u(x) + u(x) \int_{R^3} \frac{u^2(y)}{|x - y|} dy,$$

where $\triangle = \sum_{i=1}^{3} \frac{\partial^2}{\partial x_i^2}$ is the Laplace operator (Laplacian) in R^3, a and b are constants, $g(x) = g_0(x) + g_1(x)$, $g_0(x) \in L^\times(R^3)$, $g_1(x) \in L^2(R^3)$. Represent A in the form $A = L + B$, where the operator L is the linear part of A (it is the Schrödinger operator) and B is defined by the last term. It is known [107] that there exists $b \geq 0$ such that L becomes positive in the domain

$$D(L) = D(\triangle) = \{u \in H \mid \nabla u \in H \times H \times H, \ \triangle u \in H\}, \ H = L^2(R^3),$$

$$\nabla u = \left(\frac{\partial u}{\partial x_1}, \frac{\partial u}{\partial x_2}, \frac{\partial u}{\partial x_3} \right).$$

It is obvious that B is hemicontinuous. Furthermore, B is the gradient of the functional

$$\Phi(u) = \frac{1}{4} \int_{R^3} \int_{R^3} \frac{u^2(x)u^2(y)}{|x - y|} dx dy, \ u \in W_1^2(R^3),$$

which is proper convex lower semicontinuous. Therefore, B is a monotone operator from H to H. This implies that $A : L^2(R^3) \to L^2(R^3)$ is also a monotone operator.

It is known that $g(x) = -2|x|^{-2}$ and $a = \frac{1}{2}$ correspond to the case of the Coulomb potential in the quantum mechanics, and $b \geq 2$ guarantees a positivity of the Schrödinger operator [169]. The operator A describes here the atom of helium in the situation when both electrons are in the same state.

Observe that

$$A_1 u(x) = |u(x)|^{p-2}u(x) \int_{R^3} \frac{|u(y)|^p}{|x - y|} dy, \ p > 2$$

is the gradient of the functional

$$\Phi_1(u) = \frac{1}{2p} \int_{R^3} \int_{R^3} \frac{|u(x)|^p |u(y)|^p}{|x-y|} dxdy$$

and acts from $L^p(R^3)$ to $L^q(R^3)$, $p^{-1} + q^{-1} = 1$. In addition, it is not difficult to make sure that the functional $\Phi_1(u)$ is convex, i.e., the operator A_1 is monotone too.

8. In the filtration theory one has to solve the following equation [143]:

$$-div\,(g(x, \nabla^2 u)\nabla u) = f(x), \quad u|_\Gamma = 0,$$

where $x \in \Omega$, $\Omega \subset R^n$ is a bounded measurable domain, $\Gamma = \partial\Omega$, $u(x)$ is a pressure, $f(x)$ is a density of sources, ∇ is the Hamilton operator (Hamiltonian) defined for a scalar function $u(x) = u(x_1, x_2, ..., x_n)$ as

$$\nabla u = \left(\frac{\partial u}{\partial x_1}, \frac{\partial u}{\partial x_2}, ..., \frac{\partial u}{\partial x_n} \right)$$

and the symbol $div\,\bar{w}$ denotes $divergence$ of the vector field $\bar{w}$ [82].

It is known that the function $g(x, \xi^2)\xi$ can be written in the form

$$g(x, \xi^2)\xi = g_0(x, \xi^2)\xi + g_1(x, \xi^2)\xi,$$

where $g_0(x, \xi^2)\xi$ is non-negative and non-decreasing with respect to ξ, $g_0(x, \xi^2) = 0$ for $\xi \leq \beta$, $g_0(x, \xi^2)$ is measurable with respect to the first argument and absolutely continuous with respect to the second argument. Beside this,

$$g_0(x, \xi^2)\xi \leq c_1 |\xi - \beta|^{p-1}, \quad p > 1, \quad \xi \leq \beta, \quad \beta = \beta(x) \in L^p(\Omega), \quad c_1 > 0,$$

$$g_1(x, \xi^2)\xi = \begin{cases} \omega > 0, & \text{if } \xi > \beta, \\ 0, & \text{if } \xi \leq \beta. \end{cases}$$

Then the functions g_0 and g_1 define, respectively, the operators A_0 and A_1 from $\overset{o}{W}{}_1^p(\Omega)$ to $(W_1^p)^*(\Omega)$:

$$\langle A_i u, v \rangle = \int_\Omega g_i(x, \nabla^2 u)(\nabla u, \nabla v)dx, \quad i = 0, 1, \quad \forall u, v \in \overset{o}{W}{}_1^p(\Omega),$$

where both operators are monotone, A_0 is continuous, A_1 is discontinuous. Here

$$(\nabla u, \nabla v) = \sum_{i=1}^n \frac{\partial u}{\partial x_i} \frac{\partial v}{\partial x_i}.$$

Furthermore, the monotone operator $A = A_0 + A_1$ is potential and its potential is defined by the following expression:

$$F(u) = \int_\Omega \left(\int_0^{|\nabla u|} g(x, \xi^2)\xi d\xi \right) dx.$$

Introduce monotone operators with stronger properties of the monotonicity.

Definition 1.3.10 *An operator $A : X \to 2^{X^*}$ is called **uniformly monotone** if there exists a continuous increasing function $\gamma(t)$ $(t \geq 0)$, $\gamma(0) = 0$, such that the inequality*

$$\langle f - g, x - y \rangle \geq \gamma(\|x - y\|) \quad \forall f \in Ax, \quad \forall g \in Ay \qquad (1.3.7)$$

*holds for all $x, y \in D(A)$. If here $\gamma(t) = ct^2$, where c is a positive constant, then A is **strongly monotone**.*

Lemma 1.3.11 *Let $\partial\varphi : X \to 2^{X^*}$ be a subdifferential of the functional $\varphi : X \to R^1$. If φ is uniformly convex on a convex closed set Ω with modulus of convexity $\gamma(t)$, then*

$$\langle f - g, x - y \rangle \geq 2\gamma(\|x - y\|) \quad \forall x, y \in \Omega, \quad f \in \partial\varphi(x), \quad g \in \partial\varphi(y),$$

that is, $\partial\varphi$ is a uniformly monotone operator.

Remark 1.3.12 *We say that a monotone operator $A : X \to 2^{X^*}$ is **properly monotone** if there is not any strengthening of (1.3.1) (for instance, up to the level of strong or uniform monotonicity).*

In Section 1.6 we shall give examples of monotone operators A that do not satisfy the inequality (1.3.7) on the whole domain $D(A)$ but they are uniformly monotone on any bounded set of $D(A)$. Therefore, it makes sense to give the following definition.

Definition 1.3.13 *An operator $A : X \to 2^{X^*}$ is called **locally uniformly monotone** if there exists an increasing continuous function $\gamma_R(t)$ $(t \geq 0, \ R > 0)$, $\gamma_R(0) = 0$ and*

$$\langle f - g, x - y \rangle \geq \gamma_R(\|x - y\|) \quad \forall f \in Ax, \quad \forall g \in Ay$$

*for x and y from $D(A)$, where $\|x\| \leq R$, $\|y\| \leq R$. If $\gamma_R(t) = C(R)t^2$, $C(R) > 0$, then A is said to be **locally strongly monotone**.*

The following lemma asserts that the class of locally uniformly monotone operators is not empty.

Lemma 1.3.14 *A continuous strictly monotone potential operator $A : R^n \to R^n$ is locally uniformly monotone.*

Proof. Define the function

$$\gamma_R^0(\tau) = \inf \left\{ (Ax - Ay, x - y) \mid \|x - y\| = \tau, \ x, y \in B(\theta_{R^n}, R) \right\}.$$

It is obvious that $\gamma_R^0(0) = 0$. Since the operator A is strictly monotone and continuous, the function $\gamma_R^0(\tau) > 0$ as $\tau > 0$. Moreover, under our conditions, there can be found $x_0, \ y_0 \in B(\theta_{R^n}, R)$ such that $\|x_0 - y_0\| = \tau$ and $(Ax_0 - Ay_0, x_0 - y_0) = \gamma_R^0(\tau)$. We will show that $\gamma_R^0(\tau)$ is an increasing function. Let $\tau_1 < \tau_2$ be given. Then there exist x_2 and y_2 such that

$$\|x_2 - y_2\| = \tau_2, \quad \gamma_R^0(\tau_2) = (Ax_2 - Ay_2, x_2 - y_2).$$

By strong monotonicity of A, we have

$$(Ax_2 - A(y_2 + t(x_2 - y_2)), x_2 - y_2) > 0, \ \ 0 \le t < 1.$$

Hence,

$$(Ax_2 - Ay_2, x_2 - y_2) > (A(y_2 + t(x_2 - y_2)) - Ay_2, x_2 - y_2). \tag{1.3.8}$$

Let

$$w = y_2 + \frac{\tau_1(x_2 - y_2)}{\tau_2}.$$

Then $\|w - y_2\| = \tau_1$. Substitute t in (1.3.8) for $t = \tau_1/\tau_2, \ \ 0 < t < 1$, then

$$\gamma_R^0(\tau_2) > \frac{\tau_2}{\tau_1}(Aw - Ay_2, w - y_2) \ge (Aw - Ay_2, w - y_2) \ge \gamma_R^0(\tau_1).$$

Hence, the function $\gamma_R^0(\tau)$ is increasing, but it is not continuous in general. Therefore, let us proceed with the constructions. It follows from the definition of the function $\gamma_R^0(\tau)$ that

$$(A(y + t(x - y)) - Ay, x - y) \ge \frac{1}{t}\gamma_R^0(t\|x - y\|). \tag{1.3.9}$$

Then

$$0 \le \frac{\gamma_R^0(\tau)}{\tau} \le \frac{1}{\|x - y\|}\Big(A\Big(y + \tau\frac{x - y}{\|x - y\|}\Big) - Ay, x - y\Big),$$

from which we conclude that $\gamma_R^0(\tau)/\tau \to 0$ as $\tau \to 0$ because of continuity of A.

Let $Ax = grad \ \varphi(x)$ be given. Taking into account Lemma 1.2.7 and integrating (1.3.9), we write the inequality

$$\varphi(x) - \varphi(y) - (Ay, x - y) \ge \int_0^1 \gamma_R^0(t\|x - y\|)\frac{dt}{t}. \tag{1.3.10}$$

Interchange x and y in (1.3.10) and add thus obtained inequality to (1.3.10). Then

$$(Ax - Ay, x - y) \ge 2\int_0^1 \gamma_R^0(t\|x - y\|)\frac{dt}{t} = 2\int_0^{\|x-y\|} \frac{\gamma_R^0(\tau)}{\tau}d\tau.$$

Hence, we have constructed the function

$$\gamma_R(t) = 2\int_0^t \frac{\gamma_R^0(\tau)}{\tau}d\tau$$

which is increasing, continuous, $\gamma_R(0) = 0$ and for all $\|x\| \le R$ and for all $\|y\| \le R$

$$(Ax - Ay, x - y) \ge \gamma_R(\|x - y\|).$$

The lemma is proven. ∎

We present the property of the local boundedness of monotone mappings.

Definition 1.3.15 *An operator* $A : X \to 2^{X^*}$ *is said to be locally bounded at a point* $x \in X$ *if there exists a neighborhood* $M = M(x)$ *of this point such that the set*

$$A(M) = \{ y \mid y \in Ax, \ x \in M \cap D(A) \}$$

is bounded in X^*.

Theorem 1.3.16 *A monotone mapping* $A : X \to 2^{X^*}$ *is locally bounded at each interior point of its domain.*

Proof. Prove this theorem by contradiction. Suppose that A is not locally bounded at a point x_0 while $x_0 \in int \, D(A)$. Then there exists a sequence $\{x_n\}$, $x_n \to x_0$, $x_n \in D(A)$, such that $\tau_n = \|y_n\|_* \to +\infty$, where $y_n \in Ax_n$. Denote

$$t_n = max \left\{ \frac{1}{\tau_n}; \ \sqrt{\|x_n - x_0\|} \right\}.$$

It is clear that $t_n > 0$, $t_n \tau_n \geq 1$, $\|x_n - x_0\| \leq t_n^2$ and $\lim_{n \to \infty} t_n = 0$. Construct an element $z_n = x_0 + t_n z$, where any $z \in X$. Since $x_0 \in int \, D(A)$ and since $t_n \to 0$, then $z_n \in D(A)$ for sufficiently large n. Let $u_n \in Az_n$ be given and r be a positive number such that $v = x_0 + rz \in D(A)$. By the monotonicity of A,

$$\langle u_n - f, z_n - v \rangle \geq 0 \quad \forall f \in Av.$$

This implies

$$(t_n - r)\langle u_n - f, z \rangle \geq 0.$$

For $t_n < r$, we have $\langle u_n, z \rangle \leq \langle f, z \rangle$. Hence, $\limsup_{n \to \infty} |\langle u_n, z \rangle| < \infty$. Then the Banach−Steinhaus theorem ensures boundedness of the sequence $\{\|u_n\|_*\}$, That is, there exists a constant $c > 0$ such that $\|u_n\|_* \leq c$ for all $n \geq 1$. Using the monotonicity condition of A again we obtain

$$\langle y_n - u_n, x_n - z_n \rangle = \langle y_n - u_n, x_n - (x_0 + t_n z) \rangle \geq 0.$$

The last inequality yields

$$
\begin{aligned}
\langle y_n, z \rangle &\leq \frac{1}{t_n} \langle y_n, x_n - x_0 \rangle - \langle u_n, \frac{x_n - x_0}{t_n} - z \rangle \\
&\leq \frac{1}{t_n} \|y_n\|_* \|x_n - x_0\| + c(t_n + \|z\|) \leq t_n \tau_n + c(t_n + \|z\|).
\end{aligned}
$$

Then

$$\limsup_{n \to \infty} \frac{|\langle y_n, z \rangle|}{t_n \tau_n} < \infty.$$

Applying the Banach−Steinhaus theorem again one gets the inequality $(t_n \tau_n)^{-1} \|y_n\| < \infty$. At the same time, we have

$$(t_n \tau_n)^{-1} \|y_n\| = \frac{1}{t_n} \to \infty.$$

Thus, we have arrived at the contradiction. The theorem is proved. ∎

Corollary 1.3.17 *Suppose that X is a real Banach space and $A : X \to 2^{X^*}$ is a monotone operator. Let Ω be an open subset of $\overline{D(A)}$. If A is locally bounded at some point of Ω, then it is locally bounded at every point of Ω.*

Corollary 1.3.18 *If an operator $A : X \to 2^{X^*}$ is monotone and $x \in int\ D(A)$, then the set Ax is bounded in X^*.*

Observe that a monotone operator is unbounded, in general (see the example in [128]). However, the following assertion is true [1].

Theorem 1.3.19 *If X is a finite-dimensional Banach space, then any monotone operator $A : X \to 2^{X^*}$ with $D(A) = X$ is bounded.*

Proof. Let $x_n \to \bar{x}$ and $\|y_n\|_* \to \infty$, where $y_n \in Ax_n$. Since X is finite-dimensional, there exists a subsequence $\{x_k\} \subseteq \{x_n\}$ such that $y_k\|y_k\|_*^{-1} \to z$. It is obvious that $\|z\|_* = 1$. Then by the monotonicity of A, we can write

$$\|y_k\|_*^{-1}\langle y_k - y, x_k - x \rangle \geq 0 \quad \forall x \in X, \quad \forall y \in Ax.$$

Now we turn k to ∞ and obtain

$$\langle z, \bar{x} - x \rangle \geq 0 \quad \forall x \in X.$$

Assuming $x = \bar{x} + z$ in the previous inequality we come to the equality $z = \theta_{X^*}$, which contradicts the fact that $\|z\|_* = 1$. ∎

It has been noted in Section 1.1 that hemicontinuity of an operator A follows from its demicontinuity. If A is monotone then the following converse assertion is also fulfilled.

Theorem 1.3.20 *Any monotone hemicontinuous operator $A : X \to X^*$ is demicontinuous on int $D(A)$.*

Proof. Let A be hemicontinuous on $int\ D(A)$, $\{x_n\} \subset int\ D(A)$ be a sequence such that $x_n \to x \in int\ D(A)$. Then by virtue of the local boundedness of A at x, the sequence $\{Ax_n\}$ is bounded beginning with a large enough n. Therefore, we conclude that there exists some subsequence $Ax_{n_k} \rightharpoonup f \in X^*$. Write the monotonicity condition of A:

$$\langle Ax_{n_k} - Ay, x_{n_k} - y \rangle \geq 0 \quad \forall y \in D(A).$$

Passing to the limit in the previous inequality we obtain

$$\langle f - Ay, x - y \rangle \geq 0 \quad \forall y \in D(A). \tag{1.3.11}$$

Since $int\ D(A)$ is an open set, then for all $u \in X$ there exists $\bar{t}$ such that elements $y_t = x + tu \in D(A)$ as $0 \leq t \leq \bar{t}$. If we replace y in (1.3.11) by y_t then

$$\langle f - Ay_t, u \rangle \leq 0 \quad \forall u \in X.$$

Let $t \to 0$ be given. Then the inequality

$$\langle f - Ax, u \rangle \leq 0$$

holds for all $u \in X$ because of the hemicontinuity of A. Hence, $Ax = f$. Thus, $Ax_n \rightharpoonup Ax$, i.e., A is demicontinuous at a point x. ∎

Corollary 1.3.21 *If $A : X \to X^*$ is a monotone hemicontinuous operator, $D(A) = X$, X is a finite-dimensional space, then A is continuous.*

Corollary 1.3.22 *Every linear monotone operator $A : X \to X^*$ with $D(A) = X$ is continuous.*

Proof. Since A is linear, we have $A(x + ty) = Ax + tAy$ for all $t \in (-\infty, +\infty)$. Hence, A is hemicontinuous. It follows from Theorem 1.3.20 that A is demicontinuous. Prove by contradiction that A is continuous. Let $x, x_n \in X$, $x_n \to x$, and $\|Ax_n - Ax\|_* \geq \tau > 0$ be given for all $n \geq 0$. If $t_n = \|x_n - x\|^{-1/2}$ and $y_n = x + t_n(x_n - x)$ we obtain $y_n \to x$ and

$$\|Ay_n - Ax\|_* = t_n \|Ax_n - Ax\|_* \geq t_n \tau \to +\infty,$$

which contradicts the demicontinuity of A. ■

1.4 Maximal Monotone Operators

Let X be a reflexive Banach space and X^* its dual space.

Definition 1.4.1 *A monotone set $G \subseteq X \times X^*$ is called maximal monotone if it is not a proper subset of any monotone set in $X \times X^*$.*

Definition 1.4.2 *An operator $A : X \to 2^{X^*}$ with $D(A) \subseteq X$ is called maximal monotone if its graph is a maximal monotone set of $X \times X^*$.*

From this definition, immediately follows

Proposition 1.4.3 *A monotone operator $A : X \to 2^{X^*}$ is maximal on $D(A)$ if and only if the inequality*

$$\langle g - f, y - x_0 \rangle \geq 0 \quad \forall (y, g) \in grA, \tag{1.4.1}$$

implies the inclusions $x_0 \in D(A)$ and $f \in Ax_0$.

Since the graphs of the operator A and its inverse A^{-1} coincide, then the maximal monotonicity of A^{-1} follows from the maximal monotonicity of A and conversely.

Definition 1.4.4 *A set $G \subseteq X \times X^*$ is called demiclosed if the conditions $x_n \to x$, $y_n \rightharpoonup f$ or $x_n \rightharpoonup x$, $y_n \to f$, where $(x_n, f_n) \in G$, imply that $(x, f) \in G$.*

The following assertion is established by the definition of the maximal monotonicity of operators.

Lemma 1.4.5 *The graph of any maximal monotone operator $A : X \to 2^{X^*}$ is demiclosed.*

Theorem 1.4.6 *Any monotone hemicontinuous operator $A : X \to X^*$ with $D(A) = X$ is maximal monotone.*

Proof. It suffices to prove that the equality $f = Ay$ follows from the inequality

$$\langle f - Ax, y - x \rangle \geq 0 \quad \forall x \in X. \tag{1.4.2}$$

Since $D(A) = X$, it is possible to take in (1.4.2) $x = x_t = y + tz$, where $z \in X$ and $t > 0$. Then

$$\langle f - Ax_t, z \rangle \leq 0$$

for all $z \in X$. Letting $t \to 0$, by hemicontinuity of A on X, we obtain

$$\langle f - Ay, z \rangle \leq 0 \quad \forall z \in X.$$

Hence $f = Ay$. ∎

Theorem 1.4.7 *Let $A : X \to 2^{X^*}$ be a demiclosed single-valued monotone operator with $D(A) = X$. Then A is maximal monotone.*

Proof. We shall prove that the inequality (1.4.1) with any $x \in X$ and any $f \in X^*$ implies $(x, f) \in gr A$. Indeed, fix an element $z \in X$ and take $z_t = x + tz \in X$ for any $t > 0$. Then (1.4.1) with $y = z_t$ gives

$$\langle f - g_t, x - z_t \rangle \geq 0 \quad \forall g_t \in Az_t,$$

that is,

$$\langle g_t - f, z \rangle \geq 0 \quad \forall g_t \in Az_t, \quad \forall z \in X. \tag{1.4.3}$$

If $t \to 0$ then $z_t \to x$. By virtue of the local boundedness of A at a point x, it is possible to assert that $g_t \rightharpoonup \bar{g}(z) \in X^*$. Then we conclude for a demiclosed operator A that $\bar{g}(z) \in Ax$ for all $z \in X$. By (1.4.3),

$$\langle \bar{g}(z) - f, z \rangle \geq 0 \quad \forall z \in X.$$

Since A is single-valued, we have $\bar{g}(z) = Ax$. Thus,

$$\langle Ax - f, z \rangle \geq 0 \quad \forall z \in X.$$

The last inequality asserts that $Ax = f$. Then Proposition 1.4.3 proves the claim. ∎

Theorem 1.4.8 *Let $A : X \to 2^{X^*}$ be a monotone demiclosed operator such that $D(A) = X$ and for each $x \in X$ the image Ax is a nonempty convex subset of X^*. Then A is maximal monotone.*

Proof. As in the previous theorem, we prove that the inequality (1.4.1) with any $x \in X$ and any $f \in X^*$ implies $(x, f) \in gr A$. Suppose that, on the contrary, $f \notin Ax$. Since Ax is convex and A is demiclosed, Ax is weakly closed and, therefore, closed. According to the strong separation theorem, there exists an element $z \in X$ such that

$$\langle f, z \rangle > \sup \{ \langle g, z \rangle \mid g \in Ax \}. \tag{1.4.4}$$

Take $z_t = x + tz \in X$ and let $g_t \in Az_t$ for any $t > 0$. Similarly to Theorem 1.4.7, as $t \to 0$, one gets

$$\langle \bar{g}(z) - f, z \rangle \geq 0, \quad \bar{g}(z) \in Ax.$$

This contradicts (1.4.4). Consequently, $f \in Ax$ and the conclusion follows from Proposition 1.4.3 again. ∎

Theorem 1.4.9 *If $A : X \to 2^{X^*}$ is a maximal monotone operator, then the set $\{f \mid f \in Ax\}$ for every $x \in D(A)$ is convex and closed in X^*.*

Proof. Let $f_1 \in Ax$ and $f_2 \in Ax$. It is clear that the monotonicity of A yields the inequalities

$$\langle f_1 - g, x - y \rangle \geq 0 \qquad (1.4.5)$$

and

$$\langle f_2 - g, x - y \rangle \geq 0 \qquad (1.4.6)$$

for all $(y, g) \in grA$. Let $f = tf_1 + (1 - t)f_2$, where $t \in [0, 1]$. Multiply (1.4.5) and (1.4.6) by t and $1 - t$, respectively, and add them. Then we get the following inequality:

$$\langle f - g, x - y \rangle \geq 0 \quad \forall (y, g) \in grA.$$

The maximal monotonicity of the operator A implies then the inclusion $f \in Ax$, i.e., the set $\{f \mid f \in Ax\}$ is convex.

Let $f_n \to \bar{f}$, $f_n \in Ax$. We have

$$\langle f_n - g, x - y \rangle \geq 0 \quad \forall (y, g) \in grA,$$

and the limit passing as $n \to \infty$ gives

$$\langle \bar{f} - g, x - y \rangle \geq 0 \quad \forall (y, g) \in grA.$$

Hence, $\bar{f} \in Ax$, and the theorem is proved. ∎

Corollary 1.4.10 *If $A : X \to 2^{X^*}$ is a maximal monotone operator, then the set $\{x \mid f \in Ax\}$ for every $f \in R(A)$ is convex and closed in X.*

Proof. The proof follows from Theorem 1.4.9 applied to the inverse mapping A^{-1}. ∎

Consider a linear, possibly multiple-valued, operator $L : X \to 2^{X^*}$, for which grL is a linear subspace of $X \times X^*$. The monotonicity condition for it can be written in the following form:

$$\langle f, x \rangle \geq 0 \quad \forall (x, f) \in grL.$$

Let $L^* : X \to 2^{X^*}$ be the adjoint (conjugate) operator to L. It is defined as follows:

$$g \in L^*y \quad \text{implies} \quad \langle g, x \rangle = \langle f, y \rangle \quad \forall (x, f) \in gr\, L. \qquad (1.4.7)$$

An operator L is self-adjoint if $L^* = L$.

Theorem 1.4.11 *Assume that X is a reflexive strictly convex Banach space together with its dual space X^*. A linear monotone operator $L : X \to 2^{X^*}$ is maximal monotone if and only if it is closed and L^* is a monotone mapping.*

Proof. Let L be maximal monotone. Then, by Lemma 1.4.5, it is closed. We will prove that the conjugate operator L^* is monotone. Indeed, in view of (1.4.7) and the monotonicity of L, we have for all $(x, f) \in grL$ and for all $(y, g) \in grL^*$,

$$\langle f + g, x - y \rangle = \langle f, x \rangle - \langle f, y \rangle + \langle g, x \rangle - \langle g, y \rangle$$
$$= \langle f, x \rangle - \langle g, y \rangle \geq -\langle g, y \rangle. \tag{1.4.8}$$

If $\langle g, y \rangle < 0$ then $\langle f + g, x - y \rangle > 0$ for all $(x, f) \in grL$. Since L is maximal monotone, $(y, -g) \in grL$. Then assuming in (1.4.8) that $x = y$ and $f = -g$ we arrive at the contradiction. Thus, L^* is monotone. The first part of the lemma is proved. Let L be closed and L^* be a monotone operator. In order to prove the maximal monotonicity of L, it is sufficient to show that the inequality

$$\langle f - h, x - z \rangle \geq 0 \quad \forall (x, f) \in grL, \ z \in X, \ h \in X^*, \tag{1.4.9}$$

implies that $(z, h) \in grL$. Introduce on the set $grL \subseteq X \times X^*$ the norm by the formula

$$\|(x, f)\| = \|x\| + \|f\|_*, \quad (x, f) \in grL.$$

Note that grL is a strictly convex space with respect to this norm. Construct in grL the functional

$$\varphi(x, f) = 2^{-1}\|f - h\|_*^2 + 2^{-1}\|x - z\|^2 + \langle f - h, x - z \rangle, \quad (x, f) \in grL.$$

This functional is continuous, convex and due to (1.4.9), $\varphi(x, f) \to +\infty$ as $\|(x, f)\| \to \infty$. Then, by Theorem 1.1.23, there exists a point $(z_0, h_0) \in grL$ in which $\varphi(x, f)$ reaches a minimum, that is, $\varphi'(z_0, h_0) = (\theta_{X^*}, \theta_X)$. Consequently, $\langle \varphi'(z_0, h_0), (x, f) \rangle = 0$ for all $(x, f) \in grL$. It can be written in the expanded form: For all $(x, f) \in grL$,

$$\langle f, J^*(h_0 - h) \rangle + \langle J(z_0 - z), x \rangle + \langle f, z_0 - z \rangle + \langle h_0 - h, x \rangle = 0,$$

that is,

$$\langle f, J^*(h_0 - h) + z_0 - z \rangle = \langle J(z - z_0) + h - h_0, x \rangle.$$

Then by (1.4.7), we deduce that

$$J(z - z_0) + h - h_0 \in L^*(J^*(h_0 - h) + z_0 - z).$$

Now the monotonicity of L^* leads to the following inequality:

$$\langle J(z - z_0) + h - h_0, J^*(h_0 - h) + z_0 - z \rangle \geq 0.$$

Here $J : X \to X^*$ and $J^* : X^* \to X$ are normalized duality mappings in X and X^*, respectively. Taking into account the definitions of duality mappings one gets

$$\|h - h_0\|_*^2 + \|z - z_0\|^2 + \langle h - h_0, z - z_0 \rangle \leq \|z - z_0\|\|h - h_0\|_*,$$

where $(z_0, h_0) \in grL$. In view of (1.4.9), the last expression implies

$$\|h - h_0\|_*^2 + \|z - z_0\|^2 \leq \|z - z_0\|\|h - h_0\|_*.$$

It follows that $h = h_0$, $z = z_0$, hence, we have proved the inclusion $(z, h) \in grL$. ∎

Theorem 1.4.12 *If $L : X \to X^*$ is a linear single-valued maximal monotone operator, then $D(L)$ is dense in X.*

Proof. Indeed, let

$$\langle g, x \rangle = 0 \quad \forall x \in D(L), \ g \in X^*.$$

Then

$$\langle Lx - g, x - \theta_X \rangle = \langle Lx, x \rangle \geq 0$$

for every $x \in D(L)$. Due to the maximal monotonicity of L, we conclude that $g = L(\theta_X) = \theta_{X^*}$. ∎

The additional properties of maximal monotone operators are also given in Section 1.8.

1.5   Duality Mappings

In Section 1.1 we introduced the definition of the normalized duality mapping. The more general concept of a duality mapping is given by

Definition 1.5.1 *Let $\mu(t)$ be a continuous increasing function for $t \geq 0$ such that $\mu(0) = 0$, $\mu(t) \to \infty$ as $t \to \infty$. An operator $J^\mu : X \to 2^{X^*}$ is called a duality mapping with the gauge function $\mu(t)$ if the equalities*

$$\|y\|_* = \mu(\|x\|), \ \langle y, x \rangle = \mu(\|x\|)\|x\|$$

are satisfied for all $x \in X$ and for all $y \in J^\mu x$.

If $\mu(t) = t$ then J^μ coincides with the normalized duality mapping J.

Lemma 1.5.2 *A duality mapping $J^\mu : X \to 2^{X^*}$ exists in any Banach space and its domain is all of X.*

Proof. Take an arbitrary $x \in X$. Due to Corollary 1.1.2 of the Hahn−Banach theorem, there exists at least one element $\phi \in X^*$ such that $\|\phi\|_* = 1$ and $\langle \phi, x \rangle = \|x\|$. Then it is not difficult to verify that $\psi = \mu(\|x\|)\phi$ satisfies the inclusion $\psi \in J^\mu x$. ∎

Lemma 1.5.2 implies

Corollary 1.5.3 *If $J^\mu : X \to 2^{X^*}$ is a one-to-one operator at a point x, then*

$$J^\mu x = \mu(\|x\|)e^*,$$

where $e^* \in X^*$, $\langle e^*, x \rangle = \|x\|$, $\|e^*\|_* = 1$.

We present the other properties of the operator J^μ.

Lemma 1.5.4 *In any Banach space X, the duality mapping J^μ is a bounded monotone and coercive operator and it satisfies the inequality*

$$\langle u - v, x - y \rangle \geq \Big(\mu(\|x\|) - \mu(\|y\|)\Big)(\|x\| - \|y\|) \tag{1.5.1}$$

for all $u \in J^\mu x$, $v \in J^\mu y$ and for all $x, y \in X$.

If J^μ is single-valued then

$$\langle J^\mu y - J^\mu y, x - y \rangle \geq \Big(\mu(\|x\|) - \mu(\|y\|)\Big)(\|x\| - \|y\|) \quad \forall x, y \in X. \tag{1.5.2}$$

In this case, Lemma 1.5.4 gives the following important estimate for normalized duality mapping J :

$$\langle Jx - Jy, x - y \rangle \geq (\|x\| - \|y\|)^2 \quad \forall x, y \in X. \tag{1.5.3}$$

Lemma 1.5.5 *If X is a strictly convex space, then J^μ is a strictly monotone operator. If X^* is strictly convex, then J^μ is single-valued. If X is reflexive and dual space X^* is strictly convex, then J^μ is demicontinuous.*

Thus, if X is a reflexive strictly convex Banach space with strictly convex dual space X^* then J^μ is a single-valued demicontinuous (hence, hemicontinuous) and strictly monotone operator.

Lemma 1.5.6 *If the space X is reflexive and strongly smooth, then J^μ is continuous.*

Introduce in X the functional

$$\Phi(\|x\|) = \int_0^{\|x\|} \mu(t)dt, \tag{1.5.4}$$

and prove the following result:

Lemma 1.5.7 *The functional Φ defined by (1.5.4) in a strictly convex Banach space X with strictly convex X^* has the Gâteaux derivative Φ' and $\Phi'(\|x\|) = J^\mu x$ for all $x \in X$.*

Proof. Indeed, (1.5.4) implies

$$\Phi(\|y\|) - \Phi(\|x\|) = \int_{\|x\|}^{\|y\|} \mu(t) \; dt.$$

Since $\mu(t)$ is an increasing function, one gets

$$\int_{\|x\|}^{\|y\|} \mu(t) \; dt \geq \mu(\|x\|)(\|y\| - \|x\|).$$

By Definition 1.5.1, we have

$$\mu(\|x\|)\|y\| - \mu(\|x\|)\|x\| \geq \langle J^\mu x, y \rangle - \langle J^\mu x, x \rangle = \langle J^\mu x, y - x \rangle.$$

Thus,

$$\Phi(\|y\|) - \Phi(\|x\|) \geq \langle J^\mu x, y - x \rangle.$$

According to Definition 1.2.1, this means that $J^\mu x \in \partial\Phi(\|x\|)$, where $\partial\Phi(t)$ is a subgradient of $\Phi(t)$.

Let now $u \in \partial\Phi(\|x\|)$. Then we have

$$\Phi(\|y\|) - \Phi(\|x\|) \geq \langle u, y - x \rangle \quad \forall y \in X. \tag{1.5.5}$$

Choose in (1.5.5) an element $y \in X$ such that $\|x\| = \|y\|$. Then $\langle u, y - x \rangle \leq 0$, i.e.,

$$\langle u, y \rangle \leq \langle u, x \rangle \quad \forall y \in X. \tag{1.5.6}$$

It follows from strict convexity of the space X that

$$sup \{\langle u, y \rangle \mid \|y\| = r\} = \langle u, z \rangle = \|u\|_* \|z\|.$$

Therefore, by (1.5.6), if $r = \|x\|$ then $z = x$. Hence,

$$\langle u, x \rangle = \|u\|_* \|x\|. \tag{1.5.7}$$

Assume that in (1.5.5) $x = tv$, $y = sv$, $\|v\| = 1$, $t, s \in R_+^1$. Then

$$\Phi(s) - \Phi(t) \geq (s - t)\langle u, v \rangle = \frac{s - t}{t}\langle u, x \rangle = \frac{s - t}{t}\|u\|_* \|x\| = (s - t)\|u\|_*.$$

Hence, the following limit-relation is valid:

$$\lim_{s \to t} \frac{\Phi(s) - \Phi(t)}{s - t} = \|u\|_*,$$

which implies that $\mu(\|x\|) = \|u\|_*$. By (1.5.7) and by the definition of J^μ, we conclude that $u = J^\mu x$. Consequently, $\Phi'(\|x\|) = J^\mu x$. The lemma is proved. ∎

Corollary 1.5.8 *In a smooth Banach space, any normalized duality mapping is defined by the formula (1.1.12). Moreover,*

$$J^\mu x = \frac{\mu(\|x\|)}{\|x\|} Jx. \tag{1.5.8}$$

If X is a Hilbert space H then J is the identity operator I, which is linear in Hilbert spaces. Inversely, if the operator J is linear in X then X is a Hilbert space. Indeed, let $J : X \to X^*$ be a linear operator, $f = Jx$, $g = Jy$. Then

$$f + g = J(x + y), \quad f - g = J(x - y).$$

We deduce

$$\|x + y\|^2 = \langle f + g, x + y \rangle = \|x\|^2 + \langle f, y \rangle + \langle g, x \rangle + \|y\|^2,$$
$$\|x - y\|^2 = \langle f - g, x - y \rangle = \|x\|^2 - \langle f, y \rangle - \langle g, x \rangle + \|y\|^2.$$

Summing up those inequalities, we obtain the parallelogram equality

$$\|x + y\|^2 + \|x - y\|^2 = 2\|x\|^2 + 2\|y\|^2 \quad \forall x, y \in X,$$

which is possible only in a Hilbert space. Thus, the following assertion has been proven:

Theorem 1.5.9 *A normalized duality map* $J : X \to X^*$ *is linear if and only if* X *is a Hilbert space.*

The analytical representations of dual mappings are known in a number of Banach spaces. For instance, in the spaces l^p, $L^p(G)$ and $W_m^p(G)$, $p \in (1, \infty)$, respectively,

$$Jx = \|x\|_{l^p}^{2-p} y \in l^q, \quad y = \{|x_1|^{p-2}x_1, |x_2|^{p-2}x_2, ...\}, \quad x = \{x_1, x_2, ...\},$$

$$Jx = \|x\|_{L^p}^{2-p}|x(s)|^{p-2}x(s) \in L^q(G), \quad s \in G,$$

and

$$Jx = \|x\|_{W_m^p}^{2-p} \sum_{|\alpha| \leq m} (-1)^{|\alpha|} D^\alpha(|D^\alpha x(s)|^{p-2} D^\alpha x(s)) \in W_{-m}^q(G), \quad m > 0, \quad s \in G,$$

where $p^{-1} + q^{-1} = 1$. If $\mu(t) = t^{p-1}$ then the duality mappings $J^\mu = J^p$ with a gauge function $\mu(t)$ have a simpler form. Namely,
In l^p :

$$J^p x = \{|x_1|^{p-2}x_1, |x_2|^{p-2}x_2, ...\};$$

In $L^p(G)$:

$$J^p x = |x(s)|^{p-2}x(s);$$

In $W_m^p(G)$:

$$J^p x = \sum_{|\alpha| \leq m} (-1)^{|\alpha|} D^\alpha(|D^\alpha x(s)|^{p-2} D^\alpha x(s)).$$

If X is a reflexive strictly convex Banach space together with its dual space X^*, then the duality mapping J^* in X^* is an operator acting from X^* into X. It is obvious that $J^* = J^{-1}$, i.e., for all $x \in X$ and for all $\phi \in X^*$, the equalities $JJ^*\phi = \phi$ and $J^*Jx = x$ hold. At the same time, a more general assertion is true:

Lemma 1.5.10 *Let* X *be a reflexive strictly convex Banach space with strictly convex dual space* X^*. *If* $J^\mu : X \to X^*$ *and* $(J^\nu)^* : X^* \to X$ *are duality mappings with gauge functions* $\mu(t)$ *and* $\nu(s)$, *respectively, and* $\nu(s) = \mu^{-1}(t)$, *then* $(J^\nu)^* = (J^\mu)^{-1}$.

Corollary 1.5.11 *Let* X *be a reflexive strictly convex Banach space with strictly convex dual space* X^*. *If* $J^p : X \to X^*$ *and* $(J^q)^* : X^* \to X$ *are the duality mapping with gauge functions* $\mu(t) = t^{p-1}$ *and* $\nu(s) = s^{q-1}$, $p^{-1} + q^{-1} = 1$, *respectively, then* $(J^q)^* = (J^p)^{-1}$.

For instance, if in the space $L^p(G)$ the duality mapping is J^p, then $(J^q)^*$ is expressed in the explicit form as follows:

$$(J^q)^* y = |y(s)|^{q-2}y(s) \in L^p(G), \quad s \in G, \quad (J^p)^{-1} = (J^q)^*, \quad [(J^q)^*]^{-1} = J^p.$$

However, there are Banach spaces in which $(J^\nu)^*$ is not explicitly known but it can be found by means of solving some boundary value problem. Observe that such a situation arises in Sobolev spaces $W_m^p(G)$ [83].

Remark 1.5.12 *In spaces l^p, $p > 1$, the duality mapping J^p is weak-to-weak continuous, while in spaces $L^p(G)$, $p > 1$, $p \neq 2$, any J^μ has not this property. In a Hilbert space $Jx = x$, therefore, the weak-to-weak continuity of J is obvious.*

Next we will present results which are of interest for applications and prove them by making use of duality mappings.

Lemma 1.5.13 (Opial) *If in a Banach space X having a weak-to-weak continuous duality mapping J^μ the sequence $\{x_n\}$ is weakly convergent to x, then for any $y \in X$,*

$$\liminf_{n \to \infty} \|x_n - x\| \leq \liminf_{n \to \infty} \|x_n - y\|. \tag{1.5.9}$$

If, in addition, the space X is uniformly convex, then the equality in (1.5.9) occurs if and only if $x = y$.

Proof. Since J^μ is weak-to-weak continuous and $\{x_n\}$ weakly converges to x,

$$\lim_{n \to \infty} \langle J^\mu(x_n - x), y - x \rangle = 0.$$

Then

$$\lim_{n \to \infty} \langle J^\mu(x_n - x), x_n - x \rangle = \lim_{n \to \infty} \langle J^\mu(x_n - x), x_n - y \rangle.$$

Therefore, by Definition 1.5.1,

$$\begin{aligned}
\liminf_{n \to \infty} \mu(\|x_n - x\|)\|x_n - x\| &= \liminf_{n \to \infty} |\langle J^\mu(x_n - x), x_n - y \rangle| \\
&\leq \liminf_{n \to \infty} \|J^\mu(x_n - x)\|_* \|x_n - y\| \\
&= \liminf_{n \to \infty} \mu(\|x_n - x\|)\|x_n - y\|.
\end{aligned}$$

This implies (1.5.9). Let now in (1.5.9) $x \neq y$ and both limits be equal. Then for any point $z \in x + t(y - x)$, $0 < t < 1$, we would have

$$\liminf_{n \to \infty} \|x_n - z\| < \liminf_{n \to \infty} \|x_n - y\|$$

which is impossible in a uniformly convex Banach space X. ∎

Lemma 1.5.14 *Let $A : X \to 2^{X^*}$ be a monotone operator and some $x_0 \in int\, D(A)$. Then there exists a constant $r_0 > 0$ and a ball $B(x_0, r_0)$ such that for any $x \in D(A)$ and $y \in Ax$ the following inequality holds:*

$$\langle y, x - x_0 \rangle \geq \|y\|_* r_0 - c_0(\|x - x_0\| + r_0), \tag{1.5.10}$$

where

$$c_0 = \sup \{\|y\|_* \mid y \in Ax, \ x \in B(x_0, r_0)\} < \infty.$$

Proof. Since $x_0 \in int \, D(A)$, we conclude by Theorem 1.3.16 that there exists $r_0 > 0$ such that c_0 is defined. Choose an element $x \in D(A)$ and construct $z_0 = x_0 + r_0 J^* h$, where $h = \|y\|_*^{-1} y$ and $y \in Ax$. The monotonicity condition of A yields the inequality

$$\langle y - y_0, x - z_0 \rangle \geq 0 \quad \forall y \in Ax, \quad y_0 \in Az_0.$$

Therefore,

$$\frac{\langle h, x - z_0 \rangle}{\|x - z_0\|} = \left(\langle y - y_0, x - z_0 \rangle + \langle y_0, x - z_0 \rangle \right) \frac{1}{\|x - z_0\| \|y\|_*} \geq -\frac{c_0}{\|y\|_*}.$$

Hence,

$$\langle r_0 h, x - z_0 \rangle \geq -\frac{c_0 r_0}{\|y\|_*} \|x - z_0\| \geq -\frac{c_0 r_0}{\|y\|_*} (\|x - x_0\| + r_0).$$

Taking into consideration that $z_0 = x_0 + r_0 J^* h$, one gets $J(z_0 - x_0) = r_0 h$, and thus the following calculations are valid:

$$\langle r_0 h, x - x_0 \rangle = \langle r_0 h, z_0 - x_0 \rangle + \langle r_0 h, x - z_0 \rangle \geq r_0^2 - c_0 r_0 \|y\|_*^{-1} (\|x - x_0\| + r_0). \quad (1.5.11)$$

Multiplying (1.5.11) by $r_0^{-1} \|y\|_*$, we deduce (1.5.10). The lemma is completely proved. ∎

Lemma 1.5.15 *Assume that X is a reflexive strictly convex Banach space and a sequence $\{x_n\} \subset X$. If for some $x \in X$,*

$$\langle Jx_n - Jx, x_n - x \rangle \to 0 \quad \text{as} \quad n \to \infty,$$

then $x_n \rightharpoonup x$. In addition, if X is E-space, then $x_n \to x$ in X.

Proof. Since

$$\langle Jx_n - Jx, x_n - x \rangle \geq (\|x_n\| - \|x\|)^2,$$

it follows from the conditions of the lemma that $\|x_n\| \to \|x\|$. The following equality is obvious:

$$\langle Jx_n - Jx, x_n - x \rangle = (\|x_n\| - \|x\|)^2 + (\|x\| \|x_n\| - \langle Jx, x_n \rangle) + (\|x_n\| \|x\| - \langle Jx_n, x \rangle),$$

from which we have that $\langle Jx, x_n \rangle \to \langle Jx, x \rangle$. It is well known that in this case the sequence $\{x_n\}$ is bounded. Then there exists a subsequence $\{x_m\} \subseteq \{x_n\}$ such that $x_m \rightharpoonup \bar{x} \in X$. Thus, $\langle Jx, x_m \rangle \to \langle Jx, \bar{x} \rangle$ and $\langle Jx, \bar{x} \rangle = \langle Jx, x \rangle$. Since the duality map J^* is single-valued in X^*, it follows from the latter equality that $\bar{x} = x$, that is, $x_m \rightharpoonup x$. This means that any subsequence weakly converges to x. Consequently, the whole sequence $x_n \rightharpoonup x$. The last assertion of the lemma results now from the definition of the E-space. ∎

Corollary 1.5.16 *If X^* is E-space and X and X is strictly convex, then the normalized duality mapping J is continuous.*

Proof. Let $x_n \to x$. Then $\langle Jx_n - Jx, x_n - x \rangle \to 0$, hence, $\langle y_n - y, J^* y_n - J^* y \rangle \to 0$, where $y_n = Jx_k$, $y = Jx$. By Lemma 1.5.15, one gets that $y_n \to y$, i.e., $Jx_n \to Jx$. ∎

In Section 1.3 we proved that the metric projection operator P_Ω onto closed convex subset $\Omega \subset H$ satisfies the condition

$$\langle x - P_\Omega x, P_\Omega x - y \rangle \geq 0 \quad \forall y \in \Omega, \quad \forall x \in H.$$

Show that P_Ω has the similar property also in a Banach space. To this end, we now apply the normalized duality mapping J. As above, assume that X is reflexive and strictly convex together with dual space X^*. Under these conditions, the metric projection operator P_Ω is well defined for all $x \in X$ and single-valued.

Lemma 1.5.17 *For all $x \in X$, the element $z = P_\Omega x$ if and only if*

$$\langle J(x - z), z - y \rangle \geq 0 \quad \forall y \in \Omega. \tag{1.5.12}$$

Proof. Since $y, z \in \Omega$, we conclude that $(1 - t)z - ty \in \Omega$ for all $t \in [0, 1]$. By the definition of the metric projection $z = P_\Omega x$, one gets

$$\|x - z\| \leq \|x - (1 - t)z - ty\| \quad \forall y \in \Omega. \tag{1.5.13}$$

Then Corollary 1.5.8 implies

$$2 \langle J(x - z - t(y - z)), t(y - z) \rangle \leq \|x - z\|^2 - \|x - (1 - t)z - ty\|^2.$$

In view of (1.5.13), there holds the inequality

$$\langle J(x - z - t(y - z)), y - z \rangle \leq 0.$$

Letting here $t \to 0$ and using Lemma 1.5.5 we come to (1.5.12).

Suppose that (1.5.12) is now satisfied. Then

$$\|x - y\|^2 - \|x - z\|^2 \geq 2 \langle J(x - z), z - y \rangle \geq 0 \quad \forall y \in \Omega.$$

Hence, $\|x - z\| \leq \|x - y\|$ for all $y \in \Omega$, that is, $z = P_\Omega x$ by definition of P_Ω. ∎

Observe that unlike a Hilbert space (see Example 4 in Section 1.3) the metric projection operator onto a convex closed subset of a Banach space X is not necessarily contractive. Besides, since this operator acts from X to X, thus, the monotonicity property for it does not make sense. However, the following assertion is still valid:

Lemma 1.5.18 *The operator $A = J(I - P_\Omega) : X \to X^*$ is monotone, bounded and demi-continuous.*

Proof. The boundedness and demicontinuity of A is easily proved by using the properties of the operators J and P_Ω. Show that A is monotone. Take arbitrary elements $x, y \in X$. Then, by Lemma 1.5.17, we can write the inequality

$$\langle J(x - P_\Omega x), u - P_\Omega x \rangle \leq 0 \quad \forall u \in \Omega$$

and

$$\langle J(y - P_\Omega y), u - P_\Omega y \rangle \leq 0 \quad \forall u \in \Omega.$$

Substituting $u = P_\Omega y$ for the first inequality and $u = P_\Omega x$ for the second one, we sum them and thus obtain

$$\langle J(x - P_\Omega x) - J(y - P_\Omega y), P_\Omega x - P_\Omega y \rangle \geq 0. \tag{1.5.14}$$

It is easy to verify that

$$
\begin{aligned}
\langle Ax - Ay, x - y \rangle &= \langle J(x - P_\Omega x) - J(y - P_\Omega y), x - P_\Omega x - y + P_\Omega y \rangle \\
&+ \langle J(x - P_\Omega x) - J(y - P_\Omega y), P_\Omega x - P_\Omega y \rangle.
\end{aligned}
$$

It implies the monotonicity of A because of (1.5.14) and monotonicity of J. ∎

Definition 1.5.19 *Let Ω be a non-empty closed convex subset of X. A mapping $Q_\Omega : X \to \Omega$ is called*
(i) a retraction onto Ω if $Q_\Omega^2 = Q_\Omega$;
(ii) a nonexpansive retraction if it also satisfies the inequality

$$\|Q_\Omega x - Q_\Omega y\| \leq \|x - y\| \quad \forall x, y \in X;$$

(iii) a sunny retraction if for all $x \in X$ and for all $0 \leq t < \infty$,

$$Q_\Omega(Q_\Omega x + t(x - Q_\Omega x)) = Q_\Omega x.$$

Proposition 1.5.20 *Let Ω be a non-empty closed convex subset of X, J^μ be a duality mapping with gauge function $\mu(t)$. A mapping $Q_\Omega : X \to \Omega$ is a sunny nonexpansive retraction if and only if for all $x \in X$ and for all $\xi \in \Omega$,*

$$\langle J^\mu(Q_\Omega x - \xi), x - Q_\Omega x \rangle \geq 0.$$

1.6   Banach Spaces Geometry and Related Duality Estimates

The properties of duality mappings are defined by the properties of spaces X and X^*. In particular, it is well known that a duality mapping is uniformly continuous on every bounded set in a uniformly smooth Banach space X, that is, for every $R > 0$ and arbitrary $x, y \in X$ with $\|x\| \leq R$, $\|y\| \leq R$, there exists a real non-negative and continuous function $\omega_R : [0, \infty) \to R^1$ such that $\omega_R(t) > 0$ if $t > 0$ and $\omega_R(0) = 0$, for which the inequality

$$\|Jx - Jy\|_* \leq \omega_R(\|x - y\|) \tag{1.6.1}$$

holds. Furthermore, a duality mapping is uniformly monotone on every bounded set in a uniformly convex Banach space X. In other words, for every $R > 0$ and arbitrary $x, y \in X$ with $\|x\| \leq R$, $\|y\| \leq R$, there exists a real non-negative and continuous function $\psi_R : [0, \infty) \to R^1$ such that $\psi_R(t) > 0$ for $t > 0$, $\psi(0) = 0$ and

$$\langle Jx - Jy, x - y \rangle \geq \psi_R(\|x - y\|). \tag{1.6.2}$$

Our aim is to find in the analytical form the functions $\omega_R(t)$ and $\psi_R(t)$ and also the function $\tilde{\omega}_R(t)$ evaluating the left-hand side of (1.6.2) from above such that

$$\langle Jx - Jy, x - y \rangle \leq \tilde{\omega}_R(\|x - y\|). \tag{1.6.3}$$

Estimates (1.6.1) - (1.6.3) play a fundamental role in the convergence and stability analysis of approximation methods for nonlinear problems in Banach spaces.

Recall that $\delta_X(\epsilon)$ and $\rho_X(\tau)$ denote, respectively, the modulus of convexity and the modulus of smoothness of a Banach space X. It is known that in a uniformly smooth Banach space X the following inequality holds for any $0 < \tau \leq \sigma$:

$$\tau^2 \rho_X(\sigma) \leq L\sigma^2 \rho_X(\tau), \tag{1.6.4}$$

where $1 < L < 1.7$ is the Figiel constant.

Theorem 1.6.1 *In a uniformly smooth Banach space* X, *for every* $R > 0$ *and arbitrary* $x, y \in X$ *such that* $\|x\| \leq R$, $\|y\| \leq R$, *the inequality*

$$\langle Jx - Jy, x - y \rangle \leq 8\|x - y\|^2 + c_1 \rho_X(\|x - y\|) \tag{1.6.5}$$

is satisfied with $c_1 = 8max\{L, R\}$.

Proof. Denote
$$D = 2^{-1}(\|x\|^2 + \|y\|^2 - 2^{-1}\|x + y\|^2). \tag{1.6.6}$$
1) Let $\|x + y\| \leq \|x - y\|$. Then

$$\|x\| + \|y\| \leq \|x + y\| + \|x - y\| \leq 2\|x - y\|, \tag{1.6.7}$$

from which we easily obtain the inequality

$$2^{-1}\|x\|^2 + 2^{-1}\|y\|^2 + \|x\|\|y\| \leq 2\|x - y\|^2. \tag{1.6.8}$$

Subtracting $\|2^{-1}(x + y)\|^2$ from both parts of (1.6.8), we deduce

$$D \leq 2\|x - y\|^2 - (4^{-1}\|x + y\|^2 + \|x\|\|y\|).$$

If we assume that
$$\|2^{-1}(x + y)\|^2 + \|x\|\|y\| \geq \|x - y\|^2, \tag{1.6.9}$$
then we will have at once

$$D \leq \|x - y\|^2. \tag{1.6.10}$$

Suppose that a contrary inequality to (1.6.9) holds. In this case, from the inequality

$$(\|x\| - \|y\|)^2 \leq \|x + y\|^2,$$

which can be re-written as

$$\|x\|^2 - 2\|x\|\|y\| + \|y\|^2 \leq \|x + y\|^2,$$

it immediately follows that

$$D = 2^{-1}\|x\|^2 + 2^{-1}\|y\|^2 - \|2^{-1}(x + y)\|^2 \leq \|2^{-1}(x + y)\|^2 + \|x\|\|y\| \leq \|x - y\|^2.$$

Thus, (1.6.10) is satisfied again.

2) Assume now that $\|x + y\| > \|x + y\|$ and show that

$$\|x\| + \|y\| - \|x - y\| \leq \epsilon(x, y), \tag{1.6.11}$$

where

$$\epsilon(x, y) = \|x + y\|\rho_X\Big(\frac{\|x - y\|}{\|x + y\|}\Big).$$

Indeed, making the substitutions $x = 2^{-1}(u + v)$ and $y = 2^{-1}(u - v)$ for left-hand side of (1.6.11) and after setting $u^0 = u\|u\|^{-1}$ and $v^0 = v\|u\|^{-1}$, we can write the following obvious estimates:

$$\|x\| + \|y\| \quad - \quad \|x + y\| = 2^{-1}(\|u + v\| + \|u - v\|) - \|u\|$$

$$= \quad 2^{-1}\|u\|(\|u^0 + v^0\| + \|u^0 - v^0\| - 2)$$

$$\leq \quad \|u\|\sup\{2^{-1}(\|u^0 + v^0\| + \|u^0 - v^0\|) - 1 \mid \|u^0\| = 1, \|v^0\| = \tau\}$$

$$= \quad \|u\|\rho_X(\|v^0\|).$$

Returning to the previous denotations we obtain (1.6.11) which implies

$$\Big\|\frac{x + y}{2}\Big\| \geq \frac{\|x\| + \|y\| - \epsilon(x, y)}{2}. \tag{1.6.12}$$

We assert that the right-hand side of (1.6.12) is non-negative. In fact, by the property $\rho_X(\tau) \leq \tau$ (see Section 1.1), one establishes the inequality

$$\|x\| + \|y\| - \epsilon(x, y) \geq \|x\| + \|y\| - \|x - y\| \geq 0.$$

Then

$$\Big\|\frac{x + y}{2}\Big\|^2 \geq \Big(\frac{\|x\| + \|y\|}{2}\Big)^2 - \epsilon(x, y)\frac{\|x\| + \|y\|}{2}.$$

Since $|\,\|x\| - \|y\|\,| \le \|x - y\|$, we deduce

$$D \le \Big(\frac{\|x\| - \|y\|}{2}\Big)^2 + \epsilon(x, y)\frac{\|x\| + \|y\|}{2} \le \frac{\|x - y\|^2}{4} + \epsilon(x, y)\frac{\|x\| + \|y\|}{2}. \qquad (1.6.13)$$

Suppose that $\|x + y\| \le 1$. Then $\|x + y\|^{-1}\|x - y\| \ge \|x - y\|$. By (1.6.13) and (1.6.4), we have

$$\rho_X\Big(\frac{\|x - y\|}{\|x + y\|}\Big) \le \frac{L\rho_X(\|x - y\|)}{\|x + y\|^2}.$$

This inequality yields the estimate

$$D \le 4^{-1}\|x - y\|^2 + 2^{-1}L(\|x\| + \|y\|)\|x + y\|^{-1}\rho_X(\|x - y\|).$$

Since $\|x + y\| > \|x - y\|$ by the hypothesis, one gets

$$2^{-1}\|x + y\|^{-1}(\|x\| + \|y\|) \le (2\|x + y\|)^{-1}(\|x + y\| + \|x - y\|) \le 1.$$

Therefore,

$$D \le \frac{\|x - y\|^2}{4} + L\rho_X(\|x - y\|). \qquad (1.6.14)$$

Assume next that $\|x + y\| \ge 1$. Taking into account (1.6.13) and the convexity of the function $\rho_X(\tau)$, we deduce the additional estimate of (1.6.14):

$$D \le 4^{-1}\|x - y\|^2 + 2^{-1}(\|x\| + \|y\|)\rho_X(\|x - y\|).$$

Finally, (1.6.10) implies

$$2\|x\|^2 + 2\|y\|^2 - \|x + y\|^2 \le 4\|x - y\|^2 + 2max\{2L, \|x\| + \|y\|\}\rho_X(\|x - y\|). \qquad (1.6.15)$$

Denote by $k(\|x - y\|)$ the right-hand side of the last inequality. Then

$$D \le \frac{k(\|x - y\|)}{4}. \qquad (1.6.16)$$

Put into a correspondence to the convex function $\varphi(x) = 2^{-1}\|x\|^2$ the concave function

$$\Phi(\lambda) = \lambda\varphi(x) + (1 - \lambda)\varphi(y) - \varphi(y + \lambda(x - y)), \quad 0 \le \lambda \le 1.$$

It is obvious that $\Phi(0) = 0$. Furthermore,

$$\lambda_1^{-1}\Phi(\lambda_1) \ge \lambda_2^{-1}\Phi(\lambda_2) \quad \text{as} \quad \lambda_1 \le \lambda_2,$$

that is, $(\lambda^{-1}\Phi(\lambda))' \le 0$. The last inequality implies $\Phi'(\lambda) \le \lambda^{-1}\Phi(\lambda)$. In particular, we have $\Phi'\big(\frac{1}{4}\big) \le 4\Phi\big(\frac{1}{4}\big)$. At the same time,

$$\Phi\Big(\frac{1}{4}\Big) = \frac{1}{4}\varphi(x) + \frac{3}{4}\varphi(y) - \varphi\Big(\frac{1}{4}x + \frac{3}{4}y\Big).$$

By (1.6.16), one has for any $z_1, z_2 \in X$,

$$\varphi\Big(\frac{z_1 + z_2}{2}\Big) \geq \frac{\varphi(z_1)}{2} + \frac{\varphi(z_2)}{2} - \frac{k(\|z_1 - z_2\|)}{8}.$$

Assume that $z_1 = 2^{-1}(x + y)$ and $z_2 = y$. Then, by the property $k(t/2) \leq k(t)/2$, we obtain

$$\varphi\Big(\frac{1}{4}x + \frac{3}{4}y\Big) \; = \; \varphi\Big(\frac{1}{2}\Big(\frac{1}{2}x + \frac{1}{2}y\Big) + \frac{1}{2}y\Big) \geq \frac{1}{2}\varphi\Big(\frac{x + y}{2}\Big) + \frac{1}{2}\varphi(y)$$

$$- \; \frac{1}{8}k\Big(\frac{\|x - y\|}{2}\Big) \geq \frac{1}{4}\varphi(x) + \frac{1}{4}\varphi(y) - \frac{1}{16}k(\|x - y\|) + \frac{1}{2}\varphi(y)$$

$$- \; \frac{1}{8}k\Big(\frac{\|x - y\|}{2}\Big) \geq \frac{1}{4}\varphi(x) + \frac{3}{4}\varphi(y) - \frac{1}{8}k(\|x - y\|).$$

Thus, $\Phi\big(\frac{1}{4}\big) \leq 8^{-1}k(\|x - y\|)$ and

$$\Phi'\Big(\frac{1}{4}\Big) = \varphi(x) - \varphi(y) - \langle \varphi'(y + 4^{-1}(x - y)), x - y \rangle \leq 2^{-1}k(\|x - y\|).$$

It is clear that

$$\Phi'\Big(\frac{1}{4}\Big) = \varphi(y) - \varphi(x) - \langle \varphi'(x + 4^{-1}(y - x)), y - x \rangle \leq 2^{-1}k(\|x - y\|).$$

These inequalities together give

$$\langle \varphi'(x - 4^{-1}(x - y)) - \varphi'(y + 4^{-1}(x - y)), x - y \rangle \leq k(\|x - y\|).$$

Make a non-degenerate replacement of the variables

$$z_1 = 2x - 2^{-1}(x - y), \quad z_2 = 2y + 2^{-1}(x - y)$$

and recall that $Jx = \varphi'(x)$ is a homogeneous operator. Then $z_1 - z_2 = x - y$ and

$$\|x\| + \|y\| \leq \|z_1\| + \|z_2\|.$$

Therefore,

$$\langle Jz_1 - Jz_2, z_1 - z_2 \rangle \leq 2k(\|z_1 - z_2\|).$$

Thus, for $\|z_1\| \leq R$, $\|z_2\| \leq R$, we obtain (1.6.5). The proof is accomplished. $\blacksquare$

Remark 1.6.2 *It follows from (1.6.15) that for arbitrary $x, y \in X$ it is necessary to make use of the estimate (1.6.5) with $c_1 = c_1(\|x\|, \|y\|) = 4max\{2L, \|x\| + \|y\|\}$.*

Corollary 1.6.3 *Let X be a uniformly convex and smooth Banach space. Then for any $x, y \in X$ such that $\|x\| \leq R$, $\|y\| \leq R$, the following inequality holds:*

$$\langle Jx - Jy, x - y \rangle \leq 8\|Jx - Jy\|_*^2 + c_1\rho_{X^*}(\|Jx - Jy\|_*), \qquad (1.6.17)$$

where $c_1 = 8max\{L, R\}$.

Proof. Since X^* is a uniformly smooth Banach space, by Theorem 1.6.1 for any $\phi, \psi \in X^*$ such that $\|\phi\|_* \le R$, $\|\psi\|_* \le R$, we have

$$\langle \phi - \psi, J^*\phi - J^*\psi \rangle \le 8\|\phi - \psi\|_*^2 + c_1\rho_{X^*}(\|\phi - \psi\|_*), \tag{1.6.18}$$

where $c_1 = 8max\{L, R\}$. The space X is uniformly convex and smooth, consequently, it is reflexive and strictly convex together with its dual space X^*. Therefore, the normalized duality mapping J^* is single-valued. This means that for any $\phi \in X^*$ there exists a unique $x \in X$ such that $x = J^*\phi$. Besides this, $JJ^* = I_{X^*}$ and $J^*J = I_X$. Now (1.6.17) follows from (1.6.18) by the substitution $\phi = Jx$ and $\psi = Jy$. ∎

Theorem 1.6.4 *Let X be a uniformly convex Banach space. Then for any $R > 0$ and any $x, y \in X$ such that $\|x\| \le R$, $\|y\| \le R$ the following inequality holds:*

$$\langle Jx - Jy, x - y \rangle \ge (2L)^{-1}\delta_X(c_2^{-1}\|x - y\|), \tag{1.6.19}$$

where $c_2 = 2max\{1, R\}$.

Proof. As it was shown in [127], for $x, y \in X$, the equality

$$\|x\|^2 + \|y\|^2 = 2 \tag{1.6.20}$$

implies

$$\|2^{-1}(x + y)\|^2 \le 1 - \delta_X(2^{-1}\|x - y\|). \tag{1.6.21}$$

If X is uniformly convex then the function $\delta_X(\epsilon)$ is increasing, $\delta_X(0) = 0$ and $0 \le \delta_X(\epsilon) < 1$. Denote $R_1^2 = 2^{-1}(\|x\|^2 + \|y\|^2)$ and introduce the new variables by the formulas

$$\tilde{x} = \frac{x}{R_1}, \quad \tilde{y} = \frac{y}{R_1}.$$

Then

$$\|\tilde{x}\|^2 + \|\tilde{y}\|^2 = R_1^{-2}(\|x\|^2 + \|y\|^2) = 2.$$

Hence, the inequality

$$\frac{\|\tilde{x}\|^2 + \|\tilde{y}\|^2}{2} - \left\|\frac{\tilde{x} + \tilde{y}}{2}\right\|^2 \ge \delta_X\left(\frac{\|\tilde{x} - \tilde{y}\|}{2}\right)$$

is satisfied by (1.6.21). If now we return to the old variables x and y, then we will obtain

$$D \ge R_1^2\delta_X\left(\frac{\|x - y\|}{2R_1}\right),$$

where D is defined by (1.6.6). Consider two cases.
1. Let $R_1 \ge 1$. Then for $\|x\| \le R$ and $\|y\| \le R$, one gets $R_1 \le R$ and

$$D \ge \delta_X\left(\frac{\|x - y\|}{2R_1}\right) \ge \delta_X\left(\frac{\|x - y\|}{2R}\right). \tag{1.6.22}$$

2. Let $R_1 < 1$. It is known that in a uniformly convex Banach space X for any $\eta \geq \epsilon > 0$

$$\epsilon^2 \delta_X(\eta) \geq (4L)^{-1}\eta^2 \delta_X(\epsilon), \tag{1.6.23}$$

where $1 < L < 1.7$ (cf. (1.6.4)). Hence,

$$\delta_X\left(\frac{\|x-y\|}{2R_1}\right) \geq R_1^{-2}(4L)^{-1}\delta_X\left(\frac{\|x-y\|}{2}\right) \geq (4L)^{-1}\delta_X\left(\frac{\|x-y\|}{2}\right). \tag{1.6.24}$$

Combining (1.6.22) and (1.6.24) we finally deduce

$$D \geq (4L)^{-1}\delta_X(c_2^{-1}\|x-y\|), \quad c_2 = 2max\{1, R\}. \tag{1.6.25}$$

Denote $\varphi(x) = 2^{-1}\|x\|^2$. Then

$$2^{-1}\varphi(x) + 2^{-1}\varphi(y) - \varphi(2^{-1}(x+y)) \geq (8L)^{-1}\delta_X(c_2^{-1}\|x-y\|).$$

Passing to a duality mapping and using Lemma 1.3.11 we obtain (1.6.19). The theorem is proved. ∎

Remark 1.6.5 *It follows from (1.6.22) and (1.6.24) that for arbitrary $x, y \in X$,*

$$2\|x\|^2 + 2\|y\|^2 - \|x+y\|^2 \geq L^{-1}\delta_X(c_2^{-1}\|x-y\|), \tag{1.6.26}$$

where

$$c_2 = c_2(\|x\|, \|y\|) = 2max\{1, \sqrt{2^{-1}(\|x\|^2 + \|y\|^2)}\}. \tag{1.6.27}$$

In addition, the estimate (1.6.19) is satisfied if $c_2 = 2max\{1, R\}$ is replaced by (1.6.27).

Remark 1.6.6 *It should be observed that only in a uniformly convex space is $\delta_X(\epsilon)$ a strictly increasing function and $\delta_X(0) = 0$. In an arbitrary Banach space, estimate (1.6.19) guarantees, in general, the monotonicity property of normalized duality mapping J.*

The next assertion follows from Theorem 1.6.4 (cf. Corollary 1.6.3).

Corollary 1.6.7 *Let X be a uniformly smooth and strictly convex Banach space. Then for any $x, y \in X$ such that $\|x\| \leq R$, $\|y\| \leq R$, the following inequality holds:*

$$\langle Jx - Jy, x - y \rangle \geq (2L)^{-1}\delta_{X^*}(c_2^{-1}\|Jx - Jy\|_*), \tag{1.6.28}$$

where $c_2 = 2max\{1, R\}$.

Corollary 1.6.8 *Let X be a uniformly smooth and strictly convex Banach space. Suppose that the function $g_{X^*}(\epsilon) = \epsilon^{-1}\delta_{X^*}(\epsilon)$ is increasing. Then for all $x, y \in X$, $\|x\| \leq R$, $\|y\| \leq R$ the following estimate is valid:*

$$\|Jx - Jy\|_* \leq c_2 g_{X^*}^{-1}(2Lc_2\|x - y\|). \tag{1.6.29}$$

If X is a uniformly convex and smooth Banach space, $g_X(\epsilon) = \epsilon^{-1}\delta_X(\epsilon)$ is increasing, then

$$\|x - y\| \leq c_2 g_X^{-1}(2Lc_2\|Jx - Jy\|_*). \tag{1.6.30}$$

Proof. By the Cauchy–Schwarz inequality, (1.6.28) can be estimated as follows:

$$(2L)^{-1}\delta_{X^*}(c_2^{-1}\|Jx - Jy\|_*) \le \|Jx - Jy\|_* \|x - y\|.$$

Then

$$g_{X^*}\!\left(\frac{\|Jx - Jy\|_*}{c_2}\right) = \frac{c_2\delta_{X^*}(c_2^{-1}\|Jx - Jy\|_*)}{\|Jx - Jy\|_*} \le 2Lc_2\|x - y\|,$$

and (1.6.29) is obviously satisfied. Here it is necessary to recall that $\delta_X(\epsilon) \ge 0$ for all $\epsilon \ge 0$ and $g_X(0) = 0$. The estimate (1.6.30) results from (1.6.19).   ∎

Remark 1.6.9 *Note that if $g_X(t)$ does not increase strictly for all $t \in [0,2]$ but there exists a non-negative increasing continuous function $\tilde{g}_X(t)$ such that $g_X(t) \ge \tilde{g}_X(t)$, then (1.6.30) remains still valid if $g_X^{-1}(\cdot)$ is replaced by $\tilde{g}_X^{-1}(\cdot)$. The same can be said for (1.6.29) in X^*.*

Inequality (1.6.29) defines the modulus of uniform continuity of a normalized duality mapping J on set $B(\theta_X, R)$ in a uniformly smooth Banach space X, that is, it is the function $\omega_R(t)$ in (1.6.1). Namely, $\omega_R(t) = c_2 g_{X^*}^{-1}(2Lc_2 t)$.

Estimates (1.6.5), (1.6.17), (1.6.19) and (1.6.28) are reduced to calculation of the moduli of convexity and smoothness of the spaces X and X^*. However, in practice, one usually uses upper estimates for the modulus of smoothness and lower estimates for the modulus of convexity of the spaces.

For the spaces X of type l^p and L^p, $1 < p \le 2$, the modulus of smoothness $\rho_X(\tau)$ can be calculated by the Lindenstrauss formula

$$\rho_X(\tau) = sup\left\{\frac{\tau\epsilon}{2} - \delta_{X^*}(\epsilon),\ 0 \le \epsilon \le 2\right\}. \tag{1.6.31}$$

For that, it is necessary to use Hanner's equality

$$\delta_{X^*}(\epsilon) = 1 - \left(1 - (2^{-1}\epsilon)^q\right)^{\frac{1}{q}},$$

which is true if X^* is l^q or L^q, $p^{-1} + q^{-1} = 1$. By (1.6.31), it is not difficult to obtain the equality

$$\rho_X(\tau) = (1 + \tau^p)^{\frac{1}{p}} - 1.$$

By virtue of the inequality

$$a^r - b^r \le rb^{r-1}(a - b),\ 0 \le r \le 1,\ a, b > 0,$$

we have

$$\rho_X(\tau) \le p^{-1}\tau^p.$$

It makes sense to use this estimate only if $\tau \le 1$, because the relation $\rho_{X^*}(\tau) \le \tau$ is more precise by order as $\tau > 1$.

We address now an estimate of the modulus of convexity of the spaces l^q, L^q, $1 < q \le 2$. First of all, we observe that there holds the following functional identity [94]:

$$\Phi(\epsilon, \delta_{X^*}(\epsilon)) = (1 - \delta_{X^*}(\epsilon) - 2^{-1}\epsilon)^q + (1 - \delta_{X^*}(\epsilon) + 2^{-1}\epsilon)^q - 2 \equiv 0.$$

From this, after some simple algebra, we have

$$\delta'_{X^*}(\epsilon) = -\Big(\frac{\partial \Phi}{\partial \epsilon}\Big)\Big(\frac{\partial \Phi}{\partial \delta_{X^*}}\Big)^{-1} \geq \frac{(q-1)\epsilon}{8}.$$

Integrating both parts of the previous relation, one gets

$$\delta_{X^*}(\epsilon) \geq \frac{(q-1)\epsilon^2}{16}, \quad 0 \leq \epsilon \leq 2, \ 1 < q \leq 2.$$

Note that it coincides (up to constants) with Hanner's asymptotic result in [94]. Using (1.6.31) again we obtain

$$\rho_{X^*}(\tau) \leq (p-1)\tau^2, \ p \geq 2. \tag{1.6.32}$$

Thus, in the spaces $X = l^p$ and $X = L^p$, $1 < p < \infty$, one has

$$\rho_X(\tau) \leq (p-1)\tau^2, \quad \delta_X(\epsilon) \geq p^{-1}\Big(\frac{\epsilon}{2}\Big)^p, \quad p \geq 2, \tag{1.6.33}$$

and

$$\rho_X(\tau) \leq \frac{\tau^p}{p}, \quad \delta_X(\epsilon) \geq \frac{(p-1)\epsilon^2}{16}, \quad 1 < p \leq 2. \tag{1.6.34}$$

The same upper and lower estimates are also valid in the Sobolev spaces W_m^p, $1 < p < \infty$.
By (1.6.5), (1.6.19), (1.6.33) and (1.6.34), if $\|x\| \leq R$, $\|y\| \leq R$ and if $p \geq 2$, then

$$\langle Jx - Jy, x - y\rangle \leq \Big(8 + (p-1)c_1\Big)\|x - y\|^2, \quad c_1 = 8max\{L, R\}),$$

and

$$\langle Jx - Jy, x - y\rangle \geq \frac{\|x - y\|^p}{2^{p+1}Lpc_2^p}, \quad c_2 = 2max\{1, R\}.$$

Corollary 1.6.8 implies the Lipschitz-continuity of the normalized duality mapping on each bounded set in the spaces l^p, L^p and W_m^p, when $p \geq 2$, namely,

$$\|Jx - Jy\|_* \leq C(R)\|x - y\|, \quad C(R) = 32Lc_2^2(q-1)^{-1}.$$

If $1 < p \leq 2$ then

$$\langle Jx - Jy, x - y\rangle \leq 8\|x - y\|^2 + c_1 p^{-1}\|x - y\|^p \tag{1.6.35}$$

and

$$\langle Jx - Jy, x - y\rangle \geq \frac{(p-1)\|x - y\|^2}{32Lc_2^2}.$$

The Hölder-continuity of the normalized duality mapping expressed by the inequality

$$\|Jx - Jy\|_* \leq \bar{C}(R)\|x - y\|^{p-1}, \quad \bar{C}(R) = \Big(\frac{p}{p-1}\Big)^{p-1}c_2^p L^{p-1}2^{2p-1},$$

follows again from Corollary 1.6.8.
  In a Hilbert space

$$\rho_H(\tau) = (1 + \tau^2)^{\frac{1}{2}} - 1 \leq 2^{-1}\tau^2$$

and

$$\delta_H(\epsilon) = 1 - \left[1 - \left(\frac{\epsilon}{2}\right)^2\right]^{\frac{1}{2}} \geq \frac{\epsilon^2}{8}.$$

It is known that a Hilbert space has the smallest modulus of smoothness among all uniformly smooth Banach spaces, and it has the biggest modulus of convexity among all uniformly convex Banach spaces, that is $\rho_H(\tau) \leq \rho_X(\tau)$, $\delta_X(\epsilon) \geq \delta_H(\epsilon)$. Furthermore, there is a duality relationship: if $\delta_X(\epsilon) \geq k_1\epsilon^{\gamma_1}$ then $\rho_{X^*}(\tau) \leq k_2\tau^{\gamma_2}$, where k_1 and k_2 are positive constants and $\gamma_1^{-1} + \gamma_2^{-1} = 1$. These assertions follow from (1.6.31).

Remark 1.6.10 *The estimates like (1.6.5) and (1.6.19) can be also obtained for duality mappings J^μ with the gauge function $\mu(t) = t^{s-1}$, $s > 1$.*

Suppose that X and X^* are reflexive strictly convex spaces. Introduce the Lyapunov functional $W(x,y) : X \times X \to R^1$ defined by the formula

$$W(x,y) = 2^{-1}(\|x\|^2 - 2\langle Jx, y\rangle + \|y\|^2) \tag{1.6.36}$$

and study its properties.

1. Show that $W(x,y) \geq 0$ for all $x, y \in X$. Indeed,

$$W(x,y) \geq 2^{-1}(\|Jx\|_*^2 - 2\|Jx\|_*\|y\| + \|y\|^2) = 2^{-1}(\|x\| - \|y\|)^2 \geq 0. \tag{1.6.37}$$

2. By (1.6.37), it is easy to see that $W(x,y) \to \infty$ as $\|x\| \to \infty$ or/and $\|y\| \to \infty$. On the other hand,

$$W(x,y) \leq 2^{-1}(\|x\| + \|y\|)^2.$$

The latter inequality implies the following assertion: If $W(x,y) \to \infty$ then $\|x\| \to \infty$ or/and $\|y\| \to \infty$.

3. One can verify by direct substitution that $W(x,x) = 0$.

4. Let y be a fixed element of X. Consider the general functional $W_1(\phi, y) : X^* \times X \to R^1$ such that

$$W_1(\phi, y) = 2^{-1}(\|\phi\|_*^2 - 2\langle \phi, y\rangle + \|y\|^2).$$

Since $\|x\|^2 = \|Jx\|_*^2$, (1.6.36) is presented in the equivalent form as

$$W_1(Jx, y) = 2^{-1}(\|Jx\|_*^2 - 2\langle Jx, y\rangle + \|y\|^2). \tag{1.6.38}$$

Since the space X^* is smooth, $W_1(\phi, y)$ has the Gâteaux derivative $W_1'(\phi, y)$ with respect to $\phi = Jx$. It is not difficult to be sure that

$$grad\, W_1(\phi, y) = J^*\phi - y. \tag{1.6.39}$$

According to Lemma 1.2.6, $W_1(\phi, y)$ is convex and lower semicontinuous in the whole space X^*. Hence, $grad\, W_1(\phi, y) : X^* \to X$ is a monotone and single-valued operator for all $\phi \in X^*$. Definition 1.2.2 of the subdifferential gives the relation

$$W_1(\phi, y) - W_1(\psi, y) \geq \langle \phi - \psi, J^*\psi - y\rangle. \tag{1.6.40}$$

Using the fact that there exist a unique $x \in X$ and unique $z \in X$ such that for every $\phi \in X^*$ and for every $\psi \in X^*$, respectively, $J^*\phi = x$ and $J^*\psi = z$, we must have that

$$W(x, y) - W(z, y) \geq \langle Jx - Jz, z - y \rangle. \tag{1.6.41}$$

It is clear that (1.6.41) is valid for all $x, y, z \in X$.

Analogously, considering the functional $W(x, y)$ with respect to the variable y with a fixed element x, we conclude that $W(x, y)$ is convex and lower semicontinuous, and its gradient is defined in the smooth space X as follows:

$$grad\, W(x, y) = Jy - Jx. \tag{1.6.42}$$

This yields the second relation

$$W(x, y) - W(x, z) \geq \langle Jz - Jx, y - z \rangle, \tag{1.6.43}$$

which is satisfied for all $x, y, z \in X$ again.

5. By (1.6.41), one has

$$W(y, y) - W(x, y) \geq \langle Jy - Jx, x - y \rangle \quad \forall x, y \in X.$$

Now the property 3 leads to inequality

$$W(x, y) \leq \langle Jx - Jy, x - y \rangle \quad \forall x, y \in X. \tag{1.6.44}$$

Taking now into account Theorem 1.6.1, we obtain for all $x, y \in B(\theta_X, R)$ the estimate

$$W(x, y) \leq 8\|x - y\|^2 + c_1 \rho_X(\|x - y\|), \tag{1.6.45}$$

where $c_1 = 8max\{L, R\}$.

Let X be a uniformly convex Banach space. Rewrite the inequality (1.6.25) in the form

$$\|2^{-1}(x + z)\|^2 \leq 2^{-1}\|x\|^2 + 2^{-1}\|z\|^2 - (4L)^{-1}\delta_X(c_2^{-1}\|x - z\|). \tag{1.6.46}$$

By the definition of a subdifferential, we deduce

$$\|2^{-1}(x + z)\|^2 \geq \|x\|^2 + \langle Jx, z - x \rangle,$$

and then

$$\|z\|^2 \geq \|x\|^2 + 2\langle Jx, z - x \rangle + (2L)^{-1}\delta_X(c_2^{-1}\|x - z\|),$$

in view of (1.6.46). Replace z by $2^{-1}(x + y)$ in the last inequality to obtain

$$\|2^{-1}(x + y)\|^2 \geq \|x\|^2 + \langle Jx, y - x \rangle + (2L)^{-1}\delta_X((2c_2)^{-1}\|x - y\|)$$

$$= \langle Jx, y \rangle + (2L)^{-1}\delta_X((2c_2)^{-1}\|x - y\|).$$

Thus,

$$\langle Jx, y \rangle \leq \|2^{-1}(x + y)\|^2 - (2L)^{-1}\delta_X((2c_2)^{-1}\|x - y\|).$$

It follows from (1.6.36) that

$$W(x,y) \geq 2^{-1}\|x\| - \|2^{-1}(x+y)\|^2 + 2^{-1}\|y\|^2 + (2L)^{-1}\delta_X((2c_2)^{-1}\|x-y\|),$$

and due to (1.6.46), we write down

$$W(x,y) \geq (4L)^{-1}\delta_X(c_2^{-1}\|x-y\|) + (2L)^{-1}\delta_X((2c_2)^{-1}\|x-y\|). \qquad (1.6.47)$$

Since the function $t^{-1}\delta_X(t)$ is non-decreasing, the inequality $\delta_X(2^{-1}\epsilon) \leq 2^{-1}\delta_X(\epsilon)$ holds. Then (1.6.47) yields for all $x, y \in B(\theta_X, R)$ the estimate

$$W(x,y) \geq L^{-1}\delta_X((2c_2)^{-1}\|x-y\|),$$

where $c_2 = 2\max\{1, R\}$. Combining (1.6.45) and the last inequality we obtain

$$L^{-1}\delta_X((2c_2)^{-1}\|x-y\|) \leq W(x,y) \leq 8\|x-y\|^2 + c_1\rho_X(\|x-y\|). \qquad (1.6.48)$$

6. If X is a Hilbert space then $W(x,y) = 2^{-1}\|x-y\|^2$.

We may consider that

$$\delta(\epsilon) \geq c\epsilon^\gamma, \quad \gamma \geq 0, \quad c > 0. \qquad (1.6.49)$$

This assumption is not so limiting because, on the one hand, the direct calculations of $\delta(\epsilon)$ in the spaces l^p, L^p, W_m^p, $1 < p \leq \infty$, and in the Orlich spaces with Luxemburg norm [24, 119] show that (1.6.49) is true. On the other hand, it is asserted in [168] that the same relates (to within isomorphism) to super-reflexive Banach space. In these cases, there exists a constant $c > 0$ such that $\rho_{X^*}(\tau) \leq c\tau^{\gamma/(\gamma-1)}$.

Let us make a few remarks concerning the spaces l^p and L^p. In [63] the following parallelogram inequalities are presented:

$$2\|x\|^2 + 2\|y\|^2 - \|x+y\|^2 \geq (p-1)\|x-y\|^2, \quad 1 < p \leq 2; \qquad (1.6.50)$$

$$2\|x\|^2 + 2\|y\|^2 - \|x+y\|^2 \leq (p-1)\|x-y\|^2, \quad p \geq 2. \qquad (1.6.51)$$

They imply the estimates

$$\langle Jx - Jy, x - y \rangle \geq (p-1)\|x-y\|^2, \quad 1 < p \leq 2; \qquad (1.6.52)$$

$$\|Jx - Jy\|_* \leq (p-1)\|x-y\|, \quad p \geq 2. \qquad (1.6.53)$$

The other well-known relations in spaces l^p and L^p are the Clarkson inequalities [211]:

$$\frac{\|x\|^p + \|y\|^p}{2} - \left\|\frac{x+y}{2}\right\|^p \leq \left\|\frac{x-y}{2}\right\|^p, \quad 1 < p \leq 2; \qquad (1.6.54)$$

$$\frac{\|x\|^p + \|y\|^p}{2} - \left\|\frac{x+y}{2}\right\|^p \geq \left\|\frac{x-y}{2}\right\|^p, \quad p \geq 2. \qquad (1.6.55)$$

They yield, respectively, the following estimates of duality mapping J^p with the gauge function $\mu(t) = t^{p-1}$:

$$\langle J^p x - J^p y, x - y \rangle \leq 2^{3-p}p^{-1}\|x-y\|^p, \quad 1 < p \leq 2, \qquad (1.6.56)$$

and
$$\langle J^p x - J^p y, x - y \rangle \geq 2^{2-p} p^{-1} \|x - y\|^p, \quad p \geq 2. \tag{1.6.57}$$

Suppose that there exists a strictly increasing continuous function $\tilde{\delta}_X(t)$, $\tilde{\delta}_X(0) = 0$, and a positive constant K such that $\delta_X(t) \geq K\tilde{\delta}_X(t)$. We known the following estimates:

$$\langle J^p x - J^p y, x - y \rangle \geq C^p \tilde{\delta}_X(C^{-1}\|x - y\|) \quad \forall x, y \in X \tag{1.6.58}$$

and

$$\langle J^p x - J^p y, x - y \rangle \leq C^p \rho_X(C^{-1}\|x - y\|) \quad \forall x, y \in X, \tag{1.6.59}$$

where $C(\|x\|, \|y\|) = max \{\|x\|, \|y\|\}$.

Unlike Theorems 1.6.1 and 1.6.4 (when they are applied to the spaces l^p and L^p), the parallelogram inequality (1.6.50), (1.6.51) and the Clarkson inequality (1.6.54), (1.6.55) admit only one-sided estimates for each $p \neq 2$.

By analogy with (1.6.50) and (1.6.51), the inequalities (1.6.15) and (1.6.26) can be treated, respectively, as the upper parallelogram inequality in a uniformly smooth Banach space and lower parallelogram inequality in a uniformly convex Banach space.

1.7   Equations with Maximal Monotone Operators

Assume that X is a reflexive Banach space, X and X^* are strictly convex. We begin to study solvability of equations with maximal monotone operators. First of all, we prove the following fundamental auxiliary statement:

Lemma 1.7.1 (Debrunner−Flor) *Let Ω be a convex compact set in X, G be a monotone set in the product $\Omega \times X^*$, $F : \Omega \to X^*$ be a continuous operator, $h \in X^*$. Then there exists an element $u \in \Omega$ such that the inequality*

$$\langle f + Fu - h, x - u \rangle \geq 0 \tag{1.7.1}$$

holds for all $(x, f) \in G$.

Proof. Since the set of pairs $(x, f - h)$, where $(x, f) \in G$, is also monotone, we may consider, without loss of generality, that $h = \theta_{X^*}$. We want to prove the lemma by contradiction. Suppose that (1.7.1) is not true. Then for every $u \in \Omega$, there is a pair $(x, f) \in G$ such that

$$\langle f + Fu, x - u \rangle < 0.$$

For each $(x, f) \in G$, we are able to construct the set

$$N(x, f) = \{y \in \Omega \mid \langle f + Fy, x - y \rangle < 0\}.$$

The sets $N(x, f)$ with $(x, f) \in G$ form a family of open coverings of the compact set Ω. Therefore, there exists a finite subcovering, that is, finite family $(x_i, f_i) \in G$, $1 \leq i \leq n$, such that

$$\Omega = \bigcup_{i=1}^{n} N(x_i, f_i).$$

On the basis of that finite covering, we build a continuous partition of the unit on Ω. In other words, we build n continuous functions

$$\beta_i : N(x_i, f_i) \to [0,1], \quad i = 1, 2, ..., n,$$

such that $\sum_{i=1}^{n} \beta_i(y) = 1$ for all $y \in \Omega$. Define operators $T_1 : \Omega \to X$ and $T_2 : \Omega \to X^*$ as follows:

$$T_1(y) = \sum_{i=1}^{n} \beta_i(y) x_i, \quad T_2(y) = \sum_{i=1}^{n} \beta_i(y) f_i.$$

Since Ω is convex, the operator T_1 acts from Ω to Ω. Furthermore, T_1 is continuous. Then, by the Schauder principle, there exists $y_0 \in \Omega$ such that $T_1(y_0) = y_0$, i.e., y_0 is a fixed point of T_1.

Next let

$$p(y) = \langle T_2(y) + Fy, T_1(y) - y \rangle = p_1(y) + p_2(y),$$

where

$$p_1(y) = \sum_{i=1}^{n} \beta_i^2(y) \langle f_i + Fy, x_i - y \rangle$$

and

$$p_2(y) = \sum_{1 \le i < j \le n} \beta_i(y) \beta_j(y) \Big(\langle f_i + Fy, x_j - y \rangle + \langle f_j + Fy, x_i - y \rangle \Big).$$

For every $y \in \Omega$, there exists at least one number m $(1 \le m \le n)$ such that $\beta_m(y) \ne 0$, whereas $y \in N(x_m, f_m)$. Therefore, $\langle f_m + Fy, x_m - y \rangle < 0$. Thus, $p_1(y) < 0$ for all $y \in \Omega$.

The monotonicity property of G gives further the following relations for $y \in N(x_i, f_i) \cap N(x_j, f_j)$: the products $\beta_i(y)\beta_j(y) > 0$ and

$$\langle f_i + Fy, x_j - y \rangle \; + \; \langle f_j + Fy, x_i - y \rangle = \langle f_i + Fy, x_i - y \rangle$$
$$+ \; \langle f_j + Fy, x_j - y \rangle + \langle f_i - f_j, x_j - x_i \rangle < 0.$$

Hence, $p(y) < 0$ for all $y \in \Omega$. On the other hand,

$$p(y_0) = \langle T_2(y_0) + Fy_0, T_1(y_0) - y_0 \rangle = 0.$$

This contradiction proves the lemma. ∎

We now introduce the solution concept in the sense of inclusion.

Definition 1.7.2 *An element $x^0 \in D(A)$ such that $f \in Ax^0$ is called the solution (in the sense of inclusion) of the equation $Ax = f$ with a maximal monotone operator A.*

Remark 1.7.3 *If the operator A is single-valued in a solution x^0 of the equation $Ax = f$, then $f = Ax^0$. In this case, x^0 is called the classical solution.*

Theorem 1.7.4 *Let $A : X \to 2^{X^*}$ be a maximal monotone operator with $D(A)$ and $J : X \to X^*$ be a normalized duality mapping. Then $R(A + \alpha J) = X^*$ for all $\alpha > 0$.*

Proof. It suffices to consider the case $\alpha = 1$ and show that $\theta_{X^*} \in R(A + J)$. Let X_n be an n-dimensional subspace of X, $\mathcal{E}_n : X_n \to X$ be a natural imbedding operator. Denote by $\mathcal{E}_n^* : X^* \to X_n^*$ an adjoint operator to $\mathcal{E}_n$, $F = \mathcal{E}_n^* J \mathcal{E}_n$, and

$$\Omega = \{x \in X_n \mid \|x\| \le r\}.$$

Put $h = \theta_{X^*}$ and define large enough number r satisfying condition $B(\theta_X, r) \cap D(A) \ne \emptyset$. Then, by the Debrunner−Flor Lemma, there exist $x_n^r \in \Omega$ and $y_n^r = Fx_n^r$ such that

$$\langle y + y_n^r, x - x_n^r \rangle \ge 0 \quad \forall (x, y) \in G, \tag{1.7.2}$$

where $G = gr(\mathcal{E}_n^* A \mathcal{E}_n) \cap \Omega \times X_n^*$. Hence,

$$\|x_n^r\|^2 \le \|y\|_* \|x\| + \|x_n^r\|(\|x\| + \|y\|_*) \quad \forall (x, y) \in G.$$

Thus, the sequence $\{x_n^r\}$ is bounded. We know that $x_n^r \in X_n$ for all r, therefore, $x_n^r \to x_n \in X_n$ as $r \to \infty$. Since the operator $F = \mathcal{E}_n^* J \mathcal{E}_n$ is continuous (see Section 1.5) and the subspace X_n^* is closed, we conclude that $y_n^r \to y_n \in X_n^*$ as $r \to \infty$ and $y_n = Fx_n$. Hence, (1.7.2) implies

$$\langle y + y_n, x - x_n \rangle \ge 0 \quad \forall (x, y) \in grA, \quad x \in X_n. \tag{1.7.3}$$

It results from the coerciveness of J and the latter inequality that the sequences $\{x_n\}$ and $\{y_n\}$ remain bounded in X and X^*, respectively, when X_n are running through the ordered increasing filter of finite-dimensional subspaces in X. Therefore, there exist $\bar{x} \in \Omega$ and $\bar{y} \in X^*$ such that $x_n \rightharpoonup \bar{x}$ and $y_n \rightharpoonup \bar{y}$.

Let $\varphi(x) = 2^{-1}\|x\|^2$, $x \in X$, and let $\varphi^*(x^*)$ be a conjugate to $\varphi(x)$ functional. Since $\|x\|$ is a weakly lower semicontinuous functional, by Definition 1.2.10 of the functional $\varphi^*(x^*)$ and by Theorem 1.2.11, we obtain

$$\langle \bar{y}, \bar{x} \rangle \le \varphi(\bar{x}) + \varphi^*(\bar{y}) \le \liminf_{n \to \infty} \varphi(x_n) + \liminf_{n \to \infty} \varphi^*(y_n) = \liminf_{n \to \infty} \langle y_n, x_n \rangle. \tag{1.7.4}$$

Furthermore, by (1.7.3), one has

$$\langle y_n, x_n \rangle \le \langle y, x \rangle + \langle y_n, x \rangle - \langle y, x_n \rangle.$$

Hence,

$$\liminf_{n \to \infty} \langle y_n, x_n \rangle \le \langle y, x \rangle + \langle \bar{y}, x \rangle - \langle y, \bar{x} \rangle \quad \forall (x, y) \in grA, \quad x \in X_n. \tag{1.7.5}$$

Observe that $\bigcup_{n=1}^{\infty} X_n$ is dense in X. Therefore, (1.7.5) is valid for all $(x, y) \in grA$, $x \in X$. In view of (1.7.4), we thus obtain

$$\langle -\bar{y} - y, \bar{x} - x \rangle \ge 0 \quad \forall (x, y) \in grA.$$

Since the operator A is maximal monotone, this inequality gives the inclusion: $-\bar{y} \in A\bar{x}$. Assuming $x = \bar{x}$, $y = \bar{y}$ in (1.7.5) we conclude that

$$\liminf_{n \to \infty} \langle y_n, x_n \rangle \le \langle \bar{y}, \bar{x} \rangle.$$

This fact and (1.7.4) lead to the following result:

$$\langle \bar{y}, \bar{x} \rangle = \varphi(\bar{x}) + \varphi^*(\bar{y}).$$

Hence, $\bar{y} = J\bar{x}$ (see Theorem 1.2.11), i.e., $\theta_{X^*} \in (A + J)\bar{x}$. The theorem is proved. ∎

By making use of this theorem we establish the following important result.

Theorem 1.7.5 *Let $A : X \to 2^{X^*}$ be a maximal monotone and coercive operator. Then $R(A) = X^*$.*

Proof. Choose an arbitrary element $f \in X^*$. Owing to Theorem 1.7.4, for every $\alpha > 0$, there exists $x_\alpha \in D(A)$ such that

$$y_\alpha + \alpha J x_\alpha = f, \quad y_\alpha \in A x_\alpha. \tag{1.7.6}$$

Then

$$\|f\|_* \|x_\alpha\| \geq \langle f, x_\alpha \rangle = \langle y_\alpha, x_\alpha \rangle + \alpha \|x_\alpha\|^2 \geq \langle y_\alpha, x_\alpha \rangle.$$

Therefore,

$$\frac{\langle y_\alpha, x_\alpha \rangle}{\|x_\alpha\|} \leq \|f\|_*, \quad y_\alpha \in A x_\alpha.$$

Since A is coercive, it follows from the last inequality that the sequence $\{x_\alpha\}$ is bounded. Then $x_\alpha \rightharpoonup \bar{x} \in X$ as $\alpha \to 0$. By (1.7.6), one gets $y_\alpha = f - \alpha J x_\alpha$ and then the monotonicity property of A gives

$$\langle f - \alpha J x_\alpha - y, x_\alpha - x \rangle \geq 0 \quad \forall (x, y) \in gr A.$$

Letting $\alpha \to 0$, we obtain

$$\langle f - y, \bar{x} - x \rangle \geq 0 \quad \forall (x, y) \in gr A.$$

Since the operator A is maximal monotone, we deduce by Proposition 1.4.3 that $f \in A\bar{x}$. The theorem is proved. ∎

Corollary 1.7.6 *Let $A : X \to 2^{X^*}$ be a maximal monotone operator whose domain $D(A)$ is bounded. Then $R(A) = X^*$.*

The next assertions follow from Theorems 1.4.6 and 1.7.5.

Corollary 1.7.7 *Let $A : X \to X^*$ be a monotone hemicontinuous and coercive operator with $D(A) = X$. Then $R(A) = X^*$.*

Note that this statement is known in the literature as the Minty−Browder theorem.

Corollary 1.7.8 *Let $J : X \to X^*$ and $J^* : X^* \to X$ be normalized duality mappings. Then $R(J) = X^*$ and $R(J^*) = X$.*

Theorem 1.7.9 *Suppose that $A : X \to 2^{X^*}$ is a maximal monotone operator, $f \in X^*$ and there exists a number $r > 0$ such that $\langle y - f, x \rangle \geq 0$ for all $y \in Ax$ as $\|x\| \geq r$. Then there exists an element $\bar{x} \in X$ such that $f \in A\bar{x}$ and $\|\bar{x}\| \leq r$.*

Proof. Consider again the equality (1.7.6). It is obvious that

$$\langle y_\alpha - f, x_\alpha \rangle = -\alpha \|x_\alpha\|^2. \tag{1.7.7}$$

If $x_\alpha = \theta_X$ for some $\alpha > 0$ then (1.7.6) immediately gives $y_\alpha = f$, i.e., $f \in Ax_\alpha$ and $\bar{x} = x_\alpha$. If $x_\alpha \neq \theta_X$ for all $\alpha > 0$ then, by (1.7.7), we have $\langle y_\alpha - f, x_\alpha \rangle < 0$. In this case, the conditions of the theorem imply the estimate $\|x_\alpha\| < r$ for all $\alpha > 0$. Then the inclusion $f \in A\bar{x}$ is proved following the pattern of the proof given in Theorem 1.7.5, and the estimate $\|\bar{x}\| \leq r$ is obtained by the weak lower semicontinuity of the norm in X. ∎

Remark 1.7.10 *Under the conditions of Theorems 1.7.5, 1.7.9 and Corollary 1.7.7, if operator A is strictly monotone, then the equation $Ax = f$ for all $f \in X^*$ has a unique solution.*

Lemma 1.7.11 *Let $A : X \to 2^{X^*}$ be a maximal monotone operator. Then $R(A) = X^*$ if and only if A^{-1} is locally bounded on $R(A)$.*

Proof. Let $R(A) = X^*$ be given, then $D(A^{-1}) = X^*$. Since A^{-1} is monotone, it is locally bounded on $R(A) = X^*$. Let now A^{-1} be locally bounded on $R(A)$. Prove that $R(A) = X^*$. For this aim, it is sufficient to show that the set $R(A)$ is both open and closed in X^* at the same time. Let $f_n \to f$, $f_n \in Ax_n$, i.e., $f_n \in R(A)$. Since A^{-1} is locally bounded on $R(A)$, the sequence $\{x_n\}$ is bounded at least as n is sufficiently large. Then there exists some subsequence $\{x_{n_k}\} \subset \{x_n\}$ which weakly converges to $x \in X$, and $(x, f) \in grA$ because grA is demiclosed; see Lemma 1.4.5. Hence, $f \in R(A)$, which means that $R(A)$ is a closed set.

Assume now that $(x, f) \in grA$ is given and $r > 0$ such that the operator A^{-1} is bounded on $B^*(f, r)$. Show that an element $g \in B^*\left(f, \dfrac{r}{2}\right)$ belongs to $R(A)$. Consider the equality

$$g_\alpha + \alpha J(x_\alpha - x) = g, \quad g_\alpha \in Ax_\alpha, \quad \alpha > 0. \tag{1.7.8}$$

By the monotonicity of A, we have

$$\langle g - \alpha J(x_\alpha - x) - f, x_\alpha - x \rangle \geq 0.$$

Hence, $\alpha \|x_\alpha - x\| \leq \|g - f\|_* < \dfrac{r}{2}$. Then, in view of (1.7.8), one gets

$$\|g - g_\alpha\|_* = \alpha \|x_\alpha - x\| < \dfrac{r}{2}.$$

Therefore,

$$\|g_\alpha - f\|_* \leq \|g_\alpha - g\|_* + \|g - f\|_* < r,$$

i.e., $g_\alpha \in B^*(f, r)$. Using the local boundedness of A^{-1} on $B^*(f, r)$, we conclude then that the sequence $\{x_\alpha\}$ is bounded in X. Therefore, $\|g - g_\alpha\|_* = \alpha \|x_\alpha - x\| \to 0$ as $\alpha \to 0$. We established above a closedness of $R(A)$. Thus, $g \in R(A)$, which implies that $R(A)$ is open. The proof is complete. ∎

Corollary 1.7.12 *If* $A : X \to 2^{X^*}$ *is a maximal monotone and weakly coercive operator, then* $R(A) = X^*$.

Proof. Indeed, it follows from the Definition 1.1.45 of the weak coerciveness of A that A^{-1} is bounded. Then the assertion is the consequence of Lemma 1.7.11.   ∎

Next we present the following important result.

Theorem 1.7.13 *Let* $A : X \to 2^{X^*}$ *be a monotone operator. Then* A *is a maximal monotone operator if and only if* $R(A + J) = X^*$.

Proof. By Zorn's lemma, there exists a maximal monotone extension $\bar{A} : X \to 2^{X^*}$ such that $gr A \subseteq gr \bar{A}$. Applying Theorem 1.7.4 to the operator $\bar{A}$ we deduce for all $f \in X^*$ that there exist the unique elements $x \in X$ and $y \in \bar{A}x$ such that $y + Jx = f$. Hence, $gr A = gr \bar{A}$ if and only if every element $f \in X^*$ can be presented in the form of $f = y + Jx$ for some $x \in X$, $y \in Ax$, i.e., if and only if $R(A + J) = X^*$. ∎

Corollary 1.7.14 *Suppose that* H *is a Hilbert space,* $A : H \to 2^H$, *operator* $(I + \alpha A)^{-1}$ *is defined for all* $\alpha > 0$ *on the whole space* H *and* $I - A$ *is a nonexpansive mapping. Then* A *is maximal monotone.*

Proof. It follows from Section 1.3 (Example 3) that A is monotone. Further, by the hypothesis, $R(A + I) = H$. This is enough in order to apply the previous theorem and obtain the claim.   ∎

Theorem 1.7.15 *A subdifferential* $\partial\varphi : X \to 2^{X^*}$ *of a proper convex lower semicontinuous functional* $\varphi : X \to R^1$ *is a maximal monotone operator.*

Proof. In Section 1.3 (Example 2), monotonicity of $\partial\varphi$ has been established on $D(\partial\varphi)$. According to Theorem 1.7.13, to prove the maximal monotonicity of $\partial\varphi$, we have to show that $R(\partial\varphi + J) = X^*$, that is, that for any $f \in X^*$ there is $x \in D(\partial\varphi)$ satisfying the inclusion $f \in (\partial\varphi + J)x$.

Construct a proper convex lower semicontinuous functional

$$\Phi(y) = 2^{-1}\|y\|^2 + \varphi(y) - \langle f, y \rangle \quad \forall y \in D(\varphi).$$

The definition of $\partial\varphi$ at a point $x_0 \in D(\partial\varphi)$ gives

$$\varphi(y) \geq \varphi(x_0) + \langle \partial\varphi(x_0), y - x_0 \rangle \quad \forall y \in D(\varphi).$$

Then the following inequalities hold:

$$\Phi(y) \geq 2^{-1}\|y\|^2 + \varphi(x_0) + \langle \partial\varphi(x_0), y - x_0 \rangle - \langle f, y \rangle$$

$$\geq \varphi(x_0) - \langle \partial\varphi(x_0), x_0 \rangle + 2^{-1}\|y\|(\|y\| - 2\|\partial\varphi(x_0)\|_* - 2\|f\|_*).$$

Hence, $\Phi(y) \to +\infty$ as $\|y\| \to \infty$. Thus, by Theorem 1.1.23 and Lemma 1.2.5, there exists a point $x \in X$ at which $\Phi(y)$ reaches minimum. Then we can write down that $\theta_{X^*} \in \partial\Phi(x) = (J + \partial\varphi)x - f$ or $f \in (J + \partial\varphi)x$. Actually, the operator $\partial\varphi : X \to 2^{X^*}$ is maximal monotone.   ∎

Corollary 1.7.16 *Any monotone potential operator* $A : X \to 2^{X^*}$ *is maximal monotone.*

Proof. See Lemma 1.2.6 and Theorem 1.7.15. ∎

Theorem 1.7.17 *Let* $A : X \to 2^{X^*}$ *be a maximal monotone operator. Then the set* $\overline{D(A)}$ *is convex.*

Proof. For any $x^0 \in X$, consider the equation

$$J(x - x^0) + \alpha A x = 0, \quad \alpha > 0. \tag{1.7.9}$$

By Theorem 1.7.13, it has a solution $x_\alpha \in D(A)$ which is unique because A is monotone and J is strictly monotone. Then there exists $y_\alpha \in A x_\alpha$ such that

$$J(x_\alpha - x^0) + \alpha y_\alpha = \theta_{X^*}.$$

If $(v, u) \in grA$ then the following equalities hold:

$$
\begin{aligned}
\|x_\alpha - x^0\|^2 &= \langle J(x_\alpha - x^0), x_\alpha - x^0 \rangle \\[2mm]
&= \langle J(x_\alpha - x^0), x_\alpha - v \rangle + \langle J(x_\alpha - x^0), v - x^0 \rangle \\[2mm]
&= \alpha \langle u - y_\alpha, x_\alpha - v \rangle + \alpha \langle u, v - x_\alpha \rangle + \langle J(x_\alpha - x^0), v - x^0 \rangle.
\end{aligned}
$$

Now the monotonicity of A yields the relation

$$\|x_\alpha - x^0\|^2 \leq \alpha \langle u, v - x_\alpha \rangle + \langle J(x_\alpha - x^0), v - x^0 \rangle. \tag{1.7.10}$$

By the definition of J, we obtain

$$\|x_\alpha - x^0\|^2 \leq \alpha \|u\| \|v - x_\alpha\| + \|x_\alpha - x^0\| \|v - x^0\|.$$

Hence, the sequences $\{J(x_\alpha - x^0)\}$ and $\{x_\alpha\}$ are bounded for all $\alpha > 0$. There exists a subsequence $\{\alpha_n\}$, $\alpha_n \to 0$ as $n \to \infty$, such that $J(x_{\alpha_n} - x^0) \rightharpoonup \bar{y} \in X^*$. Thus, (1.7.10) gives the estimate

$$\limsup_{\alpha_n \to 0} \|x_{\alpha_n} - x^0\|^2 \leq \langle \bar{y}, v - x^0 \rangle \quad \forall v \in D(A). \tag{1.7.11}$$

Is is obvious that (1.7.11) holds for all $v \in \overline{D(A)}$. If $x^0 \in \overline{D(A)}$ then we conclude from (1.7.11) that $x_\alpha \to x^0$ as $\alpha \to 0$.

Choose now elements x_1^0, $x_2^0 \in \overline{D(A)}$ and put $x^t = t x_1^0 + (1 - t) x_2^0$, $t \in [0, 1]$. Let $x_\alpha^t \in D(A)$ be a solution of the equation (1.7.9) with x^t in place of x^0. Then, by (1.7.11), we deduce that $x_\alpha^t \to x^t$, $x_\alpha^t \in D(A)$, i.e., $x^t \in \overline{D(A)}$. Hence, the set $\overline{D(A)}$ is convex. ∎

Corollary 1.7.18 *If an operator* A *is maximal monotone, then the set* $\overline{R(A)}$ *is convex.*

Proof. This claim can be proved by applying Theorem 1.7.17 to the inverse map A^{-1}. ∎

Consider in more detail the property of the local boundedness of a maximal monotone operator.

We proved in Theorem 1.3.16 that an arbitrary monotone operator is locally bounded at any interior point of its domain. Hence, a maximal monotone operator has the same property. Furthermore, for a maximal monotone operator the statement about the local boundedness can be specified. Namely, the following theorem holds:

Theorem 1.7.19 *If int $D(A) \neq \emptyset$, then a maximal monotone operator $A : X \to 2^{X^*}$ is unbounded at any boundary points of its domain. Moreover, the range of A has at least one semi-line at these points.*

Proof. Let $x \in \partial D(A)$. Consider the set $M = \overline{D(A)}$, the closure of $D(A)$. Recall that M is convex and closed, and $int\ M \neq \emptyset$. It is clear that $x \in \partial M$. Therefore, it is possible to construct a supporting hyperplane to the set M at the point x. In other words, we assert that there exists an element $y \in X^*$ $(y \neq \theta_{X^*})$ such that

$$\langle y, x - u \rangle \geq 0 \quad \forall u \in D(A). \tag{1.7.12}$$

Let $z \in Ax$ and $w_\lambda = z + \lambda y \in X^*$, $\lambda \geq 0$. Since A is monotone, we obtain by (1.7.12)

$$\langle v - w_\lambda, u - x \rangle = \langle v - z, u - x \rangle - \lambda \langle y, u - x \rangle \geq 0 \quad \forall u \in D(A), \ \forall v \in Au. \tag{1.7.13}$$

Since an operator A is also maximal monotone, it follows from (1.7.13) that $w_\lambda \in Ax$. Hence, the range of the maximal monotone operator A at a boundary point x of $D(A)$ contains the semi-line $\{z + \lambda y, \mid z \in Ax, \ \lambda \geq 0\}$. ∎

Corollary 1.7.20 *If $A : X \to 2^{X^*}$ is a maximal monotone operator and $R(A)$ is a bounded set in X^*, then $D(A) = X$.*

Corollary 1.7.21 *If $A : X \to 2^{X^*}$ is a monotone operator and there exist elements $y \in X^*$ and $y_n \in Ax_n$, $n = 1, 2, \ldots$ such that*

$$\lim_{n \to \infty} \|x_n\| = \infty, \quad \lim_{n \to \infty} \|y_n - y\|_* = 0,$$

then y is a boundary point of $R(A)$.

1.8 Summation of Maximal Monotone Operators

Construct the indicator function $I_r(x)$ of the ball $B(\theta_X, r) \subset X$, $r > 0$:

$$I_r(x) = \begin{cases} 0, & x \in B(\theta_X, r), \\ +\infty, & x \notin B(\theta_X, r). \end{cases}$$

By Theorem 1.7.15, its subdifferential $\partial I_r : X \to 2^{X^*}$ is a maximal monotone operator and $D(\partial I_r) = B(\theta_X, r)$.

It is obvious that $\partial I_r(x) = \theta_{X^*}$ when $\|x\| < r$ and $\partial I_r(x) = \emptyset$ for all x with $\|x\| > r$. The points $x \in X$ such that $\|x\| = r$ are the boundary for $D(\partial I_r)$, therefore, by Theorem 1.7.19, the ranges of ∂I_r at these points form semi-lines in X^*. Since $\langle Jx, x - y \rangle \geq 0$ as $\|x\| = r$ and $\|y\| < r$, a semi-line at the point $x \in \partial D(\partial I_r)$ is as follows: $\{\lambda Jx \mid \lambda \geq 0\}$. Hence, the operator $\partial I_r : X \to 2^{X^*}$ has the following representation:

$$\partial I_r(x) = \begin{cases} \theta_{X^*}, & \|x\| < r; \\ \emptyset, & \|x\| > r; \\ \lambda Jx, & \|x\| = r, \ \lambda \geq 0. \end{cases} \tag{1.8.1}$$

Lemma 1.8.1 *Let $A : X \to 2^{X^*}$ be a monotone operator, $\theta_X \in D(A)$ and let there exist a number $r_0 > 0$ such that the operator $A + \partial I_r$ is maximal monotone for all $r \geq r_0$. Then A is a maximal monotone operator too.*

Proof. Without loss of generality, we presume that $\theta_{X^*} \in A(\theta_X)$. If we shall prove that $R(A + J) = X^*$ then the claim to be proved follows from Theorem 1.7.13. Take an arbitrary element $f \in X^*$ and a number $r \geq r_0$ such that $r \geq \|f\|_*$. By the hypothesis, an operator $A + \partial I_r$ is maximal monotone as $r \geq r_0$, hence, Theorem 1.7.4 guarantees existence of an $x \in X$ such that

$$f \in (A + \partial I_r + J)x = (A + J)x + \partial I_r(x). \tag{1.8.2}$$

Since $D(A + \partial I_r) = D(A) \cap D(\partial I_r) = D(A) \cap B(\theta_X, r)$, we conclude that $\|x\| \leq r$. If $\|x\| < r$ then the conclusion of the lemma follows from (1.8.1) and (1.8.2). Consider the case when $\|x\| = r$. Taking into account (1.8.1), we may rewrite (1.8.2) in the form: $f \in (A + J)x + \lambda Jx$ where $\lambda \geq 0$. If $\lambda = 0$ then the lemma is proved. Assume that $\lambda > 0$ and let $y \in Ax$ and $f = y + (1 + \lambda)Jx$. Then

$$\langle y, x \rangle + (1 + \lambda)\|x\|^2 = \langle f, x \rangle. \tag{1.8.3}$$

Since the operator A is monotone and since $\theta_{X^*} \in A(\theta_X)$, we have the inequality $\langle y, x \rangle \geq 0$ and (1.8.3) implies

$$(1 + \lambda)\|x\|^2 \leq \langle f, x \rangle \leq \|f\|_* \|x\|.$$

It follows that

$$\|x\| \leq (1 + \lambda)^{-1}\|f\|_* < r.$$

Thus, we come to the contradiction which establishes the result. ∎

Lemma 1.8.2 *Let $A : X \to 2^{X^*}$ be a maximal monotone operator. Then*

$$A_\alpha = (A + \alpha J)^{-1} : X^* \to X, \quad \alpha > 0,$$

is a single-valued, monotone and demicontinuous mapping with $D(A_\alpha) = X^*$.

Proof. By Theorem 1.7.4, $R(A + \alpha J) = X^*$ for all $\alpha > 0$ and thus $D(A_\alpha) = X^*$. Since the operator $A + \alpha J$ is strictly monotone in view of Lemma 1.5.5 and Proposition 1.3.5, the equation $Ax + \alpha Jx = f$ is uniquely solvable for any $f \in X^*$. This proves that the operator

A_α is single-valued. To show that A_α is demicontinuous, choose $\{x_n\} \subset X$ and let $f \in X^*$ and $f_n \in X^*$, $n = 1, 2, \dots$, be such that $f \in (A + \alpha J)x$ and $f_n \in (A + \alpha J)x_n$. Suppose that $f_n \to f$. It was proved in Theorem 1.7.5 that $\{x_n\}$ is bounded. Obviously,

$$\langle y_n - y, x_n - x \rangle + \alpha \langle Jx_n - Jx, x_n - x \rangle = \langle f_n - f, x_n - x \rangle, \tag{1.8.4}$$

where $y_n \in Ax_n$, $y \in Ax$, $y_n + \alpha Jx_n = f_n$ and $y + \alpha Jx = f$. Because A and J are monotone operators, we have from (1.8.4) the limit equality

$$\lim_{n \to \infty} \langle Jx_n - Jx, x_n - x \rangle = 0. \tag{1.8.5}$$

Then Lemma 1.5.15 allows us to state that $x_n \rightharpoonup x$. Furthermore, the mapping A_α is monotone as the inverse map to the monotone operator $B = A + \alpha J$. The proof is now complete. ∎

Theorem 1.8.3 *Let A_1 and A_2 be maximal monotone operators from X to 2^{X^*} and*

$$D(A_1) \cap int \, D(A_2) \neq \emptyset. \tag{1.8.6}$$

Then their sum $A_1 + A_2$ is also a maximal monotone operator.

   Proof. Making shifts, if it is necessary, in the domains of A_1 and A_2 and in the range of A_1, we assume, without loss of generality, that

$$\theta_{X^*} \in A_1(\theta_X), \quad \theta_X \in int \, D(A_2). \tag{1.8.7}$$

We deal first with the case where $D(A_2)$ is a bounded set. For this, it is sufficient to show that $R(A_1 + A_2 + J) = X^*$. Choose an arbitrary element $f \in X^*$ and prove that $f \in R(A_1 + A_2 + J)$. We may put $f = \theta_{X^*}$, shifting the range of A_2, if it is necessary. Hence, we have to prove that there exists $x \in X$ such that

$$\theta_{X^*} \in (A_1 + A_2 + J)x. \tag{1.8.8}$$

The inclusion (1.8.8) holds if and only if there exist $x \in X$ and $y \in X^*$ such that

$$-y \in (A_1 + \frac{1}{2}J)x, \quad y \in (A_2 + \frac{1}{2}J)x.$$

Construct the maps $T_1 : X^* \to X$ and $T_2 : X^* \to X$ as follows:

$$T_1 y = -(A_1 + \frac{1}{2}J)^{-1}(-y)$$

and

$$T_2 y = (A_2 + \frac{1}{2}J)^{-1}y.$$

It is clear that (1.8.8) is equivalent to the inclusion

$$\theta_X \in T_1 y + T_2 y. \tag{1.8.9}$$

This means that, instead of (1.8.8), it is sufficient to show that $\theta_X \in R(T_1 + T_2)$.

By Lemma 1.8.2, operators T_1 and T_2 are monotone, demicontinuous and $D(T_1) = D(T_2) = X^*$. Hence, the sum $T = T_1 + T_2$ is also a monotone and demicontinuous operator with $D(T) = X^*$. Then maximal monotonicity of $T : X^* \to X$ arises from Theorem 1.4.6. Since $\theta_{X^*} = J(\theta_X)$ and $\theta_{X^*} \in A_1(\theta_X)$, we conclude that

$$\theta_{X^*} \in (A_1 + \frac{1}{2}J)(\theta_X),$$

that is, $\theta_X \in T_1(\theta_{X^*})$. Therefore,

$$\langle y, T_1 y \rangle \geq 0 \quad \forall y \in X^*. \tag{1.8.10}$$

To prove (1.8.9) we can use Theorem 1.7.9. Show that there exists a number $r > 0$ such that

$$\langle y, T_1 y + T_2 y \rangle \geq 0 \ \text{ as } \ \|y\|_* \geq r.$$

In view of (1.8.10), it is necessary to find $r > 0$ satisfying the condition

$$\langle y, T_2 y \rangle \geq 0 \ \text{ as } \ \|y\|_* \geq r. \tag{1.8.11}$$

By the definition of T_2,

$$R(T_2) = D(A_2 + \frac{1}{2}J) = D(A_2).$$

Hence, the set $R(T_2)$ is bounded. Write the monotonicity condition of T_2 :

$$\langle y - z, T_2 y - T_2 z \rangle \geq 0 \quad \forall y, z \in X^*,$$

from which we obtain the inequality

$$\langle y, T_2 y \rangle \geq \langle y, T_2 z \rangle + \langle z, T_2 y - T_2 z \rangle. \tag{1.8.12}$$

Since $R(T_2)$ is bounded, there exists a constant $c > 0$ such that

$$|\langle z, T_2 y - T_2 z \rangle| \leq c \|z\|_*. \tag{1.8.13}$$

Furthermore, $\theta_X \in int \, R(T_2)$ because of (1.8.7). Hence, by Theorem 1.3.16, the inverse to T_2 operator is locally bounded at zero. This means that there exist numbers $c_1 > 0$ and $c_2 > 0$ such that

$$\{y \in X^* \mid \|T_2 y\| \leq c_1\} \subseteq B^*(\theta_{X^*}, c_2).$$

Then (1.8.12) and (1.8.13) imply

$$\langle y, T_2 y \rangle \geq \langle y, T_2 z \rangle - cc_2 \ \text{ if } \ \|T_2 z\| \leq c_1,$$

from which we deduce

$$\langle y, T_2 y \rangle \geq sup \, \{\langle y, T_2 z \rangle \mid \|T_2 z\| \leq c_1\} - cc_2 = c_1 \|y\|_* - cc_2.$$

Therefore, $\langle y, T_2 y \rangle \geq 0$ if $\|y\|_* \geq c_1^{-1} c c_2$, that is, in (1.8.11) we can put $r = c_1^{-1} c c_2$. Thus, the theorem is proved provided that $D(A_2)$ is bounded.

We now omit this assumption. Construct the maximal monotone operator ∂I_r for any $r > 0$. It is clear that $D(\partial I_r)$ is a bounded set and

$$D(A_2) \cap int\, D(\partial I_r) \neq \emptyset.$$

On the basis of the previous proof we conclude that the operator $A_2 + \partial I_r$ is maximal monotone. Further,

$$D(A_1) \cap int\, (A_2 + \partial I_r) \neq \emptyset,$$

and

$$D(A_2 + \partial I_r) = \{ x \in D(A_2) \mid \|x\| \leq r \}$$

is a bounded set. Hence, $A_1 + A_2 + \partial I_r$ is a maximal monotone operator. It arises from Lemma 1.8.1 that the map $A_1 + A_2$ is maximal monotone. The proof is accomplished. ∎

Remark 1.8.4 *Theorem 1.8.3 is valid if one of the operators A_1 and A_2 is the subdifferential of a proper convex lower semicontinuous functional.*

Theorem 1.8.5 *Let $A : X \to 2^{X^*}$ be a maximal monotone operator. $\Omega \subseteq D(A)$ be a convex closed set. $int\, \Omega \neq \emptyset$. Then there exists a maximal monotone operator $\bar{A}$ with $D(\bar{A}) = \Omega$, where $A = \bar{A}$ on $int\, \Omega$.*

Proof. Let $I_\Omega(x)$ be the indicator function associated with the set Ω, that is,

$$I_\Omega(x) = \begin{cases} 0, & x \in \Omega, \\ +\infty, & x \notin \Omega. \end{cases}$$

Using Theorem 1.7.15, we see that its subdifferential $\partial I_\Omega : X \to 2^{X^*}$, represented by the formula

$$\partial I_\Omega(x) = \begin{cases} \theta_{X^*}, & x \in int\, \Omega; \\ \emptyset, & x \notin \Omega; \\ \lambda J x, & x \in \partial\Omega, \lambda \geq 0, \end{cases} \tag{1.8.14}$$

is maximal monotone. Due to Theorem 1.8.3, the sum $A + \partial I_\Omega$ is maximal monotone too, $D(A + \partial I_\Omega) = \Omega$ and

$$(A + \partial I_\Omega)x = Ax \quad \text{as} \quad x \in int\, \Omega.$$

Therefore, on the set Ω the maximal monotone extension $\bar{A} = A + \partial I_\Omega$. Observe that $\bar{A}$ is obtained from the original operator A by joining additional values on the boundary of Ω. ∎

Remark 1.8.6 *Theorem 1.8.5 remains still valid if the condition $int\, \Omega \neq \emptyset$ is replaced by $int\, D(A) \cap \Omega \neq \emptyset$.*

It is easy to check that if Ω is a nonempty convex and closed subset of X then the subdifferential ∂I_Ω is the normality operator N_Ω given as follows:

$$N_\Omega(x) = \begin{cases} \{\psi \in X^* \mid \langle \psi, y - x \rangle \geq 0 \ \forall y \in \Omega\} & \text{if } x \in \Omega; \\ \emptyset & \text{if } x \notin \Omega. \end{cases} \tag{1.8.15}$$

The following results arise from Theorems 1.8.3 and 1.7.5.

Theorem 1.8.7 *Let $A_1 : X \to 2^{X^*}$ and $A_2 : X \to 2^{X^*}$ be maximal monotone operators satisfying the condition (1.8.6) and the sum $A_1 + A_2$ be coercive. Then $R(A_1 + A_2) = X^*$.*

Theorem 1.8.8 *Let $A_1 : X \to 2^{X^*}$ be a maximal monotone operator, $A_2 : X \to X^*$ be a monotone hemicontinuous and coercive operator. Then $R(A_1 + A_2) = X^*$.*

Corollary 1.8.9 *If $A : X \to 2^{X^*}$ is a maximal monotone operator, $J^\mu : X \to X^*$ is a duality mapping with gauge function $\mu(t)$, then the equation $Ax + \alpha J^\mu x = 0$ has a unique solution for all $\alpha > 0$ and for all $f \in X^*$.*

As a consequence, Theorem 1.4.11 allows us to present the following assertion.

Theorem 1.8.10 *Let $L : X \to X^*$ be a linear single-valued monotone and closed operator, the adjoint operator $L^* : X \to X^*$ be monotone, $A : X \to 2^{X^*}$ be a maximal monotone and coercive operator with int $D(A) \neq \emptyset$. Then $R(L + A) = X^*$.*

Return to Example 7 of Section 1.3. There we considered the monotone operator A defined as follows:

$$Au = -a^2 \Delta u + (g(x) + a)u(x) + u(x) \int_{R^3} \frac{u^2(y)}{\mid x - y \mid} dy. \tag{1.8.16}$$

We wonder whether it is maximal monotone. As before, represent A in the form: $A = L + B$, where L is a linear part of (1.8.16) and B is defined by the last term. It is not difficult to see that

$$D(L) = \{u \in H \mid \nabla u \in H \times H \times H, \ \Delta u \in H\}, \ H = L^2(R^3),$$

that is, $D(L) = D(\Delta)$. We emphasize that the operations ∇u and Δu are regarded here in the generalized sense.

Furthermore, it is obvious that L is a self-adjoint operator. Under these conditions, there exists a positive self-adjoint operator $L^{1/2}$ defined by the following equality:

$$(Lu, v) = (L^{1/2}u, L^{1/2}v).$$

We study the domain of $L^{1/2}$. First of all, $L^{1/2}$ can be represented by the sum $L^{1/2} = L_1^{1/2} + L_2^{1/2}$, where L_1 is defined by the first summand in the right-hand side of (1.8.16) and L_2 by the second one. It is clear that $D(L_2^{1/2}) \supset H$. Further, $(L_1u, v) = (\nabla u, \nabla v)$, that is, $D(L^{1/2}) \subset W_1^2(R^3)$. The following inequality holds [44]:

$$\left\| \frac{u(y)}{\mid x - y \mid} \right\|_2 \leq 2\|\nabla u\|_2.$$

Hence,

$$\left\| \int_{R^3} \frac{u^2(y)}{|x-y|} dy \right\|_\infty \leq 2\|u\|_2 \|\nabla u\|_2.$$

Since $u \in W_1^2(R^3)$, we have

$$\|Bu\|_2 \leq 2\|u\|_2^2 \|\nabla u\|_2 < \infty,$$

where $\|v\|_p$ denotes the norm of an element v in $L^p(R^3)$, $p > 1$. Thus, the inclusion $D(L^{1/2}) \subset W_1^2(R^3) \subset D(B)$ is established.

Since the condition $int\, D(B) \cap D(L) \neq \emptyset$ of Theorem 1.8.3 is difficult to verify, we use the technique of [231] to prove that A is maximal monotone. To this end, first of all note that the operator $(I + \alpha^{-1}L)^{1/2}$ with $\alpha > 0$ is well defined on all of H, positive, self-adjoint and it has a bounded inverse. Moreover,

$$D\left((I + \alpha^{-1}L)^{1/2}\right) = D(L^{1/2}). \tag{1.8.17}$$

Indeed, we have

$$L^{1/2} = \int_0^\infty \sqrt{\lambda} dE_\lambda,$$

where $\{E_\lambda\}$ is called the identity decomposition generated by L. Then $u \in D((I+\alpha^{-1}L)^{1/2})$ if and only if

$$\int_0^\infty (1 + \alpha^{-1}\lambda)(dE_\lambda u, u) < \infty.$$

The latter is fulfilled if

$$\int_0^\infty \lambda(dE_\lambda u, u) < \infty.$$

Hence, (1.8.17) holds.

Introduce the operator

$$Q = (I + \alpha^{-1}L)^{-1/2} B (I + \alpha^{-1}L)^{-1/2}$$

with $D(Q) = H$. In virtue of the properties of L and B, it is hemicontinuous and monotone. Then Theorem 1.4.6 immediately gives the maximal monotonicity of Q. In its turn, Theorem 1.7.4 guarantees that a solution w of the equation

$$Qw + \alpha w = (I + \alpha^{-1}L)^{-1/2} v$$

exists. Hence, the equation

$$Lu + Bu + \alpha u = v, \quad \alpha > 0,$$

where $u = (I + \alpha^{-1}L)^{-1/2}w$, is solvable for any $v \in H$. By Theorem 1.7.13, we conclude that the operator $A = L + B$ is maximal monotone.

1.9   Equations with General Monotone Operators

We have seen in Section 1.7 that the maximal monotonicity of operators allow us to prove the existence theorems for equations with such operators, to study their domains and ranges and to describe the structure of their solution sets. Observe that, by Zorn's lemma, an arbitrary monotone operator has at least one maximal monotone extension. However, there is no constructive way to build such extensions. Besides, the example given in the previous section shows that the establishment of maximal monotonicity of monotone operators is often a very complicated problem.

There is another problem. The reader already knows that a maximal monotone operator is multiple-valued, in general. In practice calculations, we usually consider only certain sections of maximal monotone operators which guarantee the monotonicity of resulting discontinuous mappings. Hence, there is a necessity to analyze problems with arbitrary monotone operators that, generally speaking, do not satisfy the conditions of continuity or maximal monotonicity.

Assume that X is a reflexive Banach space, X and X^* are strictly convex. The next result will be useful in the sequel.

Theorem 1.9.1 *Suppose that a monotone hemicontinuous operator $A : X \to X^*$ is given on open or linear dense set $D(A) \subseteq X$. Then the equation*

$$Ax = f \tag{1.9.1}$$

has a solution $x_0 \in D(A)$ for any $f \in X^*$ if and only if

$$\langle Ax - f, x - x_0 \rangle \geq 0 \quad \forall x \in D(A). \tag{1.9.2}$$

Proof. Let $Ax_0 = f$. Then, by the monotonicity of A,

$$\langle Ax - f, x - x_0 \rangle = \langle Ax - Ax_0, x - x_0 \rangle + \langle Ax_0 - f, x - x_0 \rangle \geq 0,$$

because

$$\langle Ax_0 - f, x - x_0 \rangle = 0.$$

Let there exist $x_0 \in D(A)$ such that (1.9.2) is valid. Take any $w \in D(A)$ and put $x_t = x_0 + tw \in D(A)$ with $t \geq 0$. Substitute x_t for x in (1.9.2). Then

$$\langle Ax_t - f, x_t - x_0 \rangle = \langle Ax_t - f, tw \rangle \geq 0,$$

that is,

$$\langle Ax_t - f, w \rangle \geq 0 \quad \forall w \in D(A).$$

By the hemicontinuity of the operator A, we have in a limit as $t \to 0$:

$$\langle Ax_0 - f, w \rangle \geq 0 \quad \forall w \in D(A).$$

This means that $Ax_0 = f$.   ∎

Observe that the second part of this theorem does not need the monotonicity property of A. Therefore, it is useful to state the following result:

Corollary 1.9.2 *Suppose that hemicontinuous operator $A : X \to X^*$ is defined on open or linear dense set $D(A) \subseteq X$ and let $x_0 \in D(A)$. Then the inequality (1.9.2) implies $f = Ax_0$.*

It has happened that a solution of the equation (1.9.1) with an arbitrary monotone operator $A : X \to 2^{X^*}$ does not exist both in the classical sense and in the sense of inclusions. For this reason, using Theorem 1.9.1, we introduce

Definition 1.9.3 *An element $x_0 \in X$ is said to be a generalized solution of the equation (1.9.1) if for all $x \in D(A)$ and for all $y \in Ax$ it satisfies the inequality*

$$\langle y - f, x - x_0 \rangle \geq 0. \tag{1.9.3}$$

It follows from (1.9.3) that if a generalized solution set of (1.9.1) is not empty then it is convex and closed. For equations with monotone hemicontinuous operators, generalized and classical solutions coincide.

We are interested in the generalized solvability of the equation

$$Ax + \alpha Jx = f, \quad \alpha > 0, \quad x \in D(A), \tag{1.9.4}$$

with a monotone operator A.

Denote by $R(Ax - f)$ a convex closed hull of the weak limits of all subsequences of the sequences $\{Ax_n - f\}$ when $x_n \to x$, $x \in X$, $x_n \in D(A)$.

Definition 1.9.4 *A point $x_0 \in X$ is called an sw-generalized solution of the equation (1.9.1) if $\theta_{X^*} \in R(Ax_0 - f)$.*

Lemma 1.9.5 *If $\operatorname{int} D(A) \neq \emptyset$, then for any point $x_0 \in \operatorname{int} D(A)$ the inclusion*

$$\theta_{X^*} \in R(Ax_0 - f)$$

is satisfied if and only if $\theta_{X^*} \in \bar{A}x_0 - f$, where $\bar{A}$ is a maximal monotone extension of A.

Proof. Let $\theta_{X^*} \in R(Ax_0 - f)$, $x_n \in D(A)$, $x_n \to x_0$. By the monotonicity of A,

$$\langle (y - f) - (y_n - f), x - x_n \rangle \geq 0 \quad \forall x \in D(A), \quad \forall y \in Ax, \quad \forall y_n \in Ax_n.$$

According to the definition of $R(Ax_0 - f)$, we obtain

$$\langle y - f, x - x_0 \rangle \geq 0 \quad \forall x \in D(A), \quad \forall y \in Ax.$$

From this inequality, it follows that $f \in \bar{A}x_0$. Hence, $\theta_{X^*} \in \bar{A}x_0 - f$.

Let now $\theta_{X^*} \in \bar{A}x_0 - f$. Show by the contradiction that $\theta_{X^*} \in R(Ax_0 - f)$. Assume that $\theta_{X^*} \notin R(Ax_0 - f)$. Then there is an element $g \in X$ such that $\langle z, g \rangle < 0$ for all $z \in R(Ax_0 - f)$. Further, construct the sequence $x_n = x_0 + t_n g$, $t_n > 0$. If $t_n \to 0$ then $x_n \to x_0$. Since $x_0 \in \operatorname{int} D(A)$ then A is locally bounded at this point. Therefore, there is some subsequence $y_k - f \rightharpoonup f_1 \in X^*$, $y_k \in Ax_k$, that is, $f_1 \in R(Ax_0 - f)$ by the definition of the last set. Now the monotonicity of $\bar{A}$ yields the inequality

$$\langle \bar{y}_k - f, x_k - x_0 \rangle \geq 0 \quad \forall \bar{y}_k \in \bar{A}x_k.$$

Consequently,

$$\langle y_k - f, x_k - x_0 \rangle \geq 0 \quad \forall y_k \in Ax_k,$$

or

$$\langle y_k - f, g \rangle \geq 0 \quad \forall y_k \in Ax_k.$$

Setting $t_k \to 0$ we obtain $\langle f_1, g \rangle \geq 0$, which contradicts the definition of g. ∎

Observe that if A is a maximal monotone operator then (1.9.3) determines the inclusion $f \in Ax_0$. This means that the solutions of (1.9.1) in the sense of Definitions 1.7.2 and 1.9.3 coincide.

Corollary 1.9.6 *A monotone hemicontinuous operator carries any bounded weakly closed set $M \subset X$ to a closed set of X^*.*

Proof. Let $x_n \in M$, $x_n \rightharpoonup x_0 \in M$ and $Ax_n \to y_0 \in X^*$. Write the monotonicity condition of A,

$$\langle Ax - Ax_n, x - x_n \rangle \geq 0.$$

Hence,

$$\langle Ax - y_0, x - x_0 \rangle \geq 0.$$

By Theorem 1.9.1, we conclude then $y_0 = Ax_0$, that is, $y_0 \in A(M)$. ∎

Corollary 1.9.7 *If the set $D(A)$ is closed and convex and int $D(A) \neq \emptyset$, then there exists only one maximal monotone extension $\bar{A}$ with the domain $D(\bar{A}) = D(A)$.*

Proof. Indeed, it follows from Lemma 1.9.5 that at each point $x^0 \in int\ D(A)$ the set $\{y_0 \mid y_0 \in \bar{A}x_0\}$ coincides with $R(Ax_0)$. In view of Theorem 1.7.19, at boundary points of $D(A)$ the operator $\bar{A}$ is finished determining A by semi-lines, if it is required. ∎

Lemma 1.9.8 *Let $A : X \to 2^{X^*}$ be a monotone operator, $D(A)$ be a convex closed set in X, int $D(A) \neq \emptyset$. Then each generalized solution of (1.9.4) is a solution of the equation*

$$\bar{A}x + \alpha J x = f, \tag{1.9.5}$$

where $\bar{A}$ is a maximal monotone extension of A with $D(\bar{A}) = D(A)$. The converse implication is also true.

Proof. By Theorem 1.7.4, equation (1.9.5) has the unique solution $x_\alpha \in D(A)$ such that

$$f \in \bar{A}x_\alpha + \alpha J x_\alpha.$$

Then the monotonicity property of $\bar{A} + \alpha J$ allows us to write down the inequality

$$\langle y + \alpha J x - f, x - x_\alpha \rangle \geq 0 \quad \forall x \in D(A), \quad \forall y \in \bar{A}x. \tag{1.9.6}$$

Furthermore, (1.9.6) holds for all $y \in Ax$. In its turn, this means that x_α is the generalized solution of (1.9.4) too.

Now let x_α be a generalized solution of (1.9.4). Then (1.9.6) is true with $y \in Ax$ and $x \in D(A)$. Hence, by Lemma 1.9.5 and Theorem 1.8.3, $f \in \overline{(A + \alpha J)}x_\alpha = \bar{A}x_\alpha + \alpha Jx_\alpha$. Thus, x_α is a solution of (1.9.5). ∎

It is obvious that the following result holds.

Theorem 1.9.9 *Under the conditions of Lemma 1.9.8, the equation (1.9.4) has the unique generalized solution belonging to $D(A)$.*

In fact, Lemma 1.9.8 proves solvability of the equation (1.9.4) with arbitrary domain of an operator A, however, its solution does not necessarily belong to $D(A)$ and uniqueness cannot be guaranteed. Below we give one more sufficient uniqueness condition for solution of the equation (1.9.4).

Introduce the set

$$T(x_0, z) = \{x \mid x = x_0 + tz, \ t \in R^1, \ x_0 \in X, \ z \in X, \ z \neq \theta_X\}.$$

Definition 1.9.10 *We say that the set $G \subset X$ densely surrounds a point x_0 if x_0 is a bilateral boundary point of the set $G \cap T(x_0, z)$ for every $T(x_0, z)$.*

Theorem 1.9.11 *If $D(A)$ densely surrounds X, then equation (1.9.4) has a unique generalized solution for all $\alpha > 0$.*

Proof. Let x_1 and x_2 be two generalized solutions of (1.9.4) and $x_1 \neq x_2$. Then it follows from (1.9.3) that there are numbers t_1 and t_2 such that $0 < t_1, t_2 < 1$, $t_1 + t_2 - 1 > 0$ and points $x_3 = t_1x_1 + (1 - t_1)x_2$, $x_4 = t_2x_2 + (1 - t_2)x_1$ belong to $D(A)$. The following equalities can be easily verified:

$$x_3 - x_1 = (1 - t_1)(x_2 - x_1),$$

$$x_4 - x_2 = (1 - t_2)(x_1 - x_2),$$

$$x_4 - x_3 = (1 - t_1 - t_2)(x_1 - x_2).$$

The operator $A + \alpha J$ is strictly monotone. Therefore,

$$\langle y_3 - y_4, x_3 - x_4 \rangle > 0, \quad y_3 \in (A + \alpha J)x_3, \quad y_4 \in (A + \alpha J)x_4. \tag{1.9.7}$$

However,

$$\langle y_3 - y_4, x_3 - x_4 \rangle = (1 - t_1 - t_2)\left(\left\langle y_3 - f, \frac{x_3 - x_1}{1 - t_1} \right\rangle + \left\langle y_4 - f, \frac{x_4 - x_2}{1 - t_2} \right\rangle\right) \leq 0$$

which contradicts (1.9.7). ∎

1.10   Equations with Semimonotone Operators

Suppose that X is a reflexive strictly convex Banach space together with its dual space X^*.

Definition 1.10.1 *An operator $A : X \to 2^{X^*}$ is said to be semimonotone, if there exists strongly continuous operator $C : X \to X^*$ such that $T = A + C$ is a monotone map and $D(A) \subseteq D(C)$.*

It is easy to obtain the following assertions using the properties of monotone operators:

a) a semimonotone operator is locally bounded at any interior point of its domain;

b) if $T : X \to 2^{X^*}$ is a maximal monotone operator, then the set of values of a semi-monotone operator at every point of the domain is a convex and closed set.

We are interested in solvability of the equation (1.9.1) with a semimonotone operator $A : X \to 2^{X^*}$.

First of all, observe that there is the following statement which is an analogy of the Minty−Browder theorem for semimonotone operators:

Theorem 1.10.2 *Let X be a reflexive Banach space, $A : X \to X^*$ be a semimonotone hemicontinuous and coercive operator with $D(A) = X$. Then $R(A) = X^*$.*

We study the solvability problem for arbitrary, possibly multiple-valued or discontinuous, semimonotone operators A with the domain $D(A) = X$. Construct a maximal monotone extension $\bar{T}$ of a monotone operator T and let $\bar{A} = \bar{T} - C$ be given.

Definition 1.10.3 *An element $x_0 \in X$ is said to be an s-generalized solution of the equation (1.9.1) if $f \in \bar{A}x_0$.*

Definition 1.10.4 *We shall say that a space X has the M-property if there exists a sequence of projectors $\{P_n\}$, $n = 1, 2, \dots$, such that $P_n X = X_n$ and for all $x \in X$,*

$$\|P_n x - x\| \to 0 \quad \text{as} \quad n \to \infty.$$

As in Section 1.9, we denote by $R(Ax - f)$ the convex closed hull of weak limits of subsequences of $\{Ax_n - f\}$, where $x_n \to x$ as $n \to \infty$, $x_n \in X$, $x \in X$.

Lemma 1.10.5 *For all $x \in X$, the sets $\{y \mid y \in \bar{A}x - f\}$ and $R(Ax - f)$ coincide.*

Proof. See the proof of Lemma 1.9.5. ∎

Theorem 1.10.6 *Suppose that a space X possesses the M-property, $A : X \to 2^{X^*}$ is a semimonotone operator, $D(A) = X$, and there exist a constant $r > 0$ and $y \in Ax$ such that*

$$\langle y - f, x \rangle \geq 0 \quad \text{as} \quad \|x\| = r.$$

Then the equation (1.9.1) has at least one s-generalized solution x_0 with $\|x_0\| \leq r$.

Proof. There exists a sequence of the adjoint projectors $\{P_n^*\}$, $n = 1, 2, \ldots$, such that $P_n^* X^* = X_n^*$. Consider in X_n the equation

$$P_n^*(Ax - f) = 0.$$

If $x \in X_n$ and $\|x\| \leq r$ then for some $y \in Ax$ we have

$$\langle P_n^*(y - f), x \rangle = \langle y - f, x \rangle \geq 0.$$

By Theorem 1.1.62, there is at least one element $x_n \in X_n$ such that

$$\theta_{X_n^*} \in R(P_n^*(Ax_n - f)),$$

where $\|x_n\| \leq r$. Show that the operator $P_n^* \bar{T} : X_n \to 2^{X_n^*}$ is a maximal monotone extension of the operator $P_n^* T : X_n \to 2^{X_n^*}$. Recalling that $T = A + C$ and $\bar{T} = \bar{A} + C$, suppose this is not the case. Let

$$\langle \bar{y} - P_n^* \tilde{y}, \bar{x} - x \rangle \geq 0 \quad \forall x \in X_n, \quad \forall \tilde{y} \in \bar{T}x, \quad \bar{x} \in X_n, \quad \bar{y} \in X_n^*, \tag{1.10.1}$$

but $\bar{y} \notin P_n^* \bar{T} \bar{x}$. By Theorem 1.4.9, it follows that a set $\{y \mid y \in \bar{T}x\}$ is convex and closed for every $x \in X$. Then there is an element $u \in X_n$ such that $\langle y - \bar{y}, u \rangle < 0$ for all $y \in \bar{T}\bar{x}$. Introduce the sequence $\{x_k\}$, where $x_k = \bar{x} + t_k u$, $t_k > 0$, $t_k \to 0$ as $k \to \infty$. Obviously,

$$x_k \to \bar{x} \quad \text{as} \quad k \to \infty. \tag{1.10.2}$$

Due to the local boundedness of $\bar{T}$, there exists a sequence $\{y_{k_m}\}$, $k_m \to \infty$, and some $g \in X^*$ such that

$$y_{k_m} \in \bar{T} x_{k_m}, \quad \{x_{k_m}\} \subseteq \{x_k\}, \quad y_{k_m} \rightharpoonup g. \tag{1.10.3}$$

Then it results from (1.10.2), (1.10.3) and Lemma 1.4.5 that $g \in \bar{T}\bar{x}$.

Substitute $x = x_{k_m}$ and $\tilde{y} = y_{k_m} \in \bar{T}x_{k_m}$ for (1.10.1) to obtain

$$0 \leq \langle P_n^* y_{k_m} - \bar{y}, u \rangle = \langle y_{k_m} - \bar{y}, u \rangle \to \langle g - \bar{y}, u \rangle = \langle P_n^* g - \bar{y}, u \rangle < 0.$$

Therefore, for some $k > 0$, the following inequality holds:

$$\langle \bar{y} - P_n^* y_k, \bar{x} - x_k \rangle < 0, \quad x_k \in X_n, \quad y_k \in \bar{T}x_k,$$

that contradicts (1.10.1), hence, $\theta_{X_n^*} \in P_n^*(\bar{A}x_n - f)$.

So, we have constructed the sequence $\{x_n\}$, $n = 1, 2, \ldots$, such that $\|x_n\| \leq r$. Then there exists a subsequence $\{x_{n_l}\} \subseteq \{x_n\}$ which converges weakly to some element $\bar{x} \in X$. We shall prove that $\bar{x}$ is a solution of the equation (1.9.1). Let $y^{n_l} \in \bar{A}x_{n_l}$ be a sequence such that $P_n^*(y^{n_l} - f) = \theta_{X_n^*}$. We assert that $\{y^{n_l}\}$ is bounded. Indeed, let $z^{n_l} \in \bar{T}x_{n_l}$. This means that $z^{n_l} = y^{n_l} + Cx_{n_l}$. Since $\bar{T}$ is locally bounded at every point X, there exist two constants $a_1 > 0$ and $a_2 > 0$ such that the inequality $\|y\|_* \leq a_2$, $y \in \bar{T}x$ holds if $\|x\| \leq a_1$. Consider in X the sequence

$$u_{n_l} = a_1 \frac{J^* y^{n_l}}{\|y^{n_l}\|_*},$$

where $J^* : X^* \to X$ is a normalized duality mapping in X^*. Then $\|u_{n_l}\| = a_1$ and hence for all $v^{n_l} \in \bar{T}u_{n_l}$ we obtain $\|v^{n_l}\|_* \leq a_2$. Now the monotonicity property of $\bar{T}$ gives

$$\langle z^{n_l} - v^{n_l}, x_{n_l} - u_{n_l} \rangle \geq 0.$$

This implies

$$
\begin{aligned}
a_1 \|y^{n_l}\|_* &\leq \langle v^{n_l}, u_{n_l} - x_{n_l} \rangle - \langle Cx_{n_l}, u_{n_l} \rangle + \langle z^{n_l}, x_{n_l} \rangle \\
&\leq a_2(a_1 + r) + a_1 \|Cx_{n_l}\|_* + r\|Cx_{n_l} + f\|_* \leq a_3,
\end{aligned}
$$

so that $\|y^{n_l}\|_* \leq a_3 a_1^{-1}$. Again, the monotonicity of $\bar{T}$ yields the inequality

$$\langle y + Cx - y^{n_l} - Cx_{n_l}, x - x_{n_l} \rangle \geq 0 \quad \forall y \in \bar{A}x, \quad x \in X,$$

or

$$\langle y - f, x - x_{n_l} \rangle \geq \langle Cx - Cx_{n_l}, x_{n_l} - x \rangle + \langle y^{n_l} - f, x \rangle.$$

Since $\{y^{n_l}\}$ is bounded as $n_l \to \infty$ and since the space X has the M-property, one gets

$$\langle y + Cx - f - C\bar{x}, x - \bar{x} \rangle \geq 0, \quad \forall y \in \bar{A}x, \quad \forall x \in X.$$

Thus,

$$\langle z - f - C\bar{x}, x - \bar{x} \rangle \geq 0 \quad \forall z \in \bar{T}x, \quad \forall x \in X.$$

It follows from the maximal monotonicity of $\bar{T}$ that $C\bar{x} + f \in \bar{T}\bar{x}$, i.e., $f \in \bar{A}\bar{x}$. The proof is now complete. ∎

1.11   Variational Inequalities with Monotone Operators

Let X be a reflexive strictly convex Banach space together with its dual space X^*, $A : X \to 2^{X^*}$ be a maximal monotone operator with domain $D(A)$, $J : X \to X^*$ be a normalized duality mapping. In this section, we consider the following variational inequality problem: To find $x \in \Omega$ such that

$$\langle Ax - f, y - x \rangle \geq 0 \quad \forall y \in \Omega, \tag{1.11.1}$$

where $\Omega \subseteq D(A)$ is a closed and convex set of X, $f \in X^*$. We present two definitions of its solutions.

Definition 1.11.1 *An element $x^0 \in \Omega$ is called the solution of the variational inequality (1.11.1) if there is an element $z^0 \in Ax^0$ such that*

$$\langle z^0 - f, y - x^0 \rangle \geq 0 \quad \forall y \in \Omega. \tag{1.11.2}$$

A solution x^0 satisfying (1.11.2) we shall also call the classical solution of the variational inequality (1.11.1).

Definition 1.11.2 *An element $x^0 \in \Omega$ is called the solution of the variational inequality (1.11.1) if*

$$\langle z - f, y - x^0 \rangle \geq 0 \quad \forall y \in \Omega, \quad \forall z \in Ay. \tag{1.11.3}$$

Lemma 1.11.3 *If $x^0 \in \Omega$ is a solution of (1.11.1) defined by the inequality (1.11.2), then it satisfies also the inequality (1.11.3).*

Proof. Write the monotonicity condition for the operator A :

$$\langle z - z^0, y - x^0 \rangle \geq 0 \quad \forall y \in \Omega, \quad \forall z \in Ay,$$

where $z^0 \in Ax^0$ and satisfies (1.11.2). Then

$$\langle z - f, y - x^0 \rangle + \langle f - z^0, y - x^0 \rangle \geq 0 \quad \forall y \in \Omega, \quad \forall z \in Ay.$$

Taking into account (1.11.2) we obtain (1.11.3). ∎

Now we present the Minty−Browder lemma for variational inequalities with maximal monotone operators.

Lemma 1.11.4 *If $\Omega \subseteq D(A)$ and if either int $\Omega \neq \emptyset$ or int $D(A) \cap \Omega \neq \emptyset$, then Definitions 1.11.1 and 1.11.2 are equivalent.*

Proof. Let $x^0 \in \Omega$ be a solution of (1.11.1) in the sense of Definition 1.11.2 and ∂I_Ω be a subdifferential of the indicator function I_Ω associated with Ω. Since $\theta_{X^*} \in \partial I_\Omega x$ for all $x \in \Omega$ and ∂I_Ω is (maximal) monotone, we have

$$\langle \eta, y - x^0 \rangle \geq 0 \quad \forall y \in \Omega, \quad \forall \eta \in \partial I_\Omega y. \tag{1.11.4}$$

Adding (1.11.3) and (1.11.4) one gets

$$\langle z + \eta - f, y - x^0 \rangle \geq 0 \quad \forall z \in Ay, \quad \forall y \in \Omega, \quad \forall \eta \in \partial I_\Omega y. \tag{1.11.5}$$

Therefore, $z + \eta \in By$, where the operator $B = A + \partial I_\Omega : X \rightarrow 2^{X^*}$ is maximal monotone in view of Theorem 1.8.3, and $D(B) = \Omega$ by the condition $\Omega \subseteq D(A)$. Then (1.11.5) implies the inclusion $f \in Bx^0$. In other words, there exist elements $z^0 \in Ax^0$ and $\eta^0 \in \partial I_\Omega x^0$ such that $f = z^0 + \eta^0$. Consequently,

$$\langle z^0 + \eta^0 - f, y - x^0 \rangle = 0 \quad \forall y \in \Omega. \tag{1.11.6}$$

It results from this that

$$\langle z^0 - f, y - x^0 \rangle = \langle \eta^0, x^0 - y \rangle.$$

However,

$$\langle \eta^0, x^0 - y \rangle \geq 0 \quad \forall y \in \Omega,$$

because ∂I_Ω is the normality operator defined by (1.8.15). Thus,

$$\langle z^0 - f, y - x^0 \rangle \geq 0 \quad \forall y \in \Omega,$$

that is, (1.11.2) follows. Taking now into account the previous lemma we obtain the result claimed in the lemma. ∎

We prove the following important assertions.

Lemma 1.11.5 *If $A : X \to 2^{X^*}$ is a maximal monotone operator with $D(A) = \Omega$, then the inequality (1.11.1) and the equation (1.9.1) are equivalent in the sense of Definition 1.11.1.*

    Proof. Indeed, a solution x^0 of the equation (1.9.1) with a maximal monotone operator A, defined by the inclusion $f \in Ax^0$, satisfies Definition 1.11.1. Therefore, it is a solution of the inequality (1.11.1). Let now x^0 be a solution of (1.11.1) in the sense of Definition 1.11.1. Then, by Lemma 1.11.3, the inequality (1.11.3) holds. Since A is a maximal monotone operator and $D(A) = \Omega$, (1.11.3) and Proposition 1.4.3 imply the inclusion $f \in Ax^0$. Thus, x^0 is the solution of the equation $Ax = f$. ∎

    In particular, the conclusion of this lemma holds if $\Omega = X$ or x^0 is a interior point of Ω.

Lemma 1.11.6 *If $A : X \to 2^{X^*}$ is a maximal monotone operator with $D(A) \subseteq X$, $\Omega \subseteq D(A)$ is convex closed set and $x_0 \in int\, \Omega$, then*

$$\langle z - f, x - x_0 \rangle \geq 0 \quad \forall x \in \Omega, \quad \forall z \in Ax, \tag{1.11.7}$$

implies the inclusion $f \in Ax_0$. The converse is also true.

    Proof. Take an arbitrary $v \in X$. Element $x_t = x_0 + tv \in \Omega$ for sufficiently small $t > 0$ because $x_0 \in int\, \Omega$. Consequently, $x_t \to x_0$ as $t \to 0$. Since operator A is local bounded at x_0, we have $y_t \rightharpoonup y$, where $y_t \in Ax_t$. We deduce from (1.11.7) that

$$\langle y_t - f, x_t - x_0 \rangle \geq 0$$

or

$$\langle y_t - f, v \rangle \geq 0 \quad \forall v \in X.$$

Letting $t \to 0$ one gets

$$\langle y - f, v \rangle \geq 0 \quad \forall v \in X.$$

This means that $y = f$. A maximal monotone operator A is demiclosed in view of Lemma 1.4.5, that leads to the inclusion $y \in Ax_0$. Hence, $f \in Ax_0$. The converse assertion immediately follows from the monotonicity of A. ∎

Lemma 1.11.7 *Let $A : X \to 2^{X^*}$ be a maximal monotone operator. Let $\Omega \subset D(A)$ be a convex and closed set. Let ∂I_Ω be a subdifferential of the indicator function I_Ω associated with Ω. If $int\, \Omega \neq \emptyset$ or $int\, D(A) \cap \Omega \neq \emptyset$, then a solution $x^0 \in \Omega$ of the variational inequality (1.11.1) is a solution in the sense of the following inclusion:*

$$f \in Ax^0 + \partial I_\Omega(x^0). \tag{1.11.8}$$

The inverse conclusion is also true.

    Proof. Let x^0 be a solution of the variational inequality (1.11.1) in the sense of Definition 1.11.1. Then (1.11.2) and (1.11.3) are satisfied. Construct the indicator function $I_\Omega(x)$ for the set Ω and find its subdifferential $\partial I_\Omega : \Omega \subset X \to 2^{X^*}$. By definition of ∂I_Ω, one gets

$$\langle u, y - x^0 \rangle \geq 0 \quad \forall y \in \Omega, \quad \forall u \in \partial I_\Omega(y).$$

Consequently, (1.11.3) involves

$$\langle z + u - f, y - x^0 \rangle \geq 0 \quad \forall y \in \Omega, \quad \forall z \in Ay, \quad \forall u \in \partial I_\Omega(y).$$

By Theorem 1.8.3, using the conditions $int\ \Omega \neq \emptyset$ or $int\ D(A) \cap \Omega \neq \emptyset$, we conclude that $A + \partial I_\Omega$ is a maximal monotone operator with $D(A + \partial I_\Omega) = \Omega$, which implies

$$f \in Ax^0 + \partial I_\Omega(x^0).$$

Let now (1.11.8) be hold with $x^0 \in \Omega$. Then there is an element $z^0 \in Ax^0$ such that $f - z^0 \in \partial I_\Omega(x^0)$. Hence, we can write down the inequality

$$\langle z^0 - f, y - x^0 \rangle \geq 0 \quad \forall y \in \Omega.$$

Thus, x^0 is a solution of the variational inequality (1.11.1). ∎

The proved lemmas imply

Theorem 1.11.8 *Under the conditions of Lemma 1.11.7, the set of solutions of the variational inequality (1.11.1), if it is nonempty, is convex and closed.*

Note also that if maximal monotone operator A is strictly monotone then the solution x^0 of the variational inequality (1.11.1) is unique. Indeed, suppose, by contradiction, that x^1 is another solution of (1.11.1). Then x^0 and x^1 satisfy (1.11.2), so that, for some $z^0 \in Ax^0$ and for some $z^1 \in Ax^1$ we have, respectively,

$$\langle z^0 - f, y - x^0 \rangle \geq 0 \quad \forall y \in \Omega,$$

and

$$\langle z^1 - f, y - x^1 \rangle \geq 0 \quad \forall y \in \Omega.$$

It is easy to see that

$$\langle z^0 - f, x^1 - x^0 \rangle \geq 0$$

and

$$\langle z^1 - f, x^0 - x^1 \rangle \geq 0.$$

Summing up two last inequalities we deduce

$$\langle z^1 - z^0, x^0 - x^1 \rangle \geq 0.$$

Since A is monotone, this gives

$$\langle z^0 - z^1, x^0 - x^1 \rangle = 0$$

which contradicts our assumption that $x^0 \neq x^1$.

Analyze the solvability problem for the inequality (1.11.1). By Lemma 1.11.7, solvability of the variational inequality (1.11.1) with the maximal monotone operator A and solvability of the equation $Ax + \partial I_\Omega(x) = f$ with the maximal monotone operator $A + \partial I_\Omega$ are equivalent. Therefore, we can apply the existence theorems of Section 1.7 to obtain the following statements.

Theorem 1.11.9 *Suppose that $A : X \to 2^{X^*}$ is a maximal monotone and coercive operator, Ω is a convex closed set in $D(A)$, and either int $\Omega \neq \emptyset$ or int $D(A) \cap \Omega \neq \emptyset$. Then inequality (1.11.1) has at least one solution for all $f \in X^*$.*

Theorem 1.11.10 *Let $A : X \to 2^{X^*}$ be a maximal monotone operator, a set Ω satisfy the conditions of Theorem 1.11.9, and there exist a number $r > 0$ such that for all y such that $\|y\| \geq r$, $y \in D(A) \cap \Omega$, the following inequality holds:*

$$\langle z - f, y \rangle \geq 0 \quad \forall z \in Ay, \; f \in X^*. \tag{1.11.9}$$

Then there exists at least one solution x of the variational inequality (1.11.1) with $\|x\| \leq r$.

Theorem 1.11.11 *Assume that $A : X \to 2^{X^*}$ is a maximal monotone operator, Ω satisfies the conditions of Theorem 1.11.9. Then the regularized variational inequality*

$$\langle Ax + \alpha Jx - f, y - x \rangle \geq 0 \quad \forall y \in \Omega, \quad x \in \Omega, \tag{1.11.10}$$

has a unique solution for all $\alpha > 0$ and for all $f \in X^*$.

A solution x_α satisfying (1.11.10) we call the regularized solution of the variational inequality (1.11.1).

Remark 1.11.12 *Similarly to Corollary 1.7.6, if Ω in Theorems 1.11.9, 1.11.10 and 1.11.11 is bounded, then the coerciveness of A and the condition (1.11.9) are unnecessary there.*

Corollary 1.11.13 *Let $A : X \to 2^{X^*}$ be a maximal monotone operator which is coercive relative to a point $x_0 \in X$, let Ω be a convex closed set in $D(A)$ and either int $\Omega \neq \emptyset$ or int $D(A) \cap \Omega \neq \emptyset$. Then inequality (1.11.1) has at least one solution for all $f \in X^*$.*

Proof. It is sufficient to consider the operator $A_1(x) = A(x + x_0)$. ∎

The constraint minimization problems for a convex functional lead to inequalities of the type (1.11.1). Let us describe such problems in detail.

1) Let $\varphi : X \to R^1$ be a proper convex lower semicontinuous functional, $\partial\varphi : X \to 2^{X^*}$ be its subdifferential which is the maximal monotone operator in view of Theorem 1.7.15. Let Ω be a convex closed set, $\Omega \subseteq D(\partial\varphi)$. The problem is to find

$$min \; \{\varphi(y) \mid y \in \Omega\}. \tag{1.11.11}$$

It is assumed that this problem is solvable.

Theorem 1.11.14 *If int $\Omega \neq \emptyset$ or int $D(\partial\varphi) \cap \Omega \neq \emptyset$, then problem (1.11.11) is equivalent to the variational inequality*

$$\langle \partial\varphi(x), y - x \rangle \geq 0 \quad \forall y \in \Omega, \quad x \in \Omega. \tag{1.11.12}$$

Proof. Let x^0 be a solution of (1.11.12). Then, by the definition of subdifferential $\partial\varphi$ at the point x^0, we have

$$\varphi(y) - \varphi(x^0) \geq \langle \partial\varphi(x^0), y - x^0 \rangle \geq 0 \quad \forall y \in \Omega.$$

This means that $\varphi(y) \geq \varphi(x^0)$ for all $y \in \Omega$. Hence, x^0 is a solution of the problem (1.11.11).

Let now x^0 be a solution of the problem (1.11.11). Construct the functional $\Phi(y) = \varphi(y) + I_\Omega(y)$, where $I_\Omega(y)$ is the indicator function associated with the set Ω. Hence, its subdifferential is $\partial\Phi = \partial\varphi + \partial I_\Omega$ and $D(\partial\Phi) = \Omega$. It is clear that Φ reaches a minimum at the point $x^0 \in \Omega$. Then, by Lemma 1.2.5, $\theta_{X^*} \in \partial\varphi(x^0) + \partial I_\Omega(x^0)$. The converse implication arises from Lemma 1.11.7. ∎

2) Let X_1 and X_2 be reflexive strictly convex Banach spaces together with their dual spaces, a function $\Phi(u,v) : X_1 \times X_2 \to R^1$ be proper convex lower semicontinuous with respect to u and concave upper semicontinuous with respect to v. Let $G_1 \subset X_1$ and $G_2 \subset X_2$ be convex and closed sets such that $int\, G_1 \neq \emptyset$, $int\, G_2 \neq \emptyset$, $G = G_1 \times G_2$ and $G \subseteq D(\partial\Phi)$.

Definition 1.11.15 *A point $(u^*, v^*) \in G$ is said to be a saddle point of the functional $\Phi(u,v)$ on G if*

$$\Phi(u^*, v) \leq \Phi(u^*, v^*) \leq \Phi(u, v^*) \tag{1.11.13}$$

for all $(u^*, v) \in G$ and $(u, v^*) \in G$.

Consider the functional

$$\Phi_1(u) = \Phi(u, v^*) - \Phi(u^*, v^*), \quad \Phi_1 : X_1 \to R^1.$$

In view of (1.11.13), $u^* \in G_1$ is a minimum point of the convex functional $\Phi_1(u)$ on G_1. By Theorem 1.11.14, we have then

$$\langle \partial\Phi_1(u^*), u^* - u \rangle \leq 0 \quad \forall u \in G_1,$$

that is,

$$\langle \partial_u\Phi(u^*, v^*), u^* - u \rangle \leq 0 \quad \forall u \in G_1. \tag{1.11.14}$$

Applying the same arguments for the functional

$$\Phi_2(v) = \Phi(u^*, v^*) - \Phi(u^*, v), \quad \Phi_2 : X_2 \to R^1,$$

we arrive at the inequality

$$\langle \partial_v\Phi(u^*, v^*), v - v^* \rangle \leq 0 \quad \forall v \in G_2. \tag{1.11.15}$$

Thus, the saddle point (u^*, v^*) is a solution of the variational inequality (1.11.12) presented by (1.11.14) and (1.11.15) together.

3) We study the variational inequality (1.11.1) with a maximal monotone operator $A : X \to 2^{X^*}$ on $\Omega = \Omega_1 \cap \Omega_2$, where $\Omega_1 \subset X$ is a convex closed set and $\Omega_2 \subset X$ is defined by the system of inequalities

$$\varphi_i(x) \leq 0, \ i = 1, 2, ..., n.$$

Here $\varphi_i : X \to R^1$ are proper convex lower semicontinuous functionals on X having, due to Theorem 1.2.8, subdifferentials on X. Denote by R_+^n the non-negative octant in R^n. Define the quantity

$$[p, \varphi(x)] = \sum_{i=1}^{n} p_i \varphi_i(x),$$

where $p = \{p_1, p_2, ..., p_n\} \in R_+^n$.

Definition 1.11.16 *We say that the Slater condition is fulfilled if for every $p \in R_+^n$ there exists an element $\bar{x} \in X$ such that $[p, \varphi(\bar{x})] < 0$.*

Let $x^* \in \Omega$ be a solution of the variational inequality (1.11.1) in the sense of Definition 1.11.1, therefore, there exists $z^* \in Ax^*$ such that

$$\langle z^* - f, x^* - y \rangle \le 0 \quad \forall y \in \Omega. \tag{1.11.16}$$

Introduce the Lagrange function:

$$L(x, p) = \langle z^* - f, x \rangle + \sum_{i=1}^{n} p_i \varphi_i(x). \tag{1.11.17}$$

By (1.11.16), we have

$$\langle z^* - f, x^* \rangle = min \{\langle z^* - f, y \rangle \mid y \in \Omega\}.$$

Next we need the Karush−Kuhn−Tucker theorem.

Theorem 1.11.17 *Let $g(x)$ and $\varphi_i(x)$, $i = 1, ..., n$, be convex functionals on convex set $\Omega \subseteq X$. Let the Slater condition be fulfilled. Then an element $x^* \in \Omega$ is the solution of the minimization problem*

$$min \{g(x) \mid x \in \Omega, \ \varphi(x) \le 0\}, \quad \varphi(x) = \{\varphi_i(x)\},$$

if and only if there exists a vector p^* such that the pair (x^*, p^*) is the saddle point of the Lagrange function

$$L(x, p) = g(x) + [p, \varphi(x)],$$

that is,

$$L(x^*, p) \le L(x^*, p^*) \le L(x, p^*) \quad \forall x \in \Omega, \quad \forall p \in R_+^n.$$

By virtue of the Karush−Kuhn−Tucker theorem, there is a vector $p^* \in R_+^n$ such that (x^*, p^*) is the saddle point of the Lagrange function (1.11.17). According to (1.11.14), (1.11.15), this point satisfies the inequalities

$$\left\langle Ax^* - f + \sum_{i=1}^{n} p_i^* \partial \varphi_i(x^*), x^* - x \right\rangle \le 0 \quad \forall x \in \Omega, \tag{1.11.18}$$

$$[q - p^*, \varphi(x^*)] \le 0 \quad \forall q \in R_+^n. \tag{1.11.19}$$

Let now (x^*, p^*) be a solution of the system (1.11.18), (1.11.19). Rewrite (1.11.19) in the following form:

$$\sum_{i=1}^{n} q_i \varphi_i(x^*) \leq \sum_{i=1}^{n} p_i^* \varphi_i(x^*) \quad \forall q \in R_+^n.$$

By simple algebra, we come to the relations

$$\varphi_i(x^*) \leq 0, \ i = 1, 2, ..., n,$$

$$[p^*, \varphi(x^*)] = 0,$$

i.e., $x^* \in \Omega_2$. Since the functional

$$[p^*, \varphi(x)] = \sum_{i=1}^{n} p_i^* \varphi_i(x)$$

is proper convex lower semicontinuous on X, it has a subdifferential. Using Definition 1.2.2, we conclude that

$$[p^*, \varphi(x)] - [p^*, \varphi(x^*)] \geq \Big\langle \sum_{i=1}^{n} p_i^* \partial \varphi_i(x^*), x - x^* \Big\rangle,$$

that with $x \in \Omega$ gives

$$0 \geq [p^*, \varphi(x)] \geq \Big\langle \sum_{i=1}^{n} p_i^* \partial \varphi_i(x^*), x - x^* \Big\rangle.$$

Then (1.11.18) implies the fact that x^* is the solution of (1.11.1). Thus, we have proved the following theorem.

Theorem 1.11.18 *If the Slater condition is satisfied, then any solution x^* of the variational inequality (1.11.1) determines a solution (x^*, p^*), $p^* \in R_+^n$ of the system of the inequalities (1.11.18), (1.11.19), and, conversely, any solution (x^*, p^*) of (1.11.18), (1.11.19) determines a solution x^* of (1.11.1).*

1.12   Variational Inequalities with Semimonotone Operators

In this section (and also in Sections 1.13 and 1.14) we study the solution existence of the variational inequality (1.11.1) with non-monotone maps $A : X \to 2^{X^*}$. We present the sufficient existence conditions for a variational inequality with the semimonotone (possibly, unbounded) operator. However, we are interested in the solutions which are points where the operator A is single-valued.

Let X be a reflexive strictly convex Banach space together with its dual space X^*. Observe one important property of semimonotone maps.

Lemma 1.12.1 *Assume that* $A : X \rightarrow 2^{X^*}$ *is a semimonotone operator,* $D(A) = X$, $C : X \rightarrow X^*$ *is strongly continuous and* $T = A + C : X \rightarrow 2^{X^*}$ *is maximal monotone. If* $x_n \rightharpoonup x$ *as* $n \rightarrow \infty$, $y_n \in Ax_n$ *and*

$$\limsup_{n \to \infty} \langle y_n, x_n - x \rangle \leq 0, \tag{1.12.1}$$

then for every $y \in X$ *there exists an element* $z(y) \in Ax$ *such that*

$$\langle z(y), x - y \rangle \leq \liminf_{n \to \infty} \langle y_n, x_n - y \rangle.$$

Proof. Let $x_n \rightharpoonup x$, $y_n \in Ax_n$ and (1.12.1) be fulfilled. Since $C : X \rightarrow X^*$ is strongly continuous, we have

$$\langle Cx_n, x_n - x \rangle \rightarrow 0 \quad \text{as} \quad n \rightarrow \infty.$$

For $z_n = y_n + Cx_n \in Tx_n$, one gets

$$\limsup_{n \to \infty} \langle z_n, x_n - x \rangle \leq 0. \tag{1.12.2}$$

By virtue of the monotonicity of T, we write

$$\langle z, x_n - x \rangle \leq \langle z_n, x_n - x \rangle \quad \forall z \in Tx.$$

Using (1.12.2) and weak convergence of $\{x_n\}$ to x, this implies the limit equality

$$\lim_{n \to \infty} \langle z_n, x_n - x \rangle = 0. \tag{1.12.3}$$

Choose an arbitrary point $(u, v) \in grT$. By the obvious equality

$$\langle z_n, x_n - u \rangle = \langle z_n, x_n - x \rangle + \langle z_n, x - u \rangle,$$

we obtain

$$\liminf_{n \to \infty} \langle z_n, x_n - u \rangle = \liminf_{n \to \infty} \langle z_n, x - u \rangle. \tag{1.12.4}$$

Since $\langle v, x_n - u \rangle \leq \langle z_n, x_n - u \rangle$, where $v \in Tu$, we deduce that the weak convergence of $\{x_n\}$ to x leads to the inequality

$$\langle v, x - u \rangle \leq \liminf_{n \to \infty} \langle z_n, x_n - u \rangle. \tag{1.12.5}$$

Let $y \in X$, $u_t = (1 - t)x + ty$, $t > 0$ and $z_t \in Tu_t$. We replace (u, v) in (1.12.5) by (u_t, z_t). Then according to (1.12.4),

$$\langle z_t, x - y \rangle \leq \liminf_{n \to \infty} \langle z_n, x - y \rangle. \tag{1.12.6}$$

It is clear that $u_t \rightarrow x$ as $t \rightarrow 0$. By the local boundedness of the operator T on X, $z_t \rightharpoonup z(y) \in X^*$, where $z(y) \in Tx$ because T is a maximal monotone mapping. Hence, by (1.12.6), we derive the inequality

$$\langle z(y), x - y \rangle \leq \liminf_{n \to \infty} \langle z_n, x - x_n \rangle + \liminf_{n \to \infty} \langle z_n, x_n - y \rangle.$$

Taking into account (1.12.3), we obtain the conclusion of the lemma. ∎

Theorem 1.12.2 *Let X be a uniformly convex Banach space, X^* be an E-space, Ω be a convex closed set in X, $\theta_X \in \Omega$, $A : X \to 2^{X^*}$ be a semimonotone operator, $D(A) = X$. Let $T = A + C$ be a maximal monotone operator, where $C : X \to X^*$ is strongly continuous, and the following inequality hold for $\|x\| = r > 0$:*

$$\langle y - f, x \rangle \geq 0 \quad \forall y \in Ax. \tag{1.12.7}$$

Assume that $z_n \in X$, $v^n \in Az^n$, and at any point $z^0 \in \Omega$, where the operator A is multiple-valued, there exists even if one element $u \in \Omega$ such that

$$\limsup_{n \to \infty} \langle v^n - f, z^n - u \rangle > 0, \tag{1.12.8}$$

provided that $z^n \rightharpoonup z^0$. Then there exists at least one solution $\bar{x}$ of the variational inequality (1.11.1) with $\|\bar{x}\| \leq r$. Moreover, the operator A is single-valued at $\bar{x}$.

Proof. For simplicity of reasoning we consider that a space X is separable. Let $\{X_n\}$ be a sequence of finite-dimensional subspaces of X, $Q_n : X \to X_n$ be projection operators, Q_n^* be their conjugate operators. Let $\Omega_n = \Omega \cap X_n$ be a convex closed set in X_n, $P_{\Omega_n} : X \to \Omega_n$, $P_\Omega : X \to \Omega$ be projection operators on Ω_n and Ω, respectively. Let

$$F_n x = J(x - P_{\Omega_n} x) \quad \forall x \in X,$$

and

$$F x = J(x - P_\Omega x) \quad \forall x \in X,$$

where $J : X \to X^*$ is the normalized duality mapping in X. It is obvious that $\theta_{X_n} \in \Omega_n$ for all $n > 0$. An operator $F : X \to X^*$ is single-valued, bounded and monotone by Lemma 1.5.18, and $Fx = \theta_{X^*}$ if and only if $x \in \Omega$. The operators F_n have the similar properties.

Observe that the sets Ω_n Mosco-approximate Ω in the following sense: For any element $g \in \Omega$ there is a sequence $\{g_n\} \in \Omega_n$ such that $g_n \to g$, and if $h_n \rightharpoonup h$, $h_n \in \Omega_n$, then $h \in \Omega$. Choose a sequence $\{\epsilon_n\}$ such that $\epsilon_n > 0$, $\epsilon_n \to 0$, and consider in X_n the equation

$$Q_n^*(Ax + \epsilon_n^{-1} F_n x - f) = 0. \tag{1.12.9}$$

Since $\theta_{X_n} \in \Omega_n$, we have $\langle F_n x, x \rangle \geq 0$ for all $x \in X_n$. Therefore, for $x \in X_n$ with $\|x\| = r$ and for $y \in Ax$, the following relations hold:

$$\langle Q_n^*(y + \epsilon_n^{-1} F_n x - f), x \rangle = \langle y - f, x \rangle + \epsilon_n^{-1} \langle F_n x, x \rangle \geq 0.$$

Now the existence of the solution $x_n \in X_n$ of equation (1.12.9) with $\|x_n\| \leq r$ follows from Theorem 1.1.62. It is clear that $x_n \rightharpoonup \bar{x} \in X$ because the sequence $\{x_n\}$ is bounded. Besides, there is $y^n \in Ax_n$ such that

$$Q_n^*(y^n + \epsilon_n^{-1} F_n x_n - f) = \theta_{X^*}. \tag{1.12.10}$$

We prove that $\{y^n\}$ is bounded. Let

$$u^n = a_1 \frac{J^* y^n}{\|y^n\|_*}, \quad v^n \in Tu^n, \quad \|v^n\|_* \leq a_2, \quad a_1 > 0, \quad a_2 > 0, \tag{1.12.11}$$

where $J^* : X^* \to X$ is the normalized duality mapping in X^*. The local boundedness of the maximal monotone operator T guarantees existence of the sequences $\{u^n\}$ and $\{v^n\}$ satisfying (1.12.11). The monotonicity property of T gives

$$\langle y^n + Cx_n - v^n, x_n - u^n \rangle \geq 0.$$

This inequality and (1.12.11) imply the estimate

$$a_1 \|y^n\|_* \leq (\|Cx_n\|_* + a_2)a_1 + (\|f\|_* + \|Cx_n\|_* + a_2)r$$

$$+ \langle Q_n^*(y^n + \epsilon_n^{-1}F_n x_n - f), x_n \rangle - \epsilon_n^{-1}\langle F_n x_n, x_n \rangle.$$

Now we make sure that the sequence $\{y^n\}$ is bounded which follows from (1.12.10) and from the properties of the operator F_n.

Using (1.12.10) again, one gets

$$\langle \epsilon_n(y^n - f) + F_n x_n, z_n \rangle = 0 \quad \forall z_n \in X_n. \tag{1.12.12}$$

From (1.12.12) we deduce that $\langle F_n x_n, z_n \rangle \to 0$ for every bounded sequence $\{z_n\}$. In particular, $\langle F_n x_n, x_n \rangle \to 0$. It is not difficult to obtain the following equalities:

$$\langle F_n x_n, x_n \rangle = \langle J(x_n - P_{\Omega_n}x_n), x_n - P_{\Omega_n}x_n \rangle + \langle J(x_n - P_{\Omega_n}x_n), P_{\Omega_n}x_n \rangle$$

$$= \|F_n x_n\|_*^2 + \langle J(x_n - P_{\Omega_n}x_n), P_{\Omega_n}x_n \rangle. \tag{1.12.13}$$

By Lemma 1.5.17 and by the definition of $P_{\Omega_n}x_n$ with $x_n \in X_n$, we can write

$$\langle J(x_n - P_{\Omega_n}x_n), P_{\Omega_n}x_n - y \rangle \geq 0 \quad \forall y \in \Omega_n.$$

Assuming $y = \theta_{X_n}$ we obtain

$$\langle J(x_n - P_{\Omega_n}x_n), P_{\Omega_n}x_n \rangle \geq 0.$$

Then the estimate

$$\langle F_n x_n, x_n \rangle \geq \|F_n x_n\|_*^2$$

appears from (1.12.12). Thus, $F_n x_n \to \theta_{X_n^*}$ as $n \to \infty$.

Show that $F_n x \to Fx$ for any $x \in X$. Let $P_\Omega x = z \in \Omega$, and $P_{\Omega_n}x = z_n \in \Omega_n$, that is,

$$\|x - z\| = \min\{\|x - y\| \mid y \in \Omega\} \tag{1.12.14}$$

and

$$\|x - z_n\| = \min\{\|x - y\| \mid y \in \Omega_n\}. \tag{1.12.15}$$

Due to Theorem 1.11.14, (1.12.14) and (1.12.15) yield the inequalities

$$\langle J(x - z), z - y \rangle \geq 0 \quad \forall y \in \Omega, \quad z \in \Omega, \tag{1.12.16}$$

and

$$\langle J(x - z_n), z_n - y \rangle \geq 0 \quad \forall y \in \Omega_n, \quad z_n \in \Omega_n. \tag{1.12.17}$$

Since sequence $\{z_n\}$ is bounded, $z_n \rightharpoonup \bar{z} \in \Omega$. Let $\bar{u}_n \in \Omega_n$, $\bar{u}_n \to z \in \Omega$. Put $y = \bar{z}$ in (1.12.16) and $y = \bar{u}_n$ in (1.12.17). Then summation of these inequalities gives

$$\langle J(x - z_n) - J(x - z), z_n - z \rangle + \langle J(x - z_n), z - \bar{u}_n \rangle + \langle J(x - z), z_n - \bar{z} \rangle \geq 0. \tag{1.12.18}$$

The following result was obtained in Theorem 1.6.4. Let $x, y \in X$. If $\|x\| \leq R$ and $\|y\| \leq R$ then there exist constants c_1, $c_2 > 0$ such that

$$\langle Jx - Jy, x - y \rangle \geq c_1 \delta_X(c_2^{-1} \|x - y\|),$$

where $\delta_X(\epsilon)$ is the modulus of convexity of a space X. Therefore, using weak convergence of $\{z_n\}$ to $\bar{z}$ and (1.12.18), we conclude that for any fixed $x \in X$,

$$c_1 \delta_X(c_2^{-1} \|z_n - z\|) \leq \|x - z_n\| \|z - \bar{u}_n\| + \langle J(x - z), z_n - \bar{z} \rangle,$$

from which the convergence $z_n \to z$ follows. By Corollary 1.5.16, J is continuous, which guarantees the strong convergence of $F_n x$ to Fx for all $x \in X$.

Write the monotonicity condition of the operators F_n:

$$\langle F_n y - F_n x_n, y - x_n \rangle \geq 0, \quad y \in X.$$

Applying the properties of the sequence $\{F_n\}$ and setting $n \to \infty$, we obtain

$$\langle Fy, y - \bar{x} \rangle \geq 0 \quad \forall y \in X, \quad \bar{x} \in X. \tag{1.12.19}$$

Since the operator $F : X \to X^*$ is monotone and demicontinuous and $D(F) = X$, we have by Theorem 1.4.6 that F is maximal monotone. Hence, the equality $F\bar{x} = \theta_{X^*}$ is satisfied because of (1.12.19). Thus, we conclude that $\bar{x} \in \Omega$.

Let $w_n \in \Omega_n$ and $w_n \to \bar{x} \in \Omega$. Then $F_n w_n = \theta_{X^*}$ and

$$\langle y^n - f, w_n - x_n \rangle = \epsilon_n^{-1} \langle F_n w_n - F_n x_n, w_n - x_n \rangle \geq 0. \tag{1.12.20}$$

Therefore,

$$\langle y^n - f, x_n - \bar{x} \rangle \leq \langle y^n - f, w_n - \bar{x} \rangle.$$

Since the sequence $\{y^n\}$ is bounded, we have

$$\limsup_{n \to \infty} \langle y^n - f, x_n - \bar{x} \rangle \leq 0. \tag{1.12.21}$$

By Lemma 1.12.1,

$$\liminf_{n \to \infty} \langle y^n - f, x_n - v \rangle \geq \langle \bar{y} - f, \bar{x} - v \rangle, \tag{1.12.22}$$

where $\bar{y} = \bar{y}(v) \in A\bar{x}$, $v \in X$. Let now $u \in \Omega$ be an arbitrary element, $w_n \in \Omega$ and $w_n \to u$. Then like (1.12.21), we obtain the inequality

$$\limsup_{n \to \infty} \langle y^n - f, x_n - u \rangle \leq 0 \quad \forall u \in \Omega. \tag{1.12.23}$$

By the condition (1.12.8), the operator A is single-valued at the point $\bar{x} \in \Omega$. Consequently, in (1.12.22) $\bar{y} = A\bar{x}$ for all $v \in X$. Assuming $v = u \in \Omega$ in (1.12.22) and taking into account (1.12.23), we deduce for $\bar{x} \in \Omega$,

$$\langle A\bar{x} - f, \bar{x} - u \rangle \leq \liminf_{n \to \infty} \langle y^n - f, x_n - u \rangle$$

$$\leq \limsup_{n \to \infty} \langle y^n - f, x_n - u \rangle \leq 0 \quad \forall u \in \Omega.$$

The proof of the theorem is accomplished. ■

One can verify that Theorem 1.12.2 holds if we require coerciveness of A in place of (1.12.7). The condition (1.12.8) of the theorem plays the defining role in proving the facts that the operator A is single-valued on the solutions of (1.11.1) and that some subsequence of $\{x_n\}$ weakly converges to $\bar{x}$.

1.13   Variational Inequalities with Pseudomonotone Operators

The property of semimonotone mappings noted in Lemma 1.12.1 is the most essential in the definition of the wider class of the so-called pseudomonotone mappings, which, in general, are not necessarily monotone ones.

Definition 1.13.1 *Let X and Y be linear topological spaces. An operator $A : X \to 2^Y$ is said to be upper semicontinuous if for each point $x_0 \in X$ and arbitrary neighborhood $\mathcal{V}$ of Ax_0 in Y, there exists a neighborhood $\mathcal{U}$ of x_0 such that for all $x \in \mathcal{U}$ one has the inclusion: $Ax \subset \mathcal{V}$.*

Definition 1.13.2 *Let Ω be a closed convex set of a reflexive Banach space X. An operator $A : \Omega \to 2^{X^*}$ is called pseudomonotone if the following conditions are satisfied:*
a) for any $x \in \Omega$ the set Ax is non-empty, bounded, convex and closed in dual space X^*;
b) A is upper semicontinuous from each finite-dimensional subspace F of X to the weak topology on X^*, that is, to a given element $x_0 \in F$ and a weak neighborhood $\mathcal{V}$ of Ax_0 in X^* there exists neighborhood $\mathcal{U}$ of x_0 in F such that $Ax \subset \mathcal{V}$ for all $x \in \mathcal{U}$.
c) if $\{x_n\}$ is a sequence of elements from Ω that converges weakly to $x \in \Omega$, elements $z_n \in Ax_n$ such that
$$\limsup_{n \to \infty} \langle z_n, x_n - x \rangle \leq 0,$$
then for every element $y \in \Omega$ there exists $z(y) \in Ax$ such that

$$\liminf_{n \to \infty} \langle z_n, x_n - y \rangle \geq \langle z(y), x - y \rangle.$$

Assume that X is a reflexive Banach space.

Lemma 1.13.3 *Let Ω be a closed convex subset of X. Any hemicontinuous monotone operator $A : \Omega \to X^*$ is pseudomonotone.*

Proof. Suppose that the sequence $\{x_n\} \subset \Omega$ weakly converges to $x \in X$ and

$$\limsup_{n \to \infty} \langle Ax_n, x_n - x \rangle \leq 0.$$

Since A is monotone, we have

$$\lim_{n \to \infty} \langle Ax_n, x_n - x \rangle = 0. \tag{1.13.1}$$

Take an arbitrary $y \in \Omega$ and put $z_t = (1 - t)x + ty$, $t \in (0, 1]$. It is clear that $z_t \in \Omega$. The monotonicity property of A gives

$$\langle Ax_n - Az_t, x_n - z_t \rangle \geq 0.$$

Then

$$\langle Az_t, x_n - x + t(x - y) \rangle - \langle Ax_n, x_n - x \rangle \leq t \langle Ax_n, x - y \rangle.$$

Using (1.13.1), one gets

$$\langle Az_t, x - y \rangle \leq \liminf_{n \to \infty} \langle Ax_n, x - y \rangle.$$

By the hemicontinuity of A, this produces in a limit as $t \to 0$,

$$\langle Ax, x - y \rangle \leq \liminf_{n \to \infty} \langle Ax_n, x - y \rangle.$$

In view of (1.13.1) again, it is not difficult to obtain from the latter inequality that

$$\langle Ax, x - y \rangle \leq \liminf_{n \to \infty} \langle Ax_n, x_n - y \rangle.$$

The lemma is proved. ■

Consider pseudomonotone operators on the whole space when $\Omega = D(A) = X$.

Lemma 1.13.4 *Any maximal monotone operator* $A : X \to 2^{X^*}$ *with* $D(A) = X$ *is pseudomonotone.*

Proof. The condition a) of Definition 1.13.2 follows from the maximal monotonicity of the operator A. Since any maximal monotone operator is semimonotone, we conclude, by Lemma 1.12.1, that the property c) is true. Let $x_n \to x$, $x_n \in F \cap X$, $x \in F \cap X$ and let F be a finite-dimensional subspace of X. In view of the local boundedness of a maximal monotone mapping at a point x, we have the weak convergence of $y_n \in Ax_n$ to some y. It is known that the graph of A is demiclosed. This guarantees the inclusion $y \in Ax$, thus, the property b) also holds. ■

It is not difficult to verify the following assertion.

Lemma 1.13.5 *Any completely continuous operator* $A : X \to X^*$ *with* $D(A) = X$ *is pseudomonotone.*

Next we give the criteria when a sum of two operators is pseudomonotone.

Lemma 1.13.6 *Let $A : \Omega \to X^*$ be a hemicontinuous monotone operator, $B : \Omega \to 2^{X^*}$ be a pseudomonotone mapping. Then their sum $C = A + B$ is pseudomonotone.*

Proof. The requirements a) and b) of Definition 1.13.2 for C are obvious. Prove the property c). Let $x_n \in \Omega$, $x_n \rightharpoonup x$ and

$$\limsup_{n \to \infty} \langle y_n, x_n - x \rangle \leq 0, \quad y_n \in Cx_n.$$

Taking into account the monotonicity of A we have

$$\langle z_n, x_n - x \rangle \leq \langle y_n, x_n - x \rangle - \langle Ax, x_n - x \rangle, \quad z_n \in Bx_n, \quad y_n = Ax_n + z_n,$$

from which we deduce the following inequality:

$$\limsup_{n \to \infty} \langle z_n, x_n - x \rangle \leq \limsup_{n \to \infty} \langle y_n, x_n - x \rangle \leq 0. \tag{1.13.2}$$

Pseudomonotonicity of B gives now the estimate

$$\langle z(y), x - y \rangle \leq \liminf_{n \to \infty} \langle z_n, x_n - y \rangle \quad \forall y \in \Omega, \quad z(y) \in Bx. \tag{1.13.3}$$

By (1.13.3) with $y = x$, one gets

$$\liminf_{n \to \infty} \langle z_n, x_n - x \rangle \geq 0.$$

The latter relation and (1.13.2) give

$$\lim_{n \to \infty} \langle z_n, x_n - x \rangle = 0.$$

Then (1.13.2), where $y_n = Ax_n + z_n$, implies

$$\limsup_{n \to \infty} \langle Ax_n, x_n - x \rangle \leq 0.$$

Due to Lemma 1.13.3 the inequality

$$\langle Ax, x - y \rangle \leq \liminf_{n \to \infty} \langle Ax_n, x_n - y \rangle \quad \forall y \in \Omega, \tag{1.13.4}$$

holds. Summing (1.13.3) and (1.13.4), we obtain the conclusion of the lemma. ∎

Lemma 1.13.7 *Let A_1 and A_2 be two pseudomonotone mappings from X into 2^{X^*}. Then their sum is also pseudomonotone.*

Let us turn to the question on solvability of variational inequalities with pseudomonotone mappings. We give one of the main results of this direction. In order to prove it we need the following two theorems.

Theorem 1.13.8 *Let Ω be a nonempty compact and convex subset of the locally convex topological vector space X, $A : \Omega \to 2^\Omega$ be an upper semicontinuous mapping such that Ax is a nonempty closed and convex subset of X for all $x \in \Omega$. Then there exists an element $x_0 \in \Omega$ such that $x_0 \in Ax_0$.*

Theorem 1.13.9 *Let Ω be a nonempty compact and convex subset of the locally convex topological vector space X, $A : \Omega \to 2^{X^*}$ be an upper semicontinuous mapping such that Ax is a nonempty compact closed convex subset of X^* for each $x \in \Omega$. Then there exists $x_0 \in \Omega$ and $y_0 \in Ax_0$ such that*

$$\langle y_0, y - x_0 \rangle \leq 0 \quad \forall y \in \Omega.$$

In the sequel, a solution of the variational inequality (1.11.1) is understood in the sense of Definition 1.11.1. The main result of this section is the following

Theorem 1.13.10 *Suppose that X is a reflexive Banach space, Ω is a closed convex subset of X, $\theta_X \in \Omega$, $A : \Omega \to 2^{X^*}$ is a pseudomonotone and coercive operator on Ω. Then for any $f \in X^*$ there exists a solution of the variational inequality (1.11.1).*

Proof. Without loss of generality, assume that $f = \theta_{X^*}$, and prove that there exists a pair $(x, z) \in grA$ such that $\langle z, x - y \rangle \leq 0$ for all $y \in \Omega$. Let Λ be a family of all finite-dimensional subspaces $F \subset X$, ordered by inclusion. For each $R > 0$, let

$$\Omega_F^R = \Omega_F \cap B(\theta_X, R), \quad \Omega_F = \Omega \cap F.$$

Then Ω_F^R is a nonempty compact convex subset of X. Hence, we may apply Theorem 1.13.9 to the mapping $-A : \Omega_F^R \to 2^{X^*}$, which allows us to assert that there exist elements $x_F^R \in \Omega_F^R$ and $z_F^R \in Ax_F^R$ such that

$$\langle z_F^R, x_F^R - y \rangle \leq 0 \quad \forall y \in \Omega_F^R. \tag{1.13.5}$$

In particular, we may take $y = \theta_F \in \Omega_F^R$ for each $R > 0$ and obtain the inequality

$$\langle z_F^R, x_F^R \rangle \leq 0, \quad z_F^R \in Ax_F^R.$$

Since the operator A is coercive, it follows that $\|x_F^R\| \leq \lambda$, $\lambda > 0$, for all $R > 0$.

Consider $R > \lambda$. Then $\|x_F^R\| \leq \lambda < R$. For any $y \in \Omega_F$, we put $x_t = (1 - t)x_F^R + ty$, $t \in (0, 1)$. By the convexity of set Ω_F, the element $x_t \in \Omega_F$. If t is sufficiently small then $\|x_t\| \leq R$. This means that $x_t \in \Omega_F^R$, hence, by (1.13.5) we have

$$\langle z_F^R, x_F^R - x_t \rangle \leq 0$$

or

$$\langle z_F^R, x_F^R - y \rangle \leq 0 \quad \forall y \in \Omega_F.$$

Thus, there exists a solution of the variational inequality

$$\langle Ax, x - y \rangle \leq 0 \quad \forall y \in \Omega_F, \quad x \in \Omega_F. \tag{1.13.6}$$

Let x_F be some solution of (1.13.6). Then for all $F \in \Lambda$ there exists an element $z_F \in Ax_F$ such that

$$\langle z_F, x_F - y \rangle \leq 0 \quad \forall y \in \Omega_F, \quad x_F \in \Omega_F.$$

Since $\theta_F \in \Omega_F$ for all $F \in \Lambda$ and A is coercive, we have

$$c(\|x_F\|)\|x_F\| \leq \langle z_F, x_F \rangle \leq 0,$$

where $c(t) \to +\infty$ as $t \to +\infty$. Hence, there exists $M > 0$ such that $\|x_F\| \leq M$ for all $F \in \Lambda$. Construct the set

$$V_F = \bigcup_{F' \supset F} \{x_{F'}\}, \quad F' \in \Lambda.$$

Then the family $\{V_F\}_{F \in \Lambda}$ has the finite intersection property. Indeed, consider $F_1 \in \Lambda$ and $F_2 \in \Lambda$. If $F_3 \in \Lambda$ is so that $F_1 \cup F_2 \subseteq F_3$, then $V_{F_1} \cap V_{F_2} \supseteq V_{F_3}$. One can show that the weak closure of V_F in Ω (for short, $weakcl\ V_F$) is weakly compact, and the set

$$G = \bigcap_{F \in \Lambda} weakcl\ V_F \neq \emptyset.$$

Let an element $x_0 \in \Omega \cap G$ be given. Choose some fixed element $y \in \Omega$ and a set $F \in \Lambda$ such that it contains the elements x_0 and y. Since $x_0 \in weakcl\ V_F$, there exists a sequence $\{x_j\} \in V_F$ such that $x_j \rightharpoonup x_0$ as $j \to \infty$. For every j, we denote by F_j just the set of Λ for which the inclusion $x_j \in \Omega_{F_j}$ holds. By the definition of the element x_j, we have

$$\langle z_j, x_j - u \rangle \leq 0 \quad \forall u \in \Omega_{F_j}, \quad z_j \in Ax_j. \tag{1.13.7}$$

Since $y \in F_j$ for all j we can write

$$\langle z_j, x_j - y \rangle \leq 0, \quad z_j \in Ax_j. \tag{1.13.8}$$

Then (1.13.7) for $u = x_0$ gives

$$\langle z_j, x_j - x_0 \rangle \leq 0$$

because $x_0 \in F_j$ for all $j \geq 1$. Therefore,

$$\limsup_{j \to \infty} \langle z_j, x_j - x_0 \rangle \leq 0.$$

By the pseudomonotonicity of A, we conclude now that there exists an element $z(y) \in Ax_0$ such that

$$\liminf_{j \to \infty} \langle z_j, x_j - y \rangle \geq \langle z(y), x_0 - y \rangle, \quad y \in \Omega.$$

In view of (1.13.8), the latter inequality implies

$$\langle z(y), x_0 - y \rangle \leq 0 \quad \forall y \in \Omega, \quad x_0 \in \Omega, \quad z(y) \in Ax_0. \tag{1.13.9}$$

In order to finish the proof of the theorem it suffices to show that there exists a unique element $z \in Ax_0$ such that (1.13.9) is satisfied and the inequality $\langle z, x_0 - y \rangle \leq 0$ holds for all

$y \in \Omega$. We shall prove this claim by contradiction. Assume that for every element $z \in Ax_0$ there exists $y(z) \in \Omega$ such that

$$\langle z, x_0 - y(z) \rangle > 0. \tag{1.13.10}$$

Construct the set

$$N_y = \{ z \mid z \in Ax_0, \ \langle z, x_0 - y \rangle > 0 \} \quad \forall y \in \Omega.$$

Since $N_y \subset Ax_0$ for all $y \in \Omega$, it follows from (1.13.10) and from Definition 1.13.2 that N_y is a nonempty open and bounded set. Besides, the family of sets $\{ N_y \mid y \in \Omega \}$ is an open covering of the weakly compact set Ax_0. Hence, there exists a finite covering $\{ N_{y_1}, N_{y_2}, ..., N_{y_s} \}$ of the set Ax_0 and a corresponding decomposition of the unit on this set. The latter is defined by the family of functions $\{ \alpha_1, \alpha_2, ..., \alpha_s \}$, where $\alpha_j : Ax_0 \to [0, 1]$ are strongly continuous,

$$0 \le \alpha_j(z) \le 1 \quad \forall z \in Ax_0,$$

$$\alpha_j(z) = 0 \quad \text{if} \quad z \notin N_{y_j}, \quad 1 \le j \le s,$$

$$\alpha_j(z) > 0 \quad \text{if} \quad z \in N_{y_j}, \quad 1 \le j \le s,$$

$$\sum_{j=1}^{s} \alpha_j(z) = 1 \quad \forall z \in Ax_0.$$

Define the map $B : Ax_0 \to \Omega$ by the equality

$$Bz = \sum_{j=1}^{s} \alpha_j(z) y_j \quad \forall z \in Ax_0.$$

Obviously, it is strongly continuous. Then, by the definitions of $\alpha_j(z)$ and y_j, we have

$$\langle z, x_0 - Bz \rangle = \sum_{j=1}^{s} \alpha_j(z) \langle z, x_0 - y_j \rangle > 0 \quad \forall z \in Ax_0, \tag{1.13.11}$$

because z belongs to N_{y_j} and $\langle z, x_0 - y_j \rangle > 0$ for any j satisfying the inequality $\alpha_j(z) > 0$. On the other hand, taking into account (1.13.9), it is possible to construct the following nonempty closed and convex set:

$$S_y = \{ z \mid z \in Ax_0, \ \langle z, x_0 - y \rangle \le 0 \},$$

so that it is a closed and convex subset of Ax_0. Thus, we defined the mapping $S : \Omega \to 2^{Ax_0}$ which has the property of upper semicontinuity. Then we consider the operator $C = SB :$ $Ax_0 \to 2^{Ax_0}$ with the following properties: (i) C is upper semicontinuous, (ii) Cz is a non-empty convex closed set for any $z \in Ax_0$. Hence, by Theorem 1.13.8, operator C has a fixed point $z_1 \in Ax_0$, i.e., $z_1 \in Cz_1$. Thus, there exists $z_1 \in Ax_0$ such that

$$\langle z_1, x_0 - Bz_1 \rangle \le 0, \quad z_1 \in Ax_0,$$

which contradicts (1.13.11). The proof is accomplished.   ∎

Remark 1.13.11 *If $\Omega = X$, then Theorem 1.13.10 gives the sufficient conditions for equations with pseudomonotone operators to be solvable.*

1.14   Variational Inequalities with Quasipotential Operators

In this section we present one more class of non-monotone mappings for which the variational inequality (1.11.1) is solvable. Let X be a reflexive strictly convex Banach space together with its dual space X^*.

Definition 1.14.1 *An operator $A : X \rightarrow X^*$ is called radially summable on Ω if for all x and y from Ω the function $\psi_{x.y}(t) = \langle A(y + t(x - y)), x - y \rangle$ is Lebesgue integrable on the interval $[0, 1]$.*

Definition 1.14.2 *A radially summable on Ω operator A is called quasipotential on Ω if there exists a functional $\varphi : \Omega \rightarrow R^1$ satisfying the equality*

$$\varphi(x) - \varphi(y) = \int_0^1 \langle A(y + t(x - y)), x - y \rangle dt \qquad (1.14.1)$$

for all $x, y \in \Omega$. A functional φ is said to be the potential of the mapping A.

Note that hemicontinuous quasipotential mappings are potential.

Definition 1.14.3 *An operator $A : X \rightarrow X^*$ is said to be upper h-semicontinuous on Ω if*

$$\limsup_{t \to 0+} \langle A(y + t(x - y)) - Ay, x - y \rangle \leq 0 \quad \forall x, y \in \Omega.$$

Lemma 1.14.4 *Let $A : X \rightarrow X^*$ denote a quasipotential upper h-semicontinuous on Ω operator, and let its potential $\varphi(x)$ have the minimum point $x \in \Omega$. Then x is a solution of the variational inequality*

$$\langle Ax, x - y \rangle \leq 0 \quad \forall y \in \Omega, \quad x \in \Omega. \qquad (1.14.2)$$

Proof. Since x is the minimum point of the functional $\varphi(x)$ on Ω, there is a number $r > 0$ such that for all $y \in \Omega$ and for all $0 < \tau < \tau_0(y)$, where $\tau_0(y) = min\{1, \dfrac{r}{\|x - y\|}\}$, the following inequality holds:

$$
\begin{aligned}
0 \quad &\leq \quad \varphi(x + \tau(y - x)) - \varphi(x) = \tau \int_0^1 \langle A(x + \tau t(y - x)), y - x \rangle dt \\
&= \quad \tau \int_0^1 \langle A(x + \tau t(y - x)) - Ax, y - x \rangle dt + \tau \langle Ax, y - x \rangle. \qquad (1.14.3)
\end{aligned}
$$

Assume that x is not a solution of the variational inequality (1.14.2). Then for some $y \in \Omega$, we have

$$\langle Ax, y - x \rangle = -\epsilon < 0.$$

By virtue of the upper h-semicontinuity of the operator A on Ω, there exists $0 < \delta < 1$ such that

$$\langle A(x + \theta(y - x)) - Ax, y - x \rangle \leq \frac{\epsilon}{2}$$

for all $0 < \theta < \delta$. Hence, if $0 < \tau < min\{\delta, \tau_0(y)\}$ then

$$\varphi(x + \tau(y - x)) - \varphi(x) \leq \frac{\tau\epsilon}{2} - \tau\epsilon = -\frac{\tau\epsilon}{2} < 0,$$

which contradicts (1.14.3). ■

Theorem 1.14.5 *Let $A : X \to X^*$ denote a quasipotential upper h-semicontinuous on Ω mapping, and let its potential φ be weakly lower semicontinuous on Ω. Suppose that either Ω is bounded or it is unbounded and*

$$\lim_{\|x\| \to +\infty} \varphi(x) = +\infty, \quad x \in \Omega. \tag{1.14.4}$$

Then the variational inequality (1.14.2) has at least one solution.

Proof. By the Weierstrass generalized theorem and by the condition (1.14.4), it follows that there exists an element $x \in \Omega$ such that $\varphi(x) = min \{f(y) \mid y \in \Omega\}$. Then Lemma 1.14.4 ensures a validity of the assertion being proved. ■

Theorem 1.14.6 *Let $A = A_1 + A_2$, where operators $A_i : X \to X^*$, $i = 1, 2$, are quasipotential on Ω. Let A_1 be a monotone mapping and A_2 be strongly continuous, A be upper h-semicontinuous on Ω, a set Ω be unbounded. If the condition (1.14.4) is satisfied, then the variational inequality (1.14.2) has a solution.*

Proof. If we establish the weak lower semicontinuity on Ω of the functional $\varphi = \varphi_1 + \varphi_2$, where φ_1 and φ_2 are potentials of the operators A_1 and A_2, respectively, then the assertion follows from Theorem 1.14.5. Let $x \in \Omega$, $x_n \in \Omega$, $x_n \rightharpoonup x$. Then the monotonicity property of the operator A_1 on Ω gives the following relations:

$$
\begin{aligned}
\varphi_1(x_n) - \varphi_1(x) &= \int_0^1 \langle A_1(x + t(x_n - x)), x_n - x \rangle dt \\
&= \int_0^1 \langle A_1(x + t(x_n - x)) - A_1 x, x_n - x \rangle dt + \langle A_1 x, x_n - x \rangle \\
&\geq \langle A_1 x, x_n - x \rangle.
\end{aligned}
$$

Therefore,

$$\liminf_{n \to \infty} \varphi_1(x_n) \geq \varphi_1(x). \tag{1.14.5}$$

Since the mapping A_2 is strongly continuous on Ω, we have for all $t \in [0, 1]$,

$$\lim_{n \to \infty} \langle A_2(x + t(x_n - x)), x_n - x \rangle = 0.$$

Furthermore, by the boundedness of A_2, there exists a constant $M > 0$ such that

$$|\langle A_2(x + t(x_n - x)), x_n - x \rangle| \leq M$$

for all $t \in [0, 1]$ and for all $n > 0$. Then we may write

$$\lim_{n \to \infty} \left(\varphi_2(x_n) - \varphi_2(x) \right) = \lim_{n \to \infty} \int_0^1 \langle A_2(x + t(x_n - x)), x_n - x \rangle dt$$

$$= \int_0^1 \lim_{n \to \infty} \langle A_2(x + t(x_n - x)), x_n - x \rangle dt = 0.$$

In view of (1.14.5), this implies

$$\liminf_{n \to \infty} \varphi(x_n) = \liminf_{n \to \infty} [\varphi_1(x_n) + \varphi_2(x_n)] = \liminf_{n \to \infty} \varphi_1(x_n) + \varphi_2(x)$$

$$\geq \varphi_1(x) + \varphi_2(x) = \varphi(x).$$

The theorem is proved. ■

Remark 1.14.7 *If $\Omega = X$, then Theorems 1.14.5, and 1.14.6 give the solvability conditions of equations with quasipotential mappings. Furthermore, if the potential of the operator A is non-negative, then the potential of $A + \alpha J (\alpha > 0)$ actually has the properties (1.14.4).*

Let us present an example. Consider the problem: To find a function $x \in \Omega$ satisfying for any $y \in \Omega$ the following inequality:

$$\sum_{1 \leq |\alpha| = |\beta| \leq m} \int_G a_{\alpha\beta}(s) D^\alpha x(s) D^\beta(y(s) - x(s)) ds + \int_G g(s, x(s))(y(s) - x(s)) ds \geq 0. \quad (1.14.6)$$

Assume that $G \subset R^n$ is a bounded measurable set with the piecewise smooth boundary ∂G, $\Omega \subset \overset{\circ}{W}_m^2(G)$, functions $a_{\alpha\beta}(s)$ are continuous in $\overline{G}$, where $a_{\alpha\beta}(s) = a_{\beta\alpha}(s)$ on $\overline{G}$, $1 \leq |\alpha| = |\beta| \leq m$, and for arbitrary $s \in G$, $\xi \in R^n$ the inequality

$$\sum_{|\alpha| = |\beta| = m} a_{\alpha\beta}(s) \xi^{\alpha+\beta} \geq \kappa |\xi|^{2m}, \quad \kappa > 0$$

holds. Let a function $g : G \times R^1 \to R^1$ be measurable superposition.

Suppose that the following conditions are satisfied:
1) the function $g(s, x)$ is continuous to the right for almost all $s \in G$ and has discontinuities just of the first kind. Moreover, if x is a discontinuity point of $g(s, x)$, then

$$g(s, x - 0) > g(s, x),$$

and for all $x \in R^1$ the function $g(s, x)$ is measurable on Ω;
2) $g(s, x)$ has $(p - 1)$-order of growth with respect to x :

$$|g(s, x)| \leq a(s) + b|x|^{p-1}$$

for all $x \in R^1$ and for almost all $s \in G$, where $a(s) \in L^q(G)$, $b > 0$, $p^{-1} + q^{-1} = 1$. Here $1 < p < \dfrac{2n}{n - 2m}$ if $n > 2m$, and $p > 1$ if $n < 2m$;
3)

$$\sum_{1 \leq |\alpha| = |\beta| \leq m} \int_G a_{\alpha\beta}(s) D^\alpha x(s) D^\beta x(s) ds \geq M \sum_{|\alpha| = m} \|D^\alpha x(s)\|_{L^2}^2,$$

for all $x \in \overset{\circ}{W}{}^2_m (G)$, $M > 0$;

4)

$$2 \int_0^x g(s,y) dy \geq -kx^2 - d(x)|x|^\lambda - c(s)$$

for all $x \in R^1$ and for almost all $s \in G$, $0 < \lambda < 2$, $d(x) \in L^\gamma(G)$, $\gamma = \dfrac{2}{2-\lambda}$, $k > 0$, $kC < M$, where C is a constant in the inequality

$$\|x\|^2_{L_2} \leq C \sum_{|\alpha|\leq m} \|D^\alpha x(s)\|^2_{L_2(G)} \quad \forall x(s) \in \overset{\circ}{W}{}^2_m (G),$$

and $c(s)$ is the Lebesgue absolutely integrable function;

5) $\Omega \subset \overset{\circ}{W}{}^2_m (G)$ is an unbounded convex closed set.

Show that the inequality (1.14.6) has a solution if all these conditions hold. Let $X = \overset{\circ}{W}{}^2_m$ (G) with the norm

$$\|u\| = \Big(\sum_{|\alpha|=m} \int_G |D^\alpha u(x)|^2 dx \Big)^{1/2}.$$

Consider an operator $A_1 : X \to X^*$ generated by the dual form

$$\langle A_1 x, y \rangle = \sum_{1\leq|\alpha|=|\beta|\leq m} \int_G a_{\alpha\beta}(s) D^\alpha x(s) D^\beta y(s) ds,$$

where x and y are elements from X. By the properties of the functions $a_{\alpha\beta}(s)$, the operator A_1 is bounded, self-adjoint and monotone. The condition 3) guarantees the estimate

$$\langle A_1 x, x \rangle \geq M\|x\|^2 \quad \forall x \in X. \tag{1.14.7}$$

Hence, A_1 is a potential mapping and its potential $\varphi_1(x) = 2^{-1}\langle A_1 x, x \rangle$. By condition 1), if $x(s)$ is measurable on G then superposition $g(s,x(s))$ is also measurable on G [213].

Define an operator $A_2 : X \to X^*$ by the formula

$$\langle A_2 x, y \rangle = \int_G g(s,x(s))y(s) ds \quad \forall x, y \in X.$$

By the condition 2), the imbedding operator $\mathcal{E}$ of the space $\overset{\circ}{W}{}^2_m (G)$ into $L^p(G)$ is completely continuous and the operator

$$A_3 x(s) = g(s,x(s)) \quad \forall x(s) \in L^p(G)$$

is bounded. It is clear that $A_3 : L^p(G) \to L^q(G)$. Denote the adjoint operator to $\mathcal{E}$ by $\mathcal{E}^*$. Now compactness of A_2 follows from the equality $A_2 = \mathcal{E}^* A_3 \mathcal{E}$.

Define functional φ_2 on X as

$$\varphi_2(x) = \int_G ds \int_0^{x(s)} g(s,y) dy.$$

Then for any $x, h \in X$ we deduce

$$
\begin{aligned}
\varphi_2(x+h) - \varphi_2(x) &= \int_G ds \int_{x(s)}^{x(s)+h(s)} g(s,y)dy \\
&= \int_G ds \int_0^1 \frac{d}{dt} \int_0^{x(s)+th(s)} g(s,y)dy dt \\
&= \int_0^1 dt \int_G g(s, x(s)+th(s))h(s)ds \\
&= \int_0^1 \langle A_2(x+th), h \rangle dt.
\end{aligned}
$$

Hence, A_2 is quasipotential on X and φ_2 is its potential. If $A = A_1 + A_2$ then

$$
\begin{aligned}
\langle A(x+th) - Ax, h \rangle &= t \sum_{1 \le |\alpha| = |\beta| \le m} \int_G a_{\alpha\beta}(s) D^\alpha h(s) D^\beta h(s) ds \\
&\quad + \int_G \Big(g(s, x(s)+th(s)) - g(s, x(s)) \Big) h(s) ds.
\end{aligned}
$$

By the conditions 1) and 2), one gets

$$
\lim_{t \to 0+} \langle A(x+th) - Ax, h \rangle \le 0.
$$

Thus, the operator A is upper h-semicontinuous on X. Now (1.14.7) and the condition 4) produce for the potential $\varphi = \varphi_1 + \varphi_2$ of operator A the following inequalities:

$$
\begin{aligned}
2\varphi(x) &\ge M\|x\|^2 + \int_G (-kx^2(s) - d(s) \mid x(s) \mid^\lambda - c(s)) ds \\
&\ge (M - kC)\|x\|^2 - \Big(\int_G \mid d(s) \mid^\gamma ds \Big)^{1/\gamma} C^{\lambda/2} \|x\|^\lambda - \int_G \mid c(s) \mid ds.
\end{aligned}
$$

Since $M - kC > 0$,

$$
\lim_{\|x\| \to +\infty} \varphi(x) = +\infty.
$$

We see that all the conditions of Theorem 1.14.6 are satisfied for an operator A. Therefore, there exists a function $x(s) \in \Omega$ such that

$$
\langle Ax, y - x \rangle \ge 0 \quad \forall y \in \Omega.
$$

The latter relation is identical to (1.14.6) in view of the definitions of A_1 and A_2.

Remark 1.14.8 *Consider the equation*

$$
\sum_{1 \le |\alpha| = |\beta| \le m} (-1)^{|\beta|} D^\beta (a_{\alpha\beta}(s) D^\alpha x(s)) + g(s, x(s)) = 0, \quad s \in G, \tag{1.14.8}
$$

with the boundary condition

$$\partial_\nu^r(\partial G)x(s) = 0 \quad \forall x \in \partial G, \quad 0 \le r \le m - 1,$$

where $\partial_\nu^r(\partial G)$ is a derivative in the interior normal direction to ∂G. A function $x \in \overset{\circ}{W}_m^2 (G)$ is said to be the solution of (1.14.8) if for all $y \in \overset{\circ}{W}_m^2 (G)$ the following equality holds:

$$\sum_{1 \le |\alpha|=|\beta| \le m} \int_G a_{\alpha\beta}(s)D^\alpha x(s)D^\beta y(s)ds + \int_G g(s,x(s))y(s)ds = 0.$$

Solvability of the equation (1.14.8) is obtained by the same arguments as in the previous example.

1.15   Equations with Accretive Operators

In this section we study the class of nonlinear operators acting from a Banach space X to X. We assume that X is reflexive and strictly convex together with its dual space X^*.

Definition 1.15.1 *An operator $A : X \to 2^X$ is called accretive if*

$$\langle J(x_1 - x_2), y_1 - y_2 \rangle \ge 0 \tag{1.15.1}$$

for all $x_1, x_2 \in D(A)$, $y_1 \in Ax_1$, $y_2 \in Ax_2$.

If an operator A is Gâteaux differentiable, then this definition is equivalent to the following one.

Definition 1.15.2 *A Gâteaux differentiable operator $A : X \to X$ is called accretive if*

$$\langle Jh, A'(x)h \rangle \ge 0 \quad \forall x, \ h \in X.$$

Remark 1.15.3 *The properties of monotonicity and accretiveness of an operator coincide in a Hilbert space.*

We present one more definition of an accretive operator.

Definition 1.15.4 *An operator $A : X \to 2^X$ is said to be accretive if*

$$\|x_1 - x_2\| \le \|x_1 - x_2 + \lambda(y_1 - y_2)\|, \quad \lambda > 0, \tag{1.15.2}$$

for all $x_1, x_2 \in D(A)$, $y_1 \in Ax_1$, $y_2 \in Ax_2$.

Theorem 1.15.5 *Definitions 1.15.1 and 1.15.4 are equivalent.*

Proof. Indeed, let (1.15.1) hold. Then the inequality

$$\langle J(x_1 - x_2), x_1 - x_2 + \lambda(y_1 - y_2)\rangle \geq \|x_1 - x_2\|^2, \ \lambda > 0,$$

is valid and (1.15.2) immediately follows.

Further, we know that if X^* is strictly convex then X is smooth and $Jx = 2^{-1}grad\|x\|^2$. By convexity of the functional $\|x\|^2$, we may write the inequality

$$\|x_1 - x_2\|^2 \geq \|x_1 - x_2 + \lambda(y_1 - y_2)\|^2 - 2\lambda\langle J(x_1 - x_2 + \lambda(y_1 - y_2)), y_1 - y_2\rangle.$$

If (1.15.2) holds then
$$\langle J(x_1 - x_2 + \lambda(y_1 - y_2)), y_1 - y_2\rangle \geq 0.$$

Letting $\lambda \to 0$ and using the hemicontinuity of J, we obtain (1.15.1). ∎

We present some properties of accretive mappings.

Definition 1.15.6 *Accretive operator $A : X \to 2^X$ is said to be coercive if*

$$\langle Jx, y\rangle \geq c(\|x\|)\|x\| \quad \forall y \in Ax,$$

where $c(t) \to +\infty$ as $t \to +\infty$.

Definition 1.15.7 *An operator $A : X \to 2^X$ is called locally bounded in a point $x \in D(A)$ if there exists a neighborhood M of that point such that the set*

$$A(M) = \{y \mid y \in Ax, \quad x \in M \cap D(A)\}$$

is bounded in X.

Theorem 1.15.8 *Let $A : X \to 2^X$ be an accretive operator, dual mappings $J : X \to X^*$ and $J^* : X^* \to X$ be continuous in X and X^*, respectively. Then A is locally bounded at every point $x \in int\ D(A)$.*

Proof. Suppose that is not the case. Let $x_0 \in int\ D(A)$, $x_n \in D(A)$, $n = 1, 2, \ldots$, $x_n \to x_0$, and $\|y_n\| \to \infty$, where $y_n \in Ax_n$. Let $t_n = \|x_n - x_0\|^{1/2}$. For any $z \in X^*$, we construct the sequence $\{z_n\}$ by the formula

$$z_n = J(x_n - x_0 - t_n w) + t_n z,$$

where $w = J^* z$. The elements z_n are well defined because of $D(J) = X$. In view of continuity of J and since $t_n \to 0$, it follows that $z_n \to \theta_{X^*}$ and the elements $x_0 + t_n w \in D(A)$ for all $w \in X$ provided that $n > 0$ is sufficiently large. If $f \in Av$, $v = x_0 + \sigma w$, $u_n \in A(x_0 + t_n w)$, then the accretiveness of A gives the inequality

$$(t_n - \sigma)\langle z, u_n - f\rangle \geq 0,$$

from which we obtain $\langle z, u_n \rangle \leq \|z\|_* \|f\|$ for $t_n < \sigma$. Since $R(J) = X^*$, one gets

$$\limsup_{n \to \infty} \langle z, u_n \rangle < \infty \quad \forall z \in X^*.$$

Now the Banach–Steinhaus theorem implies the boundedness of the sequence $\{u_n\}$, say, $\|u_n\| \leq C$ for all $n > 0$. Further, since A is accretive, we write

$$\langle z_n - t_n z, y_n - u_n \rangle \geq 0.$$

Therefore,

$$\langle z, y_n \rangle \leq \frac{1}{t_n} \langle z_n, y_n \rangle - \frac{1}{t_n} \langle z_n - t_n z, u_n \rangle$$

$$\leq \frac{\|z_n\|_*}{t_n} \|y_n\| + C \Big(\frac{\|z_n\|_*}{t_n} + \|z\|_* \Big).$$

Denoting

$$\tau_n(z) = \frac{\|z_n\|_*}{t_n} = \Big\| J \Big(\frac{x_n - x_0}{t_n} - J^* z \Big) + z \Big\|_*,$$

we deduce by the latter inequality that

$$\limsup_{n \to \infty} \frac{\langle z, y_n \rangle}{1 + \|y_n\| \tau_n(z)} < \infty.$$

The functional

$$\varphi(z) = \frac{\langle z, y_n \rangle}{1 + \|y_n\| \tau_n(z)}$$

is continuous with respect to z because J and J^* are continuous. Then, as in the proof of the Banach–Steinhaus theorem, we assert that there are constants $C_1 > 0$ and $r > 0$ such that

$$\frac{\langle z, y_n \rangle}{1 + \|y_n\| \tau_n(z)} \leq C_1, \quad n = 1, 2, \dots, \tag{1.15.3}$$

if $z \in X^*$ with $\|z\|_* \leq r$. Take in (1.15.3) $z = \bar{z}_n \in B^*(\theta_{X^*}, r)$ satisfying the condition $\langle \bar{z}_n, y_n \rangle = r\|y_n\|$. Then

$$\frac{\|y_n\|}{1 + \|y_n\| \tau_n(\bar{z}_n)} \leq \frac{C_1}{r}.$$

Hence,

$$\|y_n\| \leq \frac{C_1}{r} \Big(1 - \frac{C_1}{r} \tau_n(\bar{z}_n) \Big)^{-1}.$$

The continuity of J yields the relation

$$\tau_n(\bar{z}_n) = \Big\| J \Big(\frac{x_n - x_0}{t_n} - J^* \bar{z}_n \Big) + J J^* \bar{z}_n \Big\|_* \to 0, \quad n \to \infty.$$

Therefore, for $\epsilon > 0$ such that $1 - C_1 r^{-1} \epsilon > 0$ and for sufficiently large $n > 0$, the following inequality holds:

$$\|y_n\| \leq \frac{C_1}{r} \Big(1 - \frac{C_1}{r} \epsilon \Big)^{-1}.$$

That contradicts the assumption that $\|y_n\| \to \infty$ as $n \to \infty$. ∎

Corollary 1.15.9 *If $A : X \to 2^X$ is an accretive operator, J is continuous, $D(A) = X$ and X is finite-dimensional, then A is bounded.*

Proof. See Theorem 1.3.19. ∎

Lemma 1.15.10 *If an operator $T : X \to X$ is nonexpansive in $D(A)$, then $A = I - T$ is accretive.*

Proof. We have for all x, $y \in D(A)$,

$$\langle J(x - y), Ax - Ay \rangle = -\langle J(x - y), Tx - Ty \rangle + \langle J(x - y), x - y \rangle$$

$$\geq \|x - y\|^2 - \|Tx - Ty\|\|x - y\| \geq \|x - y\|^2 - \|x - y\|^2 = 0. \quad ∎$$

Definition 1.15.11 *An accretive operator $A : X \to 2^X$ is called maximal accretive if its graph is not the right part of the graph of any other accretive operator $B : X \to 2^X$.*

The following lemma shows that the graph of any maximal accretive operator is demi-closed (cf. Lemma 1.4.5).

Lemma 1.15.12 *Let $A : X \to 2^X$ be a maximal accretive operator. Let $x_n \in D(A)$, $y_n \in Ax_n$. Suppose that either $x_n \to x$, $y_n \rightharpoonup y$ and the duality mapping J is continuous, or $x_n \rightharpoonup x$, $y_n \to y$ and J is weak-to-weak continuous. Then $x \in D(A)$ and $y \in Ax$.*

Proof. The accretiveness A produces the inequality

$$\langle J(x_n - u), y_n - v \rangle \geq 0 \quad \forall u \in D(A), \quad \forall v \in Au.$$

Let $n \to \infty$. Under the hypotheses of the theorem, we have in a limit

$$\langle J(x - u), y - v \rangle \geq 0 \quad \forall u \in D(A), \quad \forall v \in Au.$$

Then maximal accretiveness of A implies inclusions: $x \in D(A)$ and $y \in Ax$. ∎

Lemma 1.15.13 *The value set of a maximal accretive operator $A : X \to 2^X$ at any point of its domain is convex and closed.*

Proof. This property of maximal accretive operators is easily obtained by Definition 1.15.1. ∎

Theorem 1.15.14 *Let $A : X \to X$ be a hemicontinuous accretive operator with $D(A) = X$. Then A is maximal accretive.*

Proof. It is enough to show (see Theorem 1.4.6) that the inequality

$$\langle J(x - y), Ax - f \rangle \geq 0 \quad \forall x \in X, \tag{1.15.4}$$

implies $f = Ay$. Since $D(A) = X$, we may put in (1.15.4)

$$x = x_t = y + tz \quad \forall z \in X, \ t > 0.$$

Then

$$\langle Jz, Ax_t - f \rangle \geq 0.$$

Letting $t \to 0$ and using the hemicontinuity property of A, we obtain

$$\langle Jz, Ay - f \rangle \geq 0 \quad \forall z \in X.$$

Since $R(J) = X^*$, we conclude that $f = Ay$. ∎

Theorem 1.15.15 *Let $A : X \to X$ be an accretive operator with $D(A) = X$, duality mappings J and J^* be continuous. Then the properties of hemicontinuity and demicontinuity of A on int $D(A)$ are equivalent.*

Proof. See Theorem 1.3.20. ∎

Definition 1.15.16 *An operator $A : X \to 2^X$ is called strictly accretive if the equality (1.15.1) holds only with $x_1 = x_2$.*

Definition 1.15.17 *An operator $A : X \to 2^X$ is called uniformly accretive if there exists an increasing function $\gamma(t)$, $t \geq 0$, $\gamma(0) = 0$, such that*

$$\langle J(x_1 - x_2), y_1 - y_2 \rangle \geq \gamma(\|x_1 - x_2\|),$$

where x_1, $x_2 \in D(A)$, $y_1 \in Ax_1$, $y_2 \in Ax_2$. An operator A is called strongly accretive if $\gamma(t) = ct^2$, where $c > 0$.

Remark 1.15.18 *We say that an operator $A : X \to 2^X$ is properly accretive if (1.15.1) is fulfilled and there is not any strengthening of (1.15.1), for instance, up to the level of strong or uniform accretiveness.*

Definition 1.15.19 *An accretive operator $A : X \to 2^X$ is said to be m-accretive if*

$$R(A + \alpha I) = X$$

for all $\alpha > 0$, where I is the identity operator in X.

Lemma 1.15.20 *If operator A is m-accretive, then it is maximal accretive.*

Proof. By Definition 1.15.19, $R(A + I) = X$. Since $A + I$ is the strongly accretive operator, then there exists a unique pair $(x, y) \in grA$ such that $y + x = f$ for every $f \in X$. Let $\bar{A}$ be a maximal accretive extension of the operator A (it exists by Zorn's lemma). If we admit that there exists a pair $(\bar{x}, \bar{y})$, which belongs to the graph of $\bar{A}$ and at the same time does not belong to the graph of A, then we arrive at a contradiction with the solution uniqueness of the equation $\bar{A}x + x = f$. Thus, $\bar{A} = A$. ∎

The converse assertion to Lemma 1.15.20 is not true. However, the following statement holds.

Theorem 1.15.21 *Assume that X^* is an uniformly convex Banach space, the duality mapping J^* is continuous, an operator $A : X \to 2^X$ is accretive and $D(A)$ is an open set. Then A is m-accretive if and only if A is maximal accretive.*

Next we give one important result concerning the sum of m-accretive operators.

Theorem 1.15.22 *Let X and X^* be uniformly convex Banach spaces, $A : X \to 2^X$ and $B : X \to 2^X$ be m-accretive operators in X, $D(A) \cap D(B) \neq \emptyset$ and one of them is locally bounded. Then $A + B$ is an m-accretive mapping.*

The solvability of equations with accretive operators essentially depends on the properties both of operator A and duality mapping J. Recall that the properties of duality mappings are defined by the geometric characteristics of spaces X and X^* (see Sections 1.5 and 1.6).

Next we present the existence theorem for the equation

$$Ax = f \tag{1.15.5}$$

with the accretive operator A.

Theorem 1.15.23 *Assume that X and X^* are strictly convex spaces, X possesses an approximation, operator $A : X \to X$ is accretive and demicontinuous with a domain $D(A) = X$, duality mapping $J : X \to X^*$ is continuous and weak-to-weak continuous and there exist $r > 0$ such that for all x with $\|x\| = r$,*

$$\langle Jx, Ax - f \rangle \geq 0.$$

Then equation (1.15.5) has at least one classical solution $\bar{x}$ with $\|\bar{x}\| \leq r$.

Consider the solvability problem for the equation (1.15.5) when $A : X \to 2^X$ is an arbitrary accretive operator and $D(A) = X$. First prove the following auxiliary lemma.

Lemma 1.15.24 *Assume that X_n is an n-dimensional Banach space, $P_n : X \to X_n$ is a projection operator with the norm $|P_n| = 1$ and $P_n^* : X^* \to X_n^*$ is conjugate operator to P_n. Then the equality $P_n^* Jx = Jx$ holds for every $x \in X_n$.*

Proof. It is easy to see that for all $x \in X_n$,

$$\langle P_n^* Jx, x \rangle = \langle Jx, P_n x \rangle = \langle Jx, x \rangle = \|Jx\|_* \|x\| = \|x\|^2. \tag{1.15.6}$$

Since $|P_n^*| = |P_n| = 1$, we have

$$\|P_n^* Jx\|_* \leq |P_n^*| \|Jx\|_* = \|Jx\|_* = \|x\|.$$

On the other hand, by (1.15.6),

$$\|x\|^2 \leq \|P_n^* Jx\|_* \|x\|,$$

that is, $\|x\| \leq \|P_n^* Jx\|_*$. Thus, $\|x\| = \|P_n^* Jx\|_*$. Together with (1.15.6), this means that $P_n^* J$ is a duality mapping in X and $P_n^* Jx = Jx$ because J is a single-valued operator. ∎

As in Section 1.9, denote by $R(Ax - f)$ a convex closed hull of the weak limits of all subsequences of the sequences $\{Ax_n - f\}$ when $x_n \to x$ for $x_n, x \in X$.

Definition 1.15.25 *A point $x_0 \in X$ is called an sw-generalized solution of the equation (1.15.5) with an accretive operator if $\theta_X \in R(Ax_0 - f)$.*

Observe that this definition of solution coincides with the classical one if J is continuous and A is hemicontinuous at the point x_0 (see Lemma 1.15.12).

Show that sw-generalized solution of (1.15.5) exists in a finite-dimensional space X_n.

Lemma 1.15.26 *Let $A : X_n \to 2^{X_n}$ be defined on $B(\theta_{X_n}, r)$ and the inequality*

$$\langle Jx, z - f \rangle \geq 0$$

holds for all $x \in S(\theta_{X_n}, r)$ and for some $z \in Ax$. Then there exists sw-generalized solution $x_0 \in B(\theta_{X_n}, r)$ of the equation (1.15.5).

Proof. Consider $B^*(\theta_{X_n^*}, r) = JB(\theta_{X_n}, r)$, where $J : X_n \to X_n^*$ is a duality mapping. For every $x \in B(\theta_{X_n}, r)$, there is $y \in B^*(\theta_{X_n^*}, r)$ such that $y = Jx$. At the points y of the sphere $S^*(\theta_{X_n^*}, r)$, there exists $z \in AJ^*y$ such that the inequality

$$\langle y, z - f \rangle \geq 0$$

holds because, by Lemma 1.5.10 and Corollary 1.7.8, $J^{-1} = J^*$, $(J^*)^{-1} = J$, $R(J) = X^*$ and $R(J^*) = X$. Hence, in view of Theorem 1.1.62, we are able to find an element $y_0 \in B^*(\theta_{X_n^*}, r)$ such that $\theta_{X_n} \in R(AJ^*y_0 - f)$ (spaces X_n and X_n^* can be identified). Since $J^* : X_n^* \to X_n$ is a continuous operator, $\theta_{X_n} \in R(Ax_0 - f)$, where $x_0 = J^*y_0$. ∎

A solution of the equation (1.15.5) with an arbitrary accretive operator can be also defined in the following way:

Definition 1.15.27 *A point $x_0 \in X$ is said to be a generalized solution of the equation (1.15.5) with accretive operator A if the inequality*

$$\langle J(x - x_0), y - f \rangle \geq 0 \quad \forall y \in Ax$$

is satisfied for every $x \in D(A)$.

Repeating, in fact, the proof of Lemma 1.9.5 we obtain

Lemma 1.15.28 *Suppose that the operator $A : X \to 2^X$ with $D(A) = X$ is accretive and duality mappings J and J^* are continuous. Then generalized and sw-generalized solutions of the equation (1.15.5) coincide.*

Under the conditions of this lemma, the operator A has a unique maximal accretive extension $\bar{A}$ (cf. Corollary 1.9.7).

Theorem 1.15.29 *Assume that spaces X and X^* are uniformly convex, X possesses an approximation, duality mapping J is weak-to-weak continuous, operator $A : X \to 2^X$ is accretive with a domain $D(A) = X$ and there exists $r > 0$ such that for every x with $\|x\| = r$ there is a $y \in Ax$ such that*

$$\langle Jx, y - f \rangle \geq 0.$$

Then equation (1.15.5) has at least one generalized solution $\bar{x}$ such that $\|\bar{x}\| \leq r$.

Proof. Let X_n be finite-dimensional subspaces defining the M-property of X. Fixed X_n and consider the equation

$$P_n(Ax - f) = 0,$$

where $P_n : X \to X_n$ is the projector, $|P_n| = 1$. By Lemma 1.15.24,

$$\langle Jx, P_n(y - f) \rangle = \langle Jx, y - f \rangle \geq 0, \quad y \in Ax,$$

for all $x \in X_n$ with $\|x\| = r$. Then, in view of Lemma 1.15.26, there exists at least one element $x_n \in X$ such that $\|x_n\| \leq r$ and $\theta_X \in R(P_n(Ax_n - f))$. Repeating the arguments of Theorem 1.10.6 we may verify that the operator $P_n \bar{A} : X_n \to 2^{X_n}$ is a maximal accretive extension of $P_n A : X_n \to 2^{X_n}$. Hence, there exists an element $y^n \in \bar{A}x_n$ such that

$$P_n(y^n - f) = \theta_X.$$

We are going to show that the sequence $\{y^n\}$ is bounded. Since an operator $\bar{A}$ is locally bounded, there exist numbers $a_1 > 0$ and $a_2 > 0$ such that $\|v^n\| \leq a_2$ for all $v^n \in \bar{A}q^n$ if $\|q^n\| \leq a_1$.

Suppose that the element q^n is defined by the equation

$$J(q^n - x_n) = w^n - Jx_n,$$

where $w^n = a_3 \|y^n\|^{-1} Jy^n$ and $a_3 \leq r$. Since X^* is uniformly smooth, the duality mapping $J^* : X^* \to X$ is uniformly continuous on every bounded set (see Section 1.6). In other words, there exists an increasing continuous function $\omega_R^*(t)$ for $t \geq 0$ such that $\omega_R(0) = 0$ and if $\phi_1, \phi_2 \in X^*$, $\|\phi_1\|_* \leq R$, $\|\phi_2\|_* \leq R$ then

$$\|J^*\phi_1 - J^*\phi_2\| \leq \omega_R^*(\|\phi_1 - \phi_2\|_*).$$

Put $R = 2r$. Since $J^* = J^{-1}$, we have

$$\|q^n\| = \|J^*(w^n - Jx_n) + J^*Jx_n\| \leq \omega_R^*(\|w^n - Jx_n + Jx_n\|_*) = \omega_R^*(\|w^n\|_*) = \omega_R^*(a_3).$$

Choose a_3 so small as $\omega_R^*(a_3) \leq a_1$. Clearly such a definition of the sequence $\{w^n\}$ gives the estimate $\|v^n\| \leq a_2$, where $v^n \in \bar{A}q^n$.

By the accretiveness of the operator $\bar{A}$ we can write

$$\langle J(x_n - q^n), y^n - v^n \rangle = \langle Jx_n - w^n, y^n - v^n \rangle \geq 0,$$

from which it follows that

$$\langle w^n, y^n \rangle \leq \langle w^n - Jx_n, v^n \rangle + \langle Jx_n, y^n \rangle. \tag{1.15.7}$$

Since

$$\langle Jx_n, y^n \rangle = \langle Jx_n, P_n y^n \rangle = \langle Jx_n, f \rangle,$$

one gets from (1.15.7) that

$$a_3 \|y^n\| \leq r\|f\| + a_2(\tilde{a}_1 + r) = a_4.$$

Thus, $\|y^n\| \le a_3^{-1} a_4$.

Next, let $z \in X$ and $P_n z = z_n$. Then, on account of the M-property of X, the relation $\|P_n z - z\| \to 0$ holds as $n \to \infty$. It is obvious that

$$\langle J(z - x_n), y^n - f \rangle = \langle J(z - x_n) - J(z_n - x_n), y^n - f \rangle + \langle J(z_n - x_n), y^n - f \rangle.$$

In its turn,

$$\langle J(z_n - x_n), y^n - f \rangle = \langle J(z_n - x_n), P_n(y^n - f) \rangle = 0.$$

Consequently,

$$\langle J(z - x_n), y^n - f \rangle = \langle J(z - x_n) - J(z_n - x_n), y^n - f \rangle. \qquad (1.15.8)$$

Since X is uniformly smooth, the duality mapping $J : X \to X^*$ is uniformly continuous on every bounded set, that is, there exists an increasing continuous function $\omega_R(t)$ for $t \ge 0$ such that $\omega_R(0) = 0$ and if $x_1, x_2 \in X$, $\|x_1\| \le R$, $\|x_2\| \le R$, then

$$\|Jx_1 - Jx_2\|_* \le \omega_R(\|x_1 - x_2\|).$$

This implies

$$\|J(z - x_n) - J(z_n - x_n)\|_* \le \omega_{\bar{R}}(\|z - z_n\|),$$

where $\bar{R} = r + \|z\|$. Thus,

$$\lim_{n \to \infty} \|J(z - x_n) - J(z_n - x_n)\|_* = 0.$$

We proved above that the sequence $\{y^n\}$ is bounded, therefore, by (1.15.8), we obtain

$$\lim_{n \to \infty} \langle J(z - x_n), y^n - f \rangle = 0 \quad \forall z \in X. \qquad (1.15.9)$$

Furthermore, let $\{x_k\} \subseteq \{x_n\}$ and $x_k \rightharpoonup \bar{x} \in X$. Then the accretiveness of $\bar{A}$ gives

$$\langle J(z - x_k), y - y^k \rangle \ge 0 \quad \forall y \in \bar{A}z.$$

By (1.15.9) and by weak-to-weak continuity of the duality mapping J, we have in a limit as $k \to \infty$,

$$\langle J(z - \bar{x}), y - f \rangle \ge 0 \quad \forall y \in \bar{A}x,$$

that is, $f \in \bar{A}\bar{x}$. Hence, $\bar{x}$ is the solution of (1.15.5) in the sense of Definition 1.15.27. Since $x_n \in B(\theta_X, r)$ for all $n > 0$ and $x_k \rightharpoonup \bar{x}$, we conclude by the weak lower semicontinuity of norm in X that $\|\bar{x}\| \le r$. The proof is accomplished. ∎

Remark 1.15.30 *All the conditions of Theorem 1.15.29 are satisfied, for example, in Banach spaces $X = l^p$, $p > 1$.*

Remark 1.15.31 *If an operator A in Theorem 1.15.29 is strictly accretive, then the corresponding operator equation has a unique solution.*

Consider in X the equation

$$Ax + \alpha x = f, \ \ \alpha > 0, \ \ f \in X$$

with a maximal accretive operator $A : X \to 2^X$. The following relations are easily verified:

$$\langle Jx, y + \alpha x - f \rangle \ \geq \ \alpha \|x\|^2 - \|f\|\|x\| - \|A(\theta_X)\|\|x\|$$

$$= \ \|x\|(\alpha\|x\| - \|f\| - \|A(\theta_X)\|) \ \ \forall y \in Ax.$$

Hence, there exists a number $r > 0$ such that

$$\langle Jx, y + \alpha x - f \rangle \geq 0 \ \ \forall y \in Ax \ \ \text{as} \ \ \|x\| = r.$$

Therefore, if the spaces X and X^* and operator J satisfy the conditions of Theorem 1.15.29 and if $D(A) = X$, then $R(A + \alpha I) = X$. Hence, a maximal accretive operator A is m-accretive (see also Theorem 1.15.21).

Theorem 1.15.32 *Let $A : X \to 2^X$ be a coercive and m-accretive operator. Then $\overline{R(A)} = X$.*

Proof. By the definition of m-accretiveness of A, it results for y_1, $y_2 \in X$ that there exist $x_1 \in X$ and $x_2 \in X$ such that

$$y_1 \in (A + \alpha I)x_1 \ \ \text{and} \ \ y_2 \in (A + \alpha I)x_2.$$

Applying Definition 1.15.4 to A we can write for any $\eta > 0$

$$\|x_1 - x_2 + \eta(y_1 - y_2)\| \ = \ \|(1 + \alpha\eta)(x_1 - x_2) + \eta[(y_1 - \alpha x_1) - (y_2 - \alpha x_2)]\|$$
$$\geq \ (1 + \alpha\eta)\|x_1 - x_2\|.$$

Hence, the mapping

$$C = \left(I + \eta(A + \alpha I)\right)^{-1}$$

satisfies the Lipschitz condition with the constant $L = (1 + \alpha\eta)^{-1} < 1$. Consequently, the operator C is strongly contractive. Therefore, it has a fixed point x_α which is a solution of the equation

$$x = \left(I + \eta(A + \alpha I)\right)^{-1}x.$$

It follows from this that $y_\alpha = -\alpha x_\alpha \in Ax_\alpha$.

Since A is accretive, we have for $\beta > \alpha$,

$$\alpha\|x_\alpha - x_\beta\|^2 \ \leq \ \alpha\langle J(x_\alpha - x_\beta), x_\alpha - x_\beta \rangle - \langle J(x_\alpha - x_\beta), \alpha x_\alpha - \beta x_\beta \rangle$$

$$= \ (\beta - \alpha)\langle J(x_\alpha - x_\beta), x_\beta \rangle \leq (\beta - \alpha)\|x_\alpha - x_\beta\|\|x_\beta\|.$$

This yields the inequality

$$\alpha \|x_\alpha - x_\beta\| \le (\beta - \alpha)\|x_\beta\|,$$

which implies $\alpha \|x_\alpha\| \le \beta \|x_\beta\|$. Thus, the sequence $\{\alpha x_\alpha\}$ is bounded as $\alpha \to 0$. Since

$$\langle Jx_\alpha, y_\alpha \rangle = -\alpha \|x_\alpha\|^2$$

and A is coercive, the boundedness of $\{x_\alpha\}$ follows then. Hence, $\alpha x_\alpha \to \theta_X$ as $\alpha \to 0$, i.e., $\theta_X \in \overline{R(A)}$. Now we choose an arbitrary element $f \in X$ and apply the proved assertion to the shift operator $A - f$. Finally, we obtain $\theta_X \in \overline{R(A)} - f$, thus, $f \in \overline{R(A)}$ for all $f \in X$. The theorem is proved. ∎

Theorem 1.15.33 *If an operator $A : X \to 2^X$ is m-accretive, a duality mapping J is weak-to-weak continuous and A^{-1} is locally bounded, then $R(A) = X$.*

Proof. Prove that $R(A)$ is a closed and open set at the same time and then the claim follows. Let $f_n \in R(A)$ and $f_n \to f$, $n = 1, 2, \dots$. Then $x_n \in A^{-1} f_n$ is bounded in X, that is, there exists $c > 0$ such that $\|x_n\| \le c$ for all $n > 0$. Therefore, we may consider that $x_n \rightharpoonup \bar{x} \in X$. By Lemma 1.15.20, operator A is maximal accretive, then $f \in A\bar{x}$ because of Lemma 1.15.12. Thus, the set $R(A)$ is closed.

Next we shall establish that $R(A)$ is open. Let $(x, f) \in grA$ be given. Since A^{-1} is locally bounded, there exists $r > 0$ such that the set $\{x \mid u \in Ax\}$ is bounded in X if $\|u - f\| \le r$. Take $g \in B(f, \frac{r}{2})$ and show that $g \in R(A)$. Since A is m-accretive, the equation

$$Ay + \alpha(y - x) = g$$

has a solution x_α, that is, there exists $g_\alpha \in Ax_\alpha$ such that

$$g_\alpha + \alpha(x_\alpha - x) = g. \tag{1.15.10}$$

By the accretiveness of A,

$$\langle J(x_\alpha - x), g - \alpha(x_\alpha - x) - f \rangle \ge 0,$$

from which we have

$$\alpha \|x_\alpha - x\| \le \|g - f\| \le \frac{r}{2}.$$

Then (1.15.10) implies the estimate

$$\|g_\alpha - g\| = \alpha \|x_\alpha - x\| \le \frac{r}{2}, \tag{1.15.11}$$

from which one gets

$$\|g_\alpha - f\| \le \|g_\alpha - g\| + \|g - f\| \le r.$$

The boundedness of $\{x_\alpha\}$ for sufficiently small $\alpha > 0$ follows now from the local boundedness of A^{-1}. Then $x_\alpha \rightharpoonup \bar{x} \in X$ as $\alpha \to 0$. Finally, by (1.15.11), we deduce that $g_\alpha \to g$. Hence, $g \in R(A)$ in view of Lemma 1.15.12. ∎

Theorem 1.15.34 *Under the conditions of Theorem 1.15.32, if duality mapping J is weak-to-weak continuous, then $R(A) = X$.*

Proof. The coercive operator A has bounded inverse. Therefore, the assertion follows from Theorem 1.15.33. ∎

Theorem 1.15.35 *If an operator $A : X \to 2^X$ is m-accretive and duality mapping J satisfies the Lipschitz–Hölder condition*

$$\|Jx - Jy\|_* \le c\|x - y\|^\gamma, \quad c > 0, \quad 0 < \gamma \le 1, \tag{1.15.12}$$

then $\overline{R(A)}$ is a convex set in X.

Proof. Let $x_\alpha \in D(A)$ be a unique solution of the equation

$$Ax + \alpha x = x^0, \quad x^0 \in X, \quad \alpha > 0.$$

Then there is a element $y_\alpha \in Ax_\alpha$ such that

$$y_\alpha + \alpha x_\alpha = x^0$$

or

$$J(y_\alpha - x^0) = -\alpha Jx_\alpha. \tag{1.15.13}$$

Let $(u, v) \in grA$. Using (1.15.13) we can write

$$
\begin{aligned}
\|y_\alpha - x^0\|^2 &= \langle J(y_\alpha - x^0), y_\alpha - x^0 \rangle \\
&= \langle J(y_\alpha - x^0), y_\alpha - v \rangle + \langle J(y_\alpha - x^0), v - x^0 \rangle \\
&= \alpha \langle Jx_\alpha, v - y_\alpha \rangle + \langle J(y_\alpha - x^0), v - x^0 \rangle \\
&= \alpha \langle J(x_\alpha - u), v - y_\alpha \rangle + \alpha \langle Jx_\alpha - J(x_\alpha - u), v - y_\alpha \rangle \\
&\quad + \langle J(y_\alpha - x^0), v - x^0 \rangle.
\end{aligned}
$$

Taking into account (1.15.12) and accretiveness of A, the previous norm can be evaluated in the following way:

$$\|y_\alpha - x^0\|^2 \le c\alpha \|u\|^\gamma \|v - y_\alpha\| + \langle J(y_\alpha - x^0), v - x^0 \rangle \tag{1.15.14}$$

which reduce to

$$\|y_\alpha - x^0\|^2 \le c\alpha \|u\|^\gamma \|v - y_\alpha\| + \|y_\alpha - x^0\|\|v - x^0\|.$$

We now deduce that the sequence $\{y_\alpha\}$ is bounded, and as a consequence, $\{J(y_\alpha - x^0)\}$ are also bounded. Assume that $J(y_\alpha - x^0) \rightharpoonup z \in X^*$ as $\alpha \to 0$ (as a matter of fact, this weak convergence takes place on some subsequence of $\{J(y_\alpha - x^0)\}$ but we do not change its notation). Then by (1.15.14),

$$\limsup_{\alpha \to 0} \|y_\alpha - x^0\|^2 \le \langle z, v - x^0 \rangle \quad \forall v \in R(A).$$

The rest of the proof follows the pattern of Theorem 1.7.17. ∎

Corollary 1.15.36 *If the inverse operator A^{-1} is m-accretive and duality mapping J satisfies the Lipschitz–Hölder condition, then the set $\overline{D(A)}$ is convex.*

Introduce co-variational inequalities with accretive operators. Let $A : X \rightarrow 2^X$ be an maximal accretive operator with a domain $D(A) \subseteq X$. Consider the inequality

$$\langle J(y - x), Ax - f \rangle \geq 0 \quad \forall y \in \Omega, \ x \in \Omega, \tag{1.15.15}$$

where $\Omega \subseteq D(A)$ is a closed convex set and $f \in X$.

By analogy with monotone variational inequalities, we present the following two definitions:

Definition 1.15.37 *An element $x \in \Omega$ is called a solution of the co-variational inequality (1.15.15) if there exists $z \in Ax$ such that*

$$\langle J(y - x), z - f \rangle \geq 0 \quad \forall y \in \Omega. \tag{1.15.16}$$

Definition 1.15.38 *An element $x \in \Omega$ is called a solution of the co-variational inequality (1.15.15) if*

$$\langle J(y - x), u - f \rangle \geq 0 \quad \forall y \in \Omega, \ \forall u \in Ay. \tag{1.15.17}$$

Lemma 1.15.39 *If an element $x \in \Omega$ is the solution of (1.15.15) defined by the inequality (1.15.16), then it satisfies also the inequality (1.15.17).*

Proof. Write down the property of accretivness of A,

$$\langle J(y - x), u - z \rangle \geq 0 \quad \forall x, y \in \Omega, \ \forall u \in Ay, \quad \forall z \in Ax.$$

Using (1.15.16) we obtain (1.15.17). ∎

Lemma 1.15.40 *If $A : X \rightarrow X^*$ is a hemicontinuous operator and $\Omega \subset int \ D(A)$, then Definitions 1.15.37 and 1.15.38 are equivalent.*

Proof. Let x be a solution of (1.15.15) in the sense of Definition 1.15.37. Since the set Ω is convex, the element $y_t = (1 - t)x + ty \in \Omega$ for $t \in [0, 1]$ and all $y \in \Omega$. Then (1.15.17) with $y = y_t$ gives

$$\langle J(y_t - x), Ay_t - f \rangle \geq 0 \quad \forall y \in \Omega.$$

It results from this that

$$\langle J(y - x), Ay_t - f \rangle \geq 0 \quad \forall y \in \Omega. \tag{1.15.18}$$

If $t \rightarrow 0$ then $y_t \rightarrow x$. Now the hemicontinuity of A at x implies: $Ay_t \rightharpoonup \bar{z} \in X$. In view of (1.15.18), it follows that (1.15.16) holds, that is, x is a solution of (1.15.15) in the sense of Definition 1.15.37. Joining to this result the previous lemma we obtain the necessary assertion. ∎

1.16   Equations with d-Accretive Operators

Let X be a reflexive and strictly convex Banach space and X^* also be strictly convex. In this section we study a class of d-accretive operators.

Definition 1.16.1 *An operator $A : X \to 2^X$ with $D(A) \subseteq X$ is said to be d-accretive if*

$$\langle Jx - Jy, u - v \rangle \geq 0 \quad \forall x, y \in D(A), \ \forall u \in Ax, \ \forall v \in Ay. \tag{1.16.1}$$

Present several examples of d-accretive operators.

Example 1. If F is a monotone operator from X^* to X, then the operator $A = FJ$ with $D(A) = \{x \in X \mid Jx \in D(F)\}$ is a d-accretive operator from X to X. Indeed, since F satisfies the condition

$$\langle \varphi_1 - \varphi_2, \psi_1 - \psi_2 \rangle \geq 0 \quad \forall \varphi_1, \varphi_2 \in D(F) \subset X^*, \ \forall \psi_1 \in F\varphi_1, \ \forall \psi_2 \in F\varphi_2,$$

we can also write

$$\langle JJ^*\varphi_1 - JJ^*\varphi_2, \psi_1 - \psi_2 \rangle \geq 0 \quad \forall \psi_1 \in F\varphi_1, \ \forall \psi_2 \in F\varphi_2,$$

because $JJ^* = I_{X^*}$. Setting $J^*\varphi_1 = x$ and $J^*\varphi_2 = y$, one has

$$\langle Jx - Jy, \psi_1 - \psi_2 \rangle \geq 0 \quad \forall \psi_1 \in Ax, \ \forall \psi_2 \in Ay.$$

Thus, the claim holds. As a matter of fact, it is also true that if $A : X \to 2^X$ is d-accretive, then $AJ^* : X^* \to 2^X$ is monotone.

Example 2. Suppose that the operator $T : X \to X$ satisfies the inequality

$$\|Tx - Ty\| \leq \frac{\langle Jx - Jy, x - y \rangle}{\|Jx - Jy\|_*} \quad \forall x, y \in D(T). \tag{1.16.2}$$

Then the operator $A = I - T$, where I is the identity mapping in X, is d-accretive.

Indeed,

$$\begin{aligned}
\langle Jx - Jy, Ax - Ay \rangle &= \langle Jx - Jy, (I - T)x - (I - T)y \rangle \\
&= \langle Jx - Jy, x - y \rangle - \langle Jx - Jy, Tx - Ty \rangle \\
&\geq \langle Jx - Jy, x - y \rangle - \|Tx - Ty\|\|Jx - Jy\|_* \geq 0.
\end{aligned}$$

Note that (1.16.2) implies

$$\|Tx - Ty\| \leq \|x - y\| \quad \forall x, y \in D(T),$$

i.e., the operator T is nonexpansive. The inverse assertion is not held in general. Observe also that in a Hilbert space the right-hand side of (1.16.2) is $\|x - y\|$.

Example 3. Let Ω be a nonempty closed convex subset of X, and consider the functional $W : X \times X \to R^1_+$ defined in Section 1.6 by

$$W(x, \xi) = 2^{-1}(\|x\|^2 - 2\langle Jx, \xi\rangle + \|\xi\|^2). \qquad (1.16.3)$$

By virtue of the properties of $W(x, \xi)$, for each $x \in X$ there is a unique $\hat{x} \in \Omega$ which solves the minimization problem

$$min\ \{W(x, \xi) \mid \xi \in \Omega\}.$$

Denoting $\hat{x}$ by $\Pi_\Omega x$ we define a generalized projection operator $\Pi_\Omega : X \to \Omega \subseteq X$. An element $\hat{x}$ is called a generalized projection of x onto Ω.

We claim that the operator Π_Ω is d-accretive in the space X. Indeed, with the denotations $\hat{x}_1 = \Pi_\Omega x_1$ and $\hat{x}_2 = \Pi_\Omega x_2$, the inequalities

$$W(x_1, \xi) \geq W(x_1, \hat{x}_1) \quad \text{and} \quad W(x_2, \eta) \geq W(x_2, \hat{x}_2)$$

are satisfied for all $\xi, \eta \in \Omega$ and for all $x_1, x_2 \in X$. Assume $\xi = \hat{x}_2$ and $\eta = \hat{x}_1$. Then

$$W(x_1, \hat{x}_2) \geq W(x_1, \hat{x}_1) \quad \text{and} \quad W(x_2, \hat{x}_1) \geq W(x_2, \hat{x}_2).$$

We now deduce

$$\|x_1\|^2 - 2\langle Jx_1, \hat{x}_2\rangle + \|\hat{x}_2\|^2 + \|x_2\|^2 - 2\langle Jx_2, \hat{x}_1\rangle + \|\hat{x}_1\|^2$$
$$\geq \|x_1\|^2 - 2\langle Jx_1, \hat{x}_1\rangle + \|\hat{x}_1\|^2 + \|x_2\|^2 - 2\langle Jx_2, \hat{x}_2\rangle + \|\hat{x}_2\|^2.$$

Hence,

$$\langle Jx_1, \hat{x}_1\rangle + \langle Jx_2, \hat{x}_2\rangle \geq \langle Jx_1, \hat{x}_2\rangle + \langle Jx_2, \hat{x}_1\rangle$$

and

$$\langle Jx_1 - Jx_2, \Pi_\Omega x_1 - \Pi_\Omega x_2\rangle \geq 0 \quad \forall x_1, x_2 \in X.$$

This means that Π_Ω is d-accretive operator.

Note that generalized projection operators have important applications in the theory of approximation and optimization.

We say that an operator $A : X \to 2^X$ with a domain $D(A)$ is gauge d-accretive if in Definition 1.16.1 the normalized duality mapping is replaced by the duality mapping with a gauge function. For example, using the duality mapping J^p with the gauge function $\mu(t) = t^{p-1}$, $p > 1$, we obtain instead of (1.16.1) the inequality

$$\langle J^p x - J^p y, u - v\rangle \geq 0 \quad \forall x, y \in D(A), \quad \forall u \in Ax, \quad \forall v \in Ay. \qquad (1.16.4)$$

Example 4. Consider the operator $A : L^p(G) \to L^p(G)$ defined by the following equality:

$$Ay = \varphi(x, |y|)y \quad \forall x \in G, \quad \forall y \in L^p(G), \qquad (1.16.5)$$

where G is a bounded measurable domain of R^n, $p > 1$, $\varphi(x, s)$ is a non-negative measurable function with respect to x for all $s \geq 0$ and continuous with respect to s for almost all $x \in G$. Suppose that there exists a constant $C > 0$ such that

$$\varphi(x, s) \leq C \quad \forall x \in G, \quad \forall s \in [0, +\infty). \qquad (1.16.6)$$

Let $J^p : L^p(G) \rightarrow L^q(G)$ be a duality mapping with the gauge function $\mu(t) = t^{p-1}$, $p^{-1} + q^{-1} = 1$. It is known (see Section 1.5) that in this case $J^p y = |y|^{p-2} y$. One can verify that

$$\langle J^p y_1 - J^p y_2, A y_1 - A y_2 \rangle$$

$$\geq \int_g \Big(\varphi(x, |y_1|)|y_1| - \varphi(x, |y_2|)|y_2| \Big) \big(|y_1|^{p-1} - |y_2|^{p-1} \big) \, dx. \qquad (1.16.7)$$

Let the function $\varphi(x, s)s$ be non-decreasing with respect to $s \geq 0$ for each fixed $x \in G$. Then d-accretiveness of the operator A defined by (1.16.5) arises from (1.16.7). Emphasize that the inclusion $Ay \in L^p(G)$ for $y \in L^p(G)$ guarantees the assumption (1.16.6).

Definition 1.16.2 *Let $\mathcal{W}(x, \xi)$ be defined as*

$$\mathcal{W}(x, \xi) = 2^{-1}(\|x\|^2 - 2\langle J\xi, x \rangle + \|\xi\|^2).$$

We say that an operator $A : X \rightarrow 2^X$ with domain $D(A)$ is d-accretive if

$$\mathcal{W}(x_1, x_2) \leq \mathcal{W}(x_1 + \lambda(y_1 - y_2), x_2) \qquad (1.16.8)$$

for all $x_1, x_2 \in D(A)$, for all $y_1 \in A x_1$, $y_2 \in A x_2$ and $\lambda > 0$.

Theorem 1.16.3 *Definitions 1.16.1 and 1.16.2 are equivalent.*

Proof. Let ξ be fixed. It is easy to see that $grad\, \mathcal{W}(x, \xi) = Jx - J\xi$, therefore, it is a monotone operator. Consequently, $\mathcal{W}(x, \xi)$ is a convex functional for all $x \in X$. Then

$$\mathcal{W}(x_1, x_2) \geq \mathcal{W}(x_1 + \lambda(y_1 - y_2), x_2) - \lambda\langle J(x_1 + \lambda(y_1 - y_2)) - Jx_2, y_1 - y_2 \rangle.$$

This inequality and (1.16.8) imply

$$\langle J(x_1 + \lambda(y_1 - y_2)) - Jx_2, y_1 - y_2 \rangle \geq 0.$$

Setting $\lambda \rightarrow 0$ and using the fact that J is hemicontinuous, we prove d-accretivness of A in the sense of Definition 1.16.1.

Conversely, let (1.16.1) hold, $x_1, x_2 \in D(A)$ and $\lambda > 0$. Then (1.16.8) follows from the convexity inequality for $\mathcal{W}(x, \xi)$ again, namely,

$$\mathcal{W}(x_1 + \lambda(y_1 - y_2), x_2) \geq \mathcal{W}(x_1, x_2) + \lambda\langle Jx_1 - Jx_2, y_1 - y_2 \rangle.$$

The result holds in view of the monotonicity of J. ∎

Definition 1.16.4 *A d-accretive operator $A : X \rightarrow 2^X$ is said to be maximal d-accretive if its graph is not the right part of the graph of any other d-accretive operator $B : X \rightarrow 2^X$.*

Lemma 1.16.5 *The value set of the maximal d-accretive operator at any point of its domain is convex and closed.*

Proof: It follows from Definitions 1.16.1 and 1.16.4. ∎

Theorem 1.16.6 *Let $A : X \to X$ be a demicontinuous d-accretive operator with domain $D(A) = X$ and let the normalized duality mapping J^* be continuous. Then A is a maximal d-accretive operator.*

Proof. We shall show that the inequality

$$\langle Jx - Jy, Ax - f \rangle \geq 0 \quad \forall x \in X \tag{1.16.9}$$

implies $Ay = f$. Since $D(A) = X$, we may put in (1.16.9) $x = x_t = J^*(Jy + tJz)$ for any $z \in X$ and $t > 0$. Then

$$\langle Jz, Ax_t - f \rangle \geq 0.$$

Let $t \to 0$. Using the demicontinuity property of A and continuity of J^* we obtain in a limit

$$\langle Jz, Ay - f \rangle \geq 0 \quad \forall z \in X.$$

Since $R(J) = X^*$, the lemma is proved. ∎

Theorem 1.16.7 *Let $A : X \to 2^X$ be a maximal d-accretive operator. Let $x_n \in D(A)$, $y_n \in Ax_n$. Suppose that either $x_n \to x$, $y_n \rightharpoonup y$ and the duality mapping J is continuous, or $x_n \rightharpoonup x$, $y_n \to y$ and J is weak-to-weak continuous. Then $x \in D(A)$ and $y \in Ax$.*

Definition 1.16.8 *A d-accretive operator $A : X \to 2^X$ is said to be m-d-accretive if*

$$R(A + \alpha I) = X \quad \forall \alpha > 0,$$

where I is the identity map in X.

Lemma 1.16.9 *If an operator A is m-d-accretive, then it is maximal d-accretive.*

Proof: It is produced similarly to Lemma 1.15.20. ∎

Theorem 1.16.10 *Let $A : X \to 2^X$ be a d-accretive operator, the normalized duality mappings $J : X \to X^*$ and $J^* : X^* \to X$ be continuous. Then A is locally bounded at any point $x_0 \in int\, D(A)$.*

Proof. Assume the contradiction: $x_0 \in int\, D(A)$, $x_n \in D(A)$, $n = 1, 2, \ldots$, $x_n \to x_0$, but $\|y_n\| \to \infty$, where $y_n \in Ax_n$. Introduce $z_n = Jx_n - Jx_0$, $t_n = \|Jx_n - Jx_0\|_*^{1/2}$ and construct the elements $w_n = J^*(Jx_0 + t_nJw)$ with $w \in X$. It is clear that $\|z_n\| = t_n^2$. We conclude from the continuity of J that $t_n \to 0$ and $z_n \to \theta_{X^*}$ as $n \to \infty$. Since $J^*J = I_X$ and J^* is continuous, we have that $w_n \to x_0$. Consequently, $w_n \in D(A)$ for sufficiently small $t_n \leq \sigma$. Let $v = J^*(Jx_0 + \sigma Jw)$, $u_n \in Aw_n$ and $f \in Av$. Then d-accretivness of A implies

$$(t_n - \sigma)\langle Jw, u_n - f \rangle \geq 0,$$

and we obtain for $t_n \leq \sigma$ the inequality

$$\langle z, u_n - f \rangle \leq 0, \quad z = Jw. \tag{1.16.10}$$

By $R(J) = X^*$, (1.16.10) holds for all $z \in X^*$. Therefore, $\{u_n\}$ is bounded according to the Banach–Steinhaus theorem. Let $\|u_n\| \leq C$. The property of d-accretivness of A allows us to write

$$\langle z_n - t_n z, y_n - u_n \rangle \geq 0.$$

Then we have

$$\langle z, y_n \rangle \leq \frac{1}{t_n} \langle z_n, y_n \rangle - \frac{1}{t_n} \langle z_n - t_n z, u_n \rangle \leq \frac{\|z_n\|_*}{t_n} \|y_n\| + C\left(\frac{\|z_n\|_*}{t_n} + \|z\|_* \right),$$

from which the following inequality appears:

$$\limsup_{n \to \infty} \frac{\langle z, y_n \rangle}{1 + \|y_n\| \tau_n} < \infty \quad \forall z \in X^*,$$

where $\tau_n = t_n^{-1} \|z_n\|_*$. Due to the Banach–Steinhaus theorem again, there exists a constant $K > 0$ such that

$$\|y_n\| \leq K(1 + \|y_n\| \tau_n).$$

Since $\tau_n \to 0$, the estimate $K\tau_n \leq 2^{-1}$ is satisfied for sufficiently large n. Then $\|y_n\| \leq 2K$. Thus, we have established the boundedness of $\{y_n\}$ which contradicts our assumption above. The proof is accomplished. $\blacksquare$

Theorem 1.16.11 *Let $J : X \to X^*$ be a continuous and weak-to-weak continuous mapping, $J^* : X^* \to X$ be continuous, $A : X \to X$ be a d-accretive demicontinuous operator with $D(A) = X$ and $f \in X$. Assume that the space X possesses an approximation and there exists a constant $r > 0$ such that $\langle Jx, Ax - f \rangle \geq 0$ as $\|x\| = r$. Then the equation (1.15.5) has at least one classical solution $\bar{x}$, it being known that $\|\bar{x}\| \leq r$.*

Proof. As in the proof of Lemma 1.15.24,

$$\langle Jx, P_n(Ax - f) \rangle = \langle Jx, Ax - f \rangle \geq 0$$

for all $x_n \in X$ with $\|x_n\| = r$. From this, by Lemma 1.15.26 there exists an element $x_n \in X$ such that $\|x_n\| \leq r$ and $P_n(Ax_n - f) = \theta_X$. Show that the sequence $\{Ax_n\}$ is bounded. By virtue of Theorem 1.16.10, the operator A is locally bounded, therefore, there exist constants $a > 0$ and $K > 0$ such that if $y_n \in X$, $\|y_n\| \leq a$ then $\|Ay_n\| \leq K$. Take $\tilde{a} \leq min\{a, r\}$ and put $y_n = \tilde{a} \|Ax_n\|^{-1} Ax_n$. It is clear that $\|y_n\| = \tilde{a} \leq a$ and $\|Ay_n\| \leq K$ for these y_n.

Since A is d-accretive, we write down

$$\langle Jx_n - Jy_n, Ax_n - Ay_n \rangle \geq 0$$

or in the equivalent form

$$\langle Jy_n, Ax_n \rangle \leq \langle Jy_n - Jx_n, Ay_n \rangle + \langle Jx_n, Ax_n \rangle. \tag{1.16.11}$$

It is easy to see that

$$\langle Jx_n, Ax_n \rangle = \langle Jx_n, Ax_n - f \rangle + \langle Jx_n, f \rangle$$

$$= \langle Jx_n, P_n(Ax_n - f)\rangle + \langle Jx_n, f\rangle = \langle Jx_n, f\rangle.$$

Since the operator J is homogeneous, we have $\langle Jy_n, Ax_n\rangle = \tilde{a}\|Ax_n\|$. Then (1.16.11) gives

$$\tilde{a}\|Ax_n\| \leq (\|y_n\| + \|x_n\|)\|Ay_n\| + \|x_n\|\|f\|$$

$$\leq (\tilde{a} + r)K + r\|f\|.$$

We conclude from this that the sequence $\{Ax_n\}$ is bounded. Since $\{x_n\}$ is bounded, there exists a subsequence of $\{x_n\}$ (we do not change its notation) such that $x_n \rightharpoonup \bar{x} \in X$ as $n \to \infty$.

By hypothesis, the space X possesses an approximation. This fact allows us to construct for each element $z \in X$ the sequence $\{z_n\}$ such that $P_n z = z_n \in X_n$ and $z_n \to z$. Write down again the property of the d-accretivness of A,

$$\langle Jz - Jx_n, Az - Ax_n\rangle \geq 0.$$

The latter relation is identical to

$$\langle Jz - Jz_n, Az - Ax_n\rangle + \langle Jz_n - Jx_n, Az - f\rangle + \langle Jz_n - Jx_n, f - Ax_n\rangle \geq 0. \qquad (1.16.12)$$

The first term in the left part of (1.16.12) tends to zero as $n \to \infty$ because $\{Ax_n\}$ is bounded, while $Jz_n \to Jz$ by the continuity of duality mapping J. The last term equals zero due to the equalities

$$\langle Jz_n - Jx_n, f - Ax_n\rangle = \langle Jz_n - Jx_n, P_n(f - Ax_n)\rangle$$

and

$$P_n(Ax_n - f) = \theta_X.$$

Using weak-to-weak continuity of J and setting $n \to \infty$, we have in a limit

$$\langle Jz - J\bar{x}, Az - f\rangle \geq 0.$$

Put in the last inequality $z = z_t = J^*(J\bar{x} + tJv)$, where v is an arbitrary fixed element of X and $t > 0$. Then

$$\langle Jv, Az_t - f\rangle \geq 0. \qquad (1.16.13)$$

We recall that the operator A is demicontinuous and J^* is continuous. Then (1.16.13) leads as $t \to 0$ to the inequality

$$\langle Jv, A\bar{x} - f\rangle \geq 0. \qquad (1.16.14)$$

Since $R(J) = X^*$, (1.16.14) implies $A\bar{x} = f$. In addition, the estimate $\|x_n\| \leq r$ and weak lower semicontinuity of the norm in a Banach space guarantee the inequality $\|\bar{x}\| \leq r$. This completes the proof. ∎

Remark 1.16.12 *If we omit the demicontinuity property of A in the hypotheses of Theorem 1.16.11 and understand a solution of the equation (1.15.5) as the element $x_0 \in X$ such that*

$$\langle Jx - Jx_0, y - f\rangle \geq 0 \quad \forall y \in Ax, \ \forall x \in X,$$

then one can establish solvability of (1.15.5) similarly to Theorem 1.15.29.

Theorem 1.16.13 *Let* $A : X \to 2^X$ *be a coercive and m-d-accretive operator. Then* $\overline{R(A)} = X$.

Proof. The condition of m-d-accretiveness of the operator A guarantees unique solvability of the equation $Ax + \alpha x = 0$ for all $\alpha > 0$. Consequently, there exist elements $x_\alpha \in X$ and $y_\alpha \in Ax_\alpha$ such that

$$y_\alpha + \alpha x_\alpha = \theta_X. \tag{1.16.15}$$

We have

$$\langle Jx_\alpha, y_\alpha \rangle = -\alpha \|x_\alpha\|^2 \quad \forall \alpha > 0.$$

Since A is coercive, the last equality implies boundedness of $\{x_\alpha\}$ as $\alpha \to 0$. Therefore, $\alpha x_\alpha \to \theta_X$. By (1.16.15), $y_\alpha \to \theta_X$, that is, $\theta_X \in \overline{R(A)}$. The rest of the proof follows the pattern of Theorem 1.15.32. ∎

Consider the co-variational inequality with the maximal d-accretive operator A:

$$\langle Jy - Jx, Ax - f \rangle \geq 0 \quad \forall y \in \Omega, \ x \in \Omega, \tag{1.16.16}$$

where $f \in X$, $\Omega \subseteq D(A)$ and $\Omega^* = J\Omega$ is a closed and convex set in X^*.

We also present two definitions of their solutions.

Definition 1.16.14 *An element* $x \in \Omega$ *is called the solution of the co-variational inequality (1.16.16) if there is* $z \in Ax$ *such that*

$$\langle Jy - Jx, z - f \rangle \geq 0 \quad \forall y \in \Omega. \tag{1.16.17}$$

Definition 1.16.15 *An element* $x \in \Omega$ *is called the solution of the co-variational inequality (1.16.16) if*

$$\langle Jy - Jx, u - f \rangle \geq 0 \quad \forall y \in \Omega, \ \forall u \in Ay. \tag{1.16.18}$$

Lemma 1.16.16 *If* $x \in \Omega$ *is a solution of (1.16.16) defined by the inequality (1.16.17), then it also satisfies the inequality (1.16.18).*

Proof. Write the d-accretivness property of A,

$$\langle Jy - Jx, u - z \rangle \geq 0 \quad \forall x, \ y \in \Omega, \ \forall u \in Ay, \ \forall z \in Ax.$$

In view of (1.16.17), we obtain (1.16.18). ∎

Lemma 1.16.17 *If* $A : X \to X$ *is demicontinuous,* $\Omega \subset int \, D(A)$ *and duality mapping* J^* *is continuous, then Definitions 1.16.14 and 1.16.15 are equivalent.*

Proof. Let x be a solution in the sense of (1.16.18) and choose any $y \in \Omega$. Since the set $J\Omega$ is convex, the element $(1 - t)Jx + tJy \in J\Omega$ for all $t \in [0, 1]$. Then it is obvious that the element $y_t = J^*((1 - t)Jx + tJy) \in \Omega$. By (1.16.18) with y_t in place of y and Ay_t in place of u, one gets

$$\langle Jy_t - Jx, Ay_t - f \rangle \geq 0,$$

that is,

$$\langle Jy - Jx, Ay_t - f \rangle \geq 0 \quad \forall y \in \Omega. \tag{1.16.19}$$

Letting $t \to 0$, we obtain $y_t \to x$. By virtue of the demicontinuity of A at a point x and continuity of J^* at a point Jx, it is possible to assert that $Ay_t \rightharpoonup Ax \in X$. Then, by (1.16.19), it follows that (1.16.17) holds with $z = Ax$, that is, x is a solution of (1.16.16) in the sense of Definition 1.16.14. Taking into account the previous lemma, we conclude that the proof is complete. ∎

Bibliographical Notes and Remarks

The definitions and results of Section 1.1 are standard and can be found in most textbook on functional analysis. We recommend here Dunford and Schwartz [75], Hille and Fillips [95], Kantorovich and Akilov [104], Kolmogorov and Fomin [117], Liusternik and Sobolev [141], Nirenberg [159], Riesz and Sz.-Nagy [174], Rudin [182], Yosida [235]. The theory of linear topological spaces is well covered by Schaefer [205]. The methods of Hilbert, Banach and normed spaces are treated by Maurin [145], Edwards [77] and Day [69]. The necessary material on nonlinear and convex analysis and optimization can be found in Cea [64], Ekeland and Temam [79], Holmes [96], Vainberg [221]. Interesting facts and observations are also contained in [46, 83, 113, 162, 228].

Observe that the main results in the present book are stated in uniformly convex and/or uniformly smooth Banach spaces. Recall that spaces of number sequences l^p, Lebesgue spaces $L^p(G)$, Sobolev spaces $W_m^p(G)$ with $1 < p < \infty$, $m > 0$ and most of Orlicz spaces with the Luxemburg norm are uniformly convex and uniformly smooth [119, 127, 141, 210, 237]. Presenting the geometric characteristics of Banach spaces and their properties we follow Diestel [71], Figiel [81] and Lindenstrauss and Tzafriri [127]. The estimates of modulus of convexity and smoothness of Banach spaces have been obtained by Hanner [94], Alber and Notik [24, 25, 26] (see also [16, 127, 233, 236]). The upper and lower bounds of the constant L in (1.6.4) and (1.6.23) are presented in [25, 26, 81, 207].

For the proofs of the Hahn−Banach Theorem 1.1.1 and Banach−Steinhaus Theorem 1.1.3 we refer to [117, 141]. The strong separation Theorem 1.1.10 and the generalized Weierstrass Theorem 1.1.14 are stated, for instance, in [96, 224, 225, 238]. Theorem 1.1.8 has been established in [146] and Theorem 1.1.9 in [173]. Theorems 1.1.21 and 1.1.23 are proved in [221]. Theorem 1.1.62 appeared in [1]. A reader can look through differentiability problems of functionals and operators in [71, 221]. In particular, Theorem 1.1.34 is proved in [45, 112, 209] and Theorem 1.3.9 in [148]. Zorn's Lemma 1.1.61 can be found in [235]. As regards the imbedding theorem, we are mostly concerned with [104, 210].

The concept of a monotone operator has been introduced by Kachurovskii in [102]. It plays a very important role in the theory of elliptic and parabolic differential equations and in optimization theory [111, 128, 85]. The properties of monotone operators are well described in [57, 83, 103, 162, 166, 221, 237]. Lemma 1.3.14 is contained in [85]. Local boundedness of a monotone operator at interior points of its domain has been proved by Rockafellar in [177]. Kato has shown in [106] that at such points a monotone hemicontinuous mapping is demicontinuous. The examples of monotone mappings are given in [49, 57, 76, 83, 113, 120, 128, 134, 135, 142, 162, 221, 224].

Maximal monotone operators first appeared in [148]. Their properties are studied in [48, 49, 57, 83, 162]. Necessary and sufficient conditions for a linear operator to be maximal monotone are provided, for instance, in [162]. The conditions of maximal monotonicity of the sum of monotone operators were obtained in [179]. Theorems 1.7.13, 1.7.15 and 1.7.19 are due to Minty and Rockafellar [150, 175, 177, 178]. Theorem 1.7.17 was proved in [51].

A duality mapping is one of deepest subjects of Banach spaces and theory of monotone operators. Recall that the normalized duality mapping has been introduced and investigated by Vainberg in [219] and after that by many other authors (see, for instance, [38, 57, 83, 113, 128, 162, 236]). The idea of duality mapping with a gauge function belongs to Browder. The properties of duality mappings with a gauge function were established in [160, 128, 233]. In particular, Lemma 1.5.10 has been proved in [128] and [233]. The analytical representations of duality mappings in the Lebesgue, Sobolev and Orlicz spaces can be found in [24, 128, 221, 236]. Theorems 1.6.1 and 1.6.4 were proved in [13, 25, 28]. The properties of the Lyapunov functional $W(x, y)$ are also studied in [13]. The important Remark 1.5.12 can be seen in [221]. Duality mappings in non-reflexive spaces are studied in [89].

Using the concept of a duality mapping Kato introduced in [109] accretive operators which play a significant role in the fixed point theory and theory of integral equations and evolution differential equations [57, 108]. Note that Definition 1.3.6 arises from Definition 1.15.4. The general theory and examples of accretive mappings are described in [57, 65, 67, 110, 221, 234, 165]. Theorem 1.15.29 was proved in [191].

The Debrunner−Flor lemma appeared in [70]. Theorem 1.4.6 was proved by many authors. Our proof follows [89]. For Lemma 1.7.11 and Corollary 1.7.12 we refer to [162]. Lemma 1.5.14 has been established in [12]. Its accretive version was earlier proved in [165].

The main results of Sections 1.9 and 1.10 belong to Ryazantseva [185, 201]. Definition 1.9.10 has been introduced in [149]. Browder presented in [52, 53] examples of semi-monotone mappings. The sufficient solvability conditions of equations with a single-valued hemicontinuous operator are given in [52, 53, 221].

The existence theorems for variational inequalities can be found, for instance, in [54, 100, 113, 162, 128]. Equivalence of their solutions was studied by Minty [149]. Theorem 1.12.2 was proved in [203]. Solvability of variational inequalities with semimonotone bounded operators having regular discontinuous points has been established in [164].

Section 1.13 is devoted to variational inequalities with pseudomonotone operators [47] and contains the results of [58]. We emphasize that pseudomonotone operators are non-monotone, in general. There are many different definitions of pseudomonotone operators. Following [58], we present one of them. The properties of pseudomonotone operators are described in [58, 128, 162]. Variational inequalities with quasipotential maps were investigated in [163]. In addition, one can point out the papers [2, 72, 86] in which variational inequations with non-monotone mappings are also regarded.

The concept of the d-accretive operator is introduced in [29]. The results of Section 1.16 correspond to [30]. As regards the projection operator Π_Ω, we refer the reader to [13, 16].

Chapter 2

REGULARIZATION OF OPERATOR EQUATIONS

2.1 Equations with Monotone Operators in Hilbert Spaces

1. Let H be a Hilbert space, $A : H \to H$ be a hemicontinuous monotone operator, $D(A) = H$ and $f \in H$. We study the operator equation

$$Ax = f \tag{2.1.1}$$

assuming that it has a solution (in the classical sense). Denote by N its solution set, by x^* any point of N and by $\bar{x}^*$ a point in N with the minimal norm. Note that due to Theorem 1.4.6, operator A is maximal monotone. Then N is a closed and convex set and $\bar{x}^*$ is unique in N.

As it has been already mentioned, the problem (2.1.1) is ill-posed, in general. Therefore, strong convergence and stability of approximate solutions can be proved only by applying some regularization procedure. In any method of finding $x^* \in N$, the main aim is to establish a continuous dependence of approximate solutions on data perturbations. In connection with this, we assume that right-hand side and operator in (2.1.1) are given approximately, namely, instead of f and A, we have sequences $\{f^\delta\}$ and $\{A^h\}$, where hemicontinuous monotone operators $A^h : H \to H$ with $D(A^h) = D(A)$ for all $h > 0$, and $f^\delta \in H$. We define the proximity between operators A and A^h by means of the following inequality:

$$\|Ax - A^h x\| \leq h g(\|x\|), \tag{2.1.2}$$

where $g(t)$ is a non-negative continuous function for all $t \geq 0$. It is clear that if $g(t) \equiv 1$ then (2.1.2) characterizes the uniform proximity of operators A^h and A. As regards $\{f^\delta\}$, we always assume that

$$\|f - f^\delta\| \leq \delta, \quad \delta > 0. \tag{2.1.3}$$

Thus, instead of (2.1.1), the following equation is considered:

$$A^h x = f^\delta, \tag{2.1.4}$$

117

which does not necessarily have a solution. Our goal in this section is to construct, by use of approximate data $\{A^h, f^\delta\}$, a sequence $\{x^\gamma\}$, $\gamma = (\delta, h)$, which strongly converges to $\bar{x}^* \in N$ as $\gamma \to 0$.

Consider the regularization algorithm for (2.1.4) as follows:

$$A^h x + \alpha x = f^\delta, \ \alpha > 0. \tag{2.1.5}$$

Denote $T = A + \alpha I$, where $I : H \to H$ is the identity operator. Obviously, T is monotone as a sum of two monotone operators. It is coercive because

$$(Tx, x) = (Ax - A(\theta_H), x - \theta_H) + \alpha(x, x) + (A(\theta_H), x) \geq \alpha \|x\|^2 - \|A(\theta_H)\| \|x\|,$$

and then

$$\lim_{\|x\| \to \infty} \frac{(Tx, x)}{\|x\|} \geq \lim_{\|x\| \to \infty} \alpha \|x\| - \|A(\theta_H)\| = \infty.$$

By the Minty–Browder theorem, the equation (2.1.5) has a classical solution x_α^γ, that is,

$$A^h x_\alpha^\gamma + \alpha x_\alpha^\gamma = f^\delta. \tag{2.1.6}$$

Since

$$(Tx - Ty, x - y) = (Ax - Ay, x - y) + \alpha \|x - y\|^2 \geq \alpha \|x - y\|^2,$$

T is strongly monotone for all $\alpha > 0$. Therefore, x_α^γ is the unique solution. According to Definition 5, the solution x_α^γ satisfying (2.1.5) is called the regularized solution of the operator equation (2.1.4).

Remark 2.1.1 *Note that $\{x_\alpha^\gamma\}$ is often called the solution net. However, we prefer in the sequel the more usual term "the solution sequence".*

Theorem 2.1.2 *Let (2.1.2) and (2.1.3) hold. A sequence $\{x_\alpha^\gamma\}$ generated by (2.1.6) is uniformly bounded and strongly converges to the minimal norm solution $\bar{x}^*$ if*

$$\frac{\delta + h}{\alpha} \to 0 \ \text{ as } \ \alpha \to 0. \tag{2.1.7}$$

Proof. It is clear that (2.1.7) implies $\delta \to 0$ and $h \to 0$. Let $x^* \in N$. Then it follows from (2.1.1) that $Ax^* = f$ and the equation

$$A^h x_\alpha^\gamma - f + \alpha(x_\alpha^\gamma - x^*) = f^\delta - f - \alpha x^* \tag{2.1.8}$$

is equivalent to (2.1.6). The scalar product of (2.1.8) and the difference $x_\alpha^\gamma - x^*$ gives

$$(A^h x_\alpha^\gamma - f, x_\alpha^\gamma - x^*) + \alpha \|x_\alpha^\gamma - x^*\|^2 = (f^\delta - f, x_\alpha^\gamma - x^*) + \alpha(x^*, x^* - x_\alpha^\gamma). \tag{2.1.9}$$

In view of the monotonicity of A^h,

$$(A^h x_\alpha^\gamma - A^h x^*, x_\alpha^\gamma - x^*) \geq 0, \tag{2.1.10}$$

and we have

$$\alpha\|x_\alpha^\gamma - x^*\| \leq \|f^\delta - f\| + \|A^h x^* - A x^*\| + \alpha\|x^*\|. \tag{2.1.11}$$

By the hypotheses (2.1.2) and (2.1.3),

$$\|x_\alpha^\gamma - x^*\| \leq \frac{\delta}{\alpha} + \frac{h}{\alpha} g(\|x^*\|) + \|x^*\|.$$

Now we conclude from this inequality and from (2.1.7) that the sequence $\{x_\alpha^\gamma\}$ is bounded. Hence, according to (2.1.6), $A^h x_\alpha^\gamma \to f$ as $\alpha \to 0$, and there exists a subsequence $\{x_\beta^\xi\}$, where $\beta \subseteq \alpha$ and $\xi = (\delta', h') \subseteq \gamma$, which weakly converges to some element $\tilde{x} \in H$. With this, also

$$\frac{\delta' + h'}{\beta} \to 0 \quad \text{as} \quad \alpha \to 0.$$

Show that $\tilde{x} = \bar{x}^*$. First of all, establish the inclusion $\tilde{x} \in N$. Write down the monotonicity property of $A^{h'}$ for an arbitrary $x \in H$:

$$(A^{h'} x - A^{h'} x_\beta^\xi, x - x_\beta^\xi) \geq 0. \tag{2.1.12}$$

Setting $\alpha \to 0$, and as consequence $\beta \to 0$ and $\xi \to 0$, we obtain from (2.1.12) and (2.1.2) the limit inequality

$$(A x - f, x - \tilde{x}) \geq 0 \quad \forall x \in H.$$

Since A is a hemicontinuous monotone operator, this means that $\tilde{x} \in N$ (see Theorem 1.9.1).

By (2.1.9), we further have

$$(x^*, x^* - x_\beta^\xi) + \frac{\delta'}{\beta}\|x_\beta^\xi - x^*\| + \frac{h'}{\beta} g(\|x^*\|)\|x_\beta^\xi - x^*\| \geq 0.$$

If $\beta \to 0$ then

$$(x^*, x^* - \tilde{x}) \geq 0 \quad \forall x^* \in N, \tag{2.1.13}$$

because the sequence $\{x_\beta^\xi\}$ is bounded, $\dfrac{\delta'}{\beta} \to 0$ and $\dfrac{h'}{\beta} \to 0$ as $\beta \to 0$. Since N is a convex set, $x_t = t\tilde{x} + (1-t)x^* \in N$ for all $x^* \in N$ and $t \in [0,1]$. Substitute x_t into (2.1.13) in place of x^* and use the obvious fact that $1 - t \geq 0$. Then we obtain

$$(x_t, x^* - \tilde{x}) \geq 0 \quad \forall x^* \in N.$$

If $t \to 1$, we have

$$(\tilde{x}, x^* - \tilde{x}) \geq 0 \quad \forall x^* \in N,$$

that is,

$$\|\tilde{x}\| \leq \|x^*\| \quad \forall x^* \in N.$$

The last is equivalent to the relation

$$\|\tilde{x}\| = \min\{\|x^*\| \mid x^* \in N\}.$$

Thus, $\tilde{x} = \bar{x}^*$. This means that the whole sequence $\{x_\alpha^\gamma\}$ converges weakly to $\bar{x}^*$. Finally, (2.1.9) leads to the inequality

$$\|x_\alpha^\gamma - \bar{x}^*\|^2 \leq \frac{\delta}{\alpha}\|x_\alpha^\gamma - \bar{x}^*\| + \frac{h}{\alpha}g(\|\bar{x}^*\|)\|x_\alpha^\gamma - \bar{x}^*\| + (\bar{x}^*, \bar{x}^* - x_\alpha^\gamma). \tag{2.1.14}$$

Now the conclusion of the theorem follows from (2.1.7) and the weak convergence of $\{x_\alpha^\gamma\}$ to $\bar{x}^*$. The theorem is proved. ∎

2. In comparison with (2.1.5), we study the more general regularized equation

$$A^h x + \alpha S x = f^\delta, \ \alpha > 0, \tag{2.1.15}$$

where $S : H \to H$ is a strongly monotone bounded hemicontinuous operator with domain $D(S) = H$. By the obvious inequality

$$(A^h x + \alpha S x, x) \geq \alpha c\|x\|^2 - \|A^h(\theta_H)\|\|x\| - \alpha\|S(\theta_H)\|\|x\|,$$

where c is a constant of the strong monotonicity of S, it follows that $A^h + \alpha S$ is coercive. Therefore, by Theorem 1.7.5, the equation (2.1.15) is solvable. It is uniquely solvable because S is strongly monotone. We are able to prove that the solution sequence $\{x_\alpha^\gamma\}$ generated by (2.1.15) converges in the norm of H to the element $\bar{x} \in N$ satisfying the inequality

$$(Sx^*, x^* - \bar{x}) \geq 0 \ \ \forall x^* \in N. \tag{2.1.16}$$

Let S be a potential operator, i.e., there exists a functional $\varphi(x)$ such that $Sx = grad\ \varphi(x)$ for all $x \in H$. Then Theorem 1.11.14 and (2.1.16) imply the following result:

$$\varphi(\bar{x}) = min\ \{\varphi(x^*) \mid x^* \in N\}.$$

The properties of H, S and N guarantee uniqueness of $\bar{x}$.

3. Assume now that $A : H \to 2^H$ and $A^h : H \to 2^H$ are maximal monotone (possibly, multiple-valued) operators and $D(A) = D(A^h) \subseteq H$. We recall that if A and A^h are multiple-valued, then their value sets at any point $x \in D(A)$, which we denote, respectively, by Ax and $A^h x$, are convex and closed (see Theorem 1.4.9). In this case, the proximity between the sets Ax and $A^h x$ is defined by means of the Hausdorff distance as

$$\mathcal{H}_H(Ax, A^h x) \leq hg(\|x\|), \tag{2.1.17}$$

where g(t) is a continuous non-negative function for all $t \geq 0$.

Theorem 2.1.3 *Let (2.1.3), (2.1.7) and (2.1.17) hold. Then the sequence $\{x_\alpha^\gamma\}$ generated by (2.1.5) is uniformly bounded and strongly converges to the minimal norm solution $\bar{x}^*$ of the equation (2.1.1).*

Proof. Recall that solutions of (2.1.1) and (2.1.6) are understood now in the sense of inclusions. By Theorem 1.7.4, the equation (2.1.5) has a solution, i.e., there exists x_α^γ such that

$$f^\delta \in A^h x_\alpha^\gamma + \alpha x_\alpha^\gamma.$$

Hence, there exists $y_\alpha^\gamma \in A^h x_\alpha^\gamma$ such that

$$y_\alpha^\gamma + \alpha x_\alpha^\gamma = f^\delta. \tag{2.1.18}$$

Since $A^h + \alpha I$ is strictly monotone, the solution x_α^γ is unique for all fixed $\alpha > 0$ and $\gamma > 0$.
Let $x^* \in N$. Then $f \in Ax^*$. It results from (2.1.17) that there exist $y_*^h \in A^h x^*$ such that

$$\|y_*^h - f\| \le hg(\|x^*\|).$$

Using (2.1.18) and the monotonicity condition of A^h we can estimate the following scalar product:

$$\alpha(x_\alpha^\gamma, x_\alpha^\gamma - x^*) = (f^\delta - y_\alpha^\gamma, x_\alpha^\gamma - x^*)$$
$$= (f^\delta - f, x_\alpha^\gamma - x^*) - (y_\alpha^\gamma - f, x_\alpha^\gamma - x^*)$$
$$= (f^\delta - f, x_\alpha^\gamma - x^*) - (y_\alpha^\gamma - y_*^h, x_\alpha^\gamma - x^*) + (y_*^h - f, x_\alpha^\gamma - x^*)$$
$$\le (f^\delta - f, x_\alpha^\gamma - x^*) - (y_*^h - f, x_\alpha^\gamma - x^*), \quad y_*^h \in A^h x^*.$$

Therefore,

$$(x_\alpha^\gamma, x_\alpha^\gamma - x^*) \le \left(\frac{\delta}{\alpha} + \frac{h}{\alpha}g(\|x^*\|)\right)\|x_\alpha^\gamma - x^*\|, \tag{2.1.19}$$

from which we have the quadratic inequality

$$\|x_\alpha^\gamma\|^2 - \left(\|x^*\| + \frac{\delta}{\alpha} + \frac{h}{\alpha}g(\|x^*\|)\right)\|x_\alpha^\gamma\| - \left(\frac{\delta}{\alpha} + \frac{h}{\alpha}g(\|x^*\|)\right)\|x^*\| \le 0. \tag{2.1.20}$$

Without lost of generality, one can assume by (2.1.7) that there exists a constant $C > 0$ such that

$$\frac{\delta}{\alpha} + \frac{h}{\alpha}g(\|x^*\|) \le C.$$

Then (2.1.20) yields the estimate

$$\|x_\alpha^\gamma\| \le C_1, \quad \text{where} \quad C_1 = \|x^*\| + 2C. \tag{2.1.21}$$

This implies existence of a subsequence (which for simplicity is denoted as before by $\{x_\alpha^\gamma\}$) such that $x_\alpha^\gamma \rightharpoonup \tilde{x}$ as $\alpha \to 0$.

Show that $\tilde{x} \in N$. Let $x \in D(A)$. Since the operators A^h are monotone, we have for arbitrary fixed $x \in D(A)$, $y^h \in A^h x$ and $y_\alpha^\gamma \in A^h x_\alpha^\gamma$:

$$(y^h - y_\alpha^\gamma, x - x_\alpha^\gamma) \ge 0.$$

Then, by (2.1.18), for all $y \in Ax$ and $f \in Ax^*$ we deduce

$$(y^h - y_\alpha^\gamma, x - x_\alpha^\gamma) = (y^h - f^\delta + \alpha x_\alpha^\gamma, x - x_\alpha^\gamma)$$
$$= (y - f, x - x_\alpha^\gamma) + (y^h - y, x - x_\alpha^\gamma) + (f - f^\delta, x - x_\alpha^\gamma) + \alpha(x_\alpha^\gamma, x - x_\alpha^\gamma) \ge 0.$$

This leads to the inequality

$$(y - f, x - x_\alpha^\gamma) \ge -(hg(\|x\|) + \delta + \alpha\|x_\alpha^\gamma\|)\|x - x_\alpha^\gamma\|.$$

Consequently,

$$\lim_{\alpha \to 0}(y - f, x - x_\alpha^\gamma) = (y - f, x - \tilde{x}) \geq 0 \quad \forall x \in D(A), \quad \forall y \in Ax.$$

According to Lemma 1.11.6, this imposes that $\tilde{x} \in N$.

It further follows from (2.1.19) that

$$(x^*, x^* - x_\alpha^\gamma) \geq -\Big(\frac{\delta}{\alpha} + \frac{h}{\alpha}g(\|x^*\|)\Big)\|x_\alpha^\gamma - x^*\|.$$

Setting $\alpha \to 0$, we come to the inequality

$$(x^*, x^* - \tilde{x}) \geq 0 \quad \forall x^* \in N,$$

which means that $\tilde{x} = \bar{x}^*$. At the same time, the uniqueness of $\bar{x}^*$ is guaranteed by convexity of N and properties of H. Thus, the whole sequence $\{x_\alpha^\gamma\}$ weakly converges to $\bar{x}^*$.

Finally, in view of (2.1.19) and (2.1.7), we obtain

$$(x_\alpha^\gamma, x_\alpha^\gamma - x^*) \to 0 \quad \forall x^* \in N$$

because the sequence $\{x_\alpha^\gamma\}$ is bounded. Then due to the weak convergence of x_α^γ to $\bar{x}^* \in N$, one gets

$$\|\bar{x}^* - x_\alpha^\gamma\|^2 = (\bar{x}^*, \bar{x}^* - x_\alpha^\gamma) + (x_\alpha^\gamma, x_\alpha^\gamma - \bar{x}^*) \to 0$$

as $\alpha \to 0$, that is, $\lim_{\alpha \to 0} x_\alpha^\gamma = \bar{x}^*$.   ∎

The proof of Theorem 2.1.3 can be also obtained by the scheme which has been earlier applied in the linear case. Namely, introduce the regularized operator equation with unperturbed data

$$Ax + \alpha x = f, \tag{2.1.22}$$

and denote its solution by x_α^0. Similarly to Theorem 2.1.3, one can prove that the sequence $\{x_\alpha^0\}$ converges strongly to $\bar{x}^*$ as $\alpha \to 0$, and

$$\|x_\alpha^0\| \leq \|\bar{x}^*\| \tag{2.1.23}$$

because in (2.1.21) $C = 0$. Since A^h is a monotone operator, by virtue of the equality

$$(y_\alpha^\gamma - y_\alpha^0, x_\alpha^\gamma - x_\alpha^0) + \alpha\|x_\alpha^\gamma - x_\alpha^0\|^2 = (f^\delta - f, x_\alpha^\gamma - x_\alpha^0),$$

where $y_\alpha^\gamma \in A^h x_\alpha^\gamma$ and $y_\alpha^0 \in Ax_\alpha^0$, we obtain

$$\|x_\alpha^\gamma - x_\alpha^0\| \leq \frac{\delta}{\alpha} + \frac{h}{\alpha}g(\|x_\alpha^0\|) \to 0 \quad \text{as} \quad \alpha \to 0. \tag{2.1.24}$$

The conclusion of Theorem 2.1.3 follows now from the inequality

$$\|x_\alpha^\gamma - \bar{x}^*\| \leq \|x_\alpha^\gamma - x_\alpha^0\| + \|x_\alpha^0 - \bar{x}^*\|.$$

Theorem 2.1.4 *Let (2.1.17), (2.1.3) and (2.1.7) hold. If the sequence $\{x_\alpha^\gamma\}$ generated by (2.1.5) converges (even weakly) to some element $x^* \in H$, then x^* is a solution of the equation (2.1.1).*

Proof. Since $\{x_\alpha^\gamma\}$ converges to x^*, it is bounded. Then, by (2.1.18), $y_\alpha^\gamma \to f$ as $\alpha \to 0$, where $y_\alpha^\gamma \in A^h x_\alpha^\gamma$. Write the monotonicity condition for A^h as

$$(y^h - y_\alpha^\gamma, x - x_\alpha^\gamma) \geq 0 \quad \forall x \in D(A), \quad \forall y^h \in A^h x.$$

From that, after passing to the limit as $\alpha \to 0$, we obtain

$$(y - f, x - x^*) \geq 0 \quad \forall x \in D(A), \quad \forall y \in Ax.$$

Since the operator A is maximal monotone, the latter inequality means that $f \in Ax^*$. ∎

Combining Theorems 2.1.3 and 2.1.4 implies

Theorem 2.1.5 *Suppose that the conditions (2.1.17), (2.1.3) and (2.1.7) are satisfied. Then the sequence $\{x_\alpha^\gamma\}$ strongly converges to some element $\bar{x} \in H$ if and only if there exists a solution of the equation (2.1.1).*

The next two statements immediately follow from Theorem 2.1.3.

Theorem 2.1.6 *Let equation (2.1.1) have a unique solution $x_0 \in H$ and let there exist a constant $C > 0$ such that*

$$\frac{\delta + h}{\alpha} \leq C$$

as $\alpha \to 0$. Then $x_\alpha^\gamma \rightharpoonup x_0$ as $\alpha \to 0$.

Theorem 2.1.7 *If $h, \delta, \alpha \to 0$, then*

$$1) \quad \|y_\alpha^\gamma - f^\delta\| \to 0, \quad y_\alpha^\gamma \in A^h x_\alpha^\gamma$$

and

$$2) \quad \|\tilde{y}_\alpha^\gamma - f\| \to 0, \quad \tilde{y}_\alpha^\gamma \in A x_\alpha^\gamma,$$

where y_α^γ satisfies (2.1.18) and $\tilde{y}_\alpha^\gamma$ such that

$$\|y_\alpha^\gamma - \tilde{y}_\alpha^\gamma\| \leq hg(\|x_\alpha^\gamma\|).$$

2.2 Equations with Monotone Operators in Banach Spaces

1. Let X be an E-space with strictly convex dual space X^*. Consider the equation (2.1.1) with the maximal monotone operator $A : X \to 2^{X^*}$. As in the case of Hilbert spaces, assume that (2.1.1) has a nonempty solution set N and denote by $\bar{x}^* \in N$ the minimal norm solution. It is unique because the set N is convex and closed and the Banach space X is reflexive and strictly convex. Suppose that, instead of A and f, the sequences $\{f^\delta\}$ and

$\{A^h\}$ are given, where maximal monotone operators $A^h : X \to 2^{X^*}$ have $D(A^h) = D(A)$ for all $h > 0$, and $f^\delta \in X^*$ for all $\delta > 0$. Thus, in reality, we study the equation (2.1.4) with the following proximity conditions:

$$\mathcal{H}_{X^*}(Ax, A^h x) \leq hg(\|x\|), \tag{2.2.1}$$

where $g(t)$ is a continuous non-negative function for all $t \geq 0$, $\mathcal{H}_{X^*}(G_1, G_2)$ stands the Hausdorff distance between the sets G_1 and G_2 in X^* and

$$\|f - f^\delta\|_* \leq \delta, \quad \delta > 0. \tag{2.2.2}$$

If A and A^h are single-valued then the condition (2.2.1) is

$$\|Ax - A^h x\|_* \leq hg(\|x\|). \tag{2.2.3}$$

Under these circumstances, we solve the regularized operator equation

$$A^h x + \alpha Jx = f^\delta, \tag{2.2.4}$$

where $J : X \to X^*$ is the normalized duality mapping. In our conditions, J is a demi-continuous and single-valued operator, $D(J) = X$ and $R(J) = X^*$. Then Theorem 1.7.4 guarantees solvability of the equation (2.2.4) in the sense of inclusion. Let x_α^γ be a solution of (2.2.4). It is unique because the operator $A^h + \alpha J$ is strictly monotone. It is clear that there exists an element $y_\alpha^\gamma \in A^h x_\alpha^\gamma$ such that

$$y_\alpha^\gamma + \alpha Jx_\alpha^\gamma = f^\delta.$$

According to Definition 5, the solution x_α^γ satisfying (2.2.4) is also called the regularized solution of the operator equation $A^h x = f^\delta$.

Theorem 2.2.1 *Assume that (2.2.1), (2.2.2) and (2.1.7) hold. Let A and A^h be maximal monotone operators, X be an E-space, X^* be a strictly convex space. Then the sequence $\{x_\alpha^\gamma\}$ of solutions of the equation (2.2.4) converges strongly in X to the element $\bar{x}^* \in N$ as $\alpha \to 0$.*

Proof. Similarly to Theorem 2.1.3, we obtain for an arbitrary $x^* \in N$ and $f \in Ax^*$ the equality

$$\langle y_\alpha^\gamma - f, x_\alpha^\gamma - x^* \rangle \quad + \quad \alpha \langle Jx_\alpha^\gamma - Jx^*, x_\alpha^\gamma - x^* \rangle$$
$$= \langle f^\delta - f, x_\alpha^\gamma - x^* \rangle \quad + \quad \alpha \langle Jx^*, x^* - x_\alpha^\gamma \rangle. \tag{2.2.5}$$

Using the monotonicity property of A^h, hypotheses (2.2.1) and (2.2.2) and definition of J, we deduce from (2.2.5) the relation

$$\|x_\alpha^\gamma\|^2 \leq \left(\frac{\delta}{\alpha} + \frac{h}{\alpha} g(\|x^*\|) \right) \|x_\alpha^\gamma - x^*\| + \|x^*\| \|x_\alpha^\gamma\|.$$

Now by the triangle inequality, we have

$$\|x_\alpha^\gamma\|^2 - \left(\frac{\delta}{\alpha} + \frac{h}{\alpha}g(\|x^*\|) + \|x^*\|\right)\|x_\alpha^\gamma\| - \left(\frac{\delta}{\alpha} + \frac{h}{\alpha}g(\|x^*\|)\right)\|x^*\| \le 0.$$

This quadratic inequality yields the estimate

$$\|x_\alpha^\gamma\| \le \frac{\delta}{\alpha} + \frac{h}{\alpha}g(\|x^*\|) + 2\|x^*\|,$$

which implies the boundedness of $\{x_\alpha^\gamma\}$. Then, as in the proof of Theorem 2.1.3, we get that there exists a subsequence of the sequence $\{x_\alpha^\gamma\}$ (we do not change its notation) which converges weakly to an element $\tilde{x} \in N$. Using (2.2.5) and taking into account the monotonicity of J, one gets

$$\langle Jx^*, x^* - x_\alpha^\gamma\rangle + \left(\frac{\delta}{\alpha} + \frac{h}{\alpha}g(\|x^*\|)\right)\|x_\alpha^\gamma - x^*\| \ge 0 \quad \forall x^* \in N.$$

Next we pass to the limit in this inequality as $\alpha \to 0$ and thus obtain

$$\langle Jx^*, x^* - \tilde{x}\rangle \ge 0 \quad \forall x^* \in N, \quad \tilde{x} \in N.$$

Replacing $x^* \in N$ by $x_t = t\tilde{x} + (1-t)x^*$, $t \in [0,1]$, we see that $x_t \in N$ and

$$\langle Jx_t, x^* - \tilde{x}\rangle \ge 0.$$

Since J is demicontinuous, this implies as $t \to 1$,

$$\langle J\tilde{x}, x^* - \tilde{x}\rangle \ge 0 \quad \forall x^* \in N, \quad \tilde{x} \in N.$$

According to Theorem 1.11.14, we have $\tilde{x} = \bar{x}^*$. Hence, the whole sequence $x_\alpha^\gamma \to \bar{x}^*$.

By (1.5.3) and (2.2.5) with $x^* = \bar{x}^*$, we can write

$$(\|x_\alpha^\gamma\| - \|\bar{x}^*\|)^2 \le \langle J\bar{x}^*, \bar{x}^* - x_\alpha^\gamma\rangle + \left(\frac{\delta}{\alpha} + \frac{h}{\alpha}g(\|\bar{x}^*\|)\right)\|x_\alpha^\gamma - \bar{x}^*\|. \tag{2.2.6}$$

Then in view of the proved weak convergence of $\{x_\alpha^\gamma\}$ to $\bar{x}^*$, we deduce from (2.2.6) and (2.1.7) that $\|x_\alpha^\gamma\| \to \|\bar{x}^*\|$ as $\alpha \to 0$. The conclusion of the theorem follows now from the definition of E-space. The proof is complete. ∎

Remark 2.2.2 *The convergence of the operator regularization method (2.2.4) is established in Theorem 2.2.1 without any restrictions on $D(A)$. Recall that the domain of a maximal monotone operator does not necessarily coincide with the whole space X. In particular, it may be a linear everywhere dense set, a convex closed set and open set having the convex closure.*

Corollary 2.2.3 *Theorem 2.2.1 remains valid for the operator regularization method*

$$A^h x + \alpha J(x - u) = f^\delta,$$

where $u \in X$ is some fixed element. In addition, the solution $\bar{x}^* \in N$ satisfies the equality

$$\|\bar{x}^* - u\| = min\{\|x^* - u\| \mid x^* \in N\}.$$

Consider the equation (2.1.15) again. The requirement of strong monotonicity of the operator $S : X \to X^*$ can be replaced now by the condition

$$\langle Sx - Sy, x - y \rangle \geq (\mu(\|x\|) - \mu(\|y\|))(\|x\| - \|y\|), \tag{2.2.7}$$

where a function $\mu(t)$ is continuous and increasing as $t \geq 0$, $\mu(0) = 0$, $\mu(t) \to \infty$ as $t \to \infty$. In particular, the duality mapping $J^\mu : X \to X^*$ with the gauge function $\mu(t)$ satisfies (2.2.7) as we observed in Lemma 1.5.4. In this case, the limit element of the sequence $\{x_\alpha^\gamma\}$, where x_α^γ solves the equation (2.1.15) with $S = J^\mu$ and $\alpha \to 0$, is the minimal norm solution in N. In other words, replacing in (2.2.4) the normalized duality mapping J by J^μ, we find the same solution $\bar{x}^* \in N$. However, by the corresponding choice of the function $\mu(t)$, the operator J^μ may have the stronger monotonicity property in comparison with (2.2.7) (see, for instance, (1.6.57)). This fact is of great importance in Sections 6.5, 6.6, where the convergence analysis of iterative processes are given for regularized equations.

Prove that solutions x_α^γ of the equation

$$A^h x + \alpha J^\mu x = f^\delta$$

are bounded. From (2.2.1), (2.2.2) and (1.5.2) we have the inequality

$$\mu(\|x_\alpha^\gamma\|)\|x_\alpha^\gamma\| \quad - \quad \left(\frac{\delta}{\alpha} + \frac{h}{\alpha} g(\|x^*\|) \right) \|x_\alpha^\gamma\| - \mu(\|x_\alpha^\gamma\|)\|x^*\|$$

$$- \quad \frac{\delta}{\alpha}\|x^*\| - \frac{h}{\alpha} g(\|x^*\|)\|x^*\| \leq 0. \tag{2.2.8}$$

However, we are not able to evaluate $\|x_\alpha^\gamma\|$ from above by this inequality with any function $\mu(t)$. We provide such an estimate for the most important case of power functions. Let $\mu(t) = t^s$, $s \geq 1$. Then (2.2.8) takes the following form:

$$\|x_\alpha^\gamma\|^{s+1} - \left(\frac{\delta}{\alpha} + \frac{h}{\alpha} g(\|x^*\|) \right) \|x_\alpha^\gamma\| - \|x_\alpha^\gamma\|^s\|x^*\| - \frac{\delta}{\alpha}\|x^*\| - \frac{h}{\alpha} g(\|x^*\|)\|x^*\| \leq 0. \tag{2.2.9}$$

Consider the function

$$\varphi(t) = t^{s+1} - at^s - bt - ab.$$

It is not difficult to verify that if

$$\bar{t} = \tau a + b^{1/s}, \quad \tau s \geq 2, \quad \tau > 1,$$

then $\varphi(t) > 0$ as $t \geq \bar{t}$. Hence, (2.2.9) yields the estimate

$$\|x_\alpha^\gamma\| \leq \tau\|x^*\| + \left(\frac{\delta}{\alpha} + \frac{h}{\alpha} g(\|x^*\|) \right)^{1/s} \quad \forall x^* \in N. \tag{2.2.10}$$

The boundedness of x_α^γ is finally proved by (2.1.7). ∎

2. Let $A : X \to 2^{X^*}$ and $A^h : X \to 2^{X^*}$ be monotone (possibly, multiple-valued) operators which do not satisfy the condition of the maximal monotonicity. As before, we consider

the equation (2.2.4) as regularized, and we understand solutions of all corresponding equations in the generalized sense (see Definition 1.9.3). We further assume that $D(A)$ is the convex closed set, $int\ D(A) \neq \emptyset$. Then the equation (2.2.4) has a unique solution in $D(A)$, and, according to Lemma 1.9.8, it is equivalent to the equation

$$\bar{A}^h x + \alpha J x = f^\delta, \qquad (2.2.11)$$

where $\bar{A}^h$ are maximal monotone extensions of A^h, $D(\bar{A}^h) = D(\bar{A}) = D(A)$. It is obvious that a set N of the generalized solutions of the equation (2.1.1) coincides in $D(A)$ with the set of solutions of the equation $\bar{A}x = f$. The condition (2.2.1) is replaced by the following inequality:

$$\mathcal{H}_{X^*}(\bar{A}^h x, \bar{A}x) \leq g(\|x\|)h. \qquad (2.2.12)$$

Then Theorems 2.2.1, 2.1.4 and 2.1.6 can be formulated for equations with an arbitrary monotone operator.

Theorem 2.2.4 *Let* $A : X \to 2^{X^*}$ *and* $A^h : X \to 2^{X^*}$ *be monotone operators,* $D(A) = D(A^h)$ *be a convex closed set,* $int\ D(A) \neq \emptyset$. *Let (2.2.2), (2.2.12) and (2.1.7) hold. Assume that* $x_\alpha^\gamma \in D(A)$ *is a solution of the regularized equation (2.2.4). Then the sequence* $\{x_\alpha^\gamma\}$ *strongly converges in* X *as* $\alpha \to 0$ *to the solution* $\bar{x}^*$ *of the equation (2.1.1) with the minimal norm.*

Theorem 2.2.5 *If the conditions of Theorem 2.2.4 are satisfied, then convergence of the regularization method (2.2.4) is equivalent to solvability of the equation (2.1.1).*

Theorem 2.2.6 *Assume that the conditions of Theorem 2.2.4 are fulfilled,* $\delta = O(\alpha)$, $h = O(\alpha)$ *as* $\alpha \to 0$, *and the equation (2.1.1) has a unique solution* x^*. *Then the sequence* $\{x_\alpha^\gamma\}$ *of solutions of the regularized equation (2.2.4) weakly converges to* x^* *as* $\alpha \to 0$.

   3. Let us present examples of operators satisfying the conditions (2.2.3) and (2.2.12).

Example 2.2.7 Suppose that in Example 6 of Section 1.3 the functions $a_i^h(x, s)$ are given instead of $a_i(x, s)$ with the same properties, and

$$|a_i(x, s) - a_i^h(x, s)| \leq h \quad \forall x \in G, \quad \forall s \geq 0.$$

Then the following calculations can be verified for any functions u, $v \in \overset{0}{W}{}_1^p(G)$:

$$
\begin{aligned}
\langle Au - A^h u, v \rangle &= \int_G \sum_{i=1}^n \left[a_i\left(x, \left|\frac{\partial u}{\partial x_i}\right|^{p-1}\right) - a_i^h\left(x, \left|\frac{\partial u}{\partial x_i}\right|^{p-1}\right) \right] \left|\frac{\partial u}{\partial x_i}\right|^{p-2} \frac{\partial u}{\partial x_i} \frac{\partial v}{\partial x_i} dx \\
&+ \int_G \left[a_0(x, |u|^{p-1}) - a_0^h(x, |u|^{p-1}) \right] |u|^{p-2} uv dx \\
&\leq h \int_G \left(\sum_{i=1}^n \left|\frac{\partial u}{\partial x_i}\right|^{p-1} \frac{\partial v}{\partial x_i} + |u|^{p-1} uv \right) dx \\
&\leq h \left(\sum_{i=1}^n \left\|\frac{\partial u}{\partial x_i}\right\|_{L^p}^{p-1} \left\|\frac{\partial v}{\partial x_i}\right\|_{L^p} + \|u\|_{L^p}^{p-1} \|v\|_{L^p} \right) \\
&\leq h \|u\|_{1,p}^{p-1} \|v\|_{1,p},
\end{aligned}
$$

where $\|u\|_{1,p}$ is the norm of $u \in \overset{0}{W}{}^{p}_{1}(G)$. By Corollary 1.1.2, from the Hahn$-$Banach theorem, there exists $v \in \overset{0}{W}{}^{p}_{1}(G)$ with $\|v\|_{1,p} = 1$ giving the estimate

$$\|Au - A^h u\|_{-1,q} \leq h\|u\|_{1,p}^{p-1}, \quad p^{-1} + q^{-1} = 1,$$

where $\|\cdot\|_{-1,q}$ is the norm in the space $\left(\overset{0}{W}{}^{p}_{1}(G)\right)^*$. Then we obtain that in (2.2.3) $g(t) = t^{p-1}$.

Example 2.2.8 Suppose that in Example 8 of Section 1.3, in place of the functions g_0 and g_1, their approximations g_0^h and g_1^h determining, respectively, monotone operators A_0^h and A_1^h are known. Moreover,

$$|g_0^h(x, \xi^2) - g_0(x, \xi^2)| \leq c|\xi|^{p-2}h, \quad c > 0,$$

and

$$g_1^h(x, \xi^2)\xi = \begin{cases} \omega^h, & \text{if} \quad \xi > \beta, \\ 0, & \text{if} \quad \xi \leq \beta, \end{cases}$$

where $\omega^h > 0$ and $|\omega - \omega^h| \leq h$. Repeating almost word for word all arguments given in Example 2.2.7, we come to the estimate

$$\mathcal{H}_{X^*}(\bar{A}^h u, \bar{A}u) \leq \left(c\|u\|_{1,p}^{p-1} + 1\right)h,$$

where $X^* = \left(\overset{0}{W}{}^{p}_{1}(G)\right)^*$, $\bar{A} = A_0 + \bar{A}_1$, $\bar{A}^h = A_0^h + \bar{A}_1^h$. With this, the norm in the space $\overset{0}{W}{}^{p}_{1}(G)$ is defined as

$$\|u\|_{1,p} = \left(\int_G |\nabla u|^p dx\right)^{1/p}.$$

Since the duality mapping

$$J^p u = -\sum_{i=1}^{n} \frac{\partial}{\partial x_i}\left(|\nabla u|^{p-2}\frac{\partial u}{\partial x_i}\right) = -div\left(|\nabla u|^{p-2}\nabla u\right),$$

the regularized problem has the following form:

$$-div\left(g^h(x, \nabla^2 u)\nabla u + \alpha|\nabla u|^{p-2}\nabla u\right) = f^\delta(x), \quad u\,|_{\partial\Omega} = 0, \qquad (2.2.13)$$

where

$$g^h(x, \xi^2)\xi = g_0^h(x, \xi^2)\xi + g_1^h(x, \xi^2)\xi$$

and $f^\delta(x)$ is a δ-approximation of $f(x)$, that is,

$$\|f(x) - f^\delta(x)\|_{-1,q} \leq \delta, \quad \delta > 0, \quad p^{-1} + q^{-1} = 1.$$

2.3 Estimates of the Regularized Solutions

Let the conditions of Section 2.2 hold. For simplicity of calculation, assume first that an operator A in (2.1.1) is given exactly. Consider the regularized equation

$$Ax + \alpha J^\mu x = f^\delta \tag{2.3.1}$$

with the duality mapping $J^\mu : X \to X^*$, where $\mu(t)$ is some gauge function. Let x_α^δ be a solution of (2.3.1) and x_α be a solution of the regularized equation

$$Ax + \alpha J^\mu x = f \tag{2.3.2}$$

with exact right-hand side f. We already know that the sequences $\{x_\alpha^\delta\}$ and $\{x_\alpha\}$ are bounded as $\alpha \to 0$ and $\dfrac{\delta}{\alpha} \to 0$. Let $\|x_\alpha\| \le d$ and $\|x_\alpha^\delta\| \le d$. Assume that the duality operator J^μ satisfies the following condition:

$$\langle J^\mu x - J^\mu y, x - y \rangle \ge C(R)\|x - y\|^s \quad \forall x, y \in X, \tag{2.3.3}$$

where $s \ge 2$ and $C(R)$ is a non-negative and non-increasing function of the variable

$$R = max\{\|x\|, \|y\|\}. \tag{2.3.4}$$

By (2.3.1) - (2.3.3) and by the monotonicity of the operator A, one can deduce the estimate

$$\|x_\alpha^\delta - x_\alpha\| \le \left(\frac{\delta}{\alpha C(d)}\right)^\kappa, \quad \kappa = \frac{1}{s-1}. \tag{2.3.5}$$

We are going now to appraise from above the norm $\|x_\alpha^\delta - \bar{x}^*\|$, where $\bar{x}^* \in N$. As follows from (2.3.5), it is enough for this to evaluate the norm $\|x_\alpha - \bar{x}^*\|$. Describe the conditions on the operator A and geometry of the spaces X and X^*, which allow us to solve this problem. Assume that
i) A is Fréchet differentiable and Fréchet derivative $A'(x)$ satisfies the Lipschitz−Hölder condition

$$\big|\, A'(x) - A'(y)\,\big| \le L(R)\|x - y\|^\sigma, \quad 0 < \sigma \le 1, \tag{2.3.6}$$

where $L(R)$ is a non-negative and non-decreasing function for all $R \ge 0$.
ii) There exists an element $v \in X$ such that

$$J^\mu \bar{x}^* = A'(\bar{x}^*)v, \tag{2.3.7}$$

where $\bar{x}^*$ is the minimal norm solution of the equation (2.1.1).
    Construct the linear operator $A_\alpha^\mu(x, y) : X \to X^*$ by the equality

$$A_\alpha^\mu(x, y) = A(x, y) + \alpha J^\mu(x, y), \tag{2.3.8}$$

where $x, y \in X$, $A(x, y)$ and $J^\mu(x, y)$ are linear symmetric operators from X to X^* defined as follows:

$$A(x, y)(x - y) = Ax - Ay$$

and

$$J^\mu(x,y)(x-y) = J^\mu x - J^\mu y.$$

In other words, $A(x,y)$ and $J^\mu(x,y)$ are the first order divided differences of the operators A and J^μ, respectively. We suppose that the inverse operator $[A_\alpha^\mu(x,y)]^{-1}$ exists. It is obvious that

$$\langle A_\alpha^\mu(x,y)(x-y), x-y\rangle \geq \alpha C(R)\|x-y\|^s.$$

Hence,

$$\alpha C(R)\|x-y\|^{s-2} \leq \big| A_\alpha^\mu(x,y) \big|$$

or

$$\big| [A_\alpha^\mu(x,y)]^{-1} \big| \leq \frac{1}{\alpha C(R)}\|x-y\|^{2-s}. \qquad (2.3.9)$$

Using (2.1.1), (2.3.2) and (2.3.8), it is not difficult to verify that

$$A_\alpha^\mu(x_\alpha, \bar{x}^*)(x_\alpha - \bar{x}^*) = -\alpha J^\mu \bar{x}^*.$$

Now the latter equality implies

$$\|x_\alpha - \bar{x}^*\| = \|\alpha[A_\alpha^\mu(x_\alpha, \bar{x}^*)]^{-1} A'(\bar{x}^*)v\|$$

in view of the condition (2.3.7). Let

$$\big| J^\mu(x,y) \big| \leq M(R)\|x-y\|^{-\gamma}, \quad \gamma > 0, \qquad (2.3.10)$$

where $M(R)$ is a non-negative and non-decreasing function for all $R > 0$. By (2.3.6), (2.3.9) and (2.3.10), we deduce

$$\|x_\alpha - \bar{x}^*\| = \alpha\|[A_\alpha^\mu(x_\alpha, \bar{x}^*)]^{-1}(A_\alpha^\mu(x_\alpha, \bar{x}^*) - A'(\bar{x}^*) - A_\alpha^\mu(x_\alpha, \bar{x}^*))v\|$$

$$\leq \alpha\|v\| + \frac{\|x_\alpha - \bar{x}^*\|^{2-s}}{C(r)}\Big(\big| A'(\bar{x}^* + t(x_\alpha - \bar{x}^*)) - A'\bar{x}^* \big|$$

$$+ \alpha\big| J^\mu(x_\alpha, \bar{x}^*) \big| \Big)\|v\|$$

$$\leq \alpha\|v\| + \frac{L(r)}{C(r)}\|v\|\|x_\alpha - \bar{x}^*\|^{2-s+\sigma} + \frac{\alpha\|v\|}{C(r)}M(r)\|x_\alpha - \bar{x}^*\|^{2-s-\gamma},$$

where $0 \leq t \leq 1$ is some number and $r = max\{d, \|\bar{x}^*\|\}$. Thus, we establish the estimate

$$\|x_\alpha - \bar{x}^*\| \leq \|v\|\left(\alpha + \frac{L(r)}{C(r)}\|x_\alpha - \bar{x}^*\|^{2-s+\sigma} + \frac{\alpha M(r)}{C(r)}\|x_\alpha - \bar{x}^*\|^{2-s-\gamma}\right). \qquad (2.3.11)$$

Note that if X is a Hilbert space, then

$$C(r) \equiv 1, \ s = 2, \ \mu(t) \equiv t, \ \sigma = 1, \ \gamma = 0, \ M(r) \equiv 1,$$

and if $1 - \|v\|L(r) > 0$, then (2.3.11) gives the following inequality:

$$\|x_\alpha - \bar{x}^*\| \leq \frac{2\|v\|\alpha}{1 - \|v\|L(r)}.$$

We emphasize that, in the general case, it is impossible to obtain effective estimates of $\|x_\alpha - \bar{x}^*\|$ if we use the relation (2.3.11). Therefore, we introduce some additional assumptions. Let $\sigma = s - 1$. It is clear that $\sigma \in (0,1]$ because $s \in (1,2]$. Suppose first that $s + \gamma > 2$ and that

$$a(r) = 1 - \frac{L(r)}{C(r)}\|v\| > 0. \tag{2.3.12}$$

Then (2.3.11) yields the inequality

$$\|x_\alpha - \bar{x}^*\|^{s+\gamma-1} - \frac{\alpha\|v\|}{a(r)}\|x_\alpha - \bar{x}^*\|^{s+\gamma-2} - \frac{\alpha\|v\|M(r)}{a(r)C(r)} \leq 0. \tag{2.3.13}$$

Introduce the function

$$\varphi(t) = t^\beta - a_1 t^{\beta-1} - a_2,$$

where $\beta > 1$, $t \geq 0$, $a_1 > 0$ and $a_2 > 0$. The following simple properties hold for this function: 1) $\varphi(0) = -a_2 < 0$ and 2) if $t = \bar{t} = a_2^{1/\beta} + a_1$ then

$$\varphi(\bar{t}) = \bar{t}^{\beta-1}(\bar{t} - a_1) - a_2 = \bar{t}^{\beta-1}a_2^{1/\beta} - a_2 = (a_2^{1/\beta} + a_1)^{\beta-1}a_2^{1/\beta} - a_2.$$

It is not difficult to see that if $a_1 = 0$ then $\varphi(\bar{t}) = 0$, and if $a_1 > 0$ then $\varphi(\bar{t}) > 0$. Besides, a derivative of the function $\varphi(t)$ vanishes only if $t = t_0 = a_1(\beta - 1)\beta^{-1}$. Moreover, the function $\varphi(t)$ increases for all $t > t_0$ and $\bar{t} > t_0$. Hence, by (2.3.13), we have

$$\|x_\alpha - \bar{x}^*\| \leq \left(\frac{\alpha\|v\|M(r)}{a(r)C(r)}\right)^\tau + \frac{\alpha\|v\|}{a(r)},$$

where

$$\tau = \frac{1}{s + \gamma - 1}.$$

Taking into account the estimate (2.3.5) we obtain

$$\|x_\alpha^\delta - \bar{x}^*\| \leq \left(\frac{\delta}{\alpha C(r)}\right)^\kappa + \left(\frac{\alpha\|v\|M(r)}{a(r)C(r)}\right)^\tau + \frac{\alpha\|v\|}{a(r)}. \tag{2.3.14}$$

Let now $s + \gamma < 2$. The inequality (2.3.13) can be rewritten as

$$\|x_\alpha - \bar{x}^*\| - \frac{\alpha\|v\|M(r)}{a(r)C(r)}\|x_\alpha - \bar{x}^*\|^{1-(s+\gamma-1)} - \frac{\alpha\|v\|}{a(r)} \leq 0.$$

Consider the function

$$\psi(t) = t - a_1 t^{1-\beta} - a_2$$

with $0 < \beta < 1$, $a_1 > 0$, $a_2 > 0$ and $\psi(0) = -a_2 < 0$. Denote $\bar{t}^\beta = a_1 + a_2^\beta$ and calculate

$$\psi(\bar{t}) = \bar{t}^{1-\beta}(\bar{t}^\beta - a_1) - a_2 = \bar{t}^{1-\beta}a_2^\beta - a_2 = (a_1 + a_2^\beta)^{(1-\beta)/\beta}a_2^\beta - a_2.$$

Obviously, $\psi(\bar{t}) = 0$ if $a_1 = 0$, and $\psi(t)$ increases with respect to parameter a_1. Moreover, $\psi(t)$ achieves a minimum when $t = t_0 = [a_1(1-\beta)]^{1/\beta}$. It is clear that $t_0 < \bar{t}$. Thus, $\psi(t) \leq 0$ if $t \leq (a_1 + a_2^\beta)^{1/\beta}$. Since $s + \gamma < 2$, (2.3.13) produces the estimate

$$\|x_\alpha - \bar{x}^*\| \leq \left[\frac{\alpha\|v\|M(r)}{a(r)C(r)} + \left(\frac{\alpha\|v\|}{a(r)} \right)^{\frac{1}{\tau}} \right]^\tau.$$

Therefore,

$$\|x_\alpha^\delta - \bar{x}^*\| \leq \left(\frac{\delta}{\alpha C(r)} \right)^\kappa + \left[\frac{\alpha\|v\|M(r)}{a(r)C(r)} + \left(\frac{\alpha\|v\|}{a(r)} \right)^{\frac{1}{\tau}} \right]^\tau. \qquad (2.3.15)$$

Finally, if $s + \gamma = 2$ then we conclude from (2.3.13) that

$$\|x_\alpha^\delta - \bar{x}^*\| \leq \left(\frac{\delta}{\alpha C(r)} \right)^\kappa + \frac{\alpha\|v\|}{a(r)} \left(1 + \frac{M(r)}{C(r)} \right). \qquad (2.3.16)$$

Thus, we have obtained the following results:

Theorem 2.3.1 *Assume that $A : X \to X^*$ is a maximal monotone and Fréchet differentiable operator, Fréchet derivative $A'(x)$ satisfies the Lipschitz–Hölder condition (2.3.6), duality mapping $J^\mu : X \to X^*$ with a gauge function $\mu(t)$ has the property (2.3.3), and there exists the inverse operator $[A_\alpha^\mu(x_\alpha, \bar{x}^*)]^{-1}$, where $A_\alpha^\mu(x, y)$ is defined by (2.3.8). Let $\bar{x}^*$ be the minimal norm solution of the equation (2.1.1), x_α^δ and x_α be solutions of the equations (2.3.1) and (2.3.2), respectively, and (2.3.7) and (2.3.10) hold. If $\sigma = s - 1$ and $a(r)$ is defined by (2.3.12) with $r = max\{d, \|x^*\|\}$ and if $\|x_\alpha^\delta\| \leq d$, $\|x_\alpha\| \leq d$, then the estimate (2.3.14) holds for all $s + \gamma > 2$. If $s + \gamma < 2$ or $s + \gamma = 2$ then, respectively, (2.3.15) or (2.3.16) are fulfilled.*

Remark 2.3.2 *The duality mapping J^μ satisfying the hypotheses of Theorem 2.3.1 exists in the spaces $L^p(G)$ $(1 < p \leq 2)$. Indeed, since $s = 2$, we have $\mu(t) = t$, $\kappa = 1$ and*

$$\langle Jx - Jy, x - y \rangle \leq M(R)\|x - y\|^p$$

(see (1.6.35)), that is, $\gamma = 2 - p$ and $\tau = \dfrac{1}{3 - p}$. Therefore, (2.3.14) in $L^p(G)$ is expressed as

$$\|x_\alpha^\delta - \bar{x}^*\| \leq O\left(\frac{\delta}{\alpha} \right) + O(\alpha^{\frac{1}{3-p}}) + O(\alpha).$$

Suppose that requirements of Theorem 2.3.1 are satisfied and, instead of the exact operator A, the sequence of monotone single-valued operators A^h are given such that $D(A^h) = D(A)$ and

$$\|Ax - A^h x\|_* \leq g(\|x\|)h \quad \forall x \in D(A),$$

where $g(t)$ is a non-negative and non-decreasing function for all $t \geq 0$. Let x_α^γ be a unique solution of the equation

$$A^h x + \alpha J^\mu x = f^\delta$$

with $\|x_\alpha^\gamma\| \leq d$. Then it is not difficult to verify that the estimates of $\|x_\alpha^\gamma - \bar{x}^*\|$ are obtained from (2.3.14) - (2.3.16) if the perturbation δ in their right-hand sides is replaced by $\delta + hg(r)$.

2.4 Equations with Domain Perturbations

As before, A and A^h denote, respectively, exact and perturbed maximal monotone (possibly, multiple-valued) operators, and the equations (2.1.1) are solved in an E-space X. Let X^* be a strictly convex space. It has been earlier everywhere assumed that $D(A) = D(A^h)$. Now we shall study $D(A)$ and $D(A^h)$ to be not coinciding sets and define the proximity between A and A^h in the following way: For any element $x \in D(A)$ and given $h > 0$, let there exist an element $x^h \in D(A^h)$ such that

$$\|x - x^h\| \leq a(\|x\|)h \qquad (2.4.1)$$

and

$$d_*(y, A^h x^h) \leq g(\|y\|_*)\xi(h) \quad \forall y \in Ax, \qquad (2.4.2)$$

where $d_*(y, A^h x^h)$ is distance in the space X^* between $y \in Ax$ and the convex closed set $A^h x^h$. We further assume that functions $a(t)$ and $g(t)$ are non-negative, $\xi(h) \to 0$ as $h \to 0$ and $\xi(0) = 0$. If x_α^γ is a solution of the regularized equation

$$A^h x + \alpha J^\mu x = f^\delta, \quad \alpha > 0, \qquad (2.4.3)$$

and if $y_\alpha^\gamma \in A^h x_\alpha^\gamma$, then

$$y_\alpha^\gamma + \alpha J^\mu x_\alpha^\gamma = f^\delta, \qquad (2.4.4)$$

where $f \in X^*$ satisfies (2.2.2). Let $N \neq \emptyset$, where N is a solution set of the equations (2.1.1), and $x^* \in N$. Thus, $f \in Ax^*$. By the conditions (2.4.1) and (2.4.2), one can find elements $x^h \in D(A^h)$ and $y^h \in A^h x^h$ such that for every $h > 0$,

$$\|x^* - x^h\| \leq a(\|x^*\|)h \qquad (2.4.5)$$

and

$$\|f - y^h\|_* \leq g(\|f\|_*)\xi(h). \qquad (2.4.6)$$

We subtract the element f from both parts of equation (2.4.4) and calculate their dual products with the difference $x_\alpha^\gamma - x^h$. We have

$$\langle y_\alpha^\gamma - f, x_\alpha^\gamma - x^h \rangle + \alpha \langle J^\mu x_\alpha^\gamma, x_\alpha^\gamma - x^h \rangle = \langle f^\delta - f, x_\alpha^\gamma - x^h \rangle. \qquad (2.4.7)$$

Rewrite (2.4.7) in the equivalent form:

$$\langle y_\alpha^\gamma - y^h, x_\alpha^\gamma - x^h \rangle \quad + \quad \langle y^h - f, x_\alpha^\gamma - x^h \rangle + \alpha \langle J^\mu x_\alpha^\gamma, x_\alpha^\gamma - x^h \rangle$$
$$= \quad \langle f^\delta - f, x_\alpha^\gamma - x^h \rangle.$$

By virtue of the monotonicity of operators A^h, definition of the duality mapping J^μ and by (2.4.5) and (2.4.6), one gets

$$\mu(\|x_\alpha^\gamma\|)\|x_\alpha^\gamma\| \quad - \quad \mu(\|x_\alpha^\gamma\|)\left[\|x^*\| + a(\|x^*\|)h\right]$$

$$- \quad \left(\frac{\delta}{\alpha} + \frac{\xi(h)}{\alpha}g(\|f\|_*)\right)(\|x_\alpha^\gamma\| + \|x^*\| + a(\|x^*\|)h) \leq 0. \qquad (2.4.8)$$

If

$$\lim_{\alpha \to 0} \frac{\delta + \xi(h)}{\alpha} = 0 \qquad (2.4.9)$$

then it results from (2.4.8) that the sequence $\{x_\alpha^\gamma\}$ is bounded as $\alpha \to 0$ and $x_\alpha^\gamma \rightharpoonup \bar{x} \in X$. The equality (2.4.4) allows us to assert that $y_\alpha^\gamma \to f$ as $\alpha \to 0$. Next, by (2.4.1) and (2.4.2), for every $z \in D(A)$ and every $y \in Az$, we find $z^h \in D(A^h)$ and $y^h \in A^h z^h$ such that

$$\|z - z^h\| \leq a(\|z\|)h$$

and

$$\|y - y^h\|_* \leq g(\|y\|_*)\xi(h).$$

Thus, $z^h \to z$ and $y^h \to y$ as $h \to 0$. The monotonicity property of A^h gives

$$\langle y_\alpha^\gamma - y^h, x_\alpha^\gamma - z^h \rangle \geq 0.$$

If $\alpha \to 0$ then by (2.4.9),

$$\langle f - y, \bar{x} - z \rangle \geq 0 \quad \forall z \in D(A), \quad \forall y \in Az.$$

Maximal monotonicity of the operator A ensures now the inclusion $\bar{x} \in N$.

Show that $\bar{x}$ is the minimal norm solution of (2.1.1), that is, $\|\bar{x}\| \leq \|x^*\|$ for any $x^* \in N$. Let $\bar{x} \neq \theta_X$. Then, due to the weak convergence of the sequence $\{x_\alpha^\gamma\}$ to $\bar{x}$, we obtain

$$\|\bar{x}\| \leq \liminf_{\alpha \to 0} \|x_\alpha^\gamma\|. \qquad (2.4.10)$$

Thus,

$$\mu(\|\bar{x}\|) \leq \mu(\liminf_{\alpha \to 0} \|x_\alpha^\gamma\|) = \liminf_{\alpha \to 0} \mu(\|x_\alpha^\gamma\|)$$

because $\mu(t)$ is increasing. Hence, $\mu(\|x_\alpha^\gamma\|) \geq c > 0$ for sufficiently small $\alpha > 0$. Next, (2.4.7) yields

$$\langle J^\mu x_\alpha^\gamma, x_\alpha^\gamma \rangle = \mu(\|x_\alpha^\gamma\|)\|x_\alpha^\gamma\| \leq \Big(\frac{\delta}{\alpha} + \frac{\xi(h)}{\alpha}g(\|f\|_*)\Big)\|x_\alpha^\gamma - x^h\| + \mu(\|x_\alpha^\gamma\|)\|x^h\|. \qquad (2.4.11)$$

Since, in view of (2.4.5), $x^h \to x^*$, we deduce from (2.4.11) the inequality

$$\limsup_{\alpha \to 0} \|x_\alpha^\gamma\| \leq \|x^*\|. \qquad (2.4.12)$$

The estimates (2.4.10) and (2.4.12) imply the fact that $\|\bar{x}\| = \min\{\|x^*\| \mid x^* \in N\}$, that is, $\bar{x} = \bar{x}^*$. The estimates (2.4.10) and (2.4.12) allow us to establish strong convergence of $\|x_\alpha^\gamma\|$ to $\|\bar{x}^*\|$.

Let $\bar{x} = \theta_X$. It is obvious in this case that $\bar{x} = \bar{x}^*$. Suppose that in (2.4.11) $x^h \to \theta_X$. Then the function $\mu(t)$ satisfies the limit relation

$$\lim_{\alpha \to 0} \mu(\|x_\alpha^\gamma\|)\|x_\alpha^\gamma\| = 0,$$

and we conclude that $\|x_\alpha^\gamma\| \to 0$ as $\alpha \to 0$. Thus, we come to the following result:

Theorem 2.4.1 *Let* $A : X \to 2^{X^*}$ *and* $A^h : X \to 2^{X^*}$ *be maximal monotone operators, X be an E-space, X^* be a strictly convex space. Let the hypotheses (2.2.2), (2.4.1), (2.4.2) and (2.4.9) hold. Then the sequence of solutions of the regularized equation (2.4.3) strongly converges in X to the minimal norm solution of the equation (2.1.1).*

Remark 2.4.2 *If $D(A) = D(A^h)$, then it is possible to assume $x = x^h$ in (2.4.1) and (2.4.2). In this case, (2.4.1) holds even if $a(t) \equiv 0$, and (2.4.2) can be given in the form*

$$d_*(y, A^h x) \leq g(\|y\|_*)\xi(h) \quad \forall y \in Ax.$$

The latter is weaker than (2.2.1).

Present the example realizing (2.4.1) and (2.4.2).

Example 2.4.3 Let a map $A : R^2 \to R^2$ be defined by the matrix

$$A = \begin{pmatrix} 1 & 2 \\ 2 & 4 \end{pmatrix}.$$

It is not difficult to verify that A is positive. Consider it on the set

$$\Omega = \{(x_1, x_2) \mid x_2 \leq k_1 x_1, \ x_2 \geq k_2 x_1\}, \ k_1 > k_2,$$

lying in the first quarter. Finish defining the operator A on a boundary $\partial\Omega$ by semi-lines (see the proof of Theorem 1.7.19) and thus obtain the maximal monotone operator $\bar{A}$ with $D(\bar{A}) = \Omega$. By analogy, we construct a perturbed maximal monotone operator $\bar{A}^h : R^2 \to 2^{R^2}$ by means of the closed matrix

$$A^h = \begin{pmatrix} (2+h)^2(4+h)^{-1} & 2+h \\ 2+h & 4+h \end{pmatrix},$$

where h is a small enough number and domain

$$D(\bar{A}^h) = \Omega^h = \{(x_1, x_2) \mid x_2 \leq k_1^h x_1, \ x_2 \geq k_2^h x_1\}, \ |k_i^h - k_i| \leq h, \ k_i^h > k_i, \ i = 1, 2.$$

We assume that Ω^h also lies in the first quarter. Show that the maps $\bar{A}$ and $\bar{A}^h$ satisfy the conditions (2.4.1) and (2.4.2). Indeed, if $x \in \Omega \cap \Omega^h \subset int \, \Omega$ then

$$x^h = x, \ d(y, \bar{A}^h x) = \|y - A^h x\| \leq h\|y\|,$$

where $y = Ax = \bar{A}x$. If still $x \in \Omega$ and at the same time $x \notin \Omega^h$ and it lies between the straight lines $x_2 = k_2 x_1, \ x_2 = k_2^h x_1$, then one can take $x^h = \bar{x}^h = \{x_1^h, \ k_2^h x_1^h\}$, where $\|x^h\| = \|x\|$. It is clear that there exists $c > 0$ such that $\|x - x^h\| \leq ch\|x\|$. Therefore, by the inequality

$$\|Ax - A^h x^h\| \leq \|Ax - A^h x\| + \|A^h x - A^h x^h\|,$$

we have that $d(y, \bar{A}^h x^h) \leq ch\|Ax\|$ with $y = Ax = \bar{A}x$. If x lies on the straight line $x_2 = k_2 x_1$ then, as before, we take $x^h = \bar{x}^h$. In this case, $\|x - x^h\| \leq ch\|x\|$,

$$\bar{A}x = \{Ax + \lambda\{k_2, \ -1\} \mid \lambda \geq 0\},$$

$$\bar{A}^h x^h = \{A^h x^h + \lambda\{k_2^h, -1\} \mid \lambda \geq 0\}.$$

Consequently, for $y = Ax + \lambda\{k_2, -1\}$ and $y^h = A^h x^h + \lambda\{k_2^h, -1\}$ with a fixed $\lambda \geq 0$, we have $\|y - y^h\| \leq c_1 h\|y\|$ with some $c_1 > 0$. Finally, if $x \in \Omega$ and it lies on the line $x_2 = k_1 x_1$ then we choose a point x^h on the line $x_2 = k_1^h x_1$ such that $\|x^h\| = \|x\|$. Thus, the conditions (2.4.1), (2.4.2) are satisfied with $a(t) = c_2 t$, $g(t) = c_3 t$ with some $c_2, c_3 > 0$ and $\xi(h) = h$.

Remark 2.4.4 *This example shows that in (2.4.2) it is impossible to replace $g(\|y\|)$ by $g(\|x\|)$ because in the general case (in particular, for unbounded A) $d(y, A^h x^h)$ can be arbitrarily large for a fixed x (see the case where points x and x^h lie on different straight lines defining the boundaries of sets Ω and Ω^h).*

2.5   Equations with Semimonotone Operators

Suppose that a Banach space X possesses the M-property, an operator $A : X \to 2^{X^*}$ in equation (2.1.1) is semimonotone, $D(A) = X$, and there exist $r > 0$ and $y \in Ax$ such that

$$\langle y - f, x \rangle \geq 0 \quad \text{as} \quad \|x\| \geq r. \tag{2.5.1}$$

Then, by virtue of Theorem 1.10.6, the equation (2.1.1) has in X a nonempty set N of s-generalized solutions (a solution $x_0 \in N$ is understood in the sense of Definition 1.10.3 as $f \in \bar{A} x_0$).

1. Consider the equation

$$Ax + \alpha J x = f^\delta \tag{2.5.2}$$

with $\alpha > 0$ and δ satisfying (2.2.2). Let $\delta = O(\alpha)$ as $\alpha \to 0$. Show that there exists a solution of equation (2.5.2). Indeed, for all $x \in X$, it is not difficult to verify that

$$\langle y + \alpha J x - f^\delta, x \rangle = \langle y - f, x \rangle + \alpha\|x\|^2 + \langle f - f^\delta, x \rangle$$

$$\geq \alpha\|x\|\Big(\|x\| - \frac{\delta}{\alpha}\Big) + \langle y - f, x \rangle, \quad y \in Ax. \tag{2.5.3}$$

Hence, if $\dfrac{\delta}{\alpha} \leq K$ and if $\|x\| \geq r_1 = max\{r, K\}$, then the inequality

$$\langle y + \alpha J x - f^\delta, x \rangle \geq 0$$

is satisfied. Since the duality mapping J is monotone, we are able to apply Theorem 1.10.6 and conclude that there exists a solution of equation (2.5.2) (which is not unique in general). Furthermore, it follows from (2.5.3) that the operator $A + \alpha J$ is coercive, therefore, it has a bounded inverse operator. Hence, (2.5.2) may be considered as a regularization of the equation (2.1.1).

Let $T = A + C$ be a monotone operator, where $C : X \to X^*$ is a strongly continuous mapping, and let $\bar{T}$ be a maximal monotone extension of T. Then the operator $\bar{F} = \bar{T} + \alpha J$ is also maximal monotone, that is, $\bar{F}$ is a maximal monotone extension of the operator

$F = T + \alpha J$. Denote by x_α^δ a solution of (2.5.2). If any $y_\alpha^\delta \in \bar{A}x_\alpha^\delta$, where $\bar{A} = \bar{T} - C$, then we have

$$y_\alpha^\delta + \alpha J x_\alpha^\delta = f^\delta. \tag{2.5.4}$$

Moreover, $\|x_\alpha^\delta\| \le r_1$ in view of Theorem 1.10.6 again. Hence, $x_\alpha^\delta \rightharpoonup \bar{x} \in X$ as $\alpha \to 0$. Using now (2.5.4) we can write

$$\langle z + \alpha J x - C x_\alpha^\delta - f^\delta, x - x_\alpha^\delta \rangle \ge 0 \quad \forall z \in \bar{T}x.$$

Setting $\alpha \to 0$ and taking into account the strong continuity of C we obtain in a limit the inequality

$$\langle z - C\bar{x} - f, x - \bar{x} \rangle \ge 0 \quad \forall z \in \bar{T}x. \tag{2.5.5}$$

Consequently, $f \in \bar{A}\bar{x}$, i.e., $\bar{x} \in N$. Thus, we have proved the following assertion:

Theorem 2.5.1 *Let X be a reflexive strictly convex Banach space together with its dual space X^* and have the M-property. Let $A : X \to 2^{X^*}$ in (2.1.1) be a semimonotone operator with domain $D(A) = X$ and the conditions (2.5.1) and (2.2.2) hold. If there is a constant $C > 0$ such that $\dfrac{\delta}{\alpha} \le C$ as $\alpha \to 0$, then there exists a subsequence of the sequence $\{x_\alpha^\delta\}$, where x_α^δ are solutions of the equation (2.5.2) with fixed α and δ, which weakly converges to some point $\bar{x} \in N$. If a solution $\bar{x}$ is unique, then the whole sequence $\{x_\alpha^\delta\}$ converges weakly to $\bar{x}$.*

2.   Suppose that in the equation (2.5.2) not only the right-hand side f but also the operator A is given approximately, that is, instead of A, we have the operators A^h which are also semimonotone for all $h > 0$ and $D(A^h) = X$. Since value sets of the operators $\bar{A} = \bar{T} - C$ and $\bar{A}^h = \bar{T}^h - C^h$ are convex and closed, the proximity between A and A^h can be defined by the Hausdorff distance as

$$\mathcal{H}_{X^*}(\bar{A}x, \bar{A}^h x) \le g(\|x\|)h \quad \forall x \in X, \tag{2.5.6}$$

where $g(t)$ is a continuous non-negative function for all $t \ge 0$. In this case, we use the duality mapping J^μ with a gauge function $\mu(t)$ satisfying the condition $\mu(t) > K(g(t) + 1)$ for $t \ge t_0 > 0$, where $\dfrac{\delta}{\alpha} \le K$, $\dfrac{h}{\alpha} \le K$. Then the equation

$$A^h x + \alpha J^\mu x = f^\delta \tag{2.5.7}$$

is considered in place of (2.5.2). Obviously, there exist $y \in Ax$ and $y^h \in \bar{A}^h x$ such that

$$\langle y^h + \alpha J^\mu x - f^\delta, x \rangle \ge \langle y - f, x \rangle + \alpha \mu(\|x\|)\|x\| - \delta\|x\| - hg(\|x\|)\|x\|$$

$$= \alpha\|x\|\Big(\mu(\|x\|) - \frac{h}{\alpha}g(\|x\|) - \frac{\delta}{\alpha}\Big) + \langle y - f, x \rangle.$$

Choosing the function $\mu(t)$ as was described above, we conclude that there is $r' > 0$ such that the inequality

$$\langle y^h + \alpha J^\mu x - f^\delta, x \rangle \ge 0$$

holds for all $x \in X$ with $\|x\| \geq r'$ and for some $y^h \in \bar{A}^h x$. Theorem 1.10.6 guarantees the existence of a solution x_α^γ of the equation (2.5.7). Let $y_\alpha^\gamma \in \bar{A}^h x_\alpha^\gamma$ such that

$$y_\alpha^\gamma + \alpha J^\mu x_\alpha^\gamma = f^\delta,$$

and let x be an arbitrary element of X. Then according to the condition (2.5.6), for any $y \in \bar{A}x$, there exists $y^h \in \bar{A}^h x$ such that

$$\|y - y^h\|_* \leq g(\|x\|)h \quad \forall x \in X. \tag{2.5.8}$$

The monotonicity of $\bar{T}^h + \alpha J^\mu$ implies the relation

$$\langle y^h + Cx + \alpha J^\mu x - Cx_\alpha^\gamma - f^\delta, x - x_\alpha^\gamma \rangle \geq 0.$$

Applying the estimate (2.5.8), we come again to (2.5.5) as $\alpha \to 0$. Hence, the conclusion of Theorem 2.5.1 holds for a solution sequence $\{x_\alpha^\gamma\}$ of the equation (2.5.7).

2.6   Equations with Non-Monotone Perturbations

The perturbed operators A^h are often approximations of an original operator A as a result of applications of numerical methods for solving the equations (2.1.1) in Hilbert and in Banach spaces. From this point of view approximations A^h should be done so that they retain basic properties of the operator A, because their violation influences qualitative characteristics of approximations x_α^γ to the exact solutions of (2.1.1). Sometimes to do this does not work, in particular, this concerns the monotonicity property of A^h. Therefore, it is necessary to investigate the case when the approximations A^h of the monotone operator A are non-monotone themselves. Observe that in this case regularized equations, generally speaking, may not have solutions.

1. We shall study convergence of the regularization method (2.2.4) for finding solutions of the operator equation (2.1.1) with a monotone operator $A : X \to 2^{X^*}$. The perturbed equation (2.1.4) is given under the condition that (2.2.2) holds and $\{A^h\}$ is the sequence of semimonotone operators and $D(A) = D(A^h) = X$. Let a Banach space X have the M-property, the Hausdorff distance

$$\mathcal{H}_{X^*}(\bar{A}x, \bar{A}^h x) \leq h \quad \forall x \in X, \tag{2.6.1}$$

and there exist $y \in \bar{A}x$ and $r > 0$ such that for $\|x\| \geq r$,

$$\langle y - f, x \rangle \geq 0 \quad \text{as} \quad \|x\| \geq r. \tag{2.6.2}$$

For some $y^h \in \bar{A}^h x$, we can write the following relations:

$$\langle y^h + \alpha J x - f^\delta, x \rangle \geq \langle y - f, x \rangle + \alpha\|x\|^2 - h\|x\| - \delta\|x\|$$

$$= \alpha\|x\|\left(\|x\| - \frac{\delta + h}{\alpha}\right) + \langle y - f, x \rangle.$$

Suppose that $\dfrac{\delta}{\alpha} \leq K$ and $\dfrac{h}{\alpha} \leq K$ as $\alpha \to 0$. If $\|x\| \geq r_1 = max\{r, 2K\}$ then the inequality

$$\langle y^h + \alpha Jx - f^\delta, x \rangle \geq 0$$

holds. By Theorem 1.10.6, a solution x_α^γ of the equation (2.2.4) satisfying the condition $\|x_\alpha^\gamma\| \leq r_1$ exists though it may be non-unique. Furthermore, one can find $y_\alpha^\gamma \in \bar{A}^h x_\alpha^\gamma$ such that

$$y_\alpha^\gamma + \alpha J x_\alpha^\gamma = f^\delta.$$

Therefore, $y_\alpha^\gamma \to f$ as $\alpha \to 0$. Since the sets $\bar{A}x$ and $\bar{A}^h x$ are convex and closed for every $x \in X$, it follows from (2.6.1) that, for every y_α^γ, there is an element $\tilde{y}_\alpha^\gamma \in \bar{A}x_\alpha^\gamma$ such that

$$\|y_\alpha^\gamma - \tilde{y}_\alpha^\gamma\|_* \leq h.$$

Therefore, $\tilde{y}_\alpha^\gamma \to f$ as $\alpha \to 0$. By Lemma 1.4.5, the graph of a maximal monotone operator is demiclosed. Then $x_\alpha^\gamma \rightharpoonup x^* \in N$. Taking into account the monotonicity of $\bar{A}$ and J we obtain from (2.2.5)

$$\langle Jx^*, x^* - x_\alpha^\gamma \rangle + \frac{\delta + h}{\alpha} \|x_\alpha^\gamma - x^*\| \geq 0 \quad \forall x^* \in N.$$

Repeating in fact the proof of Theorem 2.2.1 we come to the following result:

Theorem 2.6.1 *Suppose that X is an E-space and has the M-property, X^* is strictly convex, $A : X \to 2^{X^*}$ is a monotone operator and $A^h : X \to 2^{X^*}$ are semimonotone operators for all $h > 0$, $D(A) = D(A^h) = X$. Let the conditions (2.2.2), (2.6.1), (2.6.2) and (2.1.7) hold. Then the equation (2.2.4) is solvable and its solutions $x_\alpha^\gamma \to \bar{x}^*$, where $\bar{x}^*$ is the element of N with the minimal norm.*

Consider now convergence of the operator regularization method for semimonotone approximations A^h with the weaker condition of the proximity between A and A^h. Namely, let

$$\mathcal{H}_{X^*}(\bar{A}x, \bar{A}^h x) \leq g(\|x\|)h \quad \forall x \in X, \tag{2.6.3}$$

where $g(t)$ is a continuous nonnegative and increasing function for all $t \geq 0$. Instead of (2.2.4), we shall investigate the regularized equation (2.5.7) with duality mapping $J^\mu : X \to X^*$ whose gauge function $\mu(t)$ is defined by the inequality

$$\mu(t) > K(1 + g(t)) \quad \forall t \geq t_0. \tag{2.6.4}$$

Here t_0 is a fixed positive number and

$$max \left\{ \frac{\delta}{\alpha}, \frac{h}{\alpha} \right\} \leq K.$$

Theorem 2.6.2 *Assume that X is an E-space and has the M-property, X^* is strictly convex, $A : X \to 2^{X^*}$ is a monotone operator and $A^h : X \to 2^{X^*}$ are semimonotone operators for all $h > 0$, $D(A) = D(A^h) = X$. Let the conditions (2.1.7), (2.6.2) and (2.6.3) be fulfilled. Then the equation (2.5.7) is solvable and the sequence $\{x_\alpha^\gamma\}$ of its solutions converges strongly in X to the minimal norm solution of (2.1.1).*

Proof. The operator $J^\mu : X \to X^*$ is monotone, therefore, $A^h + \alpha J^\mu$ is semimonotone. Then the inequality

$$\langle y^h + \alpha J^\mu x - f^\delta, x \rangle \geq \alpha \|x\| \Big(\mu(\|x\|) - \frac{\delta}{\alpha} - \frac{h}{\alpha} g(\|x\|) \Big) + \langle y - f, x \rangle, \quad y^h \in \bar{A}^h x,$$

implies the estimate

$$\langle y^h + \alpha J^\mu x - f^\delta, x \rangle \geq 0$$

for all x with $\|x\| \geq r_1 = max\{r, t_0\}$. Hence, there exists a solution x_α^γ of the equation (2.5.7) satisfying the condition $\|x_\alpha^\gamma\| \leq r_1$. Thus, $x_\alpha^\gamma \rightharpoonup \bar{x} \in X$ as $\alpha \to 0$. Let $y_\alpha^\gamma \in \bar{A}^h x_\alpha^\gamma$ and

$$y_\alpha^\gamma + \alpha J^\mu x_\alpha^\gamma = f^\delta. \tag{2.6.5}$$

As in the previous theorem, we conclude that $y_\alpha^\gamma \to f$ and $\bar{x} \in N$. Using (2.1.1) and (2.6.5), we further deduce similarly to (2.2.5) the following equality:

$$\langle y_\alpha^\gamma - f, x_\alpha^\gamma - x^* \rangle + \alpha \langle J^\mu x_\alpha^\gamma - J^\mu x^*, x_\alpha^\gamma - x^* \rangle$$
$$= \langle f^\delta - f, x_\alpha^\gamma - x^* \rangle + \alpha \langle J^\mu x^*, x^* - x_\alpha^\gamma \rangle \quad \forall x^* \in N. \tag{2.6.6}$$

By the monotonicity of operators $\bar{A}$ and J^μ and by the condition (2.6.3), one has

$$\langle J^\mu x^*, x^* - x_\alpha^\gamma \rangle + \Big(\frac{\delta}{\alpha} + \frac{h}{\alpha} g(\|x_\alpha^\gamma\|) \Big) \|x_\alpha^\gamma - x^*\| \geq 0 \quad \forall x^* \in N.$$

Since $\{x_\alpha^\gamma\}$ is bounded and (2.1.7) holds, the latter inequality gives in a limit

$$\langle J^\mu x^*, x^* - \bar{x} \rangle \geq 0 \quad \forall x^* \in N.$$

As we have shown in Theorem 2.2.1, this yields the relation

$$\langle J^\mu \bar{x}, x^* - \bar{x} \rangle \geq 0 \quad \forall x^* \in N$$

because the operator J^μ is a demicontinuous. Lemma 1.5.7 asserts that J^μ is a potential operator and $J^\mu x = \Phi'(\|x\|)$, where $\Phi(\|x\|)$ is defined by (1.5.4). Therefore, it results from the previous inequality that $\bar{x} = \bar{x}^*$. Then Lemma 1.5.4 and (2.6.6) imply

$$\Big(\mu(\|x_\alpha^\gamma\|) - \mu(\|\bar{x}^*\|) \Big) (\|x_\alpha^\gamma\| - \|\bar{x}^*\|)$$

$$\leq \Big(\frac{\delta}{\alpha} + \frac{h}{\alpha} g(\|x_\alpha^\gamma\|) \Big) \|x_\alpha^\gamma - \bar{x}^*\| + \langle J^\mu \bar{x}^*, \bar{x}^* - x_\alpha^\gamma \rangle.$$

Since $x_\alpha^\gamma \rightharpoonup \bar{x}^*$ and (2.1.7) is satisfied, the convergence of $\|x_\alpha^\gamma\|$ to $\|\bar{x}^*\|$ follows. Then the proof of the theorem is accomplished because X is E-space. ∎

2. Next we present one more method for solving (2.1.1) with non-monotone mappings A^h. Let $A : X \to X^*$ be a monotone hemicontinuous operator with domain $D(A) = X$, $A^h : X \to X^*$ be hemicontinuous operators with $D(A^h) = X$. Suppose that

$$\|Ax - A^h x\|_* \leq hg(\|x\|), \tag{2.6.7}$$

where $g(t)$ is a non-negative continuous function for all $t \geq 0$.

We define the regularized solution x_α^γ as a solution of the so-called variational inequality with small offset:

$$\langle A^h z + \alpha J^\mu z - f^\delta, z - x_\alpha^\gamma \rangle \geq -\epsilon g(\|z\|)\|z - x_\alpha^\gamma\| \quad \forall z \in X, \quad \alpha > 0, \tag{2.6.8}$$

where $\epsilon \geq h$, $\mu(t) \geq g(t)$ as $t \geq t_0$, t_0 is a fixed positive number, $\|f^\delta - f\|_* \leq \delta$.

Lemma 2.6.3 *The inequality (2.6.8) has a nonempty solution set N for each $\alpha > 0$ and each $f^\delta \in X^*$.*

Proof. By Corollary 1.8.9, we deduce that the equation

$$Ax + \alpha J^\mu x = f^\delta \tag{2.6.9}$$

has a unique solution x_α^δ. Then (2.6.9) is equivalent to the variational inequality

$$\langle Az + \alpha J^\mu z - f^\delta, z - x_\alpha^\delta \rangle \geq 0 \quad \forall z \in X. \tag{2.6.10}$$

Making use of (2.6.7), from (2.6.10) one gets

$$\langle A^h z + \alpha J^\mu z - f^\delta, z - x_\alpha^\delta \rangle \geq -hg(\|z\|)\|z - x_\alpha^\delta\|.$$

Since $\epsilon \geq h$, we conclude that x_α^δ is all the more the solution of (2.6.8). ∎

Lemma 2.6.4 *If x_α^γ is a solution of (2.6.8) then it is also a solution of the inequality*

$$\langle A^h x_\alpha^\gamma + \alpha J^\mu x_\alpha^\gamma - f^\delta, z - x_\alpha^\gamma \rangle \geq -\epsilon g(\|x_\alpha^\gamma\|)\|z - x_\alpha^\gamma\| \quad \forall z \in X. \tag{2.6.11}$$

Proof. Replace z in (2.6.8) by $z_t = tx_\alpha^\gamma + (1-t)z$ with $t \in [0,1)$, and divide the obtained inequality by $1 - t > 0$. Then

$$\langle A^h z_t + \alpha J^\mu z_t - f^\delta, z - x_\alpha^\gamma \rangle \geq -\epsilon g(\|z_t\|)\|z - x_\alpha^\delta\|.$$

Setting $t \to 1$ and taking into account the properties of operators A^h, J^μ and function $g(t)$, we obtain in a limit (2.6.11). ∎

Theorem 2.6.5 *If*

$$\lim_{\alpha \to 0} \frac{\delta + \epsilon}{\alpha} = 0,$$

then $x_\alpha^\gamma \to \bar{x}^*$, where $\bar{x}^* \in N$ is the minimal norm solution of (2.1.1).

Proof. It follows from (2.1.1) that

$$\langle Ax^* - f, z - x^* \rangle = 0 \quad \forall z \in X, \forall x^* \in N. \tag{2.6.12}$$

By virtue of Lemma 2.6.4, one can put $z = x^*$ in (2.6.11) and $z = x_\alpha^\gamma$ in (2.6.12). After this, summation of the obtained inequalities gives

$$\langle A^h x_\alpha^\gamma - Ax^*, x^* - x_\alpha^\gamma \rangle \quad + \quad \alpha \langle J^\mu x_\alpha^\gamma, x^* - x_\alpha^\gamma \rangle + \langle f - f^\delta, x^* - x_\alpha^\gamma \rangle$$

$$\geq \quad -\epsilon g(\|x_\alpha^\gamma\|)\|x_\alpha^\gamma - x^*\|. \tag{2.6.13}$$

Now the monotonicity of A, the definition of J^μ and condition (2.6.7) imply from (2.6.13) the inequalities

$$\mu(\|x_\alpha^\gamma\|)\|x_\alpha^\gamma\| \quad \leq \quad \frac{\epsilon + h}{\alpha} g(\|x_\alpha^\gamma\|)\|x_\alpha^\gamma - x^*\| + \frac{\delta}{\alpha}\|x_\alpha^\gamma - x^*\| + \mu(\|x_\alpha^\gamma\|)\|x^*\|$$

$$\leq \quad \frac{2\epsilon}{\alpha} g(\|x_\alpha^\gamma\|)\|x_\alpha^\gamma - x^*\| + \frac{\delta}{\alpha}\|x_\alpha^\gamma - x^*\| + \mu(\|x_\alpha^\gamma\|)\|x^*\|. \tag{2.6.14}$$

Since $\alpha \to 0$, $(\delta + \epsilon)/\alpha \to 0$ and since $\mu(t) > g(t)$ beginning with a certain $t = t_0$, it results from (2.6.14) that the sequence $\{x_\alpha^\gamma\}$ is bounded for sufficiently small regularization parameter $\alpha > 0$. Hence, $x_\alpha^\gamma \rightharpoonup \bar{x} \in X$. Finally, we obtain from (2.6.8) as $\alpha \to 0$,

$$\langle Az - f, z - \bar{x} \rangle \geq 0 \quad \forall z \in X,$$

that is, in view of Theorem 1.9.1, $\bar{x} \in N$. Using (2.6.13), we complete the proof as in Theorem 2.2.1. ∎

Remark 2.6.6 *The convergence analysis of the regularization methods (2.2.4) and (2.3.2) can also be done when perturbed maps A^h are pseudomonotone or quasipotential.*

2.7   Equations with Accretive and d-Accretive Operators

In what follows, a Banach space X possesses an approximation, $A : X \to X$ is an accretive operator with domain $D(A)$, the normalized duality mapping $J : X \to X^*$ is continuous and at the same time weak-to-weak continuous in X. Consider the equation

$$Ax = f \tag{2.7.1}$$

with $f \in X$. Let N be its nonempty solution set. Suppose that the operator and right-hand side of (2.7.1) are given approximately. This means that in reality the equation $A^h x = f^\delta$ is solved, where $A^h : X \to X$ are accretive perturbations of A for all $h > 0$ with $D(A^h) = D(A)$, and $f^\delta \in X$ are perturbations of f for all $\delta > 0$. We assume that

$$\|f^\delta - f\| \leq \delta, \tag{2.7.2}$$

and

$$\|Ax - A^h x\| \leq g(\|x\|)h \quad \forall x \in X, \tag{2.7.3}$$

where $g(t)$ is a continuous non-negative function for all $t \geq 0$. The regularized equation with accretive operator A^h is written in the form:

$$A^h x + \alpha x = f^\delta. \tag{2.7.4}$$

1. First assume that operators A and A^h are hemicontinuous and $D(A) = D(A^h) = X$. Using the accretiveness of A^h we deduce

$$\langle Jx, A^h x + \alpha x \rangle \geq \alpha \|x\|^2 - \|A^h(\theta_X)\| \|x\| = \|x\|(\|x\| - \|A^h(\theta_X)\|). \tag{2.7.5}$$

Therefore, operator $T = A^h + \alpha I$ is coercive. Then, by Theorem 1.15.23, the equation (2.7.4) has a solution x_α^γ, $\gamma = \{\delta, h\}$, in the classical sense for any $\alpha > 0$. It is unique because T is strongly accretive (see Remark 1.15.31). Thus,

$$A^h x_\alpha^\gamma + \alpha x_\alpha^\gamma = f^\delta. \tag{2.7.6}$$

Theorem 2.7.1 *If the condition (2.1.7) holds, then $x_\alpha^\gamma \to \bar{x}^* \in N$, where $\bar{x}^*$ is a unique solution of (2.7.1) satisfying the inequality*

$$\langle J(\bar{x}^* - x^*), \bar{x}^* \rangle \leq 0 \quad \forall x^* \in N. \tag{2.7.7}$$

Proof. First of all, show that the element $\bar{x}^*$ is unique. Suppose that there exists $\bar{x}^{**} \in N$ such that $\bar{x}^{**} \neq \bar{x}^*$ and

$$\langle J(\bar{x}^{**} - x^*), \bar{x}^{**} \rangle \leq 0 \quad \forall x^* \in N. \tag{2.7.8}$$

Put $x^* = \bar{x}^{**}$ in (2.7.7) and $x^* = \bar{x}^*$ in (2.7.8) and add the obtained inequalities. Then

$$0 \geq \langle J(\bar{x}^* - \bar{x}^{**}), \bar{x}^* - \bar{x}^{**} \rangle = \|\bar{x}^* - \bar{x}^{**}\|^2.$$

This implies that $\bar{x}^* = \bar{x}^{**}$.

Take an arbitrary $x^* \in N$. By (2.7.1) and (2.7.6), it is not difficult to make sure that

$$\langle J(x_\alpha^\gamma - x^*), A^h x_\alpha^\gamma - Ax^* \rangle + \alpha \langle J(x_\alpha^\gamma - x^*), x_\alpha^\gamma - x^* \rangle$$
$$= \langle J(x_\alpha^\gamma - x^*), f^\delta - f \rangle - \alpha \langle J(x_\alpha^\gamma - x^*), x^* \rangle. \tag{2.7.9}$$

Since the operators A^h are accretive, we have by (2.7.2) and by (2.7.3) that

$$\|x_\alpha^\gamma - x^*\| \leq \frac{\delta}{\alpha} + \frac{h}{\alpha} g(\|x^*\|) + \|x^*\| \quad \forall x^* \in N,$$

or

$$\|x_\alpha^\gamma\| \leq \frac{\delta}{\alpha} + \frac{h}{\alpha} g(\|x^*\|) + 2\|x^*\| \quad \forall x^* \in N. \tag{2.7.10}$$

Hence, the sequence $\{x_\alpha^\gamma\}$ is bounded, therefore, it has a weak accumulation point $\bar{x}$, that is, $x_\alpha^\gamma \rightharpoonup \bar{x} \in X$ as $\alpha \to 0$ (as before, we do not change its denotation). Now we use again the accretiveness of the operator A^h with any $h > 0$ and (2.7.6) to get

$$\langle J(x - x_\alpha^\gamma), A^h x - A^h x_\alpha^\gamma \rangle = \langle J(x - x_\alpha^\gamma), A^h x + \alpha x_\alpha^\gamma - f^\delta \rangle \geq 0 \quad \forall x \in X.$$

Since the operator J is weak-to-weak continuous, we have in a limit when $\alpha \to 0$ the inequality

$$\langle J(x - \bar{x}), Ax - f \rangle \geq 0 \quad \forall x \in X. \tag{2.7.11}$$

According to Theorem 1.15.14, the operator A is maximal accretive. Then (2.7.11) implies $f = A\bar{x}$. Thus, the inclusion $\bar{x} \in N$ is established.

By (2.7.9) with $x^* = \bar{x}$, it is not difficult to obtain the estimate

$$\|x_\alpha^\gamma - \bar{x}\|^2 \leq \frac{\delta}{\alpha}\|x_\alpha^\gamma - \bar{x}\| + \frac{h}{\alpha}g(\|\bar{x}\|)\|x_\alpha^\gamma - \bar{x}\| - \langle J(x_\alpha^\gamma - \bar{x}), \bar{x} \rangle.$$

Since the sequence $\{x_\alpha^\gamma\}$ is bounded and weakly converges to $\bar{x} \in N$, we deduce from the last inequality the strong convergence of $\{x_\alpha^\gamma\}$ to $\bar{x}$ as $\alpha \to 0$. After this, by simple algebra, we obtain from (2.7.9),

$$\langle J(x_\alpha^\gamma - x^*), x_\alpha^\gamma \rangle \leq \Big(\frac{\delta}{\alpha} + \frac{h}{\alpha}g(\|x^*\|)\Big)\|x_\alpha^\gamma - x^*\| \quad \forall x^* \in N.$$

Going to the limit as $\alpha \to 0$ one has

$$\langle J(\bar{x} - x^*), \bar{x} \rangle \leq 0 \quad \forall x^* \in N,$$

i.e., $\bar{x} = \bar{x}^*$. Thus, the whole sequence $\{x_\alpha^\gamma\}$ strongly converges to $\bar{x}^*$. ∎

Similarly to Theorems 2.2.5 and 2.2.6, the following assertions can be proved for accretive equations.

Theorem 2.7.2 *Under the conditions of Theorem 2.7.1, convergence of the regularization method (2.7.4) is equivalent to solvability of the equation (2.7.1).*

Theorem 2.7.3 *Let equation (2.7.1) be uniquely solvable, $N = \{\bar{x}^*\}$, and assume that there exists a constant $C > 0$ such that $\dfrac{\delta + h}{\alpha} \leq C$ as $\alpha \to 0$. Then $x_\alpha^\gamma \to \bar{x}^*$.*

2. Suppose that $A : X \to 2^X$ is the maximal accretive (multiple-valued, in general) operator in a domain $D(A)$, $A^h : X \to 2^X$ is m-accretive (i.e., maximal accretive) and $D(A^h) = D(A)$. According to Lemma 1.15.13, the value set of a maximal accretive operator at any point of its domain is convex and closed. Therefore, the proximity of operators A and A^h can be defined by the relation

$$\mathcal{H}_X(Ax, A^h x) \leq g(\|x\|)h \quad \forall x \in D(A),$$

where $g(t)$ is a continuous non-negative function for all $t \geq 0$ and $\mathcal{H}_X(G_1, G_2)$ is the Hausdorff distance between the sets G_1 and G_2 in X. Then the regularized equation (2.7.4) has a unique solution which we denote, as before, by x_α^γ. This means that there exists $y_\alpha^\gamma \in A^h x_\alpha^\gamma$ such that

$$y_\alpha^\gamma + \alpha x_\alpha^\gamma = f^\delta.$$

Since the operator J is weak-to-weak continuous, it is not difficult to verify validity of all the assertions above in this section.

3. Let $A : X \to 2^X$ be an arbitrary accretive operator and $D(A) = X$. In this case, solutions of the equations (2.7.1) and (2.7.4) are understood in the sense of Definition 1.15.25. We assume that spaces X and X^* are uniformly convex, X has the M-property and duality mapping $J : X \to X^*$ is weak-to-weak continuous. Let $\bar{A}$ and $\bar{A}^h$ be maximal accretive extensions of accretive operators A and A^h, respectively, and

$$\mathcal{H}_X(\bar{A}x, \bar{A}^h x) \leq g(\|x\|)h \quad \forall x \in X,$$

where $g(t)$ is a continuous non-negative function for all $t \geq 0$.

We want to show that the equation (2.7.4) is uniquely solvable. Indeed, by the accretiveness of $\bar{A}^h$, one has

$$\langle Jx, y^h + \alpha x - f^\delta \rangle \geq \|x\|(\alpha\|x\| - \|y_0^h\| - \|f^\delta\|),$$

where $y_0^h \in \bar{A}^h(\theta_X)$, $y^h \in \bar{A}^h x$. Hence, there exists $r_1(\delta, h, \alpha) > 0$ such that

$$\langle Jx, y^h + \alpha x - f^\delta \rangle \geq 0$$

for all $x \in X$ with $\|x\| \geq r_1$. Due to Theorem 1.15.29, for every $\alpha > 0$ there exists a unique solution of the equation

$$\bar{A}^h x + \alpha x = f^\delta, \tag{2.7.12}$$

which is denoted by x_α^γ again. Hence, $\bar{A}^h$ is m-accretive. Obviously, the identity operator I is also m-accretive. Then, By Theorem 1.15.22, summary operator $\bar{A}^h + \alpha I$ is m-accretive. Consequently, it is maximal accretive in view of Lemma 1.15.20. Thus, under conditions of this subsection, the equations (2.7.4) and (2.7.12) are equivalent, that is, x_α^γ is a unique solution of the equation (2.7.4) as well.

Since x_α^γ is a solution of (2.7.12), there exists an element $\bar{y}_\alpha^\gamma \in \bar{A}^h x_\alpha^\gamma$ such that

$$\bar{y}_\alpha^\gamma + \alpha x_\alpha^\gamma = f^\delta.$$

Now the validity of Theorems 2.7.1, 2.7.2 and 2.7.3 is established by the same arguments as those above.

Corollary 2.7.4 *For the operator regularization method*

$$A^h x + \alpha(x - u) = f^\delta,$$

where $u \in X$ is some fixed element, all the assertions of this section still remain valid. Moreover, the solution $\bar{x}^* \in N$ satisfies the inequality

$$\langle J(\bar{x}^* - x^*), \bar{x}^* - u \rangle \leq 0 \quad \forall x^* \in N.$$

4. Up to the present, the operators A^h were assumed to be accretive. This requirement is one of the determining factors to prove solvability of the regularized equation (2.7.4). Let $A : X \to X$ be a hemicontinuous accretive operator, while $A^h : X \to X$ are arbitrary hemicontinuous operators. The solvability problem for the equation (2.7.4) is open in this case. Therefore, in order to construct approximations to solutions of the equation (2.7.1), as in Section 2.6, we make use of the additional assumptions.

Consider the variational inequality with small offset

$$\langle J(z - x_\alpha^\gamma), A^h z + \alpha z - f^\delta \rangle \geq -\epsilon g(\|z\|)\|z - x_\alpha^\gamma\| \quad \forall z \in X, \tag{2.7.13}$$

where $\epsilon \geq h$.

Lemma 2.7.5 *The inequality (2.7.13) has a solution x_α^γ for any $\alpha > 0$ and for any $f^\delta \in X$.*

Proof. Indeed, the equation

$$Ax + \alpha x = f$$

with accretive hemicontinuous operator A has a unique classical solution x_α^δ, i.e.,

$$Ax_\alpha^\delta + \alpha x_\alpha^\delta = f^\delta.$$

It follows from the accretiveness of the operator $A + \alpha I$ that

$$\langle J(z - x_\alpha^\delta), Az + \alpha z - f^\delta \rangle \geq 0 \quad \forall z \in X.$$

In view of the condition (2.7.3), we have

$$\langle J(z - x_\alpha^\delta), A^h z + \alpha z - f^\delta \rangle \geq -hg(\|z\|)\|z - x_\alpha^\delta\| \quad \forall z \in X.$$

Obviously, x_α^δ is a solution of (2.7.13) because $\epsilon \geq h$. ∎

Lemma 2.7.6 *The following inequality holds for a solution x_α^γ of the inequality (2.7.13):*

$$\langle J(z - x_\alpha^\gamma), A^h x_\alpha^\gamma + \alpha x_\alpha^\gamma - f^\delta \rangle \geq -\epsilon g(\|x_\alpha^\gamma\|)\|z - x_\alpha^\gamma\| \quad \forall z \in X. \tag{2.7.14}$$

Proof. Put in (2.7.13) $z_t = tx_\alpha^\gamma + (1-t)z$, $t \in [0,1)$, in place of $z \in X$, and apply the property $J(cx) = cJx$ of the normalized duality mapping J, where c is a constant. Then we come to the inequality

$$\langle J(z - x_\alpha^\gamma), A^h z_t + \alpha z_t - f^\delta \rangle \geq -\epsilon g(\|z_t\|)\|z - x_\alpha^\gamma\|$$

because $c = 1 - t > 0$. Since $g(s)$ is continuous and A^h is hemicontinuous, we have (2.7.14) as $t \to 1$. ∎

Theorem 2.7.7 *Let $\{x_\alpha^\gamma\}$ be a solution sequence of the inequality (2.7.13). Let $g(t) \leq Mt + Q$, $M > 0$, $Q > 0$, $(\delta + \epsilon)/\alpha \to 0$ as $\alpha \to 0$. Then $x_\alpha^\gamma \to \bar{x}^* \in N$, where $\bar{x}^*$ satisfies (2.7.7).*

Proof. Put in (2.7.14) $z = x^* \in N$. Since $Ax^* = f$, we have

$$\langle J(x^* - x_\alpha^\gamma), A^h x_\alpha^\gamma - Ax^* + \alpha x_\alpha^\gamma + f - f^\delta \rangle \geq -\epsilon g(\|x_\alpha^\gamma\|)\|x^* - x_\alpha^\gamma\| \quad \forall x^* \in N.$$

By the accretiveness of the operator A and by the condition of (2.7.3), we deduce from the previous inequality the following estimate:

$$\|x_\alpha^\gamma - x^*\| \leq \frac{\epsilon + h}{\alpha} g(\|x_\alpha^\gamma\|) + \frac{\delta}{\alpha} + \|x^*\|.$$

Hence,

$$\|x_\alpha^\gamma\| \leq \frac{\epsilon + h}{\alpha} g(\|x_\alpha^\gamma\|) + \frac{\delta}{\alpha} + 2\|x^*\|. \tag{2.7.15}$$

According to the hypothesis, we presume $(\epsilon + h)/\alpha \leq K$ for sufficiently small $\alpha > 0$ and $g(\|x_\alpha^\gamma\|) \leq M\|x_\alpha^\gamma\| + Q$. Then (2.7.15) implies the boundedness of $\{x_\alpha^\gamma\}$. Let $x_\alpha^\gamma \rightharpoonup \bar{x} \in X$. Since the duality mapping J is weak-to-weak continuous, we obtain from (2.7.13) as $\alpha \to 0$ the inequality

$$\langle J(z - \bar{x}), Az - f \rangle \geq 0 \quad \forall z \in X.$$

From this it follows that $A\bar{x} = f$ because A is a maximal accretive operator. Thus, $\bar{x} \in N$. Then the proof is accomplished by the same arguments as in Theorem 2.7.1. ∎

Remark 2.7.8 *If* $\dfrac{\delta}{\alpha} \leq K$, $\dfrac{\epsilon}{\alpha} \leq K$, $K > 0$, $g(t) \leq Mt + Q$, $M > 0$, $Q > 0$, $2KM < 1$, *and* x_0 *is a unique solution of (2.7.1), then the sequence* $\{x_\alpha^\gamma\}$ *weakly converges to* x_0 *as* $\alpha \to 0$.

5. Let X and X^* be uniformly convex spaces, X possess an approximation, J be a weak-to-weak continuous operator. Consider the equation (2.7.1) with hemicontinuous d-accretive operator $A : X \to X$. Suppose that its solution set $N \neq \emptyset$. As the regularized equation, we investigate (2.7.4) with demicontinuous d-accretive operator $A^h : X \to X$ such that $D(A^h) = D(A) = X$ for all $h > 0$. Similarly to (2.7.5), we can show that the operator $T = A^h + \alpha I$ is coercive. Then, by Theorem 1.16.11, the equation (2.7.4) has solutions x_α^γ for all positive α, δ and h. This means that x_α^γ satisfy (2.7.6). Let $N \neq \emptyset$ be a solution set of (2.7.1).

Theorem 2.7.9 *If the condition (2.1.7) holds, then* $x_\alpha^\gamma \to \bar{x}^* \in N$, *where* $\bar{x}^*$ *is a unique solution of (2.7.1) satisfying the inequality*

$$\langle J\bar{x}^* - Jx^*, \bar{x}^* \rangle \leq 0 \quad \forall x^* \in N. \tag{2.7.16}$$

Proof. As in Theorem 2.7.1, we establish from (2.7.16) a uniqueness of $\bar{x}^*$. Assume that there exists one more element $\bar{x}^{**} \in N$, $\bar{x}^{**} \neq \bar{x}^*$, such that

$$\langle J\bar{x}^{**} - Jx^*, \bar{x}^{**} \rangle \leq 0 \quad \forall x^* \in N. \tag{2.7.17}$$

We can put $x^* = \bar{x}^{**}$ in (2.7.16) and $x^* = \bar{x}^*$ in (2.7.17) and add the obtained inequalities. Then due to Theorem 1.6.4,

$$0 \geq \langle J\bar{x}^* - J\bar{x}^{**}, \bar{x}^* - \bar{x}^{**} \rangle \geq (2L)^{-1}\delta_X(c_2^{-1}\|\bar{x}^* - \bar{x}^{**}\|),$$

where $\delta_X(\epsilon)$ is the modulus of convexity of X, $1 < L < 1.7$ is the Figiel constant, $c_2 = 2\, max\{1, \|\bar{x}^*\|, \|\bar{x}^{**}\|\}$. Therefore, by the properties of the modulus of convexity $\delta_X(\epsilon)$, we conclude that $\bar{x}^* = \bar{x}^{**}$.

Take an arbitrary $x^* \in N$. Using (2.7.1) and (2.7.6), it is not difficult to verify that

$$\langle Jx_\alpha^\gamma - Jx^*, A^h x_\alpha^\gamma - Ax^* \rangle \;\; + \;\; \alpha \langle Jx_\alpha^\gamma - Jx^*, x_\alpha^\gamma \rangle$$

$$= \; \langle Jx_\alpha^\gamma - Jx^*, f^\delta - f \rangle. \tag{2.7.18}$$

Since the operator A^h is d-accretive, we have

$$\langle Jx_\alpha^\gamma - Jx^*, A^h x^* - Ax^* \rangle + \alpha \|x_\alpha^\gamma\|^2 - \alpha \langle Jx_\alpha^\gamma, x^* \rangle \leq (\|x_\alpha^\gamma\| + \|x^*\|) \|f^\delta - f\|.$$

By (2.7.2) and (2.7.3),

$$\alpha \|x_\alpha^\gamma\|^2 - \alpha \|x_\alpha^\gamma\| \|x^*\| \leq (\|x_\alpha^\gamma\| + \|x^*\|)(\delta + hg(\|x^*\|)).$$

Consequently,

$$\|x_\alpha^\gamma\|^2 - \Big(\frac{\delta + hg(\|x^*\|)}{\alpha} + \|x^*\|\Big)\|x_\alpha^\gamma\| - \|x^*\|\frac{\delta + hg(\|x^*\|)}{\alpha} \leq 0.$$

From this quadratic inequality, we obtain the estimate

$$\|x_\alpha^\gamma\| \leq \frac{2\delta}{\alpha} + \frac{2h}{\alpha}g(\|x^*\|) + \|x^*\| \quad \forall x^* \in N, \tag{2.7.19}$$

that is, the sequence $\{x_\alpha^\gamma\}$ is bounded in X, say $\|x_\alpha^\gamma\| \leq K$ and $x_\alpha^\gamma \rightharpoonup \bar{x} \in X$ as $\alpha \to 0$. Next we use again d-accretiveness of the operator A^h and (2.7.6) to get

$$\langle Jx - Jx_\alpha^\gamma, A^h x - A^h x_\alpha^\gamma \rangle = \langle Jx - Jx_\alpha^\gamma, A^h x + \alpha x_\alpha^\gamma - f^\delta \rangle \geq 0 \quad \forall x \in X.$$

Since J is weak-to-weak continuous, we have in a limit as $\alpha \to 0$,

$$\langle Jx - J\bar{x}, Ax - f \rangle \geq 0 \quad \forall x \in X. \tag{2.7.20}$$

According to Theorem 1.16.6, the operator A is maximal d-accretive. Then (2.7.20) implies $f = A\bar{x}$, i.e., $\bar{x} \in N$.

Rewrite (2.7.18) in the following form:

$$\langle Jx_\alpha^\gamma - Jx^*, A^h x_\alpha^\gamma - Ax^* \rangle \;\; + \;\; \alpha \langle Jx_\alpha^\gamma - Jx^*, x_\alpha^\gamma - x^* \rangle$$

$$= \; \langle Jx_\alpha^\gamma - Jx^*, f^\delta - f \rangle - \alpha \langle Jx_\alpha^\gamma - Jx^*, x^* \rangle. \tag{2.7.21}$$

Assume in (2.7.21) $x^* = \bar{x}$. It is not difficult to deduce the estimate

$$(2L)^{-1}\delta_X(c_2^{-1}\|x_\alpha^\gamma - \bar{x}\|) \leq \frac{\delta}{\alpha}\|Jx_\alpha^\gamma - J\bar{x}\|_* + \frac{h}{\alpha}g(\|\bar{x}\|)\|Jx_\alpha^\gamma - J\bar{x}\|_* - \langle Jx_\alpha^\gamma - J\bar{x}, \bar{x} \rangle,$$

where $c_2 = 2\,max\{1, K, \|\bar{x}\|\}$. By the boundedness of x_α^γ, weak-to-weak continuity of J and weak convergence of x_α^γ to $\bar{x}$ and by (2.1.7), the right-hand side of the previous inequality vanishes. The latter results from the limit relation

$$\delta_X(c_2^{-1}\|x_\alpha^\gamma - \bar{x}\|) \to 0 \quad \text{as} \quad \alpha \to 0.$$

Then $\lim\limits_{\alpha \to 0} \|x_\alpha^\gamma - \bar{x}\| = 0$, and we obtain strong convergence of $\{x_\alpha^\gamma\}$ to $\bar{x}$ as $\alpha \to 0$.

Finally, after some algebraic transformations in (2.7.21), one gets

$$\langle Jx_\alpha^\gamma - Jx^*, x_\alpha^\gamma \rangle \leq \left(\frac{\delta}{\alpha} + \frac{h}{\alpha}g(\|x^*\|)\right)\|Jx_\alpha^\gamma - Jx^*\|_* \quad \forall x^* \in N.$$

Passing here to the limit as $\alpha \to 0$ we come to the inequality

$$\langle J\bar{x} - Jx^*, \bar{x} \rangle \leq 0 \quad \forall x^* \in N.$$

This implies $\bar{x} = \bar{x}^*$. Thus, the whole sequence $\{x_\alpha^\gamma\}$ converges strongly to $\bar{x}^*$. The proof is complete. ∎

Remark 2.7.10 *All the assertions of this section take place if the normalized duality mapping J is replaced by J^μ.*

Bibliographical Notes and Remarks

The operator regularization methods were firstly studied for linear equations in a Hilbert space in [125]. Another scheme of proofs was used in [99]. The deepest results have been obtained by applying the spectral theory that is a rather powerful instrument of research into linear operators [39, 42, 99, 131]. In a Banach — but not in a Hilbert — space, duality mapping J is indeed not linear. This does not allow us to use spectral theory even for linear regularized equations of a type (2.2.4) in Banach spaces including problems with exact operators. The operator regularization methods for nonlinear equations in Banach spaces have required new approaches (see [5, 54, 232]).

The results of Section 2.1 have been developed in [31, 32]. The unperturbed case, that is, convergence of x_α^0 to $\bar{x}^*$, was considered in [5, 68]. Convergence analysis of the operator regularization methods in Banach E-spaces was conducted in [5, 34, 201]. Recall that E-spaces include Hilbert spaces and all uniformly convex Banach spaces [16, 74, 127]. The estimates of convergence rate of regularization methods for linear equations have been established, for instance, in [37, 222, 223]. A similar estimate for nonlinear equations in Hilbert spaces was obtained in [42] and for optimization problems in [226]. The condition (2.3.7) introduced there is the analogy of a sourcewise representability of solutions defined in [37]. Another sort of requirements was found in [80, 121, 156] that imply the estimates mentioned above for the Tikhonov regularization method. Theorem 2.3.1 for nonlinear equations in Banach spaces has been proved in [202].

The estimates of the norm $\|x_\alpha^\gamma - \bar{x}^*\|$ have their own importance. However, they are also essential when the problems connected with the optimality of regularizing algorithms are solved (see, for instance, in [99, 126]). One should note the necessity to find solutions with the property described in Corollary 2.2.3 appears, for example, in planning problems [217].

The results of Section 2.4 were obtained in [18]. Convergence of the regularization method for an equation with semimonotone operators was proved in [185]. Theorem 2.6.1 has been established in [202]. Variational inequality with a small offset has been constructed and studied by Liskovets in [132]. For the class of nonlinear accretive mappings the operator regularization method was investigated in [6, 170] and in [30] for d-accretive operators. The convergence problem of the operator regularization method (2.7.4) for maximal accretive operators was also solved in [171]. The method with a small offset for accretive maps was proposed and developed in [118].

The necessary and sufficient condition for convergence of solutions of the regularized equation (2.1.5) (namely, the solvability of (2.1.1)) has been found by Maslov [144] for the case of linear operators A in Hilbert spaces. Nonlinear versions of this criterion for Banach spaces (see, for example, Theorem 2.2.5) were obtained in [5, 7, 34].

Chapter 3

PARAMETERIZATION OF REGULARIZATION METHODS

It has been established in Chapter 2 that the condition (2.1.7) for positive perturbation parameters δ and h and regularization parameter α are sufficient for the operator regularization methods to be convergent to solutions of monotone and accretive operator equations. However, such a wide choice of parameters does not possess the regularizing properties in the sense of Definition 5 (see Preface). Our aim in this chapter is to indicate the ways to find the functions $\alpha = \alpha(\delta, h)$, which solve this problem when $\delta \to 0$ and $h \to 0$.

The following criteria of choosing the regularization parameters were widely studied for linear equations $Ax = f$ with perturbed right-hand side f in a Hilbert space:

(I) The residual principle: $\alpha = \bar{\alpha}(\delta)$ is defined by the equation

$$\|Ax_{\bar{\alpha}}^\delta - f^\delta\| = k\delta, \ k > 1,$$

where $x_{\bar{\alpha}}^\delta$ is a solution of the regularized equation $Ax + \bar{\alpha}x = f^\delta$.

(II) The smoothing functional principle : $\alpha = \bar{\alpha}(\delta)$ is defined by the equation

$$\|Ax_{\bar{\alpha}}^\delta - f^\delta\|^2 + \bar{\alpha}\|x_{\bar{\alpha}}^\delta\|^2 = \phi(\delta),$$

where $\phi(t)$ is a positive function of $t > 0$ and $x_{\bar{\alpha}}^\delta$ is a minimum point of the functional

$$\Phi_{\bar{\alpha}}^\delta(x) = \|Ax - f^\delta\|^2 + \bar{\alpha}\|x\|^2.$$

(III) The minimal residual principle: $\alpha(\delta)$ is defined by the equality

$$\alpha = \bar{\alpha}(\delta) \quad = \quad inf \ \{\alpha_0 \mid \varphi(\|Ax_{\alpha_0}^\delta - f^\delta\|)$$

$$= \quad inf \ \{\varphi(\|Ax_\alpha^\delta - f^\delta\|) \mid 0 < \alpha \le \alpha_1\}, \ x_\alpha^\delta \in M\},$$

where $\varphi(t)$ is a positive function for all $t > 0$, M is an admissible class of solutions.

(IV) The quasi-optimality principle: $\alpha = \bar{\alpha}(\delta)$ is chosen as a value realizing

$$inf_{\alpha > 0} \left\| \alpha \frac{dx_\alpha^\delta}{d\alpha} \right\|,$$

where x_α^δ is a regularized solution.

If we consider equations $Ax = f$ with arbitrary nonlinear operators A, then the criteria above are impossible [217]. However, it is possible to study this aspect for nonlinear problems with monotone and accretive operators. In this chapter we present the sufficient conditions for the parametric criteria (I) - (III) which guarantee the regularizing properties of the corresponding methods not only with perturbed right-hand side f but also with perturbed operator A. Furthermore, we answer the following significant question: whether or not the constructed function $\bar\alpha(\delta, h)$ satisfies the sufficient convergence conditions of the operator regularization method

$$\frac{\delta + h}{\bar\alpha(\delta, h)} \to 0 \text{ as } \delta,\ h \to 0.$$

We establish this result for the criteria (I) - (III) to the regularization methods of Chapter 2. We also introduce and investigate the new so-called generalized residual principle for nonlinear equations with multiple-valued and discontinuous operators. As regards the quasi-optimality principle (IV) for monotone equations, we refer the reader to [193] and [197].

3.1   Residual Principle for Monotone Equations

1. Let X be an E-space with a strictly convex space X^*, $A : X \to 2^{X^*}$ be a maximal monotone operator, $f \in X^*$. Let the equation

$$Ax = f \tag{3.1.1}$$

have a solution set $N \neq \emptyset$. Consider the regularized equations

$$A^h x + \alpha J x = f^\delta,\ \ \alpha > 0, \tag{3.1.2}$$

where $J : X \to X^*$ is the normalized duality mapping, $A^h : X \to 2^{X^*}$ are maximal monotone operators, $D(A^h) = D(A)$ for all $h > 0$ and $f^\delta \in X^*$ for all $\delta > 0$. Let x_α^γ be a solution of (3.1.2) with $\gamma = (\delta, h)$. Then there exists an element $y_\alpha^\gamma \in A^h x_\alpha^\gamma$ such that

$$y_\alpha^\gamma + \alpha J x_\alpha^\gamma = f^\delta. \tag{3.1.3}$$

Our aim is to study properties of the functions $\sigma(\alpha) = \|x_\alpha^\gamma\|$ and $\rho(\alpha) = \alpha\|x_\alpha^\gamma\|$ with fixed δ and h.

Lemma 3.1.1 *Let X be an E-space with a strictly convex space X^*, $A : X \to 2^{X^*}$ be a maximal monotone operator, $f \in X^*$. Then the function $\sigma(\alpha)$ is single-valued, continuous and non-increasing for $\alpha \geq \alpha_0 > 0$, and if $\theta_X \in D(A)$, then $\sigma(\alpha) \to 0$ as $\alpha \to \infty$.*

Proof. Single-valued solvability of the equation (3.1.2) implies the same continuity property of the function $\sigma(\alpha)$. We prove that $\sigma(\alpha)$ is continuous. Let x_β^γ be the solution of

(3.1.2) with $\alpha = \beta$ in the sense of inclusion. Then there exists an element $y_\beta^\gamma \in A^h x_\beta^\gamma$ such that

$$y_\beta^\gamma + \beta J x_\beta^\gamma = f^\delta. \qquad (3.1.4)$$

Write down the difference between (3.1.3) and (3.1.4) and then calculate the corresponding dual products on the element $x_\alpha^\gamma - x_\beta^\gamma$. We obtain

$$\langle y_\alpha^\gamma - y_\beta^\gamma, x_\alpha^\gamma - x_\beta^\gamma \rangle + \langle \alpha J x_\alpha^\gamma - \beta J x_\beta^\gamma, x_\alpha^\gamma - x_\beta^\gamma \rangle = 0.$$

Taking into account the monotonicity of A^h we deduce the inequality

$$\alpha \langle J x_\alpha^\gamma - J x_\beta^\gamma, x_\alpha^\gamma - x_\beta^\gamma \rangle + (\alpha - \beta)\langle J x_\beta^\gamma, x_\alpha^\gamma - x_\beta^\gamma \rangle \le 0. \qquad (3.1.5)$$

Then the property (1.5.3) of the J yields for $\alpha \ge \alpha_0 > 0$ the following relation:

$$(\|x_\alpha^\gamma\| - \|x_\beta^\gamma\|)^2 \le \frac{|\alpha - \beta|}{\alpha_0} \|x_\beta^\gamma\|(\|x_\alpha^\gamma\| + \|x_\beta^\gamma\|). \qquad (3.1.6)$$

Recall that the estimate

$$\|x_\alpha^\gamma\| \le 2\|x^*\| + \frac{\delta}{\alpha} + \frac{h}{\alpha} g(\|x^*\|) \quad \forall x^* \in N \qquad (3.1.7)$$

was obtained in Section 2.2. Therefore, $\|x_\alpha^\gamma\|$ is bounded when $\alpha \ge \alpha_0 > 0$ and γ is fixed. Now the continuity of $\sigma(\alpha)$ follows from (3.1.6).

Since the dual mapping J is monotone, we have from (3.1.5),

$$(\alpha - \beta)\langle J x_\beta^\gamma, x_\alpha^\gamma - x_\beta^\gamma \rangle \le 0. \qquad (3.1.8)$$

Suppose that $\alpha < \beta$. Then

$$\langle J x_\beta^\gamma, x_\beta^\gamma - x_\alpha^\gamma \rangle \le 0.$$

Using now the definition of J we conclude that $\|x_\beta^\gamma\| \le \|x_\alpha^\gamma\|$, that is, $\sigma(\beta) \le \sigma(\alpha)$ as $\beta > \alpha$. Thus, the function $\sigma(\alpha)$ does not increase.

We prove the last claim of the lemma. By (3.1.3), it is obvious that

$$\sigma(\alpha) = \frac{\|y_\alpha^\gamma - f^\delta\|_*}{\alpha}. \qquad (3.1.9)$$

We assert that the sequence $\{y_\alpha^\gamma\}$ is bounded as $\alpha \to \infty$. Indeed, since $\theta_X \in D(A)$, it follows from the monotonicity condition of A^h that

$$\langle y_\alpha^\gamma - y^h, x_\alpha^\gamma \rangle \ge 0, \qquad (3.1.10)$$

where $y^h \in A^h(\theta_X)$. A duality mapping J is homogeneous, therefore, from (3.1.3) one has

$$J(\alpha x_\alpha^\gamma) = f^\delta - y_\alpha^\gamma.$$

Under our conditions, the operator $J^* : X^* \to X$ is one-to-one and it is defined on the whole space X^*. According to Lemma 1.5.10, $J^*J = I_X$. Hence, the latter equality implies

$$\alpha x_\alpha^\gamma = J^*(f^\delta - y_\alpha^\gamma).$$

The inequality (3.1.10) can be now rewritten as follows:

$$\langle y_\alpha^\gamma - y^h, J^*(f^\delta - y_\alpha^\gamma)\rangle \geq 0,$$

because of $\alpha > 0$. It is easy to see that

$$\|y_\alpha^\gamma - f^\delta\|_*^2 = \langle f^\delta - y_\alpha^\gamma, J^*(f^\delta - y_\alpha^\gamma)\rangle \leq \langle f^\delta - y^h, J^*(f^\delta - y_\alpha^\gamma)\rangle,$$

and we thus have the estimate

$$\|y_\alpha^\gamma - f^\delta\|_* \leq \|y^h - f^\delta\|_*. \tag{3.1.11}$$

Hence, the sequence $\{y_\alpha^\gamma\}$ is bounded in X^*. Then the equality (3.1.9) allows us to assert that $\sigma(\alpha) \to 0$ as $\alpha \to \infty$. The proof is accomplished. ∎

Remark 3.1.2 *If $\theta_X \notin D(A)$, then we choose any element $u \in D(A)$ and consider the regularized equation in the form*

$$A^h x + \alpha J(x - u) = f^\delta. \tag{3.1.12}$$

In this case, all the conclusions of Lemma 3.1.1 are fulfilled for the function

$$\sigma(\alpha) = \|x_\alpha^\gamma - u\|,$$

where x_α^γ is a solution of the equation (3.1.12).

Proposition 3.1.3 *Under the conditions of the present subsection, if $D(A) = D(A^h)$ is a convex and closed subset in X, then $x_\alpha^\gamma \to x_*$ as $\alpha \to \infty$, where x_α^γ is a solution of (3.1.2) with fixed δ and h, and $x_* \in D(A)$ is the minimal norm vector, i.e.,*

$$\|x_*\| = min \,\{\|x\| \mid x \in D(A)\}.$$

Proof. It is not difficult to check that the vector x_* exists and it is defined uniquely. Choose any element $x \in D(A) = D(A^h)$. Let $y^h \in A^h x$ be given. Then, by (3.1.3),

$$\langle y_\alpha^\gamma - y^h, x_\alpha^\gamma - x\rangle + \alpha\langle Jx_\alpha^\gamma, x_\alpha^\gamma - x\rangle = \langle f^\delta - y^h, x_\alpha^\gamma - x\rangle.$$

Since A^h is monotone, we have

$$\|x_\alpha^\gamma\|^2 - \|x_\alpha^\gamma\|\Big(\|x\| + \frac{\|f^\delta - y^h\|_*}{\alpha}\Big) - \frac{\|f^\delta - y^h\|_*}{\alpha}\|x\| \leq 0.$$

From this quadratic inequality, the estimate

$$\|x_\alpha^\gamma\| \leq \|x\| + 2\frac{\|f^\delta - y^h\|_*}{\alpha} \tag{3.1.13}$$

follows for all $x \in D(A)$. Hence, the sequence $\{x_\alpha^\gamma\}$ is bounded when $\alpha \to \infty$ and γ is fixed. Then there exists a subsequence $\{x_\beta^\gamma\} \subseteq \{x_\alpha^\gamma\}$ such that $x_\beta^\gamma \rightharpoonup \bar{x} \in X$ as $\beta \to \infty$.

Furthermore, by the Mazur theorem, $\bar{x} \in D(A)$. Taking now into account the fact that the norm in X is weakly lower semicontinuous, we obtain from the inequality (3.1.13)

$$\|\bar{x}\| \leq \liminf_{\beta \to \infty} \|x_\beta^\gamma\| \leq \limsup_{\beta \to \infty} \|x_\beta^\gamma\| \leq \|x\| \quad \forall x \in D(A). \tag{3.1.14}$$

Consequently, $\bar{x} = x_*$. Therefore, the whole sequence $x_\alpha^\gamma \rightharpoonup x_*$ as $\alpha \to \infty$. In addition, by (3.1.14), $\|x_\alpha^\gamma\| \to \|x_*\|$ as $\alpha \to \infty$. Since X is an E-space, we finally establish the strong convergence of $\{x_\alpha^\gamma\}$ to x_* as $\alpha \to \infty$. ∎

Lemma 3.1.4 *Under the conditions of Lemma 3.1.1, the function $\rho(\alpha) = \alpha\|x_\alpha^\gamma\|$ is single-valued and continuous for $\alpha \geq \alpha_0 > 0$, and if $\theta_X \in D(A) = D(A^h)$, then*

$$\lim_{\alpha \to \infty} \rho(\alpha) = \|\bar{y}^h - f^\delta\|_*, \tag{3.1.15}$$

where $\bar{y}^h \in A^h(\theta_X)$ is defined as

$$\|\bar{y}^h - f^\delta\|_* = min\ \{\|y^h - f^\delta\|_* \mid y^h \in A^h(\theta_X)\}. \tag{3.1.16}$$

Proof. By Lemma 3.1.1, the function $\rho(\alpha)$ is single-valued and continuous. Since the value set of a maximal monotone operator at a point is convex and closed, the element $\bar{y}^h$ is defined uniquely. It was also proved in Lemma 3.1.1 that the sequence $\{y_\alpha^\gamma\}$ satisfying (3.1.3) is bounded in X^* and $\{x_\alpha^\gamma\}$ converges to θ_X as $\alpha \to \infty$. Therefore, there exists a subsequence of $\{y_\alpha^\gamma\}$ (we do not change its notation) such that $y_\alpha^\gamma \rightharpoonup g^\gamma \in X^*$ as $\alpha \to \infty$. Then the inclusion $g^\gamma \in A^h(\theta_X)$ follows from Lemma 1.4.5. By the weak convergence of y_α^γ to g^γ and by the inequality (3.1.11) for all $y^h \in A^h(\theta_X)$, we deduce

$$\|g^\gamma - f^\delta\|_* \leq \liminf_{\alpha \to \infty} \|y_\alpha^\gamma - f^\delta\|_* \leq \limsup_{\alpha \to \infty} \|y_\alpha^\gamma - f^\delta\|_* \leq \|y^h - f^\delta\|_*. \tag{3.1.17}$$

Thus, $g^\gamma = \bar{y}^h$, and the whole subsequence $y_\alpha^\gamma \rightharpoonup \bar{y}^h$ as $\alpha \to \infty$. Finally, by (3.1.17), we prove the last conclusion of the lemma. ∎

Remark 3.1.5 *If $\theta_X \notin D(A)$, then we are able to consider again the equation (3.1.12) and define*

$$\rho(\alpha) = \alpha\|x_\alpha^\gamma - z^0\|.$$

In this case, all the conclusions of Lemma 3.1.4 take place. Moreover,

$$\|\bar{y}^h - f^\delta\|_* = min\ \{\|y^h - f^\delta\|_* \mid y^h \in A^h(z^0)\}.$$

Remark 3.1.6 *If X^* is an E-space, then $y_\alpha^\gamma \to \bar{y}^h$ as $\alpha \to \infty$. This assertion follows from (3.1.17) and from the proven above weak convergence of $\{y_\alpha^\gamma\}$ to $\bar{y}^h$.*

We shall further study the behavior of the functions $\sigma(\alpha) = \|x_\alpha^\gamma\|$ and $\rho(\alpha) = \alpha\|x_\alpha^\gamma\|$ as $\alpha \to \infty$, where x_α^γ is a solution of (3.1.2) and $\theta_X \notin D(A)$. Let $\overline{D(A)}$ be a closure of $D(A)$. According to Theorem 1.7.17, $\overline{D(A)}$ is a convex set in X. Rewrite (3.1.3) in the equivalent form

$$y_\alpha^\gamma - y^h + \alpha J x_\alpha^\gamma - \alpha J x + \alpha J x = f^\delta - y^h, \tag{3.1.18}$$

where $y^h \in A^h x$, $x \in D(A) = D(A^h)$. Then

$$\langle (y_\alpha^\gamma + \alpha J x_\alpha^\gamma) - (y^h - \alpha J x), x_\alpha^\gamma - x \rangle + \alpha \langle J x, x_\alpha^\gamma - x \rangle = \langle f^\delta - y^h, x_\alpha^\gamma - x \rangle.$$

Owing to the monotonicity of $A^h + \alpha J$, it is not difficult to deduce from (3.1.18) the following estimate:

$$\langle J x, x_\alpha^\gamma - x \rangle \leq \frac{\| f^\delta - y^h \|_*}{\alpha} \| x_\alpha^\gamma - x \|. \tag{3.1.19}$$

By (3.1.7), the sequence $\{x_\alpha^\gamma\}$ is bounded as $\alpha \to \infty$, therefore $x_\alpha^\gamma \rightharpoonup \bar{x} \in X$. Then we obtain in a limit

$$\langle J x, \bar{x} - x \rangle \leq 0 \quad \forall x \in D(A). \tag{3.1.20}$$

Since, by Theorem 1.3.20, J is demicontinuous in X, the inequality (3.1.20) holds with all $x \in \overline{D(A)}$. Moreover, it results from the Mazur theorem that $\bar{x} \in \overline{D(A)}$. Then using Lemma 1.11.4 and Theorem 1.11.14, we conclude that $\bar{x}$ is the element $x_* \in \overline{D(A)}$ with minimal norm defined uniquely in X.

If $x_* \neq \theta_X$ and $x_* \in D(A)$ then similarly to the proof of Proposition 3.1.3, we find that $\| x_\alpha^\gamma \| \to \| x_* \|$ as $\alpha \to \infty$. Hence,

$$\lim_{\alpha \to \infty} \rho(\alpha) = +\infty. \tag{3.1.21}$$

Assume now that $x_* = \theta_X$ and, consequently, $x_* \notin D(A)$. Show by contradiction that in this case (3.1.21) also holds. Let the function

$$\rho(\alpha) = \alpha \| x_\alpha^\gamma \| = \| y_\alpha^\gamma - f^\delta \|_*$$

be bounded as $\alpha \to \infty$. Then $\| x_\alpha^\gamma \| \to 0$, and on account of $x_\alpha^\gamma \rightharpoonup x_* = \theta_X$, we conclude that $x_\alpha^\gamma \to \theta_X$ in an E-space. Furthermore, boundedness of the sequence $\{y_\alpha^\gamma\}$ allows us to assert that $y_\alpha^\gamma \rightharpoonup \tilde{y}^\gamma \in X^*$. Then $\tilde{y}^\gamma \in A^h(\theta_X)$ because maximal monotone operators A^h are demiclosed. This contradicts the assumption $\theta_X \notin D(A)$.

Let now $x_* \neq \theta_X$ and $x_* \notin D(A)$. Then due to the proved weak convergence of x_α^γ to $\bar{x} \in \overline{D(A)}$ and by the weak lower semicontinuity of the norm in any Banach space, we obtain

$$0 < \| \bar{x} \| \leq \liminf_{\alpha \to \infty} \| x_\alpha^\gamma \|.$$

Hence, (3.1.21) also holds in this case. Thus, the following assertion is established:

Lemma 3.1.7 *Under the conditions of this subsection, if $\theta_X \notin D(A)$, then property (3.1.21) of the function $\rho(\alpha)$ holds, and*

$$\lim_{\alpha \to \infty} \sigma(\alpha) = \| x_* \| = min\{ \| x \| \mid x \in \overline{D(A)} \}.$$

2. As it has been already mentioned, the regularized equation

$$A^h x + \alpha J^\mu x = f^\delta, \tag{3.1.22}$$

where $J^\mu : X \to X^*$ is the duality mapping with a gauge function $\mu(t)$, is of interest. We are going to study the properties of the functions $\sigma_\mu(\alpha) = \mu(\| x_\alpha^\gamma \|)$ and $\rho_\mu(\alpha) = \alpha \mu(\| x_\alpha^\gamma \|)$, where x_α^γ is a solution of (3.1.22).

Lemma 3.1.8 *The function $\sigma_\mu(\alpha)$ is single-valued, continuous and non-increasing when $\alpha \geq \alpha_0 > 0$. In addition, if $\theta_X \in D(A)$, then $\sigma_\mu(\alpha) \to 0$ as $\alpha \to \infty$, and if $\theta_X \notin D(A)$, then $\sigma_\mu(\alpha) \to \mu(\|x_*\|)$ as $\alpha \to \infty$.*

Proof. The operator J^μ is strictly monotone. Consequently, a solution x_α^γ of the equation (3.1.22) for each $\alpha > 0$ is unique. Then the function $\sigma_\mu(\alpha)$ is single-valued. The inequalities (3.1.5) and (3.1.6) accept, respectively, the following forms:

$$\alpha \langle J^\mu x_\alpha^\gamma - J^\mu x_\beta^\gamma, x_\alpha^\gamma - x_\beta^\gamma \rangle + (\alpha - \beta)\langle J^\mu x_\beta^\gamma, x_\alpha^\gamma - x_\beta^\gamma \rangle \leq 0 \tag{3.1.23}$$

and

$$\Big(\mu(\|x_\alpha^\gamma\|) - \mu(\|x_\beta^\gamma\|)\Big)(\|x_\alpha^\gamma\| - \|x_\beta^\gamma\|) \leq \frac{|\alpha - \beta|}{\alpha_0}\mu(\|x_\beta^\gamma\|)(\|x_\alpha^\gamma\| + \|x_\beta^\gamma\|),$$

because of the property (1.5.1) of J^μ. In Section 2.2 we proved from the inequality (2.2.8) that the sequence $\{x_\alpha^\gamma\}$ is bounded as $\alpha \geq \alpha_0 > 0$. Then the continuity of the function $\sigma_\mu(\alpha)$ as $\alpha \geq \alpha_0 > 0$ is guaranteed by the properties of the function $\mu(t)$. In view of the inequality (3.1.23), we establish, as in Lemma 3.1.1, that $\sigma_\mu(\alpha)$ is a non-increasing function.

Let $y_\alpha^\gamma \in A^h x_\alpha^\gamma$ satisfy the equality

$$y_\alpha^\gamma + \alpha J^\mu x_\alpha^\gamma = f^\delta. \tag{3.1.24}$$

Then

$$\sigma_\mu(\alpha) = \frac{\|f^\delta - y_\alpha^\gamma\|_*}{\alpha}. \tag{3.1.25}$$

Replacing in the proof of Lemma 3.1.1 normalized duality mappings J and J^* by J^μ and $(J^\nu)^*$, respectively (see Lemma 1.5.10), one can show that the sequence $\{y_\alpha^\gamma\}$ is bounded when $\alpha \to \infty$ and γ is fixed. Write down the monotonicity property of A^h for the points $x_\alpha^\gamma \in D(A)$ and $\theta_X \in D(A)$:

$$\langle y_\alpha^\gamma - y^h, x_\alpha^\gamma \rangle \geq 0 \quad \forall y^h \in A^h(\theta_X).$$

Then (3.1.24) yields the inequality

$$\langle (y_\alpha^\gamma - y^h), (J^\nu)^*(\alpha^{-1}(f^\delta - y_\alpha^\gamma))\rangle \geq 0, \quad \alpha > 0.$$

By making use of the definition of $(J^\nu)^*$, we obtain (3.1.11). Finally, (3.1.25) allows us to conclude that $\sigma_\mu(\alpha) \to 0$ as $\alpha \to \infty$.

Similarly to the proof of Lemma 3.1.7, one can be sure that if $\theta_X \notin D(A)$ then $\sigma_\mu(\alpha) \to \mu(\|x_*\|)$ as $\alpha \to \infty$. The lemma is proved. ∎

We now address the function $\rho_\mu(\alpha)$. First of all, the last lemma enables us to state that $\rho_\mu(\alpha)$ is single-valued and continuous. The limit relations (3.1.15) and (3.1.21) for $\rho_\mu(\alpha)$ are verified in the same way as in Lemmas 3.1.4 and 3.1.7. We present the final result.

Lemma 3.1.9 *If $\theta_X \in D(A)$, then*

$$\lim_{\alpha \to \infty} \rho_\mu(\alpha) = \|\bar{y}^h - f^\delta\|_*,$$

where $\bar{y}^h$ is the nearest point to f^δ in the set $A^h(\theta_X)$. If $\theta_X \notin D(A)$, then

$$\lim_{\alpha \to \infty} \rho_\mu(\alpha) = +\infty.$$

Remark 3.1.10 *By (1.5.8), we can rewrite (3.1.24) as follows:*

$$y_\alpha^\gamma + \alpha \frac{\mu(\|x_\alpha^\gamma\|)}{\|x_\alpha^\gamma\|} J x_\alpha^\gamma = f^\delta, \quad x_\alpha^\gamma \neq \theta_X.$$

Hence, the solution of (3.1.22) coincides with the solution of the equation

$$A^h x + \alpha' J x = f^\delta,$$

where

$$\alpha' = \alpha \frac{\mu(\|x_\alpha^\gamma\|)}{\|x_\alpha^\gamma\|}.$$

Still, as it has already been mentioned, the possibility of obtaining different estimates for regularized solutions essentially depends on a choice of the function $\mu(t)$.

3. Let $A : X \to X^*$ be a maximal monotone hemicontinuous operator and $\theta_X \in D(A)$. Suppose that the equation (3.1.1) is given with an exact operator A and perturbed right-hand side f^δ satisfying the condition

$$\|f - f^\delta\|_* \leq \delta, \tag{3.1.26}$$

such that in reality we solve the equation $Ax = f^\delta$. Consider the regularized equation

$$Ax + \alpha Jx = f^\delta. \tag{3.1.27}$$

Denote by x_α^δ its classical solution. Then

$$Ax_\alpha^\delta + \alpha Jx_\alpha^\delta = f^\delta. \tag{3.1.28}$$

Definition 3.1.11 *The value $\rho(\alpha) = \|Ax_\alpha^\delta - f^\delta\|_*$ is called the residual of the equation $Ax = f^\delta$ on the solution x_α^δ of the equation (3.1.27).*

It follows from (3.1.28) that

$$\rho(\alpha) = \|Ax_\alpha^\delta - f^\delta\|_* = \alpha \|x_\alpha^\delta\|. \tag{3.1.29}$$

By Lemma 3.1.4, the residual $\rho(\alpha)$ is single-valued, continuous and

$$\lim_{\alpha \to \infty} \rho(\alpha) = \|A(\theta_X) - f^\delta\|_*. \tag{3.1.30}$$

In view of (3.1.7), the estimate of solution x_α^δ to the equation (3.1.27) has the form

$$\|x_\alpha^\delta\| \leq 2\|x^*\| + \frac{\delta}{\alpha} \quad \forall x^* \in N.$$

Then (3.1.29) yields the relation

$$\rho(\alpha) \leq 2\alpha \|\bar{x}^*\| + \delta,$$

where $\bar{x}^* \in N$ is the minimal norm solution of the equation (3.1.1). Let α be such that

$$2\alpha \|\bar{x}^*\| < (k-1)\delta^p, \quad k > 1, \quad p \in (0,1].$$

We may consider, without loss of generality, that $\delta \leq 1$ because $\delta \to 0$. Hence,

$$\rho(\alpha) \leq (k-1)\delta^p + \delta < k\delta^p. \tag{3.1.31}$$

Assume that δ satisfies the inequality

$$\|A(\theta_X) - f^\delta\|_* > k\delta^p, \quad p \in (0,1], \quad k > 1. \tag{3.1.32}$$

Since (3.1.30) holds and since $\rho(\alpha)$ is continuous, we get from (3.1.31) and (3.1.32) that there exists at least one $\alpha = \bar{\alpha}(\delta)$ such that

$$\rho(\bar{\alpha}) = \bar{\alpha}\|x_{\bar{\alpha}}^\delta\| = k\delta^p, \quad k > 1, \quad p \in (0,1], \tag{3.1.33}$$

and

$$\bar{\alpha} > \frac{(k-1)\delta^p}{2\|\bar{x}^*\|}. \tag{3.1.34}$$

For every $\delta > 0$, we find $\alpha = \bar{\alpha}$ from the scalar equation (3.1.33) solving at the same time (3.1.27) with $\alpha = \bar{\alpha}$. Thus, we construct the sequence $\{x_{\bar{\alpha}}^\delta\}$. We study its behaviour as $\delta \to 0$ and at the same time the properties of the function $\bar{\alpha}(\delta)$.

At the beginning, let $p \in (0,1)$. By (3.1.34), it is easy to deduce the estimate

$$\|x_{\bar{\alpha}}^\delta\| = \frac{\rho(\bar{\alpha})}{\bar{\alpha}} \leq \frac{2k\delta^p\|\bar{x}^*\|}{(k-1)\delta^p} = \frac{2k\|\bar{x}^*\|}{k-1}.$$

Hence, the sequence $\{x_{\bar{\alpha}}^\delta\}$ is bounded, therefore, $x_{\bar{\alpha}}^\delta \rightharpoonup \bar{x} \in X$ as $\delta \to 0$. Furthermore, it follows from the rule of choosing $\bar{\alpha}$ and from Definition 3.1.11 that

$$\|Ax_{\bar{\alpha}}^\delta - f^\delta\|_* = k\delta^p \to 0 \quad \text{as} \quad \delta \to 0.$$

By Lemma 1.4.5, then $\bar{x} \in N$.

Taking into account (3.1.26) and (3.1.32), we obtain that $\theta_X \notin N$. Thus, $\bar{x}^* \neq \theta_X$. Now we find by (3.1.34) that

$$\frac{\delta}{\bar{\alpha}} \leq \frac{2\delta^{1-p}\|\bar{x}^*\|}{k-1}.$$

Consequently,

$$\frac{\delta}{\bar{\alpha}} \to 0 \quad \text{as} \quad \delta \to 0. \tag{3.1.35}$$

Show that $\bar{x} = \bar{x}^*$. As in the proof of Theorem 2.2.1, one gets

$$\langle Jx^*, x_{\bar{\alpha}}^\delta - x^* \rangle \leq \frac{\delta}{\bar{\alpha}}\|x_{\bar{\alpha}}^\delta - x^*\| \quad \forall x^* \in N.$$

It leads to the inequality

$$\langle Jx^*, \bar{x} - x^* \rangle \leq 0 \quad \forall x^* \in N,$$

because the sequence $\{x_{\bar{\alpha}}^{\delta}\} \rightharpoonup \bar{x} \in N$ as $\delta \to 0$. In addition, we proved in Theorem 2.2.1 that $\bar{x} = \bar{x}^*$ and $\|x_{\bar{\alpha}}^{\delta}\| \to \|\bar{x}^*\|$. This implies strong convergence of the sequence $\{x_{\bar{\alpha}}^{\delta}\}$ to $\bar{x}^*$ as $\delta \to 0$ in the E-space X.

Since $\|\bar{x}^*\| \neq 0$, there exists $\kappa > 0$ such that $\|x_{\bar{\alpha}}^{\delta}\| \geq \kappa$ for sufficiently small δ. Then from the equality

$$\bar{\alpha} = \frac{k\delta^p}{\|x_{\bar{\alpha}}^{\delta}\|} \tag{3.1.36}$$

it results that $\bar{\alpha} \leq \kappa^{-1} k\delta^p$. It is easy to see that $\bar{\alpha} \to 0$ as $\delta \to 0$.

Let now in (3.1.33) $p = 1$, that is, $\rho(\alpha) = k\delta$, $k > 1$. In this case

$$\frac{\delta}{\bar{\alpha}} \leq \frac{2\|\bar{x}^*\|}{k - 1}.$$

Assume that set N is a singleton and $N = \{x_0\}$. Repeating the previous arguments we obtain that $\{x_{\bar{\alpha}}^{\delta}\}$ weakly converges to x_0 as $\delta \to 0$. Then

$$\|x_0\| \leq \liminf_{\delta \to 0} \|x_{\bar{\alpha}}^{\delta}\|,$$

and from (3.1.32) we conclude that $x_0 \neq \theta_X$. Hence, the estimate

$$\|x_{\bar{\alpha}}^{\delta}\| \geq \kappa > 0$$

holds for all $\delta \in (0, 1]$, perhaps, excepting their finite number. Then (3.1.36) guarantees that $\bar{\alpha} \to 0$ as $\delta \to 0$. Thus, the following theorem (the residual principle for equations with maximal monotone hemicontinuous operators) has been proved:

Theorem 3.1.12 *Assume that X is an E-space with strictly convex dual space X^*, $A :$ $X \to X^*$ is a maximal monotone hemicontinuous operator, $\theta_X \in D(A)$, the equation (3.1.1) has a nonempty solution set N with unique minimal norm solution $\bar{x}^*$. Consider the regularized equation (3.1.27) with the conditions (3.1.26) and (3.1.32), $0 < \delta < 1$. Then there is at least one $\alpha = \bar{\alpha}$ satisfying (3.1.33), where $x_{\bar{\alpha}}^{\delta}$ is the solution of (3.1.27) with $\alpha = \bar{\alpha}$. In addition, let $\delta \to 0$. It results that: 1) $\bar{\alpha} \to 0$; 2) if $p \in (0, 1)$, then $x_{\bar{\alpha}}^{\delta} \to \bar{x}^*$ and $\dfrac{\delta}{\bar{\alpha}} \to 0$; 3) if $p = 1$, $N = \{x_0\}$, that is, the equation (3.1.1) has a unique solution, then $x_{\bar{\alpha}}^{\delta} \rightharpoonup x_0$ and there exists $C > 0$ such that $\dfrac{\delta}{\bar{\alpha}} \leq C$.*

Remark 3.1.13 *A similar theorem can be established for the regularized equation (3.1.2) with a perturbed operator A^h (see also Section 3.3).*

Remark 3.1.14 *Due to Lemma 3.1.7, if $\theta_X \notin D(A)$, then the assumption (3.1.32) in Theorem 3.1.12 is not necessary.*

Theorem 3.1.12 presents the residual principle for nonlinear equations with monotone hemicontinuous operators. It there asserts that the choice of regularization parameter according to the residual principle gives the regularizing algorithm. Moreover, it satisfies the sufficient convergence conditions of the operator regularization methods investigated in Chapter 2.

Let the sequence $\{x_{\bar{\alpha}}^{\delta}\}$ constructed by the residual principle converge to some element $x \in X$ as $\delta \to 0$ either strongly or weakly. Then $Ax_{\bar{\alpha}}^{\delta} \to f$ as $\delta \to 0$ in view of (3.1.28) and (3.1.33). Therefore, $f \in A\bar{x}$. This means that the equation (3.1.1) has a solution. Consequently, solvability of (3.1.1) is the necessary and sufficient convergence condition of the sequence $\{x_{\bar{\alpha}}^{\delta}\}$.

Next we show that the question arises if Theorem 3.1.12 is also true with $k = 1$. This problem is not only of theoretical but also of practical value, for instance, in the cases when the measurement accuracy of the right-hand side f of (3.1.1) cannot be made less than some $\delta_0 > 0$. We look for the conditions under which the given problem is positively solved.

Theorem 3.1.15 *Let X be an E-space with a strictly convex dual space X^*, $A : X \to X^*$ be a maximal monotone hemicontinuous operator, $\theta_X \in D(A)$ and the inequality*

$$\|A(\theta_X) - f^{\delta}\|_* > \delta \tag{3.1.37}$$

hold. Assume that there exists a number $r > 0$ such that for for all $x \in D(A)$ with $\|x\| \geq r$, one has

$$\langle Ax - f, x \rangle \geq 0. \tag{3.1.38}$$

Then there is at least one $\bar{\alpha} > 0$ satisfying the equation

$$\rho(\bar{\alpha}) = \bar{\alpha}\|x_{\bar{\alpha}}^{\delta}\| = \delta. \tag{3.1.39}$$

Moreover, if the equation (3.1.1) has a unique solution x_0, i.e., $N = \{x_0\}$ and in (3.1.38) the strict inequality appears, then $x_{\bar{\alpha}}^{\delta} \rightharpoonup x_0$ and there exists $C > 0$ such that $\dfrac{\delta}{\bar{\alpha}} \leq C$ as $\delta \to 0$.

Proof. Note first that $\theta_X \notin N$ because of (3.1.37). By (3.1.26), it is not difficult to be sure that

$$\langle Ax + \alpha Jx - f^{\delta}, x \rangle \geq \alpha\|x\|\left(\|x\| - \frac{\delta}{\alpha}\right) + \langle Ax - f, x \rangle.$$

Therefore, solutions x_{α}^{δ} of the equation (3.1.27) are bounded. Due to Theorem 1.7.9, there exists $\bar{\gamma} = max\{\delta\alpha^{-1}, r\}$ such that $\|x_{\alpha}^{\delta}\| \leq \bar{\gamma}$, and the estimate

$$\rho(\alpha) = \alpha\|x_{\alpha}^{\delta}\| \leq \alpha\bar{\gamma}$$

holds. If $\bar{\gamma} = r$ then the inequality $\rho(\alpha) < \delta$ is satisfied for sufficiently small α. If $\bar{\gamma} = \dfrac{\delta}{\alpha}$ then

$$\rho(\alpha) \leq \alpha\bar{\gamma} = \delta.$$

Since $\rho(\alpha)$ is continuous, (3.1.37) implies solvability of the equation (3.1.39). Now we may construct the sequence $\{x_{\bar{\alpha}}^{\delta}\}$ of solutions to the equation (3.1.27) with $\alpha = \bar{\alpha}$ and study its behavior as $\delta \to 0$. Show that $\{x_{\bar{\alpha}}^{\delta}\}$ is bounded. For this end, assume that $\|x_{\bar{\alpha}}^{\delta}\| \to \infty$ as $\delta \to 0$. If δ is sufficiently small then we have

$$
\begin{aligned}
0 &= \langle Ax_{\bar{\alpha}}^{\delta} + \bar{\alpha}Jx_{\bar{\alpha}}^{\delta} - f^{\delta}, x_{\bar{\alpha}}^{\delta} \rangle \\[2mm]
&= \langle Ax_{\bar{\alpha}}^{\delta} - f, x_{\bar{\alpha}}^{\delta} \rangle + \langle f - f^{\delta}, x_{\bar{\alpha}}^{\delta} \rangle + \bar{\alpha}\|x_{\bar{\alpha}}^{\delta}\|^2 \\[2mm]
&> \bar{\alpha}\|x_{\bar{\alpha}}^{\delta}\|^2 - \delta\|x_{\bar{\alpha}}^{\delta}\| = 0,
\end{aligned}
$$

where $\bar{\alpha}\|x_{\bar{\alpha}}^{\delta}\| = \delta$. We thus come to a contradiction. At the same time, the last equality means that there exists $C > 0$ such that $\dfrac{\delta}{\bar{\alpha}} \leq C$ as $\delta \to 0$. Then the weak convergence of $x_{\bar{\alpha}}^{\delta}$ to x_0 is established as in Theorem 3.1.12. ∎

Observe that under the conditions of Theorem 3.1.15 it is impossible to prove that simultaneously $\dfrac{\delta}{\bar{\alpha}} \to 0$ and $x_{\bar{\alpha}}^{\delta} \to x_0$ when $\delta \to 0$. This happens by reason of $\delta = \bar{\alpha}\|x_{\bar{\alpha}}^{\delta}\|$ and $x_0 \neq \theta_X$.

Theorem 3.1.16 *Assume that conditions of Theorem 3.1.15 hold. If $0 < \delta \leq 1$ and, instead of (3.1.37), the inequality*

$$
\|A(\theta_X) - f^{\delta}\|_* > \delta^p, \quad p \in (0, 1),
$$

is given, then the residual principle for choosing regularization parameter $\bar{\alpha}$ from the equation $\rho(\bar{\alpha}) = \delta^p$ takes place and it produces the convergence $x_{\bar{\alpha}}^{\delta} \to \bar{x}^*$ and $\dfrac{\delta}{\bar{\alpha}} \to 0$ as $\delta \to 0$.

Consider again the regularized equation

$$
Ax + \alpha J^{\mu}x = f^{\delta} \tag{3.1.40}
$$

with duality mapping J^{μ}. If x_{α}^{δ} is its solution then the residual

$$
\rho_{\mu}(\alpha) = \|Ax_{\alpha}^{\delta} - f^{\delta}\|_* = \alpha\mu(\|x_{\alpha}^{\delta}\|).
$$

We are not able to study the residual principle for the operator regularization method (3.1.40) with an arbitrary gauge function $\mu(t)$. However, we can do it if $\mu(t) = t^s$, $s \geq 1$. Denote by J^s the duality mapping with this gauge function and by $\rho_s(\alpha)$ the residual $\rho_{\mu}(\alpha)$, that is,

$$
\rho_s(\alpha) = \|Ax_{\alpha}^{\delta} - f^{\delta}\|_* = \alpha\|x_{\alpha}^{\delta}\|^s.
$$

Theorem 3.1.17 *Suppose that the conditions of Theorem 3.1.12 are fulfilled. Consider the equation (3.1.40) with $0 < \delta \leq 1$ as regularized to $Ax = f^{\delta}$. If x_{α}^{δ} is its solution, then there exists at least one value $\alpha = \bar{\alpha}$ such that*

$$
\rho_s(\bar{\alpha}) = \|Ax_{\bar{\alpha}}^{\delta} - f^{\delta}\|_* = k\delta^p, \quad k > 1, \quad p \in (0, 1]. \tag{3.1.41}
$$

Moreover, let $\delta \to 0$. It results that: 1) $\bar{\alpha} \to 0$; 2) if $p \in (0,1)$, then $x_{\bar{\alpha}}^{\delta} \to \bar{x}^*$ and $\dfrac{\delta}{\bar{\alpha}} \to 0$; 3) if $p = 1$ and the equation (3.1.1) is uniquely solvable, i.e., $N = \{x_0\}$, then $x_{\bar{\alpha}}^{\delta} \rightharpoonup x_0$ and there exists $C > 0$ such that $\dfrac{\delta}{\bar{\alpha}} \leq C$ as $\delta \to 0$.

Proof. According to (2.2.10), for all $x^* \in N$,

$$\|x_\alpha^\delta\| \leq \tau \|x^*\| + \left(\frac{\delta}{\alpha}\right)^\kappa, \quad \tau > 1, \quad \tau s \geq 2, \quad \kappa = \frac{1}{s}.$$

Then

$$\rho_s^\kappa(\alpha) = \alpha^\kappa \|x_\alpha^\delta\| \leq \tau \alpha^\kappa \|\bar{x}^*\| + \delta^\kappa.$$

Hence, the relation

$$\tau \alpha^\kappa \|\bar{x}^*\| \leq (k^\kappa - 1)\delta^{\kappa p}, \quad k > 1, \quad p \in (0,1],$$

holds with sufficiently small α, and

$$\rho_s(\alpha) < k\delta^p.$$

Then, by the condition (3.1.32) and by the continuity of $\rho_s(\alpha)$, there exists at least one $\alpha = \bar{\alpha}$ such that (3.1.41) is true (see Lemmas 3.1.8, 3.1.9). The final proof repeats the reasoning given in Theorem 3.1.12. ∎

3.2 Residual Principle for Accretive Equations

In what follows, $A : X \to X$ is a hemicontinuous accretive operator, $D(A) = X$, X is a reflexive strictly convex space together with its dual space X^*, $J : X \to X^*$ is a continuous and, at the same time, weak-to-weak continuous duality mapping and Banach space X possesses an approximation. Study the residual principle for this case. Let N be a non-empty solution set of the equation (3.1.1) in the classical sense. Assume that perturbed operators $A^h : X \to X$ are accretive and hemicontinuous, $D(A^h) = D(A)$ for all $h > 0$, $f^\delta \in X$ for all $\delta > 0$ and $\|f^\delta - f\| \leq \delta$. Consider the regularized equation

$$A^h x + \alpha x = f^\delta. \tag{3.2.1}$$

Let x_α^γ, $\gamma = (\delta, h)$, be its unique solution (see Sections 1.15 and 2.7), that is,

$$A^h x_\alpha^\gamma + \alpha x_\alpha^\gamma = f^\delta. \tag{3.2.2}$$

We study the functions $\sigma(\alpha) = \|x_\alpha^\gamma\|$ and $\rho(\alpha) = \alpha \|x_\alpha^\gamma\|$.

Lemma 3.2.1 *The function $\sigma(\alpha)$ is single-valued and continuous for $\alpha \geq \alpha_0 > 0$. Moreover, $\sigma(\alpha) \to 0$ as $\alpha \to \infty$.*

Proof. Obviously, $\sigma(\alpha)$ is single-valued because x_α^γ is a unique solution of (3.2.1) for each $\alpha \geq \alpha_0 > 0$. Let x_β^γ be a solution of (3.2.1) with $\alpha = \beta$. Then

$$A^h x_\beta^\gamma + \beta x_\beta^\gamma = f^\delta. \tag{3.2.3}$$

By (3.2.2) and (3.2.3), we have

$$\langle J(x_\alpha^\gamma - x_\beta^\gamma), A^h x_\alpha^\gamma - A^h x_\beta^\gamma \rangle + \langle J(x_\alpha^\gamma - x_\beta^\gamma), \alpha x_\alpha^\gamma - \beta x_\beta^\gamma \rangle = 0.$$

The accretiveness property of A^h implies

$$\langle J(x_\alpha^\gamma - x_\beta^\gamma), A^h x_\alpha^\gamma - A^h x_\beta^\gamma \rangle \geq 0.$$

Therefore,

$$\alpha \langle J(x_\alpha^\gamma - x_\beta^\gamma), x_\alpha^\gamma - x_\beta^\gamma \rangle + (\alpha - \beta)\langle J(x_\alpha^\gamma - x_\beta^\gamma), x_\beta^\gamma \rangle \leq 0.$$

From this inequality, one gets

$$(\|x_\alpha^\gamma\| - \|x_\beta^\gamma\|)^2 \leq \|x_\alpha^\gamma - x_\beta^\gamma\|^2 \leq \frac{|\alpha - \beta|}{\alpha_0} \|x_\beta^\gamma\|(\|x_\alpha^\gamma\| + \|x_\beta^\gamma\|).$$

We proved in Section 2.7 (see (2.7.10)) that the sequence $\{x_\alpha^\gamma\}$ is bounded when $\gamma \to 0$ and $\alpha \geq \alpha_0 > 0$. It results from the last inequality that the function $\sigma(\alpha)$ is continuous.

In its turn, the equality (3.2.2) yields

$$\sigma(\alpha) = \frac{\|A^h x_\alpha^\gamma - f^\delta\|}{\alpha} \tag{3.2.4}$$

and

$$\alpha J x_\alpha^\gamma = J(f^\delta - A^h x_\alpha^\gamma). \tag{3.2.5}$$

Since $D(A) = X$ and A^h are accretive operators, we have

$$\langle J x_\alpha^\gamma, A^h x_\alpha^\gamma - A^h(\theta_X) \rangle \geq 0. \tag{3.2.6}$$

Then

$$\langle J(f^\delta - A^h x_\alpha^\gamma), A^h x_\alpha^\gamma - A^h(\theta_X) \rangle \geq 0$$

in view of (3.2.5). It is easy to see that this inequality leads to the estimate

$$\|A^h x_\alpha^\gamma - f^\delta\| \leq \|A^h(\theta_X) - f^\delta\|, \tag{3.2.7}$$

that is, the sequence $\{A^h x_\alpha^\gamma - f^\delta\}$ is bounded. Obviously, the last assertion of the lemma arises from (3.2.4). ∎

Lemma 3.2.2 *The function $\rho(\alpha)$ is single-valued and continuous for $\alpha \geq \alpha_0 > 0$. Moreover,*

$$\lim_{\alpha \to \infty} \rho(\alpha) = \|A^h(\theta_X) - f^\delta\|. \tag{3.2.8}$$

Proof. It follows from Lemma 3.2.1 that the function $\rho(\alpha)$ is single-valued and continuous. The limit (3.2.8) is established on the basis of (3.2.7), as in Lemma 3.1.4. ∎

In the case of linear accretive operators, it is possible to obtain some important additional properties of the functions $\sigma(\alpha)$ and $\rho(\alpha)$. Indeed, a solution of (3.2.1) can be represented as $x_\alpha^\gamma = T_h f^\delta$, where

$$T_h = (A^h + \alpha I)^{-1}.$$

It is well known that the operator T_h exists for all $\alpha > 0$, continuous, bounded and $|T_h| \leq \alpha^{-1}$. Then

$$\sigma(\alpha) \leq |T_h| \|f^\delta\| \leq \alpha^{-1} \|f^\delta\| \to 0 \quad \text{as} \quad \alpha \to \infty.$$

By (3.2.2) and (3.2.3), we have

$$(A^h + \beta I)(x_\alpha^\gamma - x_\beta^\gamma) = -(\alpha - \beta)x_\alpha^\gamma,$$

from which one gets

$$\frac{x_\alpha^\gamma - x_\beta^\gamma}{\alpha - \beta} = -(A^h + \beta I)^{-1}x_\alpha^\gamma. \tag{3.2.9}$$

If $\beta \to \alpha$, then the limit of the right-hand side of (3.2.9) exists, hence, there exists a limit of the left-hand side. Therefore, as $\beta \to \alpha$, (3.2.9) implies

$$\frac{dx_\alpha^\gamma}{d\alpha} = -(A^h + \alpha I)^{-1}x_\alpha^\gamma \tag{3.2.10}$$

or

$$A^h \frac{dx_\alpha^\gamma}{d\alpha} + \alpha \frac{dx_\alpha^\gamma}{d\alpha} = -x_\alpha^\gamma.$$

Then the equality

$$\left\langle J\frac{dx_\alpha^\gamma}{d\alpha}, A^h \frac{dx_\alpha^\gamma}{d\alpha} \right\rangle + \alpha \left\langle J\frac{dx_\alpha^\gamma}{d\alpha}, \frac{dx_\alpha^\gamma}{d\alpha} \right\rangle = -\left\langle J\frac{dx_\alpha^\gamma}{d\alpha}, x_\alpha^\gamma \right\rangle$$

is satisfied. In view of the accretiveness of A^h, we deduce that

$$\left\| \frac{dx_\alpha^\gamma}{d\alpha} \right\| \leq \alpha^{-1} \|x_\alpha^\gamma\|$$

and

$$\left\langle J\frac{dx_\alpha^\gamma}{d\alpha}, x_\alpha^\gamma \right\rangle \leq -\alpha \left\| \frac{dx_\alpha^\gamma}{d\alpha} \right\|^2 \leq 0. \tag{3.2.11}$$

By (3.2.10),

$$\frac{d\sigma^2(\alpha)}{d\alpha} = -2\langle Jx_\alpha^\gamma, (A^h + \alpha I)^{-1}x_\alpha^\gamma \rangle.$$

Then the estimate

$$\frac{d\sigma^2(\alpha)}{d\alpha} \geq -\frac{2\|x_\alpha^\gamma\|^2}{\alpha} = -\frac{2\sigma^2(\alpha)}{\alpha} \tag{3.2.12}$$

holds. Consequently, the function $\sigma(\alpha)$ is continuous and if $\alpha \geq \alpha_0 > 0$ then

$$\sigma(\alpha) \geq \frac{\alpha_0 \sigma(\alpha_0)}{\alpha}.$$

Write now the obvious equalities

$$\frac{d\rho^2(\alpha)}{d\alpha} = 2\alpha\sigma^2(\alpha) + \alpha^2\frac{d\sigma^2(\alpha)}{d\alpha} = \alpha\Big(2\sigma^2(\alpha) + \alpha\frac{d\sigma^2(\alpha)}{d\alpha}\Big). \qquad (3.2.13)$$

Since (3.2.12) holds, it is clear that

$$\frac{d\rho^2(\alpha)}{d\alpha} \geq 0.$$

We finally obtain that the function $\rho(\alpha)$ for linear accretive operators A^h does not decrease as $\alpha \geq \alpha_0 > 0$.

Observe that in Hilbert space

$$\frac{d\sigma^2(\alpha)}{d\alpha} \leq -2\alpha\Big\|\frac{dx_\alpha^\gamma}{d\alpha}\Big\|^2 \leq 0.$$

This follows from (3.2.11). Hence, the function $\sigma(\alpha)$ does not increase as $\alpha \geq \alpha_0 > 0$.

Considering the regularized equation

$$Ax + \alpha x = f^\delta \qquad (3.2.14)$$

with an accretive operator A and denoting by x_α^δ its solution, one can give the same Definition 3.1.11 of the residual to the equation $Ax = f^\delta$ on a solution x_α^δ, namely, $\rho(\alpha) = \|Ax_\alpha^\delta - f^\delta\|$. The following residual principle for equations with accretive operators is valid:

Theorem 3.2.3 *Let $A : X \to X$ be a hemicontinuous accretive operator with $D(A) = X$, X be a reflexive strictly convex space with strictly convex dual space X^*, $J : X \to X^*$ be a continuous and weak-to-weak continuous duality mapping in X. Assume that Banach space X possesses an approximation, $\delta \in (0, 1]$ and*

$$\|A(\theta_X) - f^\delta\| > k\delta^p, \quad k > 1, \quad p \in (0, 1].$$

Then there exists at least one $\alpha = \bar{\alpha}$ satisfying the equation

$$\rho(\bar{\alpha}) = \|Ax_{\bar{\alpha}}^\delta - f^\delta\| = k\delta^p, \qquad (3.2.15)$$

where $x_{\bar{\alpha}}^\delta$ is the solution of the equation (3.2.14) with $\alpha = \bar{\alpha}$. Furthermore, let $\delta \to 0$. One has: 1) $\bar{\alpha} \to 0$; 2) if $p \in (0, 1)$, then $\dfrac{\delta}{\bar{\alpha}} \to 0$ and $x_{\bar{\alpha}}^\delta \to \bar{x}^*$, where $\bar{x}^* \in N$ is the unique solution of the inequality

$$\langle J(\bar{x} - x^*), \bar{x}^* \rangle \leq 0 \quad \forall x^* \in N;$$

3) if $p = 1$ and $N = \{x_0\}$, then $x_{\bar{\alpha}}^\delta \rightharpoonup x_0$ and there exists $C > 0$ such that $\dfrac{\delta}{\bar{\alpha}} \leq C$.

Proof. By the equations (3.1.1) and (3.2.14), we have for all $x^* \in N$,

$$\langle J(x_\alpha^\delta - x^*), Ax_\alpha^\delta - Ax^* \rangle \quad + \quad \alpha \langle J(x_\alpha^\delta - x^*), x_\alpha^\delta - x^* \rangle$$
$$= \quad \langle J(x_\alpha^\delta - x^*), f^\delta - f \rangle - \alpha \langle J(x_\alpha^\delta - x^*), x^* \rangle.$$

Taking into consideration the accretiveness of A, one gets

$$\|x_\alpha^\delta - x^*\|^2 \leq \frac{\delta}{\alpha}\|x_\alpha^\delta - x^*\| + \|x^*\|\|x_\alpha^\delta - x^*\|$$

or

$$\|x_\alpha^\delta - x^*\| \leq \frac{\delta}{\alpha} + \|x^*\|.$$

Consequently,

$$\|x_\alpha^\delta\| \leq \frac{\delta}{\alpha} + 2\|x^*\|.$$

Due to Lemmas 3.2.1 and 3.2.2, the theorem is proved by use of the same arguments as in Theorem 3.1.12 for the monotone case. ∎

3.3 Generalized Residual Principle

1. Let X be an E-space with strictly convex dual space X^*. Assume that the equation (3.1.1) with maximal monotone operator $A : X \to 2^{X^*}$ is solved, N is its nonempty solution set and the sequence $\{f^\delta\}$ of elements $f^\delta \in X^*$, $\delta > 0$, and the sequence $\{A^h\}$ of maximal monotone, possible multiple-valued, operators $A^h : X \to 2^{X^*}$, $h > 0$, are given. Besides, assume that $D(A^h) = D(A)$, (3.1.26) holds and

$$\mathcal{H}_{X^*}(Ax, A^h x) \leq hg(\|x\|) \quad \forall x \in D(A), \tag{3.3.1}$$

where $g(t)$ is a continuous non-negative and increasing function for all $t \geq 0$. This means that in reality we solve the equation

$$A^h x = f^\delta. \tag{3.3.2}$$

Study the regularized equation written as

$$A^h x + \alpha J x = f^\delta, \tag{3.3.3}$$

where J is the normalized duality mapping. Let x_α^γ be a unique solution of (3.3.3) with $\gamma = (\delta, h)$. Then Lemmas 3.1.1 and 3.1.4 enable us to propose the following definition:

Definition 3.3.1 *The value $\rho(\alpha) = \alpha\|x_\alpha^\gamma\|$ is called the generalized residual of the equation (3.3.2) in the solution x_α^γ of the equation (3.3.3).*

The generalized residual is a single-valued and continuous function of the parameter α, though operators A^h are not continuous, in general. Moreover, it follows from (3.3.3) that there exists an element $y_\alpha^\gamma \in A^h x_\alpha^\gamma$ such that

$$\rho(\alpha) = \|y_\alpha^\gamma - f^\delta\|_*.$$

To state the residual principle for such operators, it is necessary to evaluate the functional $\|x_\alpha^\gamma\|$ from above. In Section 2.2, the following estimate has been obtained:

$$\|x_\alpha^\gamma\| \leq 2\|x^*\| + \frac{\delta}{\alpha} + \frac{h}{\alpha}g(\|x^*\|) \quad \forall x^* \in N. \tag{3.3.4}$$

Consequently,

$$\rho(\alpha) = \|y_\alpha^\gamma - f^\delta\|_* \leq 2\alpha\|x^*\| + \delta + hg(\|x^*\|).$$

By the same way as in the proof of Theorem 3.1.12, we are able to determine $\bar{\alpha}$ from the scalar equation

$$\rho(\bar{\alpha}) = \Big(k + g(\|x^*\|)\Big)(\delta + h)^p, \quad k > 1, \quad p \in (0, 1].$$

The quantity $\|x^*\|$ is not known. However, it may happen that an estimate $\|x^*\| \leq c$ is known. In this case, using the properties of $g(t)$, we find that

$$\rho(\alpha) \leq 2\alpha c + \delta + hg(c).$$

Then it is possible to define $\alpha = \bar{\alpha}$ as a solution of the equation

$$\rho(\bar{\alpha}) = \Big(k + g(c)\Big)(\delta + h)^p, \quad k > 1, \quad p \in (0, 1].$$

If c is not known then we act in the following way. First of all, recall that the estimate (3.3.4) was obtained in Section 2.2 from the equality

$$\langle y_\alpha^\gamma - f, x_\alpha^\gamma - x^* \rangle + \alpha\langle Jx_\alpha^\gamma, x_\alpha^\gamma - x^* \rangle = \langle f^\delta - f, x_\alpha^\gamma - x^* \rangle \quad \forall x^* \in N,$$

applying the monotonicity property of A^h. If we use in the same equality the monotonicity of A, then we can write

$$\|x_\alpha^\gamma\|^2 - \|x_\alpha^\gamma\|\Big(\|x^*\| + \frac{h}{\alpha}g(\|x_\alpha^\gamma\|) + \frac{\delta}{\alpha}\Big) - \frac{h}{\alpha}g(\|x_\alpha^\gamma\|)\|x^*\| - \frac{\delta}{\alpha}\|x^*\| \leq 0.$$

From this it follows that

$$\|x_\alpha^\gamma\| \leq 2\|x^*\| + \frac{\delta}{\alpha} + \frac{h}{\alpha}g(\|x_\alpha^\gamma\|) \quad \forall x^* \in N. \tag{3.3.5}$$

This inequality plays an important role to establish the next theorem (generalized residual principle).

Theorem 3.3.2 *Let $A : X \to 2^{X^*}$ and $A^h : X \to 2^{X^*}$ be maximal monotone operators, $h > 0$, $D(A) = D(A^h)$, the conditions (3.1.26) and (3.3.1) hold and $0 < \delta + h \le 1$. If $\theta_X \in D(A)$, then additionally assume that the following inequality is fulfilled:*

$$\|y_*^h - f^\delta\|_* > \left(k + g(0)\right)(\delta + h)^p, \quad k > 1, \quad p \in (0,1], \tag{3.3.6}$$

where y_*^h is the nearest to f^δ element from $A^h(\theta_X)$ and $g(t)$ is a continuous non-negative and increasing function for $t \ge 0$. Then there exists a unique solution $\bar{\alpha}$ of the equation

$$\rho(\bar{\alpha}) = \left(k + g(\|x_{\bar{\alpha}}^\gamma\|)\right)(\delta + h)^p, \tag{3.3.7}$$

where $x_{\bar{\alpha}}^\gamma$ is the solution of (3.3.3) with $\alpha = \bar{\alpha}$. Moreover, let $\gamma \to 0$. It results: 1) $\bar{\alpha} \to 0$; 2) if $p \in (0,1)$, then $x_{\bar{\alpha}}^\gamma \to \bar{x}^*$ and $\dfrac{\delta + h}{\bar{\alpha}} \to 0$; 3) if $p = 1$ and $N = \{x_0\}$, then $x_{\bar{\alpha}}^\gamma \rightharpoonup x_0$ and there exists $C > 0$ such that $\dfrac{\delta + h}{\bar{\alpha}} \le C$.

Proof. By (3.3.5), we have

$$\rho(\alpha) = \|y_\alpha^\gamma - f^\delta\|_* = \alpha\|x_\alpha^\gamma\| \le 2\alpha\|x^*\| + \delta + hg(\|x_\alpha^\gamma\|),$$

where $y_\alpha^\gamma \in A^h x_\alpha^\gamma$ and

$$y_\alpha^\gamma + \alpha J x_\alpha^\gamma = f^\delta.$$

Take α so small that the inequality

$$2\alpha\|\bar{x}^*\| < (k-1)(\delta + h)^p, \quad k > 1, \quad p \in (0,1], \tag{3.3.8}$$

is satisfied. Then

$$
\begin{aligned}
\rho(\alpha) \;&<\; (k-1)(\delta + h)^p + \delta + hg(\|x_\alpha^\gamma\|) \\[4pt]
&\le\; (k-1)(\delta + h)^p + \left(1 + g(\|x_\alpha^\gamma\|)\right)(\delta + h)^p \\[4pt]
&=\; \left(k + g(\|x_\alpha^\gamma\|)\right)(\delta + h)^p. \tag{3.3.9}
\end{aligned}
$$

Construct the function

$$d(\alpha) = \rho(\alpha) - \left(k + g(\|x_\alpha^\gamma\|)\right)(\delta + h)^p.$$

Owing to the continuity of $g(t)$ and Lemmas 3.1.1 and 3.1.2, $d(\alpha)$ is also continuous for $\alpha \ge \alpha_0 > 0$. By the same lemmas, if $\theta_X \in D(A)$ then

$$\lim_{\alpha \to \infty} d(\alpha) = \|y_*^h - f^\delta\|_* - \left(k + g(0)\right)(\delta + h)^p.$$

Applying the condition (3.3.6), we come to the conclusion that

$$\lim_{\alpha \to \infty} d(\alpha) > 0.$$

At the same time, by (3.3.9), there exists $\alpha > 0$ for which $d(\alpha) < 0$. Since $d(\alpha)$ is continuous, there exists at least one $\bar{\alpha}$ which satisfies (3.3.7).

If $\theta_X \notin D(A)$, then the residual property (3.1.21) appears and, by virtue of Lemma 3.1.7,

$$\lim_{\alpha \to \infty} \|x_\alpha^\gamma\| = \|x_*\| = min\{\|x\| \mid x \in \overline{D(A)}\}.$$

Hence,

$$\lim_{\alpha \to \infty} d(\alpha) = +\infty.$$

Thus, in this case, as well, the existence problem of $\bar{\alpha}$ satisfying the equation (3.3.7) is solved positively.

Show by contradiction that $\bar{\alpha}$ is unique. Suppose that for given γ there are $\bar{\alpha}$ and $\bar{\beta}$ such that

$$\rho(\bar{\alpha}) = \Big(k + g(\|x_{\bar{\alpha}}^\gamma\|)\Big)(\delta + h)^p \tag{3.3.10}$$

and

$$\rho(\bar{\beta}) = \Big(k + g(\|x_{\bar{\beta}}^\gamma\|)\Big)(\delta + h)^p, \tag{3.3.11}$$

where $x_{\bar{\alpha}}^\gamma$ and $x_{\bar{\beta}}^\gamma$ are solutions of the regularized equation (3.3.3) with $\alpha = \bar{\alpha}$ and $\alpha = \bar{\beta}$, respectively. This means that the following equalities hold:

$$y_{\bar{\alpha}}^\gamma + \bar{\alpha} J x_{\bar{\alpha}}^\gamma = f^\delta, \quad y_{\bar{\alpha}}^\gamma \in A^h x_{\bar{\alpha}}^\gamma, \tag{3.3.12}$$

and

$$y_{\bar{\beta}}^\gamma + \bar{\beta} J x_{\bar{\beta}}^\gamma = f^\delta, \quad y_{\bar{\beta}}^\gamma \in A^h x_{\bar{\beta}}^\gamma. \tag{3.3.13}$$

If $x_{\bar{\alpha}}^\gamma = x_{\bar{\beta}}^\gamma$ then (3.3.10) and (3.3.11) imply $\bar{\alpha} = \bar{\beta}$. Therefore, assume further that $x_{\bar{\alpha}}^\gamma \neq x_{\bar{\beta}}^\gamma$. Then, according to Corollary 1.5.3, there are uniquely defined elements $e_{\bar{\alpha}}^*$ and $e_{\bar{\beta}}^*$ such that

$$\|e_{\bar{\alpha}}^*\|_* = \|e_{\bar{\beta}}^*\|_* = 1, \quad \langle e_{\bar{\alpha}}^*, x_{\bar{\alpha}}^\gamma \rangle = \|x_{\bar{\alpha}}^\gamma\|, \quad \langle e_{\bar{\beta}}^*, x_{\bar{\beta}}^\gamma \rangle = \|x_{\bar{\beta}}^\gamma\|. \tag{3.3.14}$$

Using (3.3.10) - (3.3.13) we calculate

$$\langle \bar{\alpha} J x_{\bar{\alpha}}^\gamma - \bar{\beta} J x_{\bar{\beta}}^\gamma, x_{\bar{\alpha}}^\gamma - x_{\bar{\beta}}^\gamma \rangle = \langle \bar{\alpha} \|x_{\bar{\alpha}}^\gamma\| e_{\bar{\alpha}}^* - \bar{\beta} \|x_{\bar{\beta}}^\gamma\| e_{\bar{\beta}}^*, x_{\bar{\alpha}}^\gamma - x_{\bar{\beta}}^\gamma \rangle$$

$$= k(\delta + h)^p \langle e_{\bar{\alpha}}^* - e_{\bar{\beta}}^*, x_{\bar{\alpha}}^\gamma - x_{\bar{\beta}}^\gamma \rangle$$

$$+ (\delta + h)^p \langle g(\|x_{\bar{\alpha}}^\gamma\|) e_{\bar{\alpha}}^* - g(\|x_{\bar{\beta}}^\gamma\|) e_{\bar{\beta}}^*, x_{\bar{\alpha}}^\gamma - x_{\bar{\beta}}^\gamma \rangle. \tag{3.3.15}$$

By (3.3.14), it is not difficult to verify that

$$\langle e_{\bar{\alpha}}^* - e_{\bar{\beta}}^*, x_{\bar{\alpha}}^\gamma - x_{\bar{\beta}}^\gamma \rangle \geq 0.$$

Since the function $g(t)$ increases, we deduce by applying again (3.3.14) that

$$\langle g(\|x_{\bar{\alpha}}^\gamma\|) e_{\bar{\alpha}}^* - g(\|x_{\bar{\beta}}^\gamma\|) e_{\bar{\beta}}^*, x_{\bar{\alpha}}^\gamma - x_{\bar{\beta}}^\gamma \rangle \geq \Big(g(\|x_{\bar{\alpha}}^\gamma\|) - g(\|x_{\bar{\beta}}^\gamma\|)\Big)(\|x_{\bar{\alpha}}^\gamma\| - \|x_{\bar{\beta}}^\gamma\|) > 0.$$

Consequently, the expression in the right-hand side of (3.3.15) is strictly positive. Next, (3.3.12) and (3.3.13) imply

$$\langle y_{\bar{\alpha}}^{\gamma} - y_{\bar{\beta}}^{\gamma}, x_{\bar{\alpha}}^{\gamma} - x_{\bar{\beta}}^{\gamma} \rangle + \langle \bar{\alpha} J x_{\bar{\alpha}}^{\gamma} - \bar{\beta} J x_{\bar{\beta}}^{\gamma}, x_{\bar{\alpha}}^{\gamma} - x_{\bar{\beta}}^{\gamma} \rangle = 0.$$

Hence,

$$\langle y_{\bar{\alpha}}^{\gamma} - y_{\bar{\beta}}^{\gamma}, x_{\bar{\alpha}}^{\gamma} - x_{\bar{\beta}}^{\gamma} \rangle < 0, \quad y_{\bar{\alpha}}^{\gamma} \in A^h x_{\bar{\alpha}}^{\gamma}, \quad x_{\bar{\beta}}^{\gamma} \in A^h x_{\bar{\beta}}^{\gamma},$$

and this contradicts the monotonicity of A^h. The uniqueness of $\bar{\alpha}$ is thus proved.

Note that the uniqueness proof of $\bar{\alpha}$ was done above, in fact, by the hypothesis that solutions $x_{\bar{\alpha}}^{\gamma}$ and $x_{\bar{\beta}}^{\gamma}$ of the regularized equation (3.3.3) are not θ_X. Let now, for example, $x_{\bar{\alpha}}^{\gamma} = \theta_X$ be given. Then, by (3.3.3), there exists $y^h \in A^h(\theta_X)$ such that $\|y^h - f^\delta\|_* = 0$. But this contradicts the condition (3.3.6) of the theorem.

Observe that (3.3.8) yields the estimate

$$\bar{\alpha} > \frac{k-1}{2\|\bar{x}^*\|}(\delta + h)^p,$$

therefore,

$$\frac{\delta + h}{\bar{\alpha}} \leq \frac{2\|\bar{x}^*\|}{k-1}(\delta + h)^{1-p}.$$

Hence, if $p \in (0, 1)$ then

$$\lim_{\gamma \to 0} \frac{\delta + h}{\bar{\alpha}} = 0.$$

In its turn, if $p = 1$ then there exists a constant $C > 0$ such that

$$\lim_{\gamma \to 0} \frac{\delta + h}{\bar{\alpha}} \leq C,$$

that is, the sequence $\left\{\dfrac{\delta + h}{\bar{\alpha}}\right\}$ is bounded. It follows from the proof of Theorem 2.2.1 that $\{x_{\bar{\alpha}}^{\gamma}\}$ is also bounded as $\gamma \to 0$. Then there exists a subsequence (we do not change its denotation) which weakly converges to some $\bar{x} \in X$. Since (3.3.12) holds, the equation (3.3.7) enables us to obtain the strong convergence of $y_{\bar{\alpha}}^{\gamma}$ to f.

Write down the monotonicity condition of A^h :

$$\langle y^h - y_{\bar{\alpha}}^{\gamma}, x - x_{\bar{\alpha}}^{\gamma} \rangle \geq 0 \quad \forall x \in D(A), \quad \forall y^h \in A^h x.$$

By (3.3.1), after passing in the latter inequality to the limit as $\gamma \to 0$, we come to the inequality

$$\langle y - f, x - \bar{x} \rangle \geq 0 \quad \forall x \in D(A), \quad \forall y \in Ax.$$

This means that $\bar{x} \in N$. Thus, the conclusion 3) of the theorem is completely proved.

If $p \in (0, 1)$ then the assertion 2) is guaranteed by the proof of Theorem 2.2.1.

Show now that $\bar{x}^* \neq \theta_X$. Indeed, assume that is not the case. Then, by the approximate data, we obtain for some $y^h \in A^h(\theta_X)$ the following:

$$\|y^h - f^\delta\|_* \leq \|y^h - f\|_* + \|f - f^\delta\|_* \leq hg(0) + \delta,$$

where $f \in A(\theta_X)$. Thus, we arrive at a contradiction with the condition (3.3.6). Since $\bar{x}^* \neq \theta_X$, there exists $c > 0$ such that $\|x_{\bar{\alpha}}^{\gamma}\| \geq c$ for sufficiently small γ. Observe that here we made use again of the weak lower semicontinuity of the norm in a Banach space. Then

$$\bar{\alpha} = \frac{k + g(\|x_{\bar{\alpha}}^{\gamma}\|)}{\|x_{\bar{\alpha}}^{\gamma}\|}(\delta + h)^p \leq \frac{k + g(\|x_{\bar{\alpha}}^{\gamma}\|)}{c}(\delta + h)^p.$$

Consequently, $\bar{\alpha} \to 0$ as $\gamma \to 0$ because $\{x_{\bar{\alpha}}^{\gamma}\}$ is bounded and $g(t)$ is continuous. The proof of the theorem is now accomplished. ∎

Remark 3.3.3 *If operator A is given exactly, i.e., $h = 0$ and if $g(t)$ is bounded, then the residual principle takes the form (3.1.33). If A is strictly monotone, then $\bar{\alpha}$ in Theorems 3.1.12, 3.1.15 and 3.1.17 is uniquely defined. In this case, the residual $\rho(\alpha) = \|y_{\alpha}^{\delta} - f^{\delta}\|_*$ is the increasing function of α.*

2. As a rule, the parameter $\bar{\alpha}$ is defined solving approximately the scalar equation (3.3.7). In this way, the value of $\bar{\alpha}$ can be found inexactly. Therefore, it is desirable to establish a continuous dependence of the regularized solution x_{α}^{γ} on perturbation of α with fixed γ. Let $\alpha \to \beta$. Using the fact that the function $\sigma(\alpha) = \|x_{\alpha}^{\gamma}\|$ is continuous, we obtain the convergence of $\|x_{\alpha}^{\gamma}\|$ to $\|x_{\beta}^{\gamma}\|$ as $\alpha \geq \alpha_0 > 0$ and $\beta \geq \alpha_0 > 0$. By Lemma 3.1.1, the sequence $\{x_{\alpha}^{\gamma}\}$ is bounded for $\alpha \geq \alpha_0 > 0$. Hence, $x_{\alpha}^{\gamma} \rightharpoonup \bar{x}_{\beta}$ as $\alpha \to \beta$. Since x_{α}^{γ} satisfies the equation (3.3.3), the monotonicity condition of the operator $A^h + \alpha J$ gives the following inequality:

$$\langle y^h + \alpha Jx - f^{\delta}, x - x_{\alpha}^{\gamma} \rangle \geq 0 \quad \forall x \in D(A), \quad \forall y^h \in A^h x.$$

Going to the limit when $\alpha \to \beta$ we deduce

$$\langle y^h + \beta Jx - f^{\delta}, x - \bar{x}_{\beta} \rangle \geq 0 \quad \forall x \in D(A), \quad \forall y^h \in A^h x.$$

It results from this relation that $\bar{x}_{\beta}$ is a solution of the equation

$$A^h x + \beta Jx = f^{\delta}. \tag{3.3.16}$$

We know that (3.3.16) has a unique solution x_{β}^{γ}. Consequently, $\bar{x}_{\beta} = x_{\beta}^{\gamma}$ and $x_{\alpha}^{\gamma} \rightharpoonup x_{\beta}^{\gamma}$. Finally, in the E-space X we have the strong convergence of x_{α}^{γ} to x_{β}^{γ} as $\alpha \to \beta$, provided that $\alpha \geq \alpha_0 > 0$ and $\beta \geq \alpha_0 > 0$. Thus, solutions of the regularized equation (3.3.3) are stable with respect to errors of the regularization parameter α.

3. Consider now the equation

$$A^h x + \alpha J^{s+1} x = f^{\delta}, \tag{3.3.17}$$

where $J^{s+1} : X \to X^*$ is a duality mapping with the gauge function $\mu(t) = t^s$, $s \geq 1$.

Theorem 3.3.4 *If the condition (3.3.6) of Theorem 3.3.2 is replaced by*

$$\|y_*^h - f^{\delta}\|_* > \left(\bar{k} + (1 + g(0))^{\kappa}\right)^s (\delta + h)^p, \quad \bar{k} > 0, \quad \kappa = 1/s, \quad p \in (0, 1],$$

then there exists a unique $\bar{\alpha}$ satisfying the equation

$$\rho_s(\bar{\alpha}) = \bar{\alpha}\|x_{\bar{\alpha}}^\gamma\|^s = \left(\bar{k} + (1 + g(\|x_{\bar{\alpha}}^\gamma\|))^\kappa\right)^s(\delta + h)^p,$$

where $x_{\bar{\alpha}}^\gamma$ is the solution of (3.3.17) with $\alpha = \bar{\alpha}$. Moreover, the conclusions 1) - 3) of Theorem 3.3.2 hold.

Proof. By analogy with (2.2.10), the following estimate is valid:

$$\|x_\alpha^\gamma\| \leq \tau\|x^*\| + \left(\frac{\delta}{\alpha} + \frac{h}{\alpha}g(\|x_\alpha^\gamma\|)\right)^\kappa \quad \forall x^* \in N,$$

where $\tau > 1$, $\tau s \geq 2$, $s \geq 1$. Hence,

$$\rho_s(\alpha) = \alpha\|x_\alpha^\gamma\|^s \leq \left(\alpha^\kappa\tau\|\bar{x}^*\| + (\delta + hg(\|x_\alpha^\gamma\|))^\kappa\right)^s.$$

If we choose small enough α to satisfy the relation

$$\alpha^\kappa\tau\|\bar{x}^*\| < \bar{k}(\delta + h)^{p\kappa}, \quad \bar{k} > 0, \quad p \in (0, 1], \tag{3.3.18}$$

then

$$\begin{aligned}
\rho_s(\alpha) \quad &< \quad \left(\bar{k}(\delta + h)^{p\kappa} + (\delta + hg(\|x_\alpha^\gamma\|))^\kappa\right)^s \\
&\leq \quad \left(\bar{k} + (1 + g(\|x_\alpha^\gamma\|))^\kappa\right)^s(\delta + h)^p.
\end{aligned}$$

Consequently, there exists $\alpha > 0$ such that

$$\rho_s(\alpha) < \left(\bar{k} + (1 + g(\|x_\alpha^\gamma\|))^\kappa\right)^s(\delta + h)^p.$$

Now the proof does not differ greatly from that of the previous theorem. ∎

4. Let $A : X \to 2^{X^*}$ and $A^h : X \to 2^{X^*}$ be arbitrary monotone operators and we assume that the conditions of Theorem 2.2.4 are satisfied excepting (2.1.7). We study the regularized equation (3.3.3). Let $\bar{A}^h : X \to 2^{X^*}$ be maximal monotone extensions of A^h. Then by Lemma 1.9.8, (3.3.3) is equivalent to the equation

$$\bar{A}^h x + \alpha J x = f^\delta, \tag{3.3.19}$$

if a solution of (3.3.3) is understood in the generalized sense.

Definition 3.3.1 allows us to construct the generalized residual principle for general monotone, even discontinuous, operators, as well. Let x_α^γ be a solution of (3.3.19) in the sense of inclusion. Then there exists an element $\bar{y}_\alpha^\gamma \in \bar{A}^h x_\alpha^\gamma$ such that

$$\bar{y}_\alpha^\gamma + \alpha J x_\alpha^\gamma = f^\delta.$$

In this case, the generalized residual for equations $A^h x = f^\delta$ in the point x_α^γ can be defined as

$$\rho(\alpha) = \alpha\|x_\alpha^\gamma\|$$

because

$$\alpha \|x_\alpha^\gamma\| = \|\bar{y}_\alpha^\gamma - f^\delta\|_*.$$

If A^h is a maximal monotone operator then $\bar{y}_\alpha^\gamma = y_\alpha^\gamma \in A^h x_\alpha^\gamma$ and $\rho(\alpha) = \|y_\alpha^\gamma - f^\delta\|_*$. Though, in general, operator A^h is discontinuous, the residual $\rho(\alpha)$ is always continuous and has all the other properties established by Lemmas 3.1.4 and 3.1.7. Moreover, $y_*^h \in \bar{A}^h(\theta_X)$ and $y_*^h - f^\delta$ is the element with minimal norm of the set $\bar{A}^h(\theta_X) - f^\delta$. Then the generalized residual principle for equations with arbitrary monotone operators follows from the theorems obtained for maximal monotone operators.

Theorem 3.3.5 *Let $A : X \to 2^{X^*}$ and $A^h : X \to 2^{X^*}$, $h > 0$, be monotone operators, $D(A)$ be a convex closed set, int $D(A) \neq \emptyset$, $\bar{A}$ and $\bar{A}^h$ be maximal monotone extensions of A and A^h, respectively. Assume that $D(\bar{A}) = D(\bar{A}^h) = D(A)$, the conditions (3.1.26) and (2.6.3) are fulfilled for all $x \in D(A)$ and $0 < \delta + h \leq 1$. If $\theta_X \in D(A)$, then assume additionally that (3.3.6) is satisfied, where y_*^h is the element nearest to f^δ in the set $\bar{A}^h(\theta_X)$. Then there exists $\bar{\alpha}$ such that (3.3.7) holds, where $x_{\bar{\alpha}}^\gamma$ is the generalized solution of (3.3.3) with $\alpha = \bar{\alpha}$. Moreover, let $\gamma \to 0$. It results: 1) $\bar{\alpha} \to 0$; 2) if $p \in (0,1)$, then $x_{\bar{\alpha}}^\gamma \to \bar{x}^*$ and $\dfrac{\delta + h}{\bar{\alpha}} \to 0$; 3) if $p = 1$ and $N = \{x_0\}$, then $x_{\bar{\alpha}}^\gamma \rightharpoonup x_0$ and there exists a constant $C > 0$ such that $\dfrac{\delta + h}{\bar{\alpha}} \leq C$.*

Consider the regularized equation (3.1.12), where some $u \in D(A)$ and $D(A) = D(\bar{A}) = D(\bar{A}^h)$. Suppose that the proximity of operators $\bar{A}$ and $\bar{A}^h$ is defined by the inequality

$$\mathcal{H}_{X^*}(\bar{A}x, \bar{A}^h x) \leq hg(\|x - u\|) \quad \forall x \in D(A). \tag{3.3.20}$$

Let in the previous theorem, in place of (3.3.6) and (3.3.7), the following relations be satisfied:

$$\|y_0^h - f^\delta\|_* > \Big(k + g(\|u\|)\Big)(\delta + h)^p, \tag{3.3.21}$$

where $y_0^h - f^\delta$ is a vector with minimal norm in the set $\{y - f^\delta \mid y \in \bar{A}^h u\}$, and

$$\rho(\bar{\alpha}) = \Big(k + g(\|x_{\bar{\alpha}}^\gamma - u\|)\Big)(\delta + h)^p.$$

Then all the conclusions 1) - 3) of Theorem 3.3.5 hold, where the element $\bar{x}^*$ is defined as

$$\|\bar{x}^* - u\| = min\{\|x - u\| \mid x \in N\}.$$

5. Realizing the operator regularization method numerically and choosing parameter α from the generalized residual principle, it is important to know an estimate of $\bar{\alpha}$ from above. We study this problem for the equation (3.1.1) with a maximal monotone operator $A : X \to 2^{X^*}$ and the regularized equation with duality mapping J^s. Consider first the equation with exact operator A :

$$Ax + \alpha J^{s+1}(x - u) = f^\delta, \quad s \geq 1, \quad u \in X. \tag{3.3.22}$$

By analogy with (3.1.41), we apply the generalized residual principle in the form

$$\rho_s(\bar{\alpha}) = \|y_{\bar{\alpha}}^\delta - f^\delta\|_* = \bar{\alpha}\|x_{\bar{\alpha}}^\delta - u\|^s = k\delta^p, \qquad (3.3.23)$$

where $x_{\bar{\alpha}}^\delta$ is the solution of (3.3.22) with $\alpha = \bar{\alpha}$, $p \in (0,1)$, $k > 1$ and $y_{\bar{\alpha}}^\delta \in Ax_{\bar{\alpha}}^\delta$ such that

$$y_{\bar{\alpha}}^\delta + \bar{\alpha}J^{s+1}(x_{\bar{\alpha}}^\delta - u) = f^\delta.$$

We suppose that all the requirements of Theorem 3.1.17 are satisfied. Then the sequence $\{x_{\bar{\alpha}}^\delta\}$ strongly converges to $\bar{x}^*$.

It was shown in [220, 221] that the condition "A acts from X to X^*" imposes a special restriction on the growth order of $\|Ax\|_*$. For example,

$$\|y\|_* \le \zeta(\|x - u\|) \quad \forall x \in D(A), \quad \forall y \in Ax, \qquad (3.3.24)$$

where $\zeta(t)$ is a non-negative continuous and increasing function for $t \ge 0$.

Theorem 3.3.6 *Suppose that $A : X \to 2^{X^*}$ is a maximal monotone operator, the assumptions (3.1.26) and (3.3.24) are carried out, the regularization equation has the form (3.3.22) and regularization parameter $\bar{\alpha}$ is defined by (3.3.23). Then the estimate*

$$\bar{\alpha} \le \frac{k\delta^p}{\left(\zeta^{-1}(|\|f^\delta\|_* - k\delta^p|)\right)^s} \qquad (3.3.25)$$

holds, where $\zeta^{-1}(s)$ is the function inverse to $\zeta(t)$.

Proof. By (3.3.23), we find that

$$\bar{\alpha} = \frac{k\delta^p}{\|x_{\bar{\alpha}}^\delta - u\|^s}. \qquad (3.3.26)$$

Evaluate $\|x_{\bar{\alpha}}^\delta - u\|$ from below. It is easy to see that the hypothesis (3.3.24) implies

$$\zeta(\|x_{\bar{\alpha}}^\delta - u\|) \ge \|y_{\bar{\alpha}}^\delta\|_* \ge |\|f^\delta\|_* - \|y_{\bar{\alpha}}^\delta - f^\delta\|_*| = |\|f^\delta\|_* - k\delta^p|.$$

Since $\zeta(t)$ increases, the latter inequality gives the estimate

$$\|x_{\bar{\alpha}}^\delta - u\| \ge \zeta^{-1}(|\|f^\delta\|_* - k\delta^p|).$$

Then the result (3.3.25) follows from (3.3.26). ∎

Remark 3.3.7 *If in the conditions of Theorem 3.3.6, X is a Hilbert space, $s = 1$ and $u = \theta_X$, then we can write (3.3.22) for an arbitrary linear operator A in the following form:*

$$A^*Ax + \alpha x = A^*f^\delta.$$

Therefore,

$$\bar{\alpha}\|x_{\bar{\alpha}}^\delta\| = \|A^*(Ax_{\bar{\alpha}}^\delta - f^\delta)\|.$$

Since (3.3.23) is true, we have

$$\bar{\alpha} \leq |A^*| \frac{\|Ax_{\bar{\alpha}}^{\delta} - f^{\delta}\|}{\|x_{\bar{\alpha}}^{\delta}\|} = |A| \frac{k\delta^p}{\|x_{\bar{\alpha}}^{\delta}\|}.$$

Hence,

$$\bar{\alpha} \leq |A| \frac{k\delta^p}{\zeta^{-1}(\|\|f^{\delta}\|_* - k\delta^p|)}.$$

6. Suppose now that a maximal monotone operator A in (3.1.1) is also given with some error depending on the parameter $h > 0$, such that in reality the equation $A^h x = f^{\delta}$ is solved, $D(A) = D(A^h)$, the conditions (3.1.26) and (3.3.1) hold, regularized solutions are found from the equation

$$A^h x + \alpha J^{s+1}(x - u) = f^{\delta}, \quad s \geq 1, \tag{3.3.27}$$

and the regularization parameter $\bar{\alpha}$ is defined as follows:

$$\rho_s(\bar{\alpha}) = \bar{\alpha}\|x_{\bar{\alpha}}^{\gamma} - u\|^s = \|y_{\bar{\alpha}}^{\gamma} - f^{\delta}\|_* = d(\|x_{\bar{\alpha}}^{\gamma} - u\|)(\delta + h)^p. \tag{3.3.28}$$

Here

$$d(t) = \left(\bar{k} + (1 + g(t))^{\kappa}\right)^s, \quad \kappa = \frac{1}{s}, \quad \bar{k} > 0, \quad p \in (0,1), \quad \gamma = (\delta, h),$$

$x_{\bar{\alpha}}^{\gamma}$ denotes a solution of (3.3.27) with $\alpha = \bar{\alpha}$, $y_{\bar{\alpha}}^{\gamma} \in A^h x_{\bar{\alpha}}^{\gamma}$ such that

$$y_{\bar{\alpha}}^{\gamma} + \bar{\alpha} J^{s+1}(x_{\bar{\alpha}}^{\gamma} - u) = f^{\delta}.$$

Assuming that

$$\|y^h\|_* \leq \zeta(\|x - u\|) \quad \forall h > 0, \quad \forall x \in D(A), \quad \forall y^h \in A^h x, \tag{3.3.29}$$

where $\zeta(t)$ is a non-negative continuous and increasing function for $t \geq 0$, and going through the same arguments, as in the proof of Theorem 3.3.6, we come to the inequality

$$\bar{\alpha} \leq \frac{d(\|x_{\bar{\alpha}}^{\gamma} - u\|)(\delta + h)^p}{\left(\zeta^{-1}(\|\|f^{\delta}\|_* - d(\|x_{\bar{\alpha}}^{\gamma} - u\|)(\delta + h)^p|)\right)^s}. \tag{3.3.30}$$

Evaluate from above the quantity of $\|x_{\bar{\alpha}}^{\gamma} - u\|$. According to (3.3.18),

$$\bar{\alpha} > \left(\frac{\bar{k}}{\tau\|\bar{x}^* - u\|}\right)^s (\delta + h)^p,$$

where $\tau > 1$ and $\tau s \geq 2$. Let

$$\|\bar{x}^* - u\| = min \left\{\|x - u\| \mid x \in N\right\} > 0$$

and $\bar{c}$ be a constant such that

$$\bar{c} \geq \|\bar{x}^* - u\|. \tag{3.3.31}$$

Then

$$\bar{\alpha} \geq \left(\frac{\bar{k}}{\tau \bar{c}} \right)^s (\delta + h)^p.$$

Since the function $\sigma(\alpha) = \|x_\alpha^\gamma - u\|$ is decreasing, one gets

$$\|x_{\bar{\alpha}}^\gamma - u\| \leq \|x_{\alpha^*}^\gamma - u\|,$$

where

$$\alpha^* = \left(\frac{\bar{k}}{\bar{c}\tau} \right)^s (\delta + h)^p.$$

If

$$d(\|x_* - u\|)(\delta + h)^p > \|f^\delta\|_*$$

then, by virtue of the equalities

$$\lim_{\alpha \to \infty} \|x_\alpha^\gamma - u\| = \|x_* - u\| = min\{\|x - u\| \mid x \in \overline{D(A)}\}$$

(see Lemma 3.1.7) and the properties of $\sigma(\alpha)$ and $g(t)$, the estimate (3.3.30) takes the following form:

$$\bar{\alpha} \leq \frac{d(\|x_{\alpha^*}^\gamma - u\|)(\delta + h)^p}{\left(\zeta^{-1}(|\|f^\delta\|_* - d(\|x_* - u\|)(\delta + h)^p|) \right)^s}. \tag{3.3.32}$$

Next, if

$$d(\|x_{\alpha^*}^\gamma - u\|)(\delta + h)^p < \|f^\delta\|_*$$

then it follows from (3.3.30) that

$$\bar{\alpha} \leq \frac{d(\|x_{\alpha^*}^\gamma - u\|)(\delta + h)^p}{\left(\zeta^{-1}(|\|f^\delta\|_* - d(\|x_{\alpha^*}^\gamma - u\|)(\delta + h)^p|) \right)^s}. \tag{3.3.33}$$

Thus, we are able to state the following theorem.

Theorem 3.3.8 *Under the conditions and denotations of this subsection, suppose that*

$$\|\tilde{y}_*^h - f^\delta\|_* > d(\|x_* - u\|)(\delta + h)^p,$$

where

$$\|\tilde{y}_*^h - f^\delta\|_* = min\{\|y - f^\delta\|_* \mid y \in A^h x_*\}.$$

If

$$d(\|x_* - u\|)(\delta + h)^p > \|f^\delta\|_*,$$

then estimate (3.3.32) holds. If

$$\|f^\delta\|_* > d(\|x_{\alpha^*}^\gamma - u\|)(\delta + h)^p,$$

then (3.3.33) is satisfied.

Observe that the estimates (3.3.32) and (3.3.33) become most effective when maximal monotone operators A and A^h with $D(A) = D(A^h)$ are close in the uniform metric, that is,

$$\mathcal{H}_{X^*}(Ax, A^h x) \leq h \quad \forall x \in D(A). \tag{3.3.34}$$

Then there exists a constant $k_1 > 2$ such that

$$\bar{\alpha} \leq \frac{k_1(\delta + h)^p}{\left(\zeta^{-1}(\|\|f^\delta\|\|_* - k_1(\delta + h)^p\|)\right)^s}.$$

7. Assume that the conditions of Theorem 3.3.5 are fulfilled, an operator A is strictly monotone and given exactly. As it was noted in Remark 3.3.3, in this case the function $\rho(\alpha)$ is single-valued, continuous and increasing, and the value of $\bar{\alpha}$ is defined by the equality

$$\rho(\bar{\alpha}) = \bar{\alpha}\|x_{\bar{\alpha}}^\delta\| = k\delta^p.$$

We introduce the function

$$\varphi(\alpha) = \frac{k\delta^p}{\|x_\alpha^\delta\|}.$$

Then $\bar{\alpha} = \varphi(\bar{\alpha})$. Due to the properties of the function $\sigma(\alpha)$, we conclude that $\varphi(\alpha)$ is a continuous, single-valued and increasing function. Thus, the equation $\varphi(\alpha) = \alpha$ has the unique root $\alpha = \bar{\alpha}$. Furthermore, $\varphi(\alpha) < \alpha$ as $\alpha > \bar{\alpha}$ and $\varphi(\alpha) > \alpha$ as $\alpha < \bar{\alpha}$. Consider the method of successive approximations

$$\alpha_n = \varphi(\alpha_{n-1}) \tag{3.3.35}$$

with an arbitrary initial approximation α_0. It is well-known that the inequality $|\varphi'(\alpha)| < 1$ is its sufficient convergence condition. Let $\sigma(\alpha)$ be a differentiable function. Then

$$\rho'(\alpha) = \sigma(\alpha) + \alpha\sigma'(\alpha) > 0,$$

from which we obtain that

$$\sigma'(\alpha) \geq -\frac{\sigma(\alpha)}{\alpha}.$$

Then

$$0 < \varphi'(\alpha) = -\frac{k\delta^p \sigma'(\alpha)}{\sigma^2(\alpha)} \leq \frac{k\delta^p}{\alpha\sigma(\alpha)} = \frac{\varphi(\alpha)}{\alpha}.$$

It is clear that $0 < \varphi'(\alpha) < 1$ as $\alpha > \bar{\alpha}$. Hence, method (3.3.35) converges.

Consider Example 2.2.8. Suppose that there is a function $g(t) = ct^{p-1} + 1$ with some constant $c > 0$. In order to find the regularization parameter $\bar{\alpha}$, we have the equation

$$\bar{\alpha}\|u_{\bar{\alpha}}^\gamma\|_{1,p}^{p-1} = \left(\bar{k} + (2 + c\|u_{\bar{\alpha}}^\gamma\|_{1,p}^{p-1})^\kappa\right)^{p-1}(\delta + h)^\sigma, \tag{3.3.36}$$

where $\bar{k} > 0$, $\kappa = \dfrac{1}{p-1}$, $\sigma \in (0,1]$, $p \geq 2$, and $u_{\bar{\alpha}}^\gamma$ is the solution of the equation (2.2.13) as $\alpha = \bar{\alpha}$. By the properties of the functions g_0^h and g_1^h, we establish the inequality

$$\|A^h u\|_{-1,q} \leq c_1\|u\|_{1,p}^{p-1} + \omega^h,$$

where a constant $c_1 > 0$ and $p^{-1} + q^{-1} = 1$. Thus,

$$d(t) = \left(\bar{k} + (2 + ct^{p-1})^\kappa\right)^{p-1}, \quad \zeta^{-1}(s) = c_1^{-\kappa}(s - \omega^h)^\kappa,$$

in (3.3.27) $u = \theta_X$, and the estimates (3.3.32) and (3.3.33) follow.

Assume that for all $x \in \Omega$ and for all $\xi \geq 0$,

$$|g_0^h(x, \xi^2)\xi - g_0(x, \xi^2)\xi| \leq \frac{h}{2}$$

and

$$|\omega - \omega^h| \leq \frac{h}{2}.$$

In this case, $g(t) \equiv 1$. If

$$|\|f^\delta\|_* - k_1(\delta + h)^\sigma| - \omega^h > 0,$$

then

$$\bar{\alpha} \leq \frac{k_1 c_1(\delta + h)^\sigma}{|\|f^\delta\|_* - k_1(\delta + h)^\sigma| - \omega^h}, \quad k_1 > 2, \quad \sigma \in (0, 1].$$

If in the equation (2.2.13), instead of $J^p u$, we take normalized dual mapping

$$Ju = \|u\|_{1,p}^{2-p} J^p u,$$

then (3.3.36) should be replaced by the following equation:

$$\bar{\alpha}\|u_{\bar{\alpha}}^\gamma\| = \left(k + c\|u_{\bar{\alpha}}^\gamma\|_{1,p}^{p-1}\right)(\delta + h)^\sigma,$$

where $k > 2$, $\sigma \in (0, 1]$, $c > 0$, $p > 1$.

Remark 3.3.9 *Using constructions of Section 3.2, we are able to obtain the generalized residual principle for multiple-valued and discontinuous accretive operators A and A^h.*

3.4 Modified Residual Principle

Up to the present, we have chosen the regularization parameter α from the equations (3.1.33) and (3.3.7). However, it is possible to choose it from some inequalities. We will show how it may be done.

Theorem 3.4.1 *Assume that the conditions of Theorem 3.1.12 for the equation (3.1.1) and (3.1.27) are fulfilled and there is a number $r > 0$ such that for all $x \in D(A)$ with $\|x\| \geq r$,*

$$\langle Ax - f, x \rangle \geq 0. \tag{3.4.1}$$

Then there exists $\bar{\alpha}$ satisfying the inequalities

$$\frac{\delta^p}{kr} \leq \bar{\alpha} \leq max\{\alpha \mid \rho(\alpha) \leq \delta^p\}, \quad k > 1, \quad p \in (0, 1), \quad 0 < \delta < 1. \tag{3.4.2}$$

Moreover, if $\delta \to 0$, then $x_{\bar{\alpha}}^\delta \to \bar{x}^*$, $\dfrac{\delta}{\bar{\alpha}} \to 0$ and $\bar{\alpha} \to 0$.

Proof. It is easy to check that

$$\langle Ax + \alpha Jx - f^\delta, x \rangle \geq \langle Ax - f, x \rangle + \alpha \|x\| \left(\|x\| - \frac{\delta}{\alpha} \right).$$

If $\|x\| \geq r_1 = max\{r, \delta/\alpha\}$, then

$$\langle Ax + \alpha Jx - f^\delta, x \rangle \geq 0 \quad \forall x \in D(A).$$

A solution of regularized equation (3.1.27) satisfies the inequality $\|x_\alpha^\delta\| \leq r_1$ in view of Theorem 1.7.9. Then there is α such that $\rho(\alpha) < \delta^p$. Beside this, if $r_1 = r$ then this inequality holds for all $\alpha < \dfrac{\delta^p}{r}$, while if $r_1 = \dfrac{\delta}{\alpha}$ then $\rho(\alpha) < \delta^p$ for all $\alpha > 0$. Consequently, we proved that the parameter $\alpha = \bar{\alpha}$ choice is possible by (3.4.2). Then the left inequality of (3.4.2) gives

$$\frac{\delta}{\bar{\alpha}} \leq kr\delta^{1-p}.$$

Hence, $\dfrac{\delta}{\bar{\alpha}} \to 0$ as $\delta \to 0$.

Show that the residual $\rho(\alpha) = \alpha\|x_\alpha^\delta\|$ is non-decreasing. Let

$$\rho(\alpha) = \alpha\|x_\alpha^\delta\| = \gamma_1, \quad \rho(\beta) = \beta\|x_\beta^\delta\| = \gamma_2,$$

$$\alpha Jx_\alpha^\delta = \gamma_1 e_\alpha^*, \quad \beta Jx_\beta^\delta = \gamma_2 e_\beta^*,$$

$$\|e_\alpha^*\|_* = \|e_\beta^*\|_* = 1,$$

$$\langle e_\alpha^*, x_\alpha^\delta \rangle = \|x_\alpha^\delta\|, \quad \langle e_\beta^*, x_\beta^\delta \rangle = \|x_\beta^\delta\|.$$

It is not difficult to verify by Corollary 1.5.3 that

$$\langle e_\alpha^* - e_\beta^*, x_\alpha^\delta - x_\beta^\delta \rangle \geq 0. \tag{3.4.3}$$

Let $\beta < \alpha$, but $\gamma_2 > \gamma_1$. As in the proof of Theorem 3.3.2, since x_α^δ and x_β^δ are solutions of the equation (3.1.27) with regularization parameters α and β, respectively, the following equality holds:

$$\langle Ax_\alpha^\delta - Ax_\beta^\delta, x_\alpha^\delta - x_\beta^\delta \rangle \quad + \quad \gamma_1 \langle e_\alpha^* - e_\beta^*, x_\alpha^\delta - x_\beta^\delta \rangle$$
$$+ \quad (\gamma_1 - \gamma_2)\langle e_\beta^*, x_\alpha^\delta - x_\beta^\delta \rangle = 0.$$

The monotonicity of A and (3.4.3) imply

$$(\gamma_1 - \gamma_2)\langle e_\beta^*, x_\alpha^\delta - x_\beta^\delta \rangle \leq 0.$$

Since $\gamma_2 > \gamma_1$, one gets

$$\langle e_\beta^*, x_\beta^\delta - x_\alpha^\delta \rangle \leq 0.$$

Hence, $\|x_\beta^\delta\| \leq \|x_\alpha^\delta\|$. By Lemma 3.1.1, we have that $\beta \geq \alpha$. Thus, we come to the contradiction. Consequently, the claim is proved. Therefore, $\rho(\bar{\alpha}) \leq \delta^p$.

Next, due to Theorem 3.1.12, we deduce that $x_{\bar{\alpha}}^{\delta} \to \bar{x}^{*}$ as $\delta \to 0$, where $\bar{x}^{*} \in N$ is a minimal norm solution of (3.1.1). Moreover, since

$$\bar{\alpha} \leq \frac{\delta^{p}}{\|x_{\bar{\alpha}}^{\delta}\|} \quad \text{and} \quad \bar{x}^{*} \neq \theta_{X},$$

we obtain convergence of $\bar{\alpha}$ to 0 as $\delta \to 0$. ∎

Remark 3.4.2 *A non-decreasing property of the residual established in the last theorem holds, as well, in the case of approximately given operators A.*

We are able to study in Theorem 3.4.1 general maximal monotone possibly multiple-valued mappings. If, in place of A, a sequence $\{A^{h}\}$, $h > 0$, of maximal monotone operators is given, $D(A) = D(A^{h})$ and (3.3.34) holds, then the solution sequence $\{x_{\bar{\alpha}}^{\gamma}\}$ of the equation

$$A^{h}x + \bar{\alpha}Jx = f^{\delta},$$

where $\bar{\alpha}$ is chosen from the inequalities

$$\frac{(\delta + h)^{p}}{kr} \leq \bar{\alpha} \leq max\{\alpha \mid \rho(\alpha) \leq (\delta + h)^{p}\}, \quad k > 1, \quad p \in (0, 1),$$

strongly converges to $\bar{x}^{*} \in N$. Moreover, if $\theta_{X} \notin N$ and $\delta, h \to 0$ then

$$\bar{\alpha} \to 0 \quad \text{and} \quad \frac{\delta + h}{\bar{\alpha}} \to 0.$$

If we omit the condition (3.4.1) in Theorem 3.4.1, then (3.4.2) is necessarily replaced as follows:

$$\frac{(k_{1} - 1)\delta^{p}}{kc} \leq \bar{\alpha} \leq max\{\alpha \mid \rho(\alpha) \leq k_{1}\delta^{p}\},$$

where $k_{1} > 1$, $k > 1$, $p \in (0, 1)$, and a constant c such that $\|\bar{x}^{*}\| \leq c$. If the proximity between operators A and A^{h} is defined by (3.3.1), then the criterion determining $\bar{\alpha}$ must be taken as

$$\frac{(k_{1} - 1)(\delta + h)^{p}}{kc} \leq \bar{\alpha} \leq max\{\alpha \mid \rho(\alpha) \leq [k_{1} + g(\|x_{\alpha}^{\gamma}\|)](\delta + h)^{p}\}.$$

3.5 Minimal Residual Principle

Let $A : X \to 2^{X^{*}}$ be a maximal monotone operator, $D(A) \subseteq X$, the equation (3.1.1) have a nonempty solution set N and $\theta_{X} \notin N$. We find approximations to an element $\bar{x}^{*} \in N$ by the regularized equation (3.1.27) with a right-hand side f^{δ}, $\delta > 0$, and consider that δ is not known. Construct the single-valued continuous residual $\rho(\alpha) = \alpha\|x_{\alpha}^{\delta}\|$, where x_{α}^{δ} is a unique solution of the equation (3.1.27). Parameter $\alpha = \alpha^{0}$ is defined by the following equality:

$$\alpha^{0} = inf\{\bar{\alpha} \mid \rho(\bar{\alpha}) = inf\{\rho(\alpha) \mid \alpha > 0, \ x_{\alpha}^{\delta} \in M\}\}, \tag{3.5.1}$$

where $M \subseteq D(A)$ is an admissible set of solutions. We assume that M is bounded and $\bar{x}^* \in int\ M$.

Suppose further that $f^\delta \notin R(A)$. Otherwise, if $f^\delta \in R(A)$ then, due to Theorem 2.2.5, the sequence $\{x_\alpha^\delta\}$ is bounded in X as $\alpha \to 0$. Therefore, by (3.5.1), $\alpha^0 = 0$. But such a choice of α is not acceptable in (3.1.27).

Observe that the set

$$\Lambda = \{\alpha \mid x_\alpha^\delta \in M,\ 0 < \delta < \bar{\delta}\}$$

is not empty. This follows from Theorem 2.2.1, which implies an existence of the sequence $\{x_\alpha^\delta\}$ such that $x_\alpha^\delta \to \bar{x}^*$. Thus, $x_\alpha^\delta \in M$ for a small enough α.

According to Lemma 3.1.7, if $\theta_X \notin D(A)$, then

$$\lim_{\alpha \to \infty} \rho(\alpha) = \infty, \tag{3.5.2}$$

and if $\theta_X \in D(A)$ then

$$\lim_{\alpha \to \infty} \rho(\alpha) = \|y_* - f^\delta\|_*,$$

where $y_* \in A(\theta_X)$ and

$$\|y_* - f^\delta\|_* = min\ \{\|y - f^\delta\|_* \mid y \in A(\theta_X)\}.$$

Prove that $\rho(\alpha) \neq 0$ as $\alpha \neq 0$. Let this claim be not true. Then the equality $\rho(\alpha) = 0$ appears only if $x_\alpha^\delta = \theta_X$. In this case, $f^\delta \in A(\theta_X)$, which contradicts the assumption that $f^\delta \notin R(A)$. Then the inequality (3.1.11) implies for $\alpha > 0$,

$$\rho(\alpha) = \|y_\alpha^\delta - f^\delta\|_* \leq \|y_* - f^\delta\|_*,$$

where $y_\alpha^\delta \in Ax_\alpha^\delta$, and

$$y_\alpha^\delta + \alpha J x_\alpha^\delta = f^\delta. \tag{3.5.3}$$

Taking into account (3.5.2), we get that the parameters $\bar{\alpha}$ and α_0 cannot approach infinity.

Lemma 3.5.1 *Under the conditions of this section, if $0 < \delta < \bar{\delta}$, then there exists a unique $\alpha^0 > 0$ satisfying (3.5.1).*

Proof. The uniqueness of α_0 is obvious. Let $\{\alpha_n\}$ be a minimizing sequence for $\rho(\alpha)$, that is,

$$\alpha_n \to \bar{\alpha}, \quad \rho(\alpha_n) \to inf\ \{\rho(\alpha) \mid \alpha > 0,\ x_\alpha^\delta \in M\}.$$

Show that $\bar{\alpha} > 0$. Indeed, if $\bar{\alpha} = 0$ then $\|x_\alpha^\delta\| \to \infty$ because $f^\delta \notin R(A)$. Hence, beginning with a certain number n, an element $x_{\alpha_n}^\delta \notin M$. This contradicts (3.5.1). Since $\rho(\alpha)$ is continuous for $\alpha > 0$, one gets that $\rho(\alpha_n) \to \rho(\bar{\alpha})$. Let now $\bar{\alpha}_n \to \alpha^0$ and

$$\rho(\bar{\alpha}_n) = inf\ \{\rho(\alpha) \mid \alpha > 0,\ x_\alpha^\delta \in M\}.$$

Reasoning by contradiction, as above, we establish that $\alpha^0 > 0$. ∎

We study the behavior of the sequences $\{\alpha^0(\delta)\}$ and $\{x_{\alpha^0}^\delta\}$ as $\delta \to 0$, where $\alpha^0(\delta)$ are defined by (3.5.1) and $x_{\alpha^0}^\delta$ are solutions of (3.1.27) with $\alpha = \alpha^0(\delta)$. It follows from (3.5.1) that $\{x_{\alpha^0}^\delta\}$ is bounded, therefore, $x_{\alpha^0}^\delta \rightharpoonup \bar{x} \in X$ as $\delta \to 0$. It is known that there exists a sequence $\{\alpha(\delta)\} \to 0$ such that $\rho(\alpha) = \alpha\|x_\alpha^\delta\| \to 0$ and $x_\alpha^\delta \to \bar{x}^*$ with $\alpha = \alpha(\delta)$. Since $\bar{x}^* \in int\, M$, we have the inclusion: $x_\alpha^\delta \in M$ if δ is small enough. Well then, all the more

$$\|y_{\alpha^0}^\delta - f^\delta\|_* \to 0 \quad \text{as} \quad \delta \to 0,$$

where $y_{\alpha^0}^\delta$ satisfies (3.5.3) with $\alpha = \alpha^0$. Thus, $y_{\alpha^0}^\delta \to f$. This limit relation together with one $x_{\alpha^0}^\delta \rightharpoonup \bar{x}$ and with demiclosedness property of A allow us to assert that $\bar{x} \in N$.

Show that $\alpha^0(\delta) \to 0$ as $\delta \to 0$. Let $\alpha^0(\delta) \to \alpha^* \neq 0$. We already proved that

$$\rho(\alpha^0) = \alpha^0\|x_{\alpha^0}^\delta\| \to 0.$$

Then $x_{\alpha^0}^\delta \to \theta_X$. Thus, $\bar{x} = \theta_X$, which contradicts the fact that $\theta_X \notin N$. Thus, we have proved the following theorem.

Theorem 3.5.2 *Under the conditions of this section, any weak accumulation point of the sequence $\{x_{\alpha^0}^\delta\}$, where α^0 is defined by (3.5.1), is a solution of the equation (3.1.1). Moreover, $\alpha^0(\delta) \to 0$ and $\rho(\alpha^0) \to 0$ as $\delta \to 0$.*

3.6 Smoothing Functional Principle

1. We solve the equation (3.1.1) with a monotone hemicontinuous and potential operator $A : X \to X^*$. Assume that $D(A) = X$, a solution set N of (3.1.1) is nonempty, $\bar{x}^*$ is its minimal norm solution, and, as before, f^δ is an approximation of f. Let $\omega(x)$ be a potential of A, i.e., $A = grad\,\omega$. Then regularized equation (3.1.27) is equivalent to the minimization problem of the functional

$$\Phi_\delta^\alpha(x) = \omega(x) - \langle f^\delta, x\rangle + \frac{\alpha}{2}\|x\|^2 \quad \forall x \in X, \quad \alpha > 0. \tag{3.6.1}$$

In the terms of variational regularization methods, (3.6.1) is a smoothing functional for the equation $Ax = f^\delta$. Let x_α^δ be a unique minimum point of (3.6.1), which coincides with solution of the equation (3.1.27). We introduce the following denotations:

$$F^\delta(x) = \omega(x) - \langle f^\delta, x\rangle,$$
$$F^0(x) = \omega(x) - \langle f, x\rangle,$$
$$m = min\,\{F^0(x) \mid x \in X\},$$
$$m_\delta(\alpha) = min\,\{\Phi_\delta^\alpha(x) \mid x \in X\},$$
$$h^\delta = inf\,\{F^\delta(x) \mid x \in X\}.$$

Observe that

$$h^\delta \leq inf\,\{F^0(x^*) + \delta\|x^*\| \mid x^* \in N\} \leq m + \delta\|\bar{x}^*\|, \tag{3.6.2}$$

and assume, without loss of generality, that $F^\delta(x) \geq 0$ for all $x \in X$.

Lemma 3.6.1 *The function $m_\delta(\alpha)$ is continuous and non-decreasing for all $\alpha \geq 0$,*

$$\lim_{\alpha \to 0+} m_\delta(\alpha) = h^\delta$$

and

$$\lim_{\alpha \to \infty} m_\delta(\alpha) = \omega(\theta_X).$$

Moreover, if $h^\delta < \omega(\theta_X)$, then this function is increasing on the interval $(0, \alpha_\delta^*)$, where

$$\alpha_\delta^* = \sup \{\alpha \mid m_\delta(\alpha) < \omega(\theta_X)\}.$$

Next we take functions $\psi(\delta)$ and $C(\delta)$ with the following properties:

$$\lim_{\delta \to 0} \psi(\delta) = m, \quad m + \delta C(\delta) \leq \psi(\delta) < \omega(\theta_X) \tag{3.6.3}$$

and

$$C(\delta) \to \infty \quad \text{as} \quad \delta \to 0. \tag{3.6.4}$$

Theorem 3.6.2 *Let*

$$\omega(\theta_X) > m + \delta \|\bar{x}^*\| \tag{3.6.5}$$

and the function $\psi(\delta)$ satisfy (3.6.3) and (3.6.4). Then there exists a unique $\bar{\alpha} = \alpha(\delta) > 0$ which is defined by the following equation of the smoothing functional principle:

$$m_\delta(\bar{\alpha}) = \psi(\delta). \tag{3.6.6}$$

Furthermore, the sequence $\{x_{\bar{\alpha}}^\delta\}$ of minimal points of the functional $\Phi_\delta^\alpha(x)$ with $\alpha = \bar{\alpha}$ converges strongly to $\bar{x}^*$ as $\delta \to 0$.

Proof. By (3.6.5), (3.6.2) and by Lemma 3.6.1, we conclude that the function $m_\delta(\alpha)$ is continuous and increasing as $\alpha \in (0, \alpha_\delta^*)$. It follows from (3.6.3) and (3.6.4) that there exists a unique positive root of equation (3.6.6).

Prove that regularized solutions $x_{\bar{\alpha}}^\delta$ strongly converge to $\bar{x}^*$. Since

$$|F^\delta(x) - F^0(x)| \leq \delta \|x\|,$$

we obtain by (3.6.6) the inequality

$$F^0(x_{\bar{\alpha}}^\delta) - \delta \|x_{\bar{\alpha}}^\delta\| + \frac{\bar{\alpha}}{2} \|x_{\bar{\alpha}}^\delta\|^2 \leq F^\delta(x_{\bar{\alpha}}^\delta) + \frac{\bar{\alpha}}{2} \|x_{\bar{\alpha}}^\delta\|^2 = \psi(\delta).$$

Owing to the inequality $F^0(x_{\bar{\alpha}}^\delta) \geq m$, one gets

$$\frac{\bar{\alpha}}{2} \|x_{\bar{\alpha}}^\delta\|^2 \leq \delta \|x_{\bar{\alpha}}^\delta\| + \psi(\delta) - m. \tag{3.6.7}$$

From this quadratic inequality, the estimate

$$\|x_{\bar{\alpha}}^\delta\| \leq \frac{\delta}{\bar{\alpha}} + \sqrt{\left(\frac{\delta}{\bar{\alpha}}\right)^2 + \frac{2(\psi(\delta) - m)}{\bar{\alpha}}} \tag{3.6.8}$$

holds. By (3.6.3) and (3.6.6), we deduce

$$m \quad + \quad \delta C(\delta) \le \psi(\delta) = m_\delta(\bar{\alpha}) \le \Phi_\delta^{\bar{\alpha}}(\bar{x}^*) = F^\delta(\bar{x}^*) + \frac{\bar{\alpha}}{2}\|\bar{x}^*\|^2$$

$$\le \quad F^0(\bar{x}^*) + \delta\|\bar{x}^*\| + \frac{\bar{\alpha}}{2}\|\bar{x}^*\|^2 = m + \delta\|\bar{x}^*\| + \frac{\bar{\alpha}}{2}\|\bar{x}^*\|^2. \tag{3.6.9}$$

Without loss of generality, one can consider that $C(\delta) - \|\bar{x}^*\| > 0$. Therefore,

$$\frac{\delta}{\bar{\alpha}} \le \frac{\|\bar{x}^*\|^2}{2(C(\delta) - \|\bar{x}^*\|)}.$$

Then, since $C(\delta) \to \infty$ as $\delta \to 0$, we have $\dfrac{\delta}{\bar{\alpha}} \to 0$ as $\delta \to 0$. Moreover, from (3.6.9) follows the inequality

$$\frac{\psi(\delta) - m}{\bar{\alpha}} \le \frac{\|\bar{x}^*\|^2}{2} + \frac{\delta}{\bar{\alpha}}\|\bar{x}^*\|.$$

By (3.6.8), we have a boundedness of the sequence $\{x_{\bar{\alpha}}^\delta\}$ as $\delta \to 0$. Show that $\bar{\alpha} \to 0$ as $\delta \to 0$. For this end, make use of inequality (3.6.8). The right-hand side of it vanishes as $\delta \to 0$ if $\bar{\alpha} \not\to 0$. We prove by contradiction that

$$\lim_{\delta \to 0} \|x_{\bar{\alpha}}^\delta\| \ne 0.$$

Let $x_{\bar{\alpha}}^\delta \to \theta_X$. Since the potential ω of A is weakly lower semicontinuous, we can write

$$F^0(\theta_X) \le \lim_{\delta \to 0} F^0(x_{\bar{\alpha}}^\delta). \tag{3.6.10}$$

On the other hand,

$$F^0(x_{\bar{\alpha}}^\delta) \le F^\delta(x_{\bar{\alpha}}^\delta) + \delta\|x_{\bar{\alpha}}^\delta\| \le \psi(\delta) + \delta\|x_{\bar{\alpha}}^\delta\|,$$

therefore,

$$\limsup_{\delta \to 0} F^0(x_{\bar{\alpha}}^\delta) \le m.$$

Then, in view of (3.6.10), we obtain that $F^0(\theta_X) \le m$. Thus, $\omega(\theta_X) \le m$. Taking into account the condition (3.6.5), we arrive at the contradiction. Consequently, $\lim_{\delta \to 0} \|x_{\bar{\alpha}}^\delta\| \ne 0$. By reason of (3.6.7), this implies the convergence $\bar{\alpha} \to 0$ as $\delta \to 0$. Then the last assertion of the theorem being proved follows from Theorem 2.2.1. ∎

Remark 3.6.3 *Under the conditions of Theorem 3.6.2, if the equation (3.1.1) has a unique solution x_0 and the relations in (3.6.3) is replaced by the inequality*

$$m + C\delta \le \psi(\delta) < \omega(\theta_X), \quad C > \|x_0\|,$$

then the weak convergence of $\{x_{\bar{\alpha}}^\delta\}$ to x_0 follows when $\delta \to 0$.

Remark 3.6.4 *The assumptions of Theorem 3.6.2 such that A is hemicontinuous and* $D(A) = X$ *may be omitted if we consider maximal monotone or arbitrary monotone operators with domains which not necessarily coincide with the whole of space X, understanding solutions in the sense of Definitions 1.7.2 and 1.9.3, respectively. Then, applying the results of Section 2.2, we may establish the smoothing functional principle for these cases, as well.*

2. We study now the smoothing functional principle with approximately given operators A. Instead of ω, let a sequence of functionals $\{\omega^h\}$ with $D(\omega) = D(\omega^h) = X$, $h > 0$, be known. Assume that ω^h have the same properties as ω and the inequality

$$|\omega(x) - \omega^h(x)| \le \eta(\|x\|)h \quad \forall x \in X \tag{3.6.11}$$

holds, where $\eta(t)$ is non-negative, continuous and increasing for all $t \ge 0$. Then the equation (3.1.2) with $A^h x = grad\, \omega^h(x)$ is equivalent to the minimization problem of the smoothing functional

$$\Phi_\gamma^\alpha(x) = \omega^h(x) - \langle f^\delta, x \rangle + \frac{\alpha}{2}\|x\|^2, \quad \gamma = (\delta, h), \quad \alpha > 0. \tag{3.6.12}$$

Denote a unique solution of the problems (3.1.2) and minimal point of (3.6.12) by x_α^γ, and assume

$$F^\gamma(x) = \omega^h(x) - \langle f^\delta, x \rangle,$$

$$m_\gamma(\alpha) = min \, \{\Phi_\gamma^\alpha(x) \mid x \in X\},$$

$$h^\gamma = inf \, \{F^\gamma(x) \mid x \in X\}.$$

Similarly to (3.6.2), we have

$$h^\gamma \ \le \ inf \, \{F^0(x^*) + h\eta(\|x^*\|) + \delta\|x^*\| \mid x^* \in N\}$$
$$\le \ m + h\eta(\|\bar{x}^*\|) + \delta\|\bar{x}^*\|.$$

The assertions like Lemma 3.6.1 and Theorem 3.6.2 are as follows:

Lemma 3.6.5 *A function* $m_\gamma(\alpha)$ *is continuous and non-decreasing for* $\alpha \ge 0$, *and*

$$\lim_{\alpha \to 0+} m_\gamma(\alpha) = h^\gamma, \qquad \lim_{\alpha \to \infty} m_\gamma(\alpha) = \omega^h(\theta_X).$$

Moreover, if $h^\gamma < \omega^h(\theta_X)$, *then* $m_\gamma(\alpha)$ *is increasing for* $\alpha \in (0, \alpha_\gamma^*)$, *where*

$$\alpha_\gamma^* = sup \, \{\alpha \mid m_\gamma(\alpha) < \omega^h(\theta_X)\}.$$

Theorem 3.6.6 *Suppose that a function* $\psi(\gamma)$ *satisfies the conditions:*

$$\lim_{\gamma \to 0} \psi(\gamma) = m, \quad m + (\delta + h)C(\gamma) \le \psi(\gamma) < \omega^h(\theta_X), \tag{3.6.13}$$

where $C(\gamma) \to \infty$ *as* $\gamma \to 0$. *Furthermore, let the inequality (5.6.3) hold, and*

$$\omega^h(\theta_X) > m + \delta\|\bar{x}^*\| + h\eta(\|\bar{x}^*\|), \tag{3.6.14}$$

where the function $\eta(t)$ possesses the property

$$\limsup_{t \to \infty} \frac{\eta(t)}{t^2} = M < \infty. \tag{3.6.15}$$

Then there exists a unique $\bar{\alpha} > 0$ such that

$$m_\gamma(\bar{\alpha}) = \psi(\gamma), \tag{3.6.16}$$

and the sequence $\{x_{\bar{\alpha}}^\gamma\}$ strongly converges to $\bar{x}^*$ as $\gamma \to 0$.

Proof. By the condition (3.6.14) and Lemma 3.6.5, the function $m_\gamma(\alpha)$ is continuous and increasing on the interval $(0, \alpha_\gamma^*)$. Hence, (3.6.13) implies the first assertion of the theorem. Similarly to (3.6.7), we deduce the estimate

$$\|x_{\bar{\alpha}}^\gamma\|^2 \leq 2\left(\frac{\psi(\gamma) - m}{\bar{\alpha}} + \frac{\delta}{\bar{\alpha}}\|x_{\bar{\alpha}}^\gamma\| + \frac{h}{\bar{\alpha}}\eta(\|x_{\bar{\alpha}}^\gamma\|)\right). \tag{3.6.17}$$

Applying (3.6.13) and (3.6.16), one gets the following relations:

$$\begin{aligned}
m + (\delta + h)C(\gamma) &\leq \psi(\gamma) = m_\gamma(\bar{\alpha}) \leq \Phi_\gamma^{\bar{\alpha}}(\bar{x}^*) \\[2mm]
&= F^\gamma(\bar{x}^*) + \frac{\bar{\alpha}}{2}\|\bar{x}^*\|^2 \leq F^0(\bar{x}^*) + \delta\|\bar{x}^*\| + h\eta(\|\bar{x}^*\|) + \frac{\bar{\alpha}}{2}\|\bar{x}^*\|^2 \\[2mm]
&= m + \delta\|\bar{x}^*\| + h\eta(\|\bar{x}^*\|) + \frac{\bar{\alpha}}{2}\|\bar{x}^*\|^2 \\[2mm]
&\leq m + (\delta + h)c_1 + \frac{\bar{\alpha}}{2}\|\bar{x}^*\|^2, \tag{3.6.18}
\end{aligned}$$

where $c_1 = max\{\|\bar{x}^*\|, \eta(\|\bar{x}^*\|)\}$. Since $\lim_{\gamma \to 0} C(\gamma) = \infty$, one can regard that $C(\gamma) - c_1 > 0$. Then the inequality

$$\frac{\delta + h}{\bar{\alpha}} \leq \frac{\|\bar{x}^*\|^2}{2(C(\gamma) - c_1)}$$

is valid, and

$$\frac{\delta + h}{\bar{\alpha}} \to 0 \quad \text{as} \quad \bar{\alpha} \to 0.$$

Moreover, it results from (3.6.18) that

$$\frac{\psi(\gamma) - m}{\bar{\alpha}} \leq \frac{\|\bar{x}^*\|^2}{2} + \frac{\delta + h}{\bar{\alpha}}c_1.$$

Taking into account the condition (3.6.15) and estimate (3.6.17), we conclude that the sequence $\{x_{\bar{\alpha}}^\gamma\}$ as $\gamma \to 0$ is bounded. As in the proof of Theorem 3.6.2, we next deduce that $\bar{\alpha} \to 0$ when $\gamma \to 0$. The last assertion follows from the sufficient convergence condition for the operator regularization method in Theorem 2.2.1. ∎

Remark 3.6.7 *Under the conditions of Theorem 3.6.6, if the equation (3.1.1) has a unique solution x_0, $C(\gamma) = C > c_1$ in (3.6.13) and*

$$M \frac{\|x^*\|^2}{C - c_1} < 1,$$

then the weak convergence of $\{x_{\tilde{\alpha}}^\gamma\}$ to x_0 follows as $\gamma \to 0$.

3. Theorem 3.6.6 imposes a growth condition of the function $\eta(t)$. It may be omitted if we consider the smoothing functional in the form

$$\Phi_\gamma^\alpha(\|x\|) = \omega^h(x) - \langle f^\delta, x \rangle + \alpha\Phi(x),$$

where $\Phi(\|x\|)$ is defined by the equality (1.5.4). In view of Lemma 1.5.7, $\Phi'(\|x\|) = J^\mu(x)$. We suppose that the gauge function $\mu(t)$ of duality mapping J^μ is such that $\Phi(t) \geq \eta(t)$. Under these conditions, strong convergence is proved following the same scheme as in Theorem 3.6.6. Weak convergence holds if a constant $C > c_1$ and

$$\frac{\Phi(\|x^*\|)}{C - c_1} < 1.$$

For instance, in Example 2.2.8, the smoothing functional principle can be written as follows:

$$\bar{\alpha}\|u_{\tilde{\alpha}}^\gamma\|_{1,p}^p + \omega^h(u_{\tilde{\alpha}}^\gamma) = m + \bar{c}(\delta + h)^\eta, \quad \eta \in (0, 1),$$

where $\bar{c} > 0$ satisfies the inequality

$$m + \bar{c}(\delta + h)^\eta < \omega^h(\theta_X). \tag{3.6.19}$$

Thus, in (3.6.13),

$$C(\gamma) = \bar{c}(\delta + h)^{\eta - 1}.$$

Observe in addition that all results of this section can be reformulated for $D(A) \subset X$ provided that $D(A)$ are convex sets.

4. We further present the equation with a potential operator describing the twisting of reinforced bars (see [101, 147]), the minimal norm solution of which can be found by the approach above. If G is a bounded convex two-dimensional domain with the boundary ∂G, then the equation and boundary condition defining the elasto-plastic twisting $x(t, s)$ are written in the form

$$Ax = -\frac{\partial}{\partial t}\Big(g(T^2)\frac{\partial x}{\partial t}\Big) - \frac{\partial}{\partial s}\Big(g(T^2)\frac{\partial x}{\partial s}\Big) = f, \tag{3.6.20}$$

$$x|_{\partial G} = 0. \tag{3.6.21}$$

Here

$$T = grad\, u = \sqrt{\Big(\frac{\partial x}{\partial t}\Big)^2 + \Big(\frac{\partial x}{\partial s}\Big)^2}$$

is a maximal tangential stress, f is an angle of twist per unit length of the bar and $g(T^2)$ is a function which is characterized by the material of the bar in the stress state such that the constraint equation $\Gamma = g(T^2)T$ is fulfilled for intensity Γ of the shear deformations. It is well known that $g(T^2) \geq C_0$, where C_0 depends on the shear modulus of the stressed material, and

$$\frac{\partial \Gamma}{\partial T} \geq 0,$$

that is,

$$g(T^2) + 2T^2 g'(T^2) \geq 0. \tag{3.6.22}$$

If we replace (3.6.22) by the stronger inequality

$$g(T^2) + 2T^2 g'(T^2) \geq C_1 > 0 \tag{3.6.23}$$

and denote by $A'(x)$ the Gâteaux derivative at a point $x \in G$ of the operator A acting in $L^2(G)$, then one can show that

$$(A'(x)h, h) \geq C \int \int_G \left(\frac{\partial x}{\partial t}\right)^2 + \left(\frac{\partial x}{\partial s}\right)^2 dt ds, \quad h|_{\partial G} = 0,$$

where $C = min\{C_0, C_1\}$. Making use of the Friedrichs' inequality with a constant k we obtain

$$(A'(x)h, h) \geq kC\|x\|^2.$$

If

$$0 < C_2 \leq \lim_{v \to \infty} g(v)v^{1-\frac{1}{2}p} \leq C_3$$

and

$$\lim_{v \to \infty} |g'(v)|v^{2-\frac{1}{2}p} \leq C_4,$$

then the problem (3.6.20), (3.6.21) is equivalent to the minimization problem of the functional

$$F^0(x) = \int \int_G dt ds \int_0^{T^2(x)} \frac{g(v)}{2} dv - f \int \int_G x dt ds. \tag{3.6.24}$$

It is proved that the functional $F^0(x)$ is well defined, uniformly convex, coercive and weakly lower semicontinuous on the space W_1^p. If $C_1 = 0$ then the problem (3.6.20), (3.6.21) is ill-posed because the convexity of $F(x)$ ceases to be uniform. In this case, for given f^δ such that $\delta \geq \|f^\delta - f\|$, the twisting $x(t, s)$ satisfying (3.6.20), (3.6.21) can be defined by the regularization methods of this section.

Bibliographical Notes and Remarks

The residual properties on solutions of the regularized linear equations with perturbed data in Hilbert spaces were studied in [152, 153]. Analogous results for nonlinear equations with monotone and accretive operators in Banach spaces have been obtained in [5, 6, 7] and [32, 33, 34]. The residual principle and generalized residual principle were stated in [11] for accretive and in [8, 33, 34] for monotone operator equations. The concept of the generalized residual given in Definition 3.3.1 is due to Alber. Theorem 3.1.15 was proved in [188]. A more general form of the residual was used in [187].

The modified residual principle was stated in [192]. The estimates (3.3.32) and (3.3.33) of $\bar{\alpha}$ have been found in [202]. The minimal residual principle for nonlinear monotone equations has been provided in [198]. The linear case can be seen in [215]. The choice of the regularization parameter according to the smoothing functional principle for potential monotone equations was investigated in [22]. It was earlier studied in [129, 130, 224] for linear equations. The proof of Lemma 3.6.1 can be found in [130].

Chapter 4

REGULARIZATION OF VARIATIONAL INEQUALITIES

4.1 Variational Inequalities on Exactly Given Sets

1. Let X be an E-space, X^* be a strictly convex space, $A : X \to 2^{X^*}$ be a maximal monotone operator with domain $D(A)$, $\Omega \subset D(A)$ be a convex closed subset in X. Let either

$$int\ \Omega \neq \emptyset \quad \text{or} \quad int\ D(A) \cap \Omega \neq \emptyset. \tag{4.1.1}$$

Consider a variational inequality problem: To find $x \in \Omega$ such that

$$\langle Ax - f, z - x \rangle \geq 0 \quad \forall z \in \Omega. \tag{4.1.2}$$

As usual, we assume that its solution set N is nonempty. Hence, by Definition 1.11.1, for every $x^* \in N$ there exists $y \in Ax^*$ such that

$$\langle y - f, z - x^* \rangle \geq 0 \quad \forall z \in \Omega. \tag{4.1.3}$$

Moreover, according to Lemma 1.11.4, x^* satisfies the inequality

$$\langle y - f, z - x^* \rangle \geq 0 \quad \forall z \in \Omega, \quad \forall y \in Az. \tag{4.1.4}$$

Suppose that f and A in (7.1.1) are given with perturbations, that is, in place of f and A, their δ-approximations f^δ and h-approximations $A^h : X \to 2^{X^*}$ are known such that A^h are maximal monotone, $D(A^h) = D(A)$ and, respectively,

$$\|f - f^\delta\|_* \leq \delta, \quad \delta > 0, \tag{4.1.5}$$

and

$$\mathcal{H}_{X^*}(A^h x, Ax) \leq g(\|x\|)h \quad \forall x \in \Omega, \quad h > 0, \tag{4.1.6}$$

where $g(t)$ is a continuous non-negative function for $t \geq 0$ and $\mathcal{H}_{X^*}(G_1, G_2)$ stands the Hausdorff distance between the sets G_1 and G_2 in X^*. Thus, in reality, we solve the following approximate variational inequality:

$$\langle A^h x - f^\delta, z - x \rangle \geq 0 \quad \forall z \in \Omega, \quad x \in \Omega. \tag{4.1.7}$$

By Theorem 1.11.8, the set N is convex and closed, therefore, there exists a unique vector $\bar{x}^* \in N$ with the minimal norm. Our aim is to prove the convergence to $\bar{x}^*$ of approximations defined by the regularized variational inequality

$$\langle A^h x + \alpha J x - f^\delta, z - x \rangle \geq 0 \quad \forall z \in \Omega, \quad x \in \Omega, \quad \alpha > 0. \tag{4.1.8}$$

By Theorem 1.11.11, the inequality (4.1.8) is uniquely solvable. Denote its solutions by x_α^γ, where $\gamma = (\delta, h) \in \Re_0$ and $\Re_0 = (0, \delta^*] \times (0, h^*]$ with positive δ^* and h^*. Observe that x_α^γ is the regularized solution of the variational inequality (7.1.1). Since A^h is a maximal monotone operator, there is a vector $y_\alpha^\gamma \in A^h x_\alpha^\gamma$ such that

$$\langle y_\alpha^\gamma + \alpha J x_\alpha^\gamma - f^\delta, z - x_\alpha^\gamma \rangle \geq 0 \quad \forall z \in \Omega. \tag{4.1.9}$$

Theorem 4.1.1 *Let all the conditions of this subsection hold and*

$$\frac{\delta + h}{\alpha} \to 0 \quad \text{as} \quad \alpha \to 0. \tag{4.1.10}$$

Then the sequence $\{x_\alpha^\gamma\}$ converges in the norm of X to $\bar{x}^*$.

Proof. Presuming $z = x_\alpha^\gamma \in \Omega$ and $z = x^* \in N$ in (4.1.3) and in (4.1.9), respectively, and summing obtained inequalities, we can write down the following result:

$$\langle y_\alpha^\gamma - y, x^* - x_\alpha^\gamma \rangle + \alpha \langle J x_\alpha^\gamma, x^* - x_\alpha^\gamma \rangle - \langle f^\delta - f, x^* - x_\alpha^\gamma \rangle \geq 0. \tag{4.1.11}$$

Since A^h are monotone operators, we have

$$\langle y_\alpha^\gamma - y^h, x^* - x_\alpha^\gamma \rangle \leq 0 \quad \forall y^h \in A^h x^*, \quad \forall x^* \in N.$$

The condition (4.1.6) enables us to choose $y^h \in A^h x^*$ such that

$$\|y^h - y\|_* \leq g(\|x^*\|)h.$$

Then, as in the proof of Theorem 2.2.1, we deduce from (4.1.11) the inequality

$$\|x_\alpha^\gamma\|^2 - \left(\frac{\delta}{\alpha} + \frac{h}{\alpha} g(\|x^*\|) + \|x^*\|\right) \|x_\alpha^\gamma\| - \frac{\delta}{\alpha}\|x^*\| - \frac{h}{\alpha}g(\|x^*\|)\|x^*\| \leq 0.$$

By (4.1.10), it implies the boundedness of $\{x_\alpha^\gamma\}$ as $\alpha \to 0$. Let $x_\alpha^\gamma \rightharpoonup \bar{x} \in X$ as $\alpha \to 0$. It results from the Mazur theorem that Ω is weakly closed. Then we establish the inclusion $\bar{x} \in \Omega$ because $x_\alpha^\gamma \in \Omega$. By virtue of Lemma 1.11.4, the variational inequality (4.1.8) is equivalent to

$$\langle y^h + \alpha J z - f^\delta, z - x_\alpha^\gamma \rangle \geq 0 \quad \forall y^h \in A^h z, \quad \forall z \in \Omega. \tag{4.1.12}$$

Passing in (4.1.12) to the limit as $\alpha \to 0$ and taking into account that $f^\delta \to f$, one gets

$$\langle y - f, z - \bar{x} \rangle \geq 0 \quad \forall y \in Az, \quad \forall z \in \Omega. \tag{4.1.13}$$

By Lemma 1.11.4 again, $\bar{x} \in N$. Using further (4.1.11), by the same arguments as in the proof of Theorem 2.2.1, we conclude that $\bar{x} = \bar{x}^*$, $\|x_\alpha^\gamma\| \to \|\bar{x}^*\|$ and $x_\alpha^\gamma \to \bar{x}^*$. The final result is established by the fact that X is the E-space. Observe, that the assumptions (4.1.1) are used in Theorem 1.11.11 and Lemma 1.11.4. ∎

Theorem 4.1.2 *Under the conditions of Theorem 4.1.1, if $N = \{x_0\}$ and there exists a constant $C > 0$ such that $\dfrac{\delta + h}{\alpha} \leq C$ as $\alpha \to 0$, then $x_\alpha^\gamma \rightharpoonup x_0$.*

Proof follows from the previous theorem. ∎

Theorem 4.1.3 *Under the conditions of Theorems 4.1.1 and 4.1.2, convergence of the operator regularization method (4.1.8) is equivalent to solvability of the variational inequality (7.1.1).*

Proof. Let $x_\alpha^\gamma \rightharpoonup \bar{x} \in X$ as $\alpha \to 0$. Then the inclusion $\bar{x} \in N$ is proved as in Theorem 4.1.1. At the same time, the inverse assertion follows from Theorems 4.1.1 and 4.1.2. ∎

2. It is not difficult to verify that Theorems 4.1.1 - 4.1.3 remain still valid for more general regularized inequality

$$\langle A^h x + \alpha J^\mu x - f^\delta, z - x \rangle \geq 0 \quad \forall z \in \Omega, \quad x \in \Omega, \tag{4.1.14}$$

where $J^\mu : X \to X^*$ is the duality mapping with a gauge function $\mu(t)$. Existence and uniqueness of its solution $x_\alpha^\gamma \in \Omega$ can be proved as in Theorem 1.11.11. In other words, we assert that there is an element $y_\alpha^\gamma \in A^h x_\alpha^\gamma$ such that

$$\langle y_\alpha^\gamma + \alpha J^\mu x_\alpha^\gamma - f^\delta, z - x_\alpha^\gamma \rangle \geq 0 \quad \forall z \in \Omega. \tag{4.1.15}$$

To prove that $\lim_{\gamma \to 0} x_\alpha^\gamma = \bar{x}^*$, we further assume that the function $g(t)$ in (4.1.6) is increasing for $t \geq 0$ and apply the generalized residual principle of Chapter 3 to the variational inequality (4.1.14), where the residual of (4.1.7) on solutions x_α^γ is understood as follows (cf. Definition 3.3.1):

$$\rho_\mu(\alpha) = \alpha \|J^\mu x_\alpha^\gamma\|_* = \alpha \mu(\|x_\alpha^\gamma\|). \tag{4.1.16}$$

We establish some important properties of the functions $\rho_\mu(\alpha)$ and $\sigma_\mu(\alpha) = \mu(\|x_\alpha^\gamma\|)$ for variational inequalities.

Lemma 4.1.4 *Let X be an E-space, X^* be strictly convex, $A : X \to 2^{X^*}$ and $A^h : X \to 2^{X^*}$ be maximal monotone operators with domains $D(A) = D(A^h)$, $\Omega \subset D(A)$ be a convex closed subset in X such that (4.1.1) holds. Then the function $\sigma_\mu(\alpha)$ is single-valued, continuous and non-increasing for $\alpha \geq \alpha_0 > 0$.*

Proof. Fix two values of the regularization parameter α_1 and α_2 and write down for them the inequality (4.1.15). We have

$$\langle y_{\alpha_1}^\gamma + \alpha_1 J^\mu x_{\alpha_1}^\gamma - f^\delta, z - x_{\alpha_1}^\gamma \rangle \geq 0 \quad \forall z \in \Omega, \quad x_{\alpha_1}^\gamma \in \Omega, \quad y_{\alpha_1}^\gamma \in A^h x_{\alpha_1}^\gamma,$$

and

$$\langle y_{\alpha_2}^\gamma + \alpha_2 J^\mu x_{\alpha_2}^\gamma - f^\delta, z - x_{\alpha_2}^\gamma \rangle \geq 0 \quad \forall z \in \Omega, \quad x_{\alpha_2}^\gamma \in \Omega, \quad y_{\alpha_2}^\gamma \in A^h x_{\alpha_2}^\gamma.$$

Assuming $z = x_{\alpha_2}^\gamma$ and $z = x_{\alpha_1}^\gamma$, respectively, in the first and in the second inequalities, and summing them, one gets

$$\langle y_{\alpha_1}^\gamma - y_{\alpha_2}^\gamma, x_{\alpha_2}^\gamma - x_{\alpha_1}^\gamma \rangle + \langle \alpha_1 J^\mu x_{\alpha_1}^\gamma - \alpha_2 J^\mu x_{\alpha_2}^\gamma, x_{\alpha_2}^\gamma - x_{\alpha_1}^\gamma \rangle \geq 0.$$

Then the monotonicity property of A^h implies

$$\langle \alpha_1 J^\mu x_{\alpha_1}^\gamma - \alpha_2 J^\mu x_{\alpha_2}^\gamma, x_{\alpha_2}^\gamma - x_{\alpha_1}^\gamma \rangle \geq 0. \tag{4.1.17}$$

The rest of the proof follows the pattern of Lemma 3.1.8. ∎

Lemma 4.1.5 *Assume the conditions of Lemma 4.1.4. Then* $\lim_{\alpha \to \infty} x_\alpha^\gamma = x_*$, *where* $x_* \in \Omega$ *and* $\|x_*\| = min \{\|x\| \mid x \in \Omega\}$.

Proof. First of all, we note that, by virtue of the convexity and closedness of Ω in E-space X, an element x_* exists and it is uniquely defined. Applying (4.1.3), (4.1.15) and the definition of duality mapping J^μ, we deduce the inequality (cf. Theorem 4.1.1):

$$\mu(\|x_\alpha^\gamma\|)(\|x_\alpha^\gamma\| - \|x^*\|) \quad - \quad \|x_\alpha^\gamma\|\Big(\frac{\delta}{\alpha} + \frac{h}{\alpha}g(\|x^*\|)\Big)$$

$$- \quad \|x^*\|\Big(\frac{\delta}{\alpha} + \frac{h}{\alpha}g(\|x^*\|)\Big) \leq 0 \quad \forall x^* \in N,$$

which implies the boundedness of the sequence $\{x_\alpha^\gamma\}$ as $\alpha \to \infty$. Hence, $x_\alpha^\gamma \rightharpoonup \bar{x} \in X$, and then $\bar{x} \in \Omega$ according to the Mazur theorem. Show that $\bar{x} = x_*$ using the properties of J^μ. By (4.1.15),

$$\langle y_\alpha^\gamma - y^h, x_\alpha^\gamma - z \rangle \quad + \quad \alpha\Big(\mu(\|x_\alpha^\gamma\|) - \mu(\|z\|)\Big)(\|x_\alpha^\gamma\| - \|z\|)$$

$$+ \quad \alpha\langle J^\mu z, x_\alpha^\gamma - z \rangle \leq \langle y^h - f^\delta, z - x_\alpha^\gamma \rangle \quad \forall y^h \in A^h z, \quad \forall z \in \Omega.$$

Now the monotonicity condition of A^h yields the inequality

$$\Big(\mu(\|x_\alpha^\gamma\|) - \mu(\|z\|)\Big)(\|x_\alpha^\gamma\| - \|z\|) + \langle J^\mu z, x_\alpha^\gamma - z \rangle \leq \frac{\|y^h - f^\delta\|_*}{\alpha}\|z - x_\alpha^\gamma\|. \tag{4.1.18}$$

Then the estimate

$$\langle J^\mu z, x_\alpha^\gamma - z \rangle \leq \frac{\|y^h - f^\delta\|_*}{\alpha}\|z - x_\alpha^\gamma\| \quad \forall z \in \Omega \tag{4.1.19}$$

holds. Passing in (4.1.19) to the limit as $\alpha \to \infty$ we obtain

$$\langle J^\mu z, \bar{x} - z \rangle \leq 0 \quad \forall z \in \Omega. \tag{4.1.20}$$

As it has been established more than once (see, e.g., Theorem 2.2.1), (4.1.20) ensures the equality $\bar{x} = x_*$. If in (4.1.18) $z = x_*$ then $\|x_\alpha^\gamma\| \to \|x_*\|$ as $\alpha \to \infty$. Since X is E-space, the lemma is proved. ∎

3. Consider the generalized residual (4.1.16). Obviously, $\rho_\mu(\alpha) = \alpha \sigma_\mu(\alpha)$. By Lemma 4.1.4, $\rho_\mu(\alpha)$ is single-valued and continuous for $\alpha \geq \alpha_0 > 0$. We study its behaviour as $\alpha \to \infty$. It follows from Lemma 4.1.5 that if $\theta_X \notin \Omega$, then

$$\lim_{\alpha \to \infty} \rho_\mu(\alpha) = \infty.$$

Let $\theta_X \in \Omega$. Then $x_\alpha^\gamma \to \theta_X$ as $\alpha \to \infty$. By Lemma 1.11.7, the variational inequality (4.1.14) is equivalent to the inclusion

$$f^\delta \in B^h x_\alpha^\gamma + \alpha J^\mu x_\alpha^\gamma$$

with the maximal monotone operator B^h such that

$$B^h = A^h + \partial I_\Omega : X \to 2^{X^*}, \tag{4.1.21}$$

where $\partial I_\Omega : X \to 2^{X^*}$ is a subdifferential of the indicator function of the set Ω. It is clear that $D(B^h) = \Omega$. Hence, there is an element $\xi_\alpha^\gamma \in B^h x_\alpha^\gamma$ such that

$$\xi_\alpha^\gamma + \alpha J^\mu x_\alpha^\gamma = f^\delta. \tag{4.1.22}$$

Show that the sequence $\{\xi_\alpha^\gamma\}$ is bounded as $\alpha \to \infty$. The monotonicity of B^h yields the inequality

$$\langle \xi_\alpha^\gamma - \xi^h, x_\alpha^\gamma \rangle \geq 0 \quad \forall \xi^h \in B^h(\theta_X). \tag{4.1.23}$$

Making use of the property $(J^\nu)^* = (J^\mu)^{-1}$ of duality mapping J^μ in an E-space X, we obtain from (4.1.22) the following formula:

$$x_\alpha^\gamma = (J^\nu)^* \left(\frac{f^\delta - \xi_\alpha^\gamma}{\alpha} \right),$$

where $(J^\nu)^*$ is a duality mapping in X^* with the gauge function $\nu(t) = \mu^{-1}(t)$. Then (4.1.23) can be rewritten as

$$\left\langle \frac{\xi_\alpha^\gamma - \xi^h}{\alpha}, (J^\nu)^* \left(\frac{f^\delta - \xi_\alpha^\gamma}{\alpha} \right) \right\rangle \geq 0 \quad \forall \xi^h \in B^h(\theta_X).$$

It is not difficult to deduce from this the inequality

$$\|\xi_\alpha^\gamma - f^\delta\|_* \leq \|\xi^h - f^\delta\|_* \quad \forall \xi^h \in B^h(\theta_X), \tag{4.1.24}$$

that guarantees the boundedness of $\{\xi_\alpha^\gamma\}$. Therefore,

$$\xi_\alpha^\gamma \rightharpoonup \xi \in X^* \quad \text{as} \quad \alpha \to \infty. \tag{4.1.25}$$

Since maximal monotone operator B^h is demiclosed, $\xi_\alpha^\gamma \in B^h(x_\alpha^\gamma)$ and $x_\alpha^\gamma \to \theta_X$ as $\alpha \to \infty$, we have the limit inclusion $\xi \in B^h(\theta_X)$. Then from the weak convergence (4.1.25) and from (4.1.24) follows the chain of inequalities:

$$\|\xi - f^\delta\|_* \leq \liminf_{\alpha \to \infty} \|\xi_\alpha^\gamma - f^\delta\|_* \leq \limsup_{\alpha \to \infty} \|\xi_\alpha^\gamma - f^\delta\|_* \leq \|\xi^h - f^\delta\|_* \quad \forall \xi^h \in B^h(\theta_X). \quad (4.1.26)$$

Hence,

$$\|\xi - f^\delta\|_* = min \{\|\zeta - f^\delta\|_* \mid \zeta \in B^h(\theta_X)\}. \quad (4.1.27)$$

Thus, the whole sequence $\{\xi_\alpha^\gamma\}$ converges weakly to ξ because ξ is uniquely defined by (4.1.27).

The representation of the operator B^h by (4.1.21) means that if $\theta_X \in int\ \Omega$, then the value sets of A^h and B^h coincide at θ_X. Therefore, in this case, $\xi \in A^h(\theta_X)$ and

$$\|\xi - f^\delta\|_* = min \{\|\zeta - f^\delta\|_* \mid \zeta \in A^h(\theta_X)\}. \quad (4.1.28)$$

The result (4.1.26) also implies the convergence of norms:

$$\|\xi_\alpha^\gamma - f^\delta\|_* \to \|\xi - f^\delta\|_* \quad \text{as} \quad \alpha \to \infty.$$

At the same time, we find from (4.1.22) that

$$\rho_\mu(\alpha) = \|\xi_\alpha^\gamma - f^\delta\|_*.$$

Consequently,

$$\lim_{\alpha \to \infty} \rho_\mu(\alpha) = \|\xi - f^\delta\|_*,$$

where $\xi \in B^h(\theta_X)$. Finally, one can prove, as before, that if X^* is an E-space, then $\xi_\alpha^\gamma \to \xi$ as $\alpha \to \infty$. We are able to state the following lemma.

Lemma 4.1.6 Let B^h be defined by (4.1.21). Under the conditions of Lemma 4.1.4:
(i) if $\theta_X \notin \Omega$ then

$$\lim_{\alpha \to \infty} \rho_\mu(\alpha) = \infty,$$

(ii) if $\theta_X \in \Omega$ then $\xi_\alpha^\gamma \rightharpoonup \xi$ as $\alpha \to \infty$, where $\xi_\alpha^\gamma \in B^h x_\alpha^\gamma$ and $\xi \in B^h(\theta_X)$. In addition,

$$\lim_{\alpha \to \infty} \rho_\mu(\alpha) = \|\xi - f^\delta\|_*.$$

Moreover,
(iii) if $\theta_X \in \partial\Omega$ then ξ satisfies (4.1.27),
(iv) if $\theta_X \in int\ \Omega$ then ξ satisfies (4.1.28),
(v) if X^* is an E-space then $\xi_\alpha^\gamma \to \xi$ as $\alpha \to \infty$.

Instead of (4.1.14), consider the following regularized variational inequality:

$$\langle A^h x + \alpha J^\mu(x - z^0) - f^\delta, z - x \rangle \geq 0 \quad \forall z \in \Omega, \quad x \in \Omega, \quad (4.1.29)$$

where z^0 is a fixed element of X.

Lemma 4.1.7 *Let x_α^γ be a solution of (4.1.29) and B^h be defined by (4.1.21). Under the conditions of Lemma 4.1.4, the function*

$$\sigma(\alpha) = \|x_\alpha^\gamma - z^0\|$$

is single-valued, continuous and non-increasing. In addition, $x_\alpha^\gamma \to x_*$ as $\alpha \to \infty$, where $x_* \in \Omega$ is defined by the following minimum problem:

$$\|x_* - z^0\| = min\ \{\|x - z^0\| \mid x \in \Omega\}.$$

The function

$$\rho_\mu(\alpha) = \alpha\mu(\|x_\alpha^\gamma - z^0\|)$$

is single-valued and continuous. Moreover,
(i) if $z^0 \notin \Omega$ then

$$\lim_{\alpha \to \infty} \rho_\mu(\alpha) = \infty,$$

(ii) if $z^0 \in \Omega$ then $\xi_\alpha^\gamma \rightharpoonup \xi_0$ as $\alpha \to \infty$, where $\xi_\alpha^\gamma \in B^h x_\alpha^\gamma$ and $\xi_0 \in B^h z^0$. In addition,

$$\lim_{\alpha \to \infty} \rho_\mu(\alpha) = \|\xi_0 - f^\delta\|_*,$$

(iii) if $z^0 \in \partial\Omega$ then $\xi_0 \in B^h z^0$ and there holds the equality

$$\|\xi_0 - f^\delta\|_* = min\{\|\zeta - f^\delta\|_* \mid \zeta \in B^h z^0\},$$

(iv) if $z^0 \in int\ \Omega$ then $\xi_0 \in A^h z^0$ and

$$\|\xi_0 - f^\delta\|_* = min\ \{\|\zeta - f^\delta\|_* \mid \zeta \in A^h z^0\}.$$

Proof is established by the same arguments as in the previous lemma.   ∎

4. By making use of Lemmas 4.1.4 - 4.1.6, state and prove the generalized residual principle for the variational inequality (4.1.14). In the beginning, we study the case of exactly given monotone operators.

Theorem 4.1.8 *Let X be an E-space with strictly convex X^*, $A : X \to X^*$ be a maximal monotone and hemicontinuous operator with domain $D(A)$, $\Omega \subset D(A)$ be a convex closed set. Let the variational inequality (7.1.1) have a nonempty solution set N and $\bar{x}^* \in N$ be a minimal norm solution. f^δ be a δ-approximation of f such that $\|f - f^\delta\|_* \leq \delta \leq 1$ and $\theta_X \notin \Omega$. Then there exists $\bar{\alpha} > 0$ such that*

$$\rho(\bar{\alpha}) = \bar{\alpha}\|x_{\bar{\alpha}}^\delta\| = k\delta^p, \quad k > 1, \quad p \in (0, 1], \tag{4.1.30}$$

where $x_{\bar{\alpha}}^\delta$ is a (classical) solution of the regularized variational inequality

$$\langle Ax + \alpha Jx - f^\delta, z - x \rangle \geq 0 \quad \forall z \in \Omega, \quad x \in \Omega, \tag{4.1.31}$$

with $\alpha = \bar{\alpha}$. Moreover,
(i) if $\delta \to 0$ then $\bar{\alpha} \to 0$,

(ii) if $\delta \to 0$ and $p \in (0,1)$ then $x_\alpha^\delta \to \bar{x}^*$ and $\dfrac{\delta}{\bar\alpha} \to 0$,

(iii) if $\delta \to 0$, $p = 1$ and $N = \{x_0\}$ then $x_{\bar\alpha}^\delta \rightharpoonup x_0$ and there exists a constant $C > 0$ such that $\dfrac{\delta}{\bar\alpha} \leq C$.

Proof. In view of Theorem 1.7.19, maximal monotone and hemicontinuous operator A is defined on an open set $D(A)$. Therefore, *int* $D(A) \cap \Omega \neq \emptyset$. Hence, in our circumstances, it is possible to apply Lemma 1.11.4 and Theorem 1.11.11. As in the proof of Theorem 4.1.1, we deduce the quadratic inequality

$$\|x_\alpha^\delta\|^2 \leq \left(\frac{\delta}{\alpha} + \|x^*\|\right)\|x_\alpha^\delta\| + \frac{\delta}{\alpha}\|x^*\| \quad \forall x^* \in N. \tag{4.1.32}$$

Then there holds the estimate

$$\|x_\alpha^\delta\| \leq 2\|x^*\| + \frac{\delta}{\alpha} \quad \forall x^* \in N. \tag{4.1.33}$$

Fix some $x^* \in N$ and choose α such that

$$2\alpha\|x^*\| < (k-1)\delta^p, \quad k > 1, \quad p \in (0,1]. \tag{4.1.34}$$

This allows us to find the estimate for the residual, namely,

$$\rho(\alpha) = \alpha\|x_\alpha^\delta\| \leq 2\alpha\|x^*\| + \delta < (k-1)\delta^p + \delta \leq k\delta^p. \tag{4.1.35}$$

Furthermore, by Lemma 4.1.6,

$$\lim_{\alpha \to \infty} \rho(\alpha) = \infty.$$

Then the existence of $\bar\alpha$ follows from the continuity of $\rho(\alpha)$.

Next, by (4.1.34), we have the inequality

$$\bar\alpha > \frac{(k-1)\delta^p}{2\|x^*\|}.$$

Consequently,

$$\frac{\delta}{\bar\alpha} \leq \frac{2\|x^*\|\delta^{1-p}}{k-1}.$$

Therefore, $\dfrac{\delta}{\bar\alpha} \to 0$ when $\delta \to 0$ and $p \in (0,1)$, and $\dfrac{\delta}{\bar\alpha} \leq C$ with $C = 2\|x^*\|(k-1)^{-1}$ when $\delta \to 0$ and $p = 1$. It is easy to see that (4.1.33) implies the boundedness of the sequence $\{x_{\bar\alpha}^\delta\}$. Hence, we obtain that $x_{\bar\alpha}^\delta \rightharpoonup \bar{x} \in X$ as $\delta \to 0$. Since Ω is weakly closed by the Mazur theorem, the inclusion $\bar{x} \in \Omega$ holds.

Show that $\bar{x} \in N$. Construct the maximal monotone operator

$$B = A + \partial I_\Omega : X \to 2^{X^*}.$$

It is clear that $D(B) = \Omega$. By Lemma 1.11.7, from variational inequality (4.1.31) with $\alpha = \bar\alpha$, we have

$$f^\delta \in Bx_{\bar\alpha}^\delta + \bar\alpha J x_{\bar\alpha}^\delta.$$

This means that there exists $\xi_{\bar{\alpha}}^{\delta} \in Bx_{\bar{\alpha}}^{\delta}$ such that

$$\xi_{\bar{\alpha}}^{\delta} + \bar{\alpha} J x_{\bar{\alpha}}^{\delta} = f^{\delta}.$$

In view of (4.1.30), one gets

$$\rho(\bar{\alpha}) = \bar{\alpha} \|x_{\bar{\alpha}}^{\delta}\| = \|\xi_{\bar{\alpha}}^{\delta} - f^{\delta}\|_* = k\delta^p.$$

This implies the limit result: $\xi_{\bar{\alpha}}^{\delta} \to f$ as $\delta \to 0$.

Write down the monotonicity property of the operator B as

$$\langle \xi_{\bar{\alpha}}^{\delta} - Az - y, x_{\bar{\alpha}}^{\delta} - z \rangle \geq 0 \quad \forall z \in \Omega, \quad \forall y \in \partial I_{\Omega} z. \tag{4.1.36}$$

Since $\theta_{X^*} \in \partial I_{\Omega} z$ for all $z \in \Omega$ (see (1.8.14)), we may put in (4.1.36) $y = \theta_{X^*}$. After that, passing in (4.1.36) to the limit as $\delta \to 0$, we deduce

$$\langle f - Az, \bar{x} - z \rangle \geq 0 \quad \forall z \in \Omega.$$

The latter inequality is equivalent to (7.1.1) because of Lemma 1.11.4. This means that $\bar{x} \in N$.

Let $p \in (0, 1)$. Along with (4.1.33), the quadratic inequality (4.1.32) gives also the estimate

$$\|x_{\bar{\alpha}}^{\delta}\| \leq \frac{2\delta}{\bar{\alpha}} + \|x^*\| \quad \forall x^* \in N. \tag{4.1.37}$$

Together with the weak convergence of $x_{\bar{\alpha}}^{\delta}$ to $\bar{x} \in N$, (4.1.37) yields the following relations:

$$\|\bar{x}\| \leq \liminf_{\delta \to 0} \|x_{\bar{\alpha}}^{\delta}\| \leq \limsup_{\delta \to 0} \|x_{\bar{\alpha}}^{\delta}\| \leq \|x^*\| \quad \forall x^* \in N. \tag{4.1.38}$$

Therefore, $\bar{x} = \bar{x}^*$ because the minimal norm solution $\bar{x}^* \in N$ is unique. Moreover, the convergence of $\|x_{\bar{\alpha}}^{\delta}\|$ to $\|\bar{x}^*\|$ as $\delta \to 0$ follows from (4.1.38). Since X is E-space, the claim (ii) holds. By (4.1.30),

$$\bar{\alpha} = \frac{k\delta^p}{\|x_{\bar{\alpha}}^{\delta}\|}. \tag{4.1.39}$$

Observe that $\|x_{\bar{\alpha}}^{\delta}\| > 0$ for a small enough $\delta > 0$ by reason of $\bar{x}^* \neq \theta_X$. The latter follows from the hypotheses that $\theta_X \notin \Omega$. Then one gets from (4.1.39) that $\bar{\alpha} \to 0$ as $\delta \to 0$ and $p \in (0, 1)$. If $p = 1$ then $x_{\bar{\alpha}}^{\delta} \rightharpoonup x_0 \neq \theta_X$. By the property of the norm in a Banach space, we have the relation

$$\|x_0\| \leq \liminf_{\delta \to 0} \|x_{\bar{\alpha}}^{\delta}\|,$$

which enables us to conclude that $\bar{\alpha} \to 0$ as $\delta \to 0$. This completes the proof. ∎

Theorem 4.1.9 *Assume that the conditions of Theorem 4.1.8 are satisfied, $\theta_X \in \Omega$ and*

$$\|\xi_0^{\delta} - f^{\delta}\|_* > k\delta^p, \quad k > 1, \quad p \in (0, 1], \tag{4.1.40}$$

where ξ_0^{δ} is defined as follows:
(i) if $\theta_X \in int\ \Omega$ then $\xi_0^{\delta} = A(\theta_X)$,
(ii) if $\theta_X \in \partial \Omega$ then

$$\|\xi_0^{\delta} - f^{\delta}\|_* = min\ \{\|\zeta - f^{\delta}\|_* \mid \zeta \in B(\theta_X)\},$$

where $B = A + \partial I_{\Omega}$. Then all the conclusions of Theorem 4.1.8 remain still valid.

Proof. Due to Lemma 4.1.6,

$$\lim_{\alpha \to \infty} \rho(\alpha) = \|\xi_0^\delta - f^\delta\|_*.$$

As it was established in (4.1.35), there is $\alpha > 0$ such that $\rho(\alpha) < k\delta^p$. Then the condition (4.1.40) guarantees that $\bar{\alpha}$ satisfying (4.1.30) exists. Since the solvability of variational inequality (4.1.31) is equivalent to the inclusion $f \in Bx$, where $D(B) = \Omega$ and $Bx = Ax$ for $x \in int\ \Omega$, it results from (4.1.40) that $\theta_X \notin N$. The remaining assertions are proved as in Theorem 4.1.8. ■

Along with (4.1.31), it is possible to research the regularized variational inequality

$$\langle Ax_\alpha^\delta + \alpha J(x_\alpha^\delta - z^0) - f^\delta, z - x_\alpha^\delta \rangle \geq 0 \quad \forall z \in \Omega,$$

where $x_\alpha^\delta \in \Omega$ and z^0 is an arbitrary fixed point of X. Under the conditions of Theorems 4.1.8 and 4.1.9, it is not difficult to prove the following:
(i) If $z^0 \notin \Omega$ then Theorem 4.1.8 holds with $\rho(\alpha) = \alpha\|x_\alpha^\delta - z^0\|$. The vector $\bar{x}^*$ is defined there by the minimization problem:

$$\|\bar{x}^* - z^0\| = min\ \{\|x^* - z^0\| \mid x^* \in N\}.$$

(ii) If $z^0 \in \Omega$ then Theorem 4.1.9 holds, provided that: in the case of $z^0 \in int\ \Omega$, (4.1.40) is replaced by the inequality

$$\|Az^0 - f^\delta\|_* > k\delta^p.$$

In the case of $z^0 \in \partial\Omega$, the element ξ_0^δ is defined as

$$\|\xi_0^\delta - f^\delta\|_* = min\ \{\|\zeta - f^\delta\|_* \mid \zeta \in Bz^0\},$$

where $B = A + \partial I_\Omega$.

Next we omit the hemicontinuity property of A and present the following theorem.

Theorem 4.1.10 *Assume that A is a maximal monotone (possibly, set-valued) operator. Then all the conclusions of Theorems 4.1.8 and 4.1.9 remain still valid if (4.1.1) is satisfied. Moreover, if $\theta_X \in int\ \Omega$, then ξ_0^δ is defined by the minimization problems:*

$$\|\xi_0^\delta - f^\delta\|_* = min\{\|\zeta - f^\delta\|_* \mid \zeta \in A(\theta_X)\}. \tag{4.1.41}$$

By analogy, if $z^0 \in int\ \Omega$, then ξ_δ^0 is defined as

$$\|\xi_\delta^0 - f^\delta\|_* = min\{\|\zeta - f^\delta\|_* \mid \zeta \in Az^0\}. \tag{4.1.42}$$

If $\theta_X \in \partial\Omega$ (respectively, $z^0 \in \partial\Omega$), then ξ_δ^0 is defined by (4.1.41) (respectively, (4.1.42)), where A is replaced by $B = A + \partial I_\Omega$.

5. We further discuss the variational inequalities with approximately given operators.

Theorem 4.1.11 *Suppose that X is an E-space with strictly convex X^*, $A : X \to 2^{X^*}$ is a maximal monotone operator with domain $D(A)$, $\Omega \subset D(A)$ is a convex closed subset, condition (4.1.1) holds, variational inequality (7.1.1) has a nonempty solution set N and $\bar{x}^* \in N$ is the minimal norm solution. Let $A^h : X \to 2^{X^*}$ with $h > 0$ be maximal monotone operators. $\Omega \subset D(A^h)$, and if int $\Omega = \emptyset$, then int $D(A^h) \cap \Omega \neq \emptyset$. Let conditions (4.1.5) and (4.1.6) hold, where $g(t)$ is a non-negative, continuous and increasing function and $0 < \delta + h \leq 1$. Furthermore, assume that in the case of $\theta_X \in \Omega$, the additional inequality*

$$\|\xi_0 - f^\delta\|_* > \big(k + g(0)\big)(\delta + h)^p, \quad k > 1, \quad p \in (0, 1], \tag{4.1.43}$$

is satisfied, where ξ_0 is defined by the following minimization problems:
1) if $\theta_X \in int\ \Omega$ then

$$\|\xi_0 - f^\delta\|_* = min\ \{\|\zeta - f^\delta\|_* \mid \zeta \in A^h(\theta_X)\},$$

2) if $\theta_X \in \partial\Omega$ then

$$\|\xi_0 - f^\delta\|_* = min\ \{\|\zeta - f^\delta\|_* \mid \zeta \in B^h(\theta_X)\},$$

where $B^h = A^h + \partial I_\Omega$. Then there exists a unique $\bar{\alpha}$ satisfying the equation

$$\rho(\bar{\alpha}) = \bar{\alpha}\|x_{\bar{\alpha}}^\gamma\| = \big(k + g(\|x_{\bar{\alpha}}^\gamma\|)\big)(\delta + h)^p, \quad k > 1, \quad p \in (0, 1], \tag{4.1.44}$$

where $x_{\bar{\alpha}}^\gamma$ is a solution of the regularized variational inequality (4.1.8) with $\alpha = \bar{\alpha}$ and $\gamma = (\delta, h)$. Moreover,
(i) if $\gamma \to 0$ then $\bar{\alpha} \to 0$,
(ii) if $\gamma \to 0$ and $p \in (0, 1)$ then $x_{\bar{\alpha}}^\gamma \to \bar{x}^*$ and $\dfrac{\delta + h}{\bar{\alpha}} \to 0$,
(iii)) if $\gamma \to 0$, $p = 1$ and $N = \{x_0\}$ then $x_{\bar{\alpha}}^\gamma \rightharpoonup x_0$ and there exists a constant $C > 0$ such that $\dfrac{\delta + h}{\bar{\alpha}} \leq C$.

Proof. Using the monotonicity of A, inequality (4.1.11) and conditions (4.1.5) and (4.1.6) of the theorem, we calculate the estimate

$$\|x_\alpha^\gamma\|^2 \leq \Big(\frac{\delta}{\alpha} + \frac{h}{\alpha}g(\|x_\alpha^\gamma\|) + \|x^*\|\Big)\|x_\alpha^\gamma\| + \Big(\frac{\delta}{\alpha} + \frac{h}{\alpha}g(\|x_\alpha^\gamma\|)\Big)\|x^*\| \quad \forall x^* \in N. \tag{4.1.45}$$

It implies

$$\|x_\alpha^\gamma\| \leq 2\|x^*\| + \frac{\delta}{\alpha} + \frac{h}{\alpha}g(\|x_\alpha^\gamma\|) \quad \forall x^* \in N.$$

Hence, for the residual function, one gets

$$\rho(\alpha) = \alpha\|x_\alpha^\gamma\| \leq 2\alpha\|x^*\| + \delta + hg(\|x_\alpha^\gamma\|).$$

If, for some $x^* \in N$, the parameter α is such that

$$2\alpha\|x^*\| < (k - 1)(\delta + h)^p, \quad k > 1, \quad p \in (0, 1],$$

that is,

$$\alpha < \frac{k-1}{2\|x^*\|}(\delta + h)^p, \tag{4.1.46}$$

then

$$\rho(\alpha) < \Big(k + g(\|x_\alpha^\gamma\|)\Big)(\delta + h)^p. \tag{4.1.47}$$

Consider first the case when $\theta_X \in \Omega$. Due to Lemmas 4.1.5 and 4.1.6, we conclude that

$$\lim_{\alpha \to \infty} \|x_\alpha^\gamma\| = 0$$

and

$$\lim_{\alpha \to \infty} \rho(\alpha) = \|\xi - f^\delta\|_*, \quad \xi \in B^h(\theta_X).$$

Then, it follows from the continuity of $\rho(\alpha)$ and $g(t)$ and from (4.1.43) and (4.1.47) that there exists a solution $\bar\alpha > 0$ of the equation (4.1.44). Moreover, (4.1.46) implies

$$\bar\alpha > \frac{k-1}{2\|x^*\|}(\delta + h)^p. \tag{4.1.48}$$

As in the proof of Theorem 4.1.9, assertions (i) and (ii) for the sequence $\left\{\dfrac{\delta + h}{\bar\alpha}\right\}$ can be deduced from (4.1.48), separately for $p = 1$ and $p \in (0,1)$. If the monotonicity condition of operators A^h is used in (4.1.11) with $\alpha = \bar\alpha$ then we come to the inequality similar to (4.1.45):

$$\|x_{\bar\alpha}^\gamma\|^2 \le \Big(\frac{\delta}{\bar\alpha} + \frac{h}{\bar\alpha}g(\|x^*\|) + \|x^*\|\Big)\|x_{\bar\alpha}^\gamma\| + \Big(\frac{\delta}{\bar\alpha} + \frac{h}{\bar\alpha}g(\|x^*\|)\Big)\|x^*\| \quad \forall x^* \in N.$$

As before, there holds the estimate

$$\|x_{\bar\alpha}^\gamma\| \le \|x^*\| + 2\frac{\delta}{\bar\alpha} + 2\frac{h}{\bar\alpha}g(\|x^*\|). \tag{4.1.49}$$

Therefore, $\{x_{\bar\alpha}^\gamma\}$ is bounded, hence, $x_{\bar\alpha}^\gamma \rightharpoonup \bar x \in \Omega$ as $\gamma \to 0$.

Show that $\bar x \in N$. Indeed, variational inequality (4.1.8) is reduced to the inclusion

$$f^\delta \in B^h x_{\bar\alpha}^\gamma + \bar\alpha J x_{\bar\alpha}^\gamma,$$

where $B^h = A^h + \partial I_\Omega$. This means that there exists $\xi_{\bar\alpha}^\gamma \in B^h x_{\bar\alpha}^\gamma$ such that

$$\xi_{\bar\alpha}^\gamma + \bar\alpha J x_{\bar\alpha}^\gamma = f^\delta.$$

The latter equality and (4.1.44) induce the strong convergence of $\{\xi_{\bar\alpha}^\gamma\}$ to f as $\gamma \to 0$. By (4.1.6), for every $x \in \Omega$ and every $y \in Ax$, one can construct a sequence $\{y^h\}$, $y^h \in A^h x$, such that $y^h \to y$ as $h \to 0$. Then the monotonicity of operator B^h yields the following inequality:

$$\langle \xi_{\bar\alpha}^\gamma - y^h, x_{\bar\alpha}^\gamma - x \rangle \ge 0 \quad \forall x \in \Omega.$$

Letting $\gamma \to 0$ one has

$$\langle f - y, \bar{x} - x \rangle \geq 0 \quad \forall x \in \Omega, \quad \forall y \in Ax.$$

Now, by Lemma 1.11.4, we conclude that $\bar{x} \in N$, and the assertions of the theorem for $p = 1$ are proved completely.

Let $p \in (0, 1)$. Using the weak convergence of $x_{\bar{\alpha}}^{\gamma}$ to $\bar{x} \in N$ and inequality (4.1.49), we make sure, as in Theorem 4.1.8, that $\bar{x} = \bar{x}^*$ and $\|x_{\bar{\alpha}}^{\gamma}\| \to \|\bar{x}^*\|$ as $\gamma \to 0$. Thus, the strong convergence $x_{\bar{\alpha}}^{\gamma} \to \bar{x}^*$ is established in an E-space X.

Prove that $\bar{\alpha} \to 0$ as $\gamma \to 0$. For this end, show that condition (4.1.43) ensures that $\theta_X \notin N$. If $\theta_X \in N$ then $f \in B(\theta_X)$, where $B = A + \partial I_{\Omega}$. By virtue of the definitions of operators B and B^h, there exist $\xi^h \in B^h(\theta_X)$ such that $\|\xi^h - f\|_* \leq hg(0)$. Then

$$\|\xi^h - f\|_* \leq \|\xi^h - f^{\delta}\|_* + \|f - f^{\delta}\|_* \leq hg(0) + \delta \leq \big(k + g(0)\big)(\delta + h)^p,$$

which contradicts (4.1.43). Then the claim is established in the same way as in Theorem 3.3.2.

Prove that $\bar{\alpha}$ involving (4.1.44) is unique. Assume that (4.1.44) has two solutions $\bar{\alpha}$ and $\bar{\beta}$ such that

$$\rho(\bar{\alpha}) = \bar{\alpha}\|x_{\bar{\alpha}}^{\gamma}\| = \big(k + g(\|x_{\bar{\alpha}}^{\gamma}\|)\big)(\delta + h)^p \tag{4.1.50}$$

and

$$\rho(\bar{\beta}) = \bar{\beta}\|x_{\bar{\beta}}^{\gamma}\| = \big(k + g(\|x_{\bar{\beta}}^{\gamma}\|)\big)(\delta + h)^p, \tag{4.1.51}$$

where $x_{\bar{\alpha}}^{\gamma}$ and $x_{\bar{\beta}}^{\gamma}$ satisfy, respectively, the following inequalities:

$$\langle y_{\bar{\alpha}}^{\gamma} + \bar{\alpha}Jx_{\bar{\alpha}}^{\gamma} - f^{\delta}, z - x_{\bar{\alpha}}^{\gamma} \rangle \geq 0 \quad \forall z \in \Omega, \quad y_{\bar{\alpha}}^{\gamma} \in A^h x_{\bar{\alpha}}^{\gamma}, \quad x_{\bar{\alpha}}^{\gamma} \in \Omega, \tag{4.1.52}$$

and

$$\langle y_{\bar{\beta}}^{\gamma} + \bar{\beta}Jx_{\bar{\beta}}^{\gamma} - f^{\delta}, z - x_{\bar{\beta}}^{\gamma} \rangle \geq 0 \quad \forall z \in \Omega, \quad y_{\bar{\beta}}^{\gamma} \in A^h x_{\bar{\beta}}^{\gamma}, \quad x_{\bar{\beta}}^{\gamma} \in \Omega. \tag{4.1.53}$$

Put $x = x_{\bar{\beta}}^{\gamma}$ in (4.1.52) and $z = x_{\bar{\alpha}}^{\gamma}$ in (4.1.53), and add the obtained inequalities. Then we obtain

$$\langle y_{\bar{\alpha}}^{\gamma} - y_{\bar{\beta}}^{\gamma}, x_{\bar{\alpha}}^{\gamma} - x_{\bar{\beta}}^{\gamma} \rangle + \langle \bar{\alpha}Jx_{\bar{\alpha}}^{\gamma} - \bar{\beta}Jx_{\bar{\beta}}^{\gamma}, x_{\bar{\alpha}}^{\gamma} - x_{\bar{\beta}}^{\gamma} \rangle \leq 0.$$

Further, it is necessary to use (4.1.50) and (4.1.51) and repeat the reasoning from the proof of Theorem 3.3.2.

If $\theta_X \notin \Omega$ then, by Lemma 4.1.6, $\lim_{\alpha \to \infty} \rho(\alpha) = \infty$. Therefore the requirement (4.1.43) in this case can be omitted. The proof is accomplished. $\blacksquare$

Remark 4.1.12 *Along with the generalized residual principle, the choice of regularization parameter in the variational inequality (4.1.8) can be realized by the modified residual principle of Section 3.4 (see Theorem 3.4.1 and Remarks 3.4.2).*

Remark 4.1.13 *Some important applied problems are reduced to variational inequalities for which sets Ω are defined as follows [128]:*

$$\Omega = \{u \mid u = u(x) \in X, \ u(x) \geq 0\}$$

and

$$\Omega = \{u \mid u = u(x) \in X, \ u(x)|_{\partial\Omega} = 0\},$$

where $x \in G$, G is a bounded measurable subset in R^n and X is a Banach space. These sets in the operator regularization methods can be considered as unperturbed ones.

4.2 Variational Inequalities on Approximately Given Sets

In the previous section, we dealt with the regularizing processes for variational inequalities provided that their constraint sets Ω are given exactly. However, there are many practical variational problems with approximately given sets. In this section, we show that the operator regularization methods also solve these problems stably.

Under the conditions of Theorem 4.1.1, let $\{A^h\}$ be a sequence of maximal monotone operators, $\{\Omega_\sigma\}$ be a sequence of convex closed sets such that $\Omega_\sigma \subseteq D(A^h)$ and either

$$int \ \Omega_\sigma \neq \emptyset \quad or \quad int \ D(A^h) \cap \Omega_\sigma \neq \emptyset. \tag{4.2.1}$$

In the sequel, for some positive δ^*, h^* and σ^*, we denote $\Re = (0, \delta^*] \times (0, h^*] \times (0, \sigma^*]$. Let the inequality (4.1.6) hold for all $x \in \Omega \cap \Omega_\sigma$ and for every couple of A^h and Ω_σ.

Suppose that X is a Hilbert space H and operators A^h have the following growth order: There exists a constant $M > 0$ such that

$$\|y^h - f^\delta\| \leq M(\|x\| + 1) \quad \forall y^h \in A^h x, \quad \forall x \in \Omega_\sigma, \tag{4.2.2}$$

where $\gamma = (\delta, h, \sigma) \in \Re$. We study the strong convergence of the operator regularization method for variational inequalities with different proximity conditions between Ω_σ and Ω.

Let Ω_σ uniformly approximate Ω in the Hausdorff metric, that is,

$$\mathcal{H}_H(\Omega, \Omega_\sigma) \leq \sigma. \tag{4.2.3}$$

Consider approximations to a solution of (7.1.1) generated by the regularized variational inequality

$$\langle A^h x + \alpha x - f^\delta, z - x \rangle \geq 0 \quad \forall z \in \Omega_\sigma, \ x \in \Omega_\sigma. \tag{4.2.4}$$

As it follows from Section 1.11, this inequality has unique regularized solution x_α^γ. Therefore, there exists $y_\alpha^\gamma \in A^h x_\alpha^\gamma$ such that

$$\langle y_\alpha^\gamma + \alpha x_\alpha^\gamma - f^\delta, z - x_\alpha^\gamma \rangle \geq 0 \quad \forall z \in \Omega_\sigma. \tag{4.2.5}$$

We emphasize that each operator A^h in the variational inequality (4.2.4) is not assumed to be defined on every subset Ω_σ. However, we declare that always there is a possibility to approach the parameters h and σ to zero at the same time. This remark has to do with all the variational inequalities of the type (4.2.4).

1. In this subsection we presume that $\Omega \subseteq \Omega_\sigma$. This is the so-called exterior approximations of Ω. In the next subsection we will study the interior approximations of Ω when $\Omega_\sigma \subseteq \Omega$.

Theorem 4.2.1 *Assume that*
(i) $A : H \to 2^H$ is a maximal monotone operator;
(ii) Ω is a convex closed set in H;
(iii) $\{A^h\}$ is a sequence of the maximal monotone operators $A^h : H \to 2^H$;
(iv) $\{\Omega_\sigma\}$ is a sequence of the convex closed sets such that $\Omega_\sigma \subseteq D(A^h)$;
(v) $\Omega \subseteq \Omega_\sigma$ for all $\sigma \in (0, \sigma^*]$;
(vi) for all $x \in \Omega$, the proximity between the operators A and A^h is given by

$$\mathcal{H}_H(A^h x, Ax) \leq g(\|x\|)h \quad \forall x \in \Omega, \quad h \in (0, h^*], \tag{4.2.6}$$

where $g(t)$ is a continuous and non-negative function for $t \geq 0$;
(vii) the conditions (4.1.1), (4.1.5), (4.2.1), (4.2.2) and (4.2.3) are fulfilled;
(viii) variational inequality (7.1.1) has a nonempty solution set N.
If

$$\lim_{\alpha \to 0} \frac{\delta + h + \sigma}{\alpha} = 0, \tag{4.2.7}$$

then the solution sequence $\{x_\alpha^\gamma\}$ of the variational inequality (4.2.4) converges strongly in H to the minimal norm solution $\bar{x}^* \in N$ as $\alpha \to 0$.

Proof. According to condition (4.2.3), for all $x_\alpha^\gamma \in \Omega_\sigma$ and for all $x^* \in N \subset \Omega$, there exist respective elements $u_\alpha^\gamma \in \Omega$ and $v_\alpha^\gamma \in \Omega_\sigma$ such that

$$\|x_\alpha^\gamma - u_\alpha^\gamma\| \leq \sigma \tag{4.2.8}$$

and

$$\|x^* - v_\alpha^\gamma\| \leq \sigma. \tag{4.2.9}$$

Since $x^* \in N$, there exists $y \in Ax^*$ such that (4.1.3) holds. Presuming $z = u_\alpha^\gamma$ in (4.1.3) and $z = v_\alpha^\gamma$ in (4.2.5) and summing the obtained inequalities, one gets for $y_\alpha^\gamma \in A^h x_\alpha^\gamma$,

$$(y_\alpha^\gamma + \alpha x_\alpha^\gamma - f^\delta, v_\alpha^\gamma - x_\alpha^\gamma) + (y - f, u_\alpha^\gamma - x^*) \geq 0 \quad \forall x^* \in N, \quad y \in Ax^*. \tag{4.2.10}$$

Since A^h is monotone for each $h \in (0, h^*]$ and $\Omega \subseteq \Omega_\sigma \subseteq D(A^h)$, we have

$$(y_\alpha^\gamma - y^h, x_\alpha^\gamma - x^*) \geq 0 \quad \forall y^h \in A^h x^*.$$

Moreover, there exists $y^h \in A^h x^*$ such that

$$\|y^h - y\| \leq g(\|x^*\|)h.$$

On the basis of (4.2.10), we obtain

$$
\begin{aligned}
\alpha(x_\alpha^\gamma, x_\alpha^\gamma - v_\alpha^\gamma) &\leq (y - f, u_\alpha^\gamma - x_\alpha^\gamma) + (y_\alpha^\gamma - f^\delta, v_\alpha^\gamma - x_\alpha^\gamma) + (y - f, x_\alpha^\gamma - x^*) \\
&\leq (y - f, u_\alpha^\gamma - x_\alpha^\gamma) + (y_\alpha^\gamma - f^\delta, v_\alpha^\gamma - x^*) + (y - y_\alpha^\gamma, x_\alpha^\gamma - x^*) + (f^\delta - f, x_\alpha^\gamma - x^*) \\
&\leq (y - f, u_\alpha^\gamma - x_\alpha^\gamma) + (y_\alpha^\gamma - f^\delta, v_\alpha^\gamma - x^*) + (y^h - y, x^* - x_\alpha^\gamma) + (f - f^\delta, x^* - x_\alpha^\gamma) \\
&\leq (y - f, u_\alpha^\gamma - x_\alpha^\gamma) + (y_\alpha^\gamma - f^\delta, v_\alpha^\gamma - x^*) + \left(hg(\|x^*\|) + \delta\right)\|x_\alpha^\gamma - x^*\|. \qquad (4.2.11)
\end{aligned}
$$

Then (4.2.2), (4.2.8) and (4.2.9) imply the quadratic inequality

$$
\|x_\alpha^\gamma\|^2 \; - \; \|x_\alpha^\gamma\|\Big(\frac{\delta}{\alpha} + M\frac{\sigma}{\alpha} + \frac{h}{\alpha}g(\|x^*\|) + \sigma + \|x^*\|\Big)
$$
$$
- \; \Big(\frac{h}{\alpha}g(\|x^*\|) + \frac{\delta}{\alpha} + M\frac{\sigma}{\alpha}\Big)\|x^*\| - 2M\frac{\sigma}{\alpha} \leq 0 \quad \forall x^* \in N. \qquad (4.2.12)
$$

It yields the following estimate:

$$
\|x_\alpha^\gamma\| \leq \frac{\delta}{\alpha} + M\frac{\sigma}{\alpha} + \frac{h}{\alpha}g(\|x^*\|) + \sigma + 2(\|x^*\| + 1) \quad \forall x^* \in N, \qquad (4.2.13)
$$

which together with (4.2.7) proves that $\{x_\alpha^\gamma\}$ is bounded. Since $\|x_\alpha^\gamma - u_\alpha^\gamma\| \leq \sigma$, $\{u_\alpha^\gamma\}$ is also bounded. Therefore, $u_\alpha^\gamma \rightharpoonup \bar{x} \in H$. Moreover, $\bar{x} \in \Omega$ because $u_\alpha^\gamma \in \Omega$ and Ω is weakly closed. Thus, we have established that $x_\alpha^\gamma \rightharpoonup \bar{x} \in \Omega$ as $\alpha \to 0$.

By Lemma 1.11.7, variational inequality (4.2.4) is equivalent to the inclusion

$$
f^\delta \in B^\lambda x_\alpha^\gamma + \alpha x_\alpha^\gamma,
$$

where operator

$$
B^\lambda = A^h + \partial I_{\Omega_\sigma}, \quad \lambda = (h, \sigma),
$$

is maximal monotone with $D(B^\lambda) = \Omega_\sigma$. Recall that we denoted by $\partial I_{\Omega_\sigma}$ a subdifferential of the indicator function of the set Ω_σ. Hence, there exists an element $\xi_\alpha^\gamma \in B^\lambda x_\alpha^\gamma$ such that

$$
\xi_\alpha^\gamma + \alpha x_\alpha^\gamma = f^\delta. \qquad (4.2.14)
$$

Obviously, ξ_α^γ can be represented as

$$
\xi_\alpha^\gamma = \tilde{y}_\alpha^\gamma + z_\alpha^\gamma, \qquad (4.2.15)
$$

where $\tilde{y}_\alpha^\gamma \in A^h x_\alpha^\gamma$ and $z_\alpha^\gamma \in \partial I_{\Omega_\sigma} x_\alpha^\gamma$. It follows from (4.1.5), (4.2.14) and (4.2.15) that

$$
\lim_{\alpha \to 0}(\tilde{y}_\alpha^\gamma + z_\alpha^\gamma) = f \qquad (4.2.16)
$$

because $\{x_\alpha^\gamma\}$ is bounded. By inclusion $\Omega \subseteq \Omega_\sigma$ and by (4.2.6), for every $x \in \Omega$ and for every $y \in Ax$, we may define a sequence $\{y^h\}$ such that $y^h \in A^h x$ and $y^h \to y$ as $h \to 0$. Since B^λ is (maximal) monotone and $\Omega \subseteq \Omega_\sigma = D(B^\lambda)$, one has

$$
(\tilde{y}_\alpha^\gamma + z_\alpha^\gamma - y^h - z^\sigma, x_\alpha^\gamma - x) \geq 0 \quad \forall x \in \Omega, \quad z^\sigma \in \partial I_{\Omega_\sigma} x.
$$

It is known that $\theta_H \in \partial I_{\Omega_\sigma} x$ for all $x \in \Omega_\sigma$. Therefore, we may put in the latter inequality $z^\sigma = \theta_H$. Then, after passing to limit as $\alpha \to 0$, one gets

$$(f - y, \bar{x} - x) \geq 0 \quad \forall x \in \Omega, \quad \forall y \in Ax,$$

where $\bar{x} \in \Omega$. This means that $\bar{x} \in N$ in view of Lemma 1.11.4.

After some simple transformations, (4.2.11) is reduced to the inequality

$$\|x_\alpha^\gamma - x^*\|^2 \leq \|x_\alpha^\gamma - x^*\| \left(\frac{\delta}{\alpha} + \frac{h}{\alpha} g(\|x^*\|) \right) + M \left(\|x^*\| + \|x_\alpha^\gamma\| + 2 \right) \frac{\sigma}{\alpha}$$

$$+ \sigma \|x_\alpha^\gamma\| + (x^*, x^* - x_\alpha^\gamma) \quad \forall x^* \in N. \tag{4.2.17}$$

Put here $x^* = \bar{x}$. Then the strong convergence of x_α^γ to $\bar{x}$ follows because (4.2.7) is satisfied. Letting $\alpha \to 0$ in (4.2.17), one gets

$$(x^*, x^* - \bar{x}) \geq 0 \quad \forall x^* \in N,$$

which implies the equality $\bar{x} = \bar{x}^*$ (see, for instance, Theorem 2.1.2). Consequently, the whole sequence x_α^γ strongly converges to $\bar{x}^*$ as $\alpha \to 0$, and the proof is complete. ∎

Corollary 4.2.2 *If $N = \{x_0\}$ and there exists $C > 0$ such that*

$$\frac{\delta + h + \sigma}{\alpha} \leq C \quad as \quad \alpha \to 0,$$

then Theorem 4.2.1 guarantees the weak convergence of x_α^γ to x_0.

2. We study the interior approximations of Ω. This means that (4.2.6) holds for any $x \in \Omega_\sigma$. Furthermore, we assume that function $g(t)$ in (4.2.6) is bounded on bounded sets. The regularized solutions are found from the inequality

$$(A^h x + \alpha U^\mu x - f^\delta, z - x) \geq 0 \quad \forall z \in \Omega_\sigma, \quad x \in \Omega_\sigma, \tag{4.2.18}$$

where

$$U^\mu x = \frac{\mu(\|x\|) x}{\|x\|} \quad \text{if} \quad x \neq \theta_H,$$

and $J^\mu(\theta_H) = \theta_H$, $\mu(t)$ has the same properties as a gauge function of duality mappings J^μ. We additionally assume that there exists $t_0 > 0$ such that $\mu(t) > g(t)$ for $t \geq t_0$. Let $x_\alpha^\gamma \in \Omega_\sigma$ with $\gamma \in \Re$ be a solution of (4.2.18) and $y_\alpha^\gamma \in A^h x_\alpha^\gamma$ such that

$$(y_\alpha^\gamma + \alpha U^\mu x_\alpha^\gamma - f^\delta, z - x_\alpha^\gamma) \geq 0 \quad \forall z \in \Omega_\sigma.$$

Then (4.2.10) accepts the following form:

$$(y_\alpha^\gamma + \alpha U^\mu x_\alpha^\gamma - f^\delta, v_\alpha^\gamma - x_\alpha^\gamma) + (y - f, u_\alpha^\gamma - x^*) \geq 0 \quad \forall x^* \in N, \quad y \in Ax^*. \tag{4.2.19}$$

Since $\Omega_\sigma \subseteq \Omega$, the condition (4.2.6) asserts that there exists an element $\eta_\alpha^\gamma \in Ax_\alpha^\gamma$ such that

$$\|y_\alpha^\gamma - \eta_\alpha^\gamma\| \leq hg(\|x_\alpha^\gamma\|). \tag{4.2.20}$$

Taking into account the monotonicity of A and (4.2.19), one gets

$$
\begin{aligned}
\alpha(U^\mu x_\alpha^\gamma, x_\alpha^\gamma - v_\alpha^\gamma) &\leq \alpha(U^\mu x_\alpha^\gamma, x_\alpha^\gamma - v_\alpha^\gamma) + (y - \eta_\alpha^\gamma, x^* - x_\alpha^\gamma) \\
&\leq (\eta_\alpha^\gamma - y_\alpha^\gamma, x_\alpha^\gamma - x^*) + (y_\alpha^\gamma - f^\delta, v_\alpha^\gamma - x^*) \\
&\quad + (y - f, u_\alpha^\gamma - x_\alpha^\gamma) + (f^\delta - f, x_\alpha^\gamma - x^*) \quad \forall x^* \in N. \quad (4.2.21)
\end{aligned}
$$

Similarly to (4.2.12), we obtain with the help of (4.1.5), (4.2.2), (4.2.3), (4.2.8), (4.2.9) and (4.2.20) the following inequality:

$$
\mu(\|x_\alpha^\gamma\|)(\|x_\alpha^\gamma\| - \sigma - \|x^*\|) \;-\; \left(\frac{\delta}{\alpha} + \frac{h}{\alpha} g(\|x_\alpha^\gamma\|)\right)\|x_\alpha^\gamma - x^*\|
$$

$$
-\; \frac{\sigma}{\alpha} M(2 + \|x^*\| + \|x_\alpha^\gamma\|) \leq 0 \quad \forall x^* \in N.
$$

The properties of $\mu(t)$ and condition (4.2.7) enable us to assert that the sequence $\{x_\alpha^\gamma\}$ is bounded. Then $x_\alpha^\gamma \rightharpoonup \bar{x}$. Since Ω_σ are convex closed sets and $x_\alpha^\gamma \in \Omega_\sigma \subseteq \Omega$, the Mazur theorem guarantees the inclusion $\bar{x} \in \Omega$.

Show that $\bar{x} \in N$. Let $\xi_\alpha^\gamma \in B^\lambda x_\alpha^\gamma$, where $B^\lambda = A^h + \partial I_{\Omega_\sigma}$, $\lambda = (h, \sigma)$, such that

$$
\xi_\alpha^\gamma + \alpha U^\mu x_\alpha^\gamma = f^\delta. \quad (4.2.22)
$$

It is clear that (4.2.15) is true. By making use of (4.2.6), which is valid for all $x \in \Omega_\sigma$, we conclude that for every element $y_\alpha^\gamma \in A^h x_\alpha^\gamma$ there exists $w_\alpha^\gamma \in A x_\alpha^\gamma$ such that

$$
\|\bar{y}_\alpha^\gamma - w_\alpha^\gamma\| \leq g(\|x_\alpha^\gamma\|)h \quad (4.2.23)
$$

is satisfied. Since $\{x_\alpha^\gamma\}$ is bounded,

$$
\|y_\alpha^\gamma - w_\alpha^\gamma\| \to 0 \quad \text{as} \quad \alpha \to 0.
$$

Then (4.2.16) and (4.2.23) imply

$$
\lim_{\alpha \to 0}(w_\alpha^\gamma + z_\alpha^\gamma) = f, \quad (4.2.24)
$$

where $z_\alpha^\gamma \in \partial I_{\Omega_\sigma} x_\alpha^\gamma$.

Write down the monotonicity property of the operator $B = A + \partial I_\Omega$ with $D(B) = \Omega$:

$$
(w_\alpha^\gamma + \tilde{z}_\alpha^\gamma - \zeta, x_\alpha^\gamma - x) \geq 0 \quad \forall x \in \Omega, \quad \forall \zeta \in Bx, \quad \forall \tilde{z}_\alpha^\gamma \in \partial I_\Omega x_\alpha^\gamma.
$$

Since $\theta_H \in \partial I_\Omega x$ for all $x \in \Omega$, we may presume in the latter inequality $\tilde{z}_\alpha^\gamma = \theta_H$. According to (4.2.3), for each $x \in \Omega$, there exists $\{x_\sigma\} \subset \Omega_\sigma$ such that $x_\sigma \to x$ as $\sigma \to 0$. By simple algebra, we now come to the relation

$$
(w_\alpha^\gamma + z_\alpha^\gamma - \zeta, x_\alpha^\gamma - x) + (z_\alpha^\gamma, x_\sigma - x_\alpha^\gamma) - (z_\alpha^\gamma, x_\sigma - x) \geq 0. \quad (4.2.25)
$$

Taking into account that $\theta_H \in \partial I_{\Omega_\sigma} z$ for all $z \in \Omega_\sigma$ and that $\partial I_{\Omega_\sigma}$ is monotone, it is easy to see that the second term in (4.2.25) is non-positive. Omitting it, we only strengthen this inequality. Further, (4.2.2) allows us to establish the boundedness of $\{\bar{y}_\alpha^\gamma\}$ as $\alpha \to 0$. In its turn, the boundedness of $\{z_\alpha^\gamma\}$ results from (4.2.23) and (4.2.24). Finally, (4.2.25) yields the inequality

$$(f - \zeta, \bar{x} - x) \geq 0 \quad \forall x \in \Omega, \quad \forall \zeta \in Ax, \quad \bar{x} \in \Omega.$$

By Lemma 1.11.4, it implies the inclusion $\bar{x} \in N$. Using (4.2.21) and following the scheme of the previous theorem, we obtain

Theorem 4.2.3 *Assume that the conditions of Theorem 4.2.1 are satisfied with the inverse inclusion $\Omega_\sigma \subseteq \Omega$ for all $\sigma \in (0, \sigma^*]$, the sequence $\{x_\alpha^\gamma\}$ is generated by the variational inequality (4.2.18), the function $g(t)$ in (4.2.6) is bounded on bounded sets and there exists $t_0 > 0$ such that $\mu(t) > g(t)$ for $t \geq t_0$. Then $\{x_\alpha^\gamma\}$ strongly converges to the minimal norm solution $\bar{x}^*$ as $\alpha \to 0$.*

Remark 4.2.4 *Theorem 4.2.1 remains still valid if the inclusion $\Omega \subseteq \Omega_\sigma$ is replaced by $N \subset \Omega_\sigma$ for all $\sigma \in (0, \sigma^*]$. Indeed, in this case the estimate (4.2.13) is satisfied and verification of the fact that $\bar{x} \in N$ can be done as in the proof of the previous theorem.*

3. Let the condition (4.2.6) hold for all $x \in \Omega \cup \Omega_\sigma$ with all $\sigma \in (0, \sigma^*]$. In other words, we suppose that $D(A)$ contains Ω and $D(A^h)$ with $h \in (0, h^*]$ contain Ω_σ. In particular, this is carried out if $D(A) = D(A^h)$ for all $h \in (0, h^*]$. In this case, the requirements $\Omega \subseteq \Omega_\sigma$ and $\Omega_\sigma \subseteq \Omega$ in Theorems 4.2.1 and 4.2.3 should be omitted. Then the proof of the fact that $\bar{x} \in N$ is simplified. Indeed, let x be any fixed element of Ω. Put in (4.2.5) $z = x_\sigma \in \Omega_\sigma$, provided that $\|x - x_\sigma\| \leq \sigma$. We have for all $\gamma \in \Re$,

$$(y_\alpha^\gamma + \alpha x_\alpha^\gamma - f^\delta, x_\sigma - x_\alpha^\gamma) \geq 0,$$

where $y_\alpha^\gamma \in A^h x_\alpha^\gamma$. Rewrite this inequality in the following form:

$$\begin{aligned}
(y_\alpha^\gamma + \alpha x_\alpha^\gamma - f^\delta, x_\sigma - x) \quad &+ \quad (y - f, x - x_\alpha^\gamma) + (y_\alpha^\gamma - w_\alpha^\gamma, x - x_\alpha^\gamma) \\
&+ \quad (w_\alpha^\gamma - y, x - x_\alpha^\gamma) + (f - f^\delta, x - x_\alpha^\gamma) \\
&+ \quad \alpha(x_\alpha^\gamma, x - x_\alpha^\gamma) \geq 0 \quad \forall y \in Ax, \quad (4.2.26)
\end{aligned}$$

where $w_\alpha^\gamma \in Ax_\alpha^\gamma$ such that

$$\|y_\alpha^\gamma - w_\alpha^\gamma\| \leq g(\|x_\alpha^\gamma\|)h.$$

Assuming that $g(t)$ is bounded on bounded sets and passing in (4.2.26) to the limit as $\alpha \to 0$, one gets

$$(y - f, x - \bar{x}) \geq 0 \quad \forall x \in \Omega, \quad \forall y \in Ax, \quad \bar{x} \in \Omega,$$

which is equivalent to (4.1.3). We come to the following statement.

Theorem 4.2.5 *If in the condition (vi) of Theorem 4.2.1, inequality (4.2.6) holds for all $h \in (0, h^*]$ and for all $x \in \Omega \cup \Omega_\sigma$ and if function $g(t)$ is bounded on bounded sets, then the solution sequences $\{x_\alpha^\gamma\}$ generated by the variational inequalities (4.2.4) and (4.2.18) converge strongly in H to the minimal norm solution $\bar{x}^*$ of the variational inequality (7.1.1) as $\alpha \to 0$.*

4. Convergence of the operator regularization methods is established in Theorems 4.2.1, 4.2.3 and 4.2.5 under the condition (4.2.3). It is natural if Ω and Ω_σ are bounded subsets in a Hilbert space. However, there are variational problems of the type (7.1.1) in which (4.2.3) may not be fulfilled. For instance, if Ω is an unbounded set in R^2 and if boundary $\partial\Omega$ of Ω is given by linear functions $y = kx + b$ with $k, b \in R^1$, then any arbitrarily small error of k implies the fact that $\mathcal{H}_H(\Omega, \Omega_\sigma) = \infty$.

Assume that Ω is unbounded. Construct the sets

$$\Omega^R = \Omega \cap B(\theta_H, R)$$

and

$$\Omega_\sigma^R = \Omega_\sigma \cap B(\theta_H, R) \quad \forall\, \sigma \in (0, \sigma^*].$$

Choose R large enough that sets $G^R = N \cap \Omega^R \neq \emptyset$ and $\Omega_\sigma^R \neq \emptyset$. Furthermore, we assume that either

$$int\, \Omega_\sigma^R \neq \emptyset \quad \text{or} \quad int\, D(A^h) \cap \Omega_\sigma^R \neq \emptyset \quad \forall\, h \in (0, h^*], \ \ \forall\, \sigma \in (0, \sigma^*],$$

and

$$\mathcal{H}_H(\Omega^R, \Omega_\sigma^R) \leq \sigma.$$

Restricting the given problem (7.1.1) on the subset Ω^R with fixed R, we obtain in Theorems 4.2.1, 4.2.3 and 4.2.5 the strong convergence of x_α^γ to $\bar{x}_R^* \in N^R$, where x_α^γ are solutions of corresponding regularized variational inequalities on the sets Ω_σ^R, and N^R is a solutions set of the variational inequality

$$(Ax - f, z - x) \geq 0 \quad \forall z \in \Omega^R, \quad x \in \Omega^R, \tag{4.2.27}$$

at that

$$\|\bar{x}_R^*\| = min\, \{\|x\| \mid x \in N^R\}.$$

The question arises: can we find in this way the solution $\bar{x}^*$ of the variational inequality (7.1.1)? The answer is affirmative. Indeed, first of all, we note that $\bar{x}^* \in N^R$, $\|\bar{x}_R^*\| \leq \|\bar{x}^*\|$ and $G^R \subseteq N^R$. In our assumptions, solutions of (4.2.27) coincide with solutions of the inequality

$$(Az - f, z - x) \geq 0 \quad \forall z \in \Omega^R, \quad x \in \Omega^R.$$

Therefore, any element $x \in N^R \backslash G^R$ cannot be an interior point of Ω^R, because the inclusion

$$f \in Ax + \partial I_{\Omega^R} x$$

is equivalent to (4.2.27). If $x \in int\, \Omega^R$, then it is transformed into the inclusion $f \in Ax$. Thus, it follows from the inclusion $x \in N^R \backslash G^R$ that $x \in \partial \Omega^R$. Further, the sets N^R and G^R are convex and closed. Therefore, if $N^R \backslash G^R \neq \emptyset$ and $int\, G^R \neq \emptyset$ then there exists $\bar{x} \in N^R \backslash G^R$ such that $\bar{x} \in int\, \Omega^R$. Hence, $G^R = N^R$ in this case.

Sets N^R and G^R also coincide if $\partial \Omega^R$ (more precisely, $\partial \Omega^R \backslash S(\theta_H, R)$) has no convex subsets. Indeed, construct maximal monotone operators

$$T = A + \partial I_\Omega \quad \text{and} \quad T^R = A + \partial I_{\Omega^R}.$$

Then inequalities (7.1.1) and (4.2.27) are equivalent to the equations $Tx = f$ and $T^R x = f$, respectively. Show that values of T and T^R coincide on the set S. To this end, it is sufficient to establish that sets $\partial I_\Omega x$ and $\partial I_{\Omega^R} x$ coincide on S. Let $\eta \in \partial I_{\Omega^R} x$ and $x \in S$. Then, by the definition of a subdifferential,

$$(\eta, x - y) \geq 0 \quad \forall y \in \Omega^R.$$

By the hypotheses, for every $y_1 \in \Omega$, there exists a constant $\lambda > 0$ and an element $y \in \Omega^R$ such that $x - y_1 = \lambda(x - y)$. Hence, η also belongs to $\partial I_\Omega x$, that is, $\partial I_{\Omega^R} x \subseteq \partial I_\Omega x$ for all $x \in S$. Taking into account the obvious contrary inclusion $\partial I_\Omega x \subseteq \partial I_{\Omega^R} x$ for $x \in S$, we finally obtain the claim. Thus, the inequality (4.2.27) can be considered in place of (7.1.1). After that, we are able to state Theorems 4.2.1, 4.2.3 and 4.2.5 for corresponding operator regularization methods. Observe that Theorem 1.11.10 may be also useful for an evaluation of R.

Let now a constant R, for which $N^R = G^R$, be unknown. In this case, define the proximity between Ω and Ω_σ in a different way than in (4.2.3). Namely, let

$$s(R, \Omega, \Omega_\sigma) = \sup \{ \inf \{ \| u - v \| \mid u \in \Omega_\sigma \} \mid v \in \Omega^R \} \quad \forall R > 0. \tag{4.2.28}$$

If $\Omega^R = \emptyset$ then we presume that $s(R, \Omega, \Omega_\sigma) = 0$. At the same time, if $\Omega^R \neq \emptyset$ or $\Omega_\sigma^R \neq \emptyset$, define

$$\tau(R, \Omega, \Omega_\sigma) = \max \{ s(R, \Omega, \Omega_\sigma), s(R, \Omega_\sigma, \Omega) \}.$$

Suppose that

$$\tau(R, \Omega, \Omega_\sigma) \leq a(R)\sigma, \tag{4.2.29}$$

where $a(R)$ is a non-negative continuous and increasing function for $R \geq 0$, $a(0) = 0$ and $a(R) \to \infty$ as $R \to \infty$. In order to introduce an operator U^μ in the regularized inequality (4.2.18), choose $\mu(t)$ such that

$$\mu(t) > \max\{a(t), g(t)\}, \quad t \geq t_0 > 0, \tag{4.2.30}$$

where $g(t)$ satisfies (4.2.6).

Theorem 4.2.6 *If the requirement (4.2.3) in Theorems 4.2.1, 4.2.3 and 4.2.5 is replaced by (4.2.29), then the solution sequence $\{x_\alpha^\gamma\}$ generated by (4.2.18) with $\mu(t)$ satisfying (4.2.30) converges strongly to the minimal norm solution $\bar{x}^*$ of the variational inequality (7.1.1) as $\alpha \to 0$.*

Proof. We shall not repeat the proofs of the mentioned theorems because their arguments remain as before. For instance, the inequality of the type (4.2.12) has the following form:

$$\mu(\|x_\alpha^\gamma\|)\Big(\|x_\alpha^\gamma\| \quad - \quad a(\|x^*\|)\sigma - \|x^*\|\Big) - \frac{\sigma}{\alpha} Ma(\|x_\alpha^\gamma\|)(\|x^*\| + 1)$$

$$- \quad \|x_\alpha^\gamma\|\Big(\frac{\sigma}{\alpha} Ma(\|x^*\|) + \frac{\delta}{\alpha} + \frac{h}{\alpha} g(\|x^*\|)\Big) - \frac{\delta}{\alpha}\|x^*\|$$

$$- \quad \frac{h}{\alpha} g(\|x^*\|)\|x^*\| - Ma(\|x^*\|)\frac{\sigma}{\alpha} \leq 0 \quad \forall x^* \in N. \tag{4.2.31}$$

We deduce from this that $\{x_\alpha^\gamma\}$ is bounded as $\alpha \to 0$. Then the rest of the proof follows the pattern of Theorem 4.2.1. ∎

5. Next we present the residual principle for regularized variational inequality on approximately given sets Ω_σ. Observe that the following Lemmas 4.2.7 - 4.2.9 are similar to Lemmas 4.1.4 - 4.1.6 of Section 4.1 and can be proved in the same manner.

Lemma 4.2.7 *A function $\sigma_\mu(\alpha) = \mu(\|x_\alpha^\gamma\|)$ is single-valued, continuous and non-increasing for $\alpha \geq \alpha_0 > 0$.*

Lemma 4.2.8 $\lim\limits_{\alpha \to \infty} x_\alpha^\gamma = x_*^\sigma$, *where* $x_*^\sigma \in \Omega_\sigma$ *and* $\|x_*^\sigma\| = min\{\|x\| \mid x \in \Omega_\sigma\}$.

Lemma 4.2.9 *Let* $B^\lambda = A^h + \partial I_{\Omega_\sigma}$ *with* $\lambda = (h, \sigma)$. *One has:*
(i) If $\theta_H \notin \Omega_\sigma$ *then*
$$\lim_{\alpha \to \infty} \rho_\mu(\alpha) = \lim_{\alpha \to \infty} \alpha\mu(\|x_\alpha^\gamma\|) = \infty,$$

(ii) if $\theta_H \in \Omega_\sigma$ *then* $\xi_\alpha^\gamma \rightharpoonup \xi$ *as* $\alpha \to \infty$, *where* $\xi_\alpha^\gamma \in B^\lambda x_\alpha^\gamma$ *in view of (4.1.22), and* $\xi \in B^\lambda(\theta_H)$. *In addition,*
$$\lim_{\alpha \to \infty} \rho_\mu(\alpha) = \|\xi - f^\delta\|.$$

Moreover,
(iii) if $\theta_H \in int\ \Omega_\sigma$ *then* $\xi \in A^h(\theta_H)$ *is defined as*
$$\|\xi - f^\delta\|_* = min\ \{\|\zeta - f^\delta\|_* \mid \zeta \in A^h(\theta_H)\}, \tag{4.2.32}$$

(iv) if $\theta_H \in \partial\Omega_\sigma$ *then* $\xi \in B^\lambda(\theta_H)$ *satisfies the equality*
$$\|\xi - f^\delta\|_* = min\ \{\|\zeta - f^\delta\|_* \mid \zeta \in B^\lambda(\theta_H)\}. \tag{4.2.33}$$

These lemmas enable us to find the residual principle for choosing the regularization parameter α in a Hilbert space.

Theorem 4.2.10 *Let* $A : H \to 2^H$ *be a maximal monotone operator with domain* $D(A)$, $\Omega \subseteq D(A)$ *be a convex closed set satisfying (4.1.1) and let the variational inequality (7.1.1) with* $f \in H$ *have a nonempty solution set* N *with the unique minimal norm solution* $\bar{x}^*$. *Assume that a sequence of operators* $\{A^h\}$, *where* $A^h : H \to 2^H$ *has domain* $D(A^h)$, *sequence of convex closed sets* $\{\Omega_\sigma\} \subseteq D(A^h)$ *and the sequence* $\{f^\delta\} \in H$ *are known in place of* A, Ω *and* f, *respectively, such that in reality the variational inequality*
$$\langle A^h x - f^\delta, z - x \rangle \geq 0 \quad \forall z \in \Omega_\sigma$$

is solved. Let $\gamma = (h, \delta, \sigma) \in \Re$ *such that* $0 < \delta + h + \sigma \leq 1$. *Using the perturbed data, construct the regularized inequality (4.2.4) with the condition (4.2.1). Denote its (unique) solution by* x_α^γ. *Suppose that (4.1.5), (4.2.2) and (4.2.3) are satisfied and (4.2.6) holds on the set* $\Omega \cup \Omega_\sigma$, *where the function* $g(t)$ *is continuous and increasing. If* $\theta_H \in \Omega_\sigma$, *then the following additional condition is given:*
$$\|\xi - f^\delta\| > \left(k + M + g(0)\right)(\delta + h + \sigma)^p, \quad k > 1, \quad p \in (0, 1], \tag{4.2.34}$$

provided that
1) *if* $\theta_H \in int\ \Omega_\sigma$ *then* $\xi \in A^h(\theta_H)$ *and satisfies (4.2.32)*;
2) *if* $\theta_H \in \partial\Omega_\sigma$ *then* $\xi \in B^\lambda(\theta_H)$ *and satisfies (4.2.33)*;
3) *if* $\theta_H \notin \Omega_\sigma$, *but* $\theta_H \in \Omega$, *then it is assumed that* $\theta_H \notin N$.
    Then there exists a unique $\bar{\alpha}$ *satisfying the equality*

$$\rho(\bar{\alpha}) = \bar{\alpha}\|x_{\bar{\alpha}}^\gamma\| = \Big(k + M + g(\|x_{\bar{\alpha}}^\gamma\|)\Big)(\delta + h + \sigma)^p, \qquad (4.2.35)$$

where $x_{\bar{\alpha}}^\gamma$ *is a solution of (4.2.4) with* $\alpha = \bar{\alpha}$. *Moreover,*
(i) if $\gamma \to 0$ *then* $\bar{\alpha} \to 0$,
(ii) if $\gamma \to 0$ *and* $p \in (0,1)$ *then* $x_{\bar{\alpha}}^\gamma \to \bar{x}^*$ *and*

$$\lim_{\gamma \to 0} \frac{\delta + h + \sigma}{\bar{\alpha}} = 0,$$

(iii) if $\gamma \to 0$, $p = 1$ *and* $N = \{x_0\}$ *then* $x_{\bar{\alpha}}^\gamma \to x_0$ *and there exists a constant* $C > 0$ *such that*

$$\limsup_{\gamma \to 0} \frac{\delta + h + \sigma}{\bar{\alpha}} \leq C. \qquad (4.2.36)$$

Proof. The monotonicity of A in the inequality (4.2.10) and the condition (4.2.2) induce the following estimate:

$$\|x_\alpha^\gamma\| \leq \frac{\delta}{\alpha} + M\frac{\sigma}{\alpha} + \frac{h}{\alpha}g(\|x_\alpha^\gamma\|) + \sigma + 2(\|x^*\| + 1) \quad \forall x^* \in N. \qquad (4.2.37)$$

Take a small enough $\alpha > 0$ such that for any fixed $x^* \in N$,

$$\alpha\Big(\sigma + 2(\|x^*\| + 1)\Big) < (k-1)(\delta + h + \sigma)^p, \quad k > 1, \quad p \in (0,1]. \qquad (4.2.38)$$

Then one gets

$$\begin{aligned}
\rho(\alpha) \ &\leq \ \delta + M\sigma + hg(\|x_\alpha^\gamma\|) + \alpha\Big(\sigma + 2(\|x^*\| + 1)\Big) \\
&< \ \delta + M\sigma + hg(\|x_\alpha^\gamma\|) + (k-1)(\delta + h + \sigma)^p \\
&\leq \ \Big(k + M + g(\|x_\alpha^\gamma\|)\Big)(\delta + h + \sigma)^p. \qquad (4.2.39)
\end{aligned}$$

Now (4.2.34) guarantees the solvability of equation (4.2.35) if $\theta_H \in \Omega_\sigma$. At the same time, if $\theta_H \notin \Omega_\sigma$ then the existence of $\bar{\alpha}$ follows from (4.2.39) and Lemma 4.2.9. By (4.2.38), we find that

$$\bar{\alpha} > \frac{k-1}{2(\|x^*\| + 1) + \sigma}(\delta + h + \sigma)^p.$$

Consequently,

$$\frac{\delta + h + \sigma}{\bar{\alpha}} < \frac{2\|x^*\| + 3}{k-1}(\delta + h + \sigma)^{1-p}.$$

Since, $\gamma \in \Re$, (4.2.36) holds in the cases of $p \in (0,1)$ and $p = 1$ at the same time. The boundedness of $\{x_{\bar{\alpha}}^\gamma\}$ results from (4.2.13). In the standard way, we establish that $x_{\bar{\alpha}}^\gamma \rightharpoonup \bar{x} \in \Omega$ as $\gamma \to 0$ and $\xi_{\bar{\alpha}}^\gamma \to f$, where $\xi_{\bar{\alpha}}^\gamma \in B^\lambda(x_{\bar{\alpha}}^\gamma)$ satisfies the equality

$$\xi_{\bar{\alpha}}^\gamma + \bar{\alpha}x_{\bar{\alpha}}^\gamma = f^\delta.$$

Other properties of $\{x_{\bar{\alpha}}^{\gamma}\}$ are proved similarly to Theorem 4.2.5.

Observe that $\theta_H \notin N$. Indeed, the inclusion $\theta_H \in N$ is possible if and only if $\theta_H \in \Omega_\sigma$ for all $\sigma \in (0, \sigma^*]$. Then it is not difficult to deduce from (4.2.3) that $\theta_H \in \Omega$. Finally, one can make certain, as in the proof of Theorem 4.1.11, that (4.2.34) implies the claim. Then the convergence of $\bar{\alpha}$ to zero follows if $\gamma \to 0$. ∎

6. Let further X be an E-space, X^* be strictly convex, $A : X \to 2^{X^*}$ and $A^h : X \to 2^{X^*}$ be maximal monotone operators. Note that condition (4.2.2) on the growth order of operators A^h is quite natural just in a Hilbert space. If we consider variational inequalities in Banach spaces, then we have to replace the linear function in the right-hand side of (4.2.2) by a nonlinear one. For a short view, we presume in Theorem 4.2.6 that

$$\|y^h - f^\delta\|_* \leq \kappa(\|x\|) \quad \forall y^h \in A^h x, \quad \forall x \in \Omega_\sigma, \tag{4.2.40}$$

where $\gamma = (\sigma, h, \delta) \in \Re$, $\kappa(t)$ is a non-negative, continuous and increasing function for $t \geq 0$. Define the function $\mu(t)$ such that

$$\mu(t) > max\{a(t), g(t), \kappa(t)\}, \quad t \geq t_0 > 0,$$

and corresponding duality mapping $J^\mu : X \to X^*$. To define approximations to the minimal norm solution of the variational inequality (7.1.1), we apply the regularization method

$$\langle A^h x + \alpha J^\mu x - f^\delta, z - x \rangle \geq 0 \quad \forall z \in \Omega_\sigma, \quad x \in \Omega_\sigma. \tag{4.2.41}$$

As usual, denote its solutions by x_α^γ.

Theorem 4.2.11 *Under the hypothesis of the present subsection, the results of Theorems 4.1.3, 4.2.1, 4.2.3, 4.2.5, 4.2.6 (and also Corollary 4.2.2) remain still valid for a solution sequence $\{x_\alpha^\gamma\}$ of the variational inequality (4.2.41).*

Proof. The inequality (4.2.31) for $\|x_\alpha^\gamma\|$ takes the following form:

$$\mu(\|x_\alpha^\gamma\|)\Big(\|x_\alpha^\gamma\| \quad - \quad a(\|x^*\|)\sigma - \|x^*\|\Big) - a(\|x_\alpha^\gamma\|)\kappa(\|x^*\|)\frac{\sigma}{\alpha}$$

$$- \quad \kappa(\|x_\alpha^\gamma\|)a(\|x^*\|)\frac{\sigma}{\alpha} - \|x_\alpha^\gamma\|\Big(\frac{\delta}{\alpha} + \frac{h}{\alpha}g(\|x^*\|)\Big)$$

$$- \quad \frac{\delta}{\alpha}\|x^*\| - \frac{h}{\alpha}g(\|x^*\|)\|x^*\| \leq 0 \quad \forall x^* \in N. \tag{4.2.42}$$

Owing to the properties of $\mu(t)$, we can be sure that the sequence $\{x_\alpha^\gamma\}$ is bounded as $\alpha \to 0$. The rest of the proof follows the same scheme as in Theorems 4.1.3, 4.2.1, 4.2.3 and 4.2.5. ∎

In order to write down the residual principle in a Banach space X, we need to obtain any upper estimate of $\|x_\alpha^\gamma\|$ in the explicit form. Emphasize that it may be done, for example, if $g(t)$ and $\kappa(t)$ are power functions of the following kind:

$$g(t) = c_1(t^{s-1} + 1), \quad \kappa(t) = c_2(t^{s-1} + 1), \quad s > 2,$$

with some positive constant c_1 and c_2, and if

$$\mu(t) = c_3 t^{s-1}, \quad c_3 > max\{c_1, c_2\}.$$

In this special case, Theorem 3.3.4 can be stated in X.

Remark 4.2.12 *The conditions (4.2.1), (4.2.2) and (4.2.40) do not need to hold for every pair (A^h, Ω_σ) of sequences $\{A^h\}$ and $\{\Omega_\sigma\}$. However, they should be fulfilled for such pairs which guarantee realization of the sufficiency criterion (4.2.7) for the strong convergence of regularized solutions.*

4.3 Variational Inequalities with Domain Perturbations

Let X be an E-space, X^* be a strictly convex space, $A : X \to 2^{X^*}$ be a maximal monotone operator, $\Omega \subset D(A)$ be a closed and convex set, the inequality (7.1.1) with $f \in X^*$ have a nonempty solution set N and the condition (4.1.1) hold. Suppose that for non-negative t, h and σ, there exist non-negative functions $a_1(t)$, $a_2(t)$, $g_1(t)$ and $\beta(h, \sigma)$ with the additional hypotheses that $a_2(t)$ is non-decreasing and

$$\lim_{t \to \infty} \frac{a_2(t)}{t\mu(t)} = 0, \tag{4.3.1}$$

where $\mu(t)$ is a gauge function of the duality mapping J^μ. For some positive δ^*, h^* and σ^*, denote $\Re = (0, \delta^*] \times (0, h^*] \times (0, \sigma^*]$. We assume that for each $\gamma = (\delta, h, \sigma) \in \Re$ one can define maximal monotone operators $A^h : X \to 2^{X^*}$, elements $f^\delta \in X^*$ and closed convex sets $\Omega_\sigma \subseteq D(A^h)$ which are $(f^\delta, A^h, \Omega_\sigma)$–approximations of (f, A, Ω), satisfying (4.1.5), (4.2.1) and the following conditions:
(i) for each $x \in \Omega$ there exists $z_\sigma \in \Omega_\sigma$ such that

$$\|x - z_\sigma\| \leq a_1(\|x\|)\sigma, \tag{4.3.2}$$

and then for each $\zeta \in Ax$,

$$d_*(\zeta, A^h z_\sigma) = inf\ \{\|\zeta - u\|_* \mid u \in A^h z_\sigma\} \leq g_1(\|\zeta\|_*)\beta(h, \sigma); \tag{4.3.3}$$

(ii) for each $w \in \Omega_\sigma$ there exists $v \in \Omega$ such that

$$\|w - v\| \leq a_2(\|w\|)\sigma. \tag{4.3.4}$$

Observe that hypotheses (i) and (ii) do not require that $D(A)$ and $D(A^h)$ coincide. Moreover, $\Omega_\sigma \nsubseteq D(A)$ in general.

We study again the regularized inequality (4.2.41). Let x_α^γ be its (unique) solution.

Theorem 4.3.1 *Under the assumptions above, let $\theta_X \notin N$ and*

$$\lim_{\alpha \to 0} \frac{\delta + \sigma + \beta(h, \sigma)}{\alpha} = 0. \tag{4.3.5}$$

Then solution sequence $\{x_\alpha^\gamma\}$ of the variational inequality (4.2.41) strongly converges, as $\alpha \to 0$, to the minimal norm solution $\bar{x}^*$ of (7.1.1).

Proof. Take $x^* \in N \subset \Omega$ and $\zeta^* \in Ax^*$ such that

$$\langle \zeta^* - f, z - x^* \rangle \geq 0 \quad \forall z \in \Omega. \tag{4.3.6}$$

By conditions (4.3.2) and (4.3.3) for $x = x^*$, there are elements $z_\sigma \in \Omega_\sigma$ and $\zeta^\lambda \in A^h z_\sigma$ with $\lambda = (h, \sigma)$ satisfying the inequalities

$$\|x^* - z_\sigma\| \leq a_1(\|x^*\|)\sigma \tag{4.3.7}$$

and

$$\|\zeta^* - \zeta^\lambda\|_* \leq g_1(\|\zeta^*\|_*)\beta(h, \sigma). \tag{4.3.8}$$

Let $\zeta_\alpha^\gamma \in A^h x_\alpha^\gamma$ be such that

$$\langle \zeta_\alpha^\gamma + \alpha J^\mu x_\alpha^\gamma - f^\delta, z - x_\alpha^\gamma \rangle \geq 0 \quad \forall z \in \Omega_\sigma. \tag{4.3.9}$$

By (4.3.4), there exists $u_\alpha^\gamma \in \Omega$ such that

$$\|x_\alpha^\gamma - u_\alpha^\gamma\| \leq a_2(\|x_\alpha^\gamma\|)\sigma. \tag{4.3.10}$$

Put $z = u_\alpha^\gamma$ in (4.3.6) and $z = z_\sigma$ in (4.3.9). Adding the obtained inequalities we have

$$\langle \zeta_\alpha^\gamma - f^\delta, x_\alpha^\gamma - z_\sigma \rangle + \alpha \langle J^\mu x_\alpha^\gamma, x_\alpha^\gamma - z_\sigma \rangle + \langle \zeta^* - f, x^* - u_\alpha^\gamma \rangle \leq 0$$

or

$$\begin{aligned}
\alpha \langle J^\mu x_\alpha^\gamma, x_\alpha^\gamma - z_\sigma \rangle \quad &+ \quad \langle \zeta_\alpha^\gamma - \zeta^\lambda, x_\alpha^\gamma - z_\sigma \rangle \\
&+ \quad \langle \zeta^\lambda - f^\delta, x_\alpha^\gamma - z_\sigma \rangle + \langle \zeta^* - f, x^* - u_\alpha^\gamma \rangle \leq 0.
\end{aligned}$$

Since A^h are monotone operators, the second term of the last inequality is non-negative. Hence,

$$\begin{aligned}
\alpha \langle J^\mu x_\alpha^\gamma, x_\alpha^\gamma \rangle \quad &\leq \quad \alpha \langle J^\mu x_\alpha^\gamma, z_\sigma \rangle + \langle \zeta^* - \zeta^\lambda, x_\alpha^\gamma - x^* \rangle \\
&+ \quad \langle \zeta^* - f, u_\alpha^\gamma - x_\alpha^\gamma \rangle + \langle f^\delta - f, x_\alpha^\gamma - x^* \rangle \\
&+ \quad \langle \zeta^\lambda - f^\delta, z_\sigma - x^* \rangle. \tag{4.3.11}
\end{aligned}$$

Further, from (4.1.5), (4.3.7), (4.3.8) and (4.3.10), one gets, respectively:

$$\|f^\delta\|_* \leq \|f\|_* + \delta,$$

$$\|z_\sigma\| \le \|x^*\| + a_1(\|x^*\|)\sigma,$$
$$\|\zeta^\lambda\|_* \le \|\zeta^*\|_* + g_1(\|\zeta^*\|_*)\beta(h, \sigma),$$
$$\|u_\alpha^\gamma\| \le \|x_\alpha^\gamma\| + a_2(\|x_\alpha^\gamma\|)\sigma.$$

Now we can evaluate all the terms in the right-hand side of (4.3.11) in the following way:

$$\alpha\langle J^\mu x_\alpha^\gamma, z_\sigma\rangle \le \alpha\mu(\|x_\alpha^\gamma\|)\Big(\|x^*\| + a_1(\|x^*\|)\sigma\Big),$$

$$\langle \zeta^* - \zeta^\lambda, x_\alpha^\gamma - x^*\rangle \le (\|x_\alpha^\gamma\| + \|x^*\|)g_1(\|\zeta^*\|_*)\beta(h, \sigma),$$

$$\langle \zeta^* - f, u_\alpha^\gamma - x_\alpha^\gamma\rangle \le \|\zeta^* - f\|_* a_2(\|x_\alpha^\gamma\|)\sigma,$$

$$\langle f^\delta - f, x_\alpha^\gamma - x^*\rangle \le (\|x_\alpha^\gamma\| + \|x^*\|)\delta,$$

$$\langle \zeta^\lambda - f^\delta, z_\sigma - x^*\rangle \le \Big(\|\zeta^*\|_* + \|f\|_* + \delta + g_1(\|\zeta^*\|_*)\beta(h, \sigma)\Big)\sigma.$$

With due regard for these facts, (4.3.11) leads to the inequality

$$\mu(\|x_\alpha^\gamma\|)\Big(\|x_\alpha^\gamma\| - \|x^*\| - a_1(\|x^*\|)\sigma\Big)$$

$$- \Big(\frac{\delta}{\alpha} + \frac{\beta(h, \sigma)}{\alpha}g_1(\|\zeta^*\|_*)\Big)(\|x_\alpha^\gamma\| + \|x^*\|)$$

$$\le \frac{\sigma}{\alpha}\Big(\|\zeta^* - f\|_* a_2(\|x_\alpha^\gamma\|) + \|\zeta^*\|_* + \|f\|_* + \delta + g_1(\|\zeta^*\|_*)\beta(h, \sigma)\Big). \quad (4.3.12)$$

Then (4.3.12) and (4.3.1) guarantee that $\{x_\alpha^\gamma\}$ is bounded and $x_\alpha^\gamma \rightharpoonup \bar{x} \in X$ as $\alpha \to 0$. Since the function $a_2(t)$ is non-decreasing, the estimate (4.3.10) implies the weak convergence of $\{u_\alpha^\gamma\}$ to $\bar{x} \in X$. We recall that $u_\alpha^\gamma \in \Omega$ for all $\alpha > 0$ and $\gamma \in \Re$. Therefore, by the Mazur theorem, $\bar{x} \in \Omega$.

We show that $\bar{x} \in N$. Due to the condition (4.2.1) and Lemma 1.11.4, solutions x_α^γ of (4.2.41) satisfy the following inequality:

$$\langle \zeta^h + \alpha J^\mu z - f^\delta, z - x_\alpha^\gamma\rangle \ge 0 \quad \forall z \in \Omega_\sigma, \quad \forall \zeta^h \in A^h z. \quad (4.3.13)$$

Choose and fix $u \in \Omega$ and $\eta \in Au$. The conditions (4.3.2) and (4.3.3) enable us to construct elements $v_\sigma \in \Omega_\sigma$ and $\eta^\lambda \in A^h v_\sigma$ such that

$$\|u - v_\sigma\| \le a_1(\|u\|)\sigma \quad (4.3.14)$$

and

$$\|\eta - \eta^\lambda\|_* \le g_1(\|\eta\|_*)\beta(h, \sigma). \quad (4.3.15)$$

Put $z = v_\sigma$ and $\zeta^h = \eta^\lambda$ in (4.3.13). Then we come to the relation

$$\langle \eta^\lambda + \alpha J^\mu v_\sigma - f^\delta, v_\sigma - x_\alpha^\gamma\rangle \ge 0.$$

Passing to the limit as $\alpha \to 0$ and taking into account (4.1.5), (4.3.5), (4.3.14), (4.3.15) and the weak convergence of $\{x_\alpha^\gamma\}$ to $\bar{x} \in \Omega$, one gets

$$\langle \eta - f, u - \bar{x} \rangle \geq 0 \quad \forall u \in \Omega, \quad \forall \eta \in Au.$$

By (4.1.1) and by Lemma 1.11.4, we conclude that $\bar{x} \in N$.

It is not difficult to see from (4.3.12) that there exists a constant $M > 0$ such that

$$\mu(\|x_\alpha^\gamma\|)\|x_\alpha^\gamma\| \leq \mu(\|x_\alpha^\gamma\|)\|x^*\| + M\frac{\delta + \sigma + \beta(h,\sigma)}{\alpha}. \qquad (4.3.16)$$

By the hypothesis, $\bar{x} \neq \theta_X$. Therefore, the weak convergence of $\{x_\alpha^\gamma\}$ to $\bar{x}$ as $\alpha \to 0$ yields

$$0 < \|\bar{x}\| \leq \liminf_{\alpha \to 0} \|x_\alpha^\gamma\|. \qquad (4.3.17)$$

Hence, if α is a sufficiently small positive parameter, then there exists $\tau > 0$ such that $\tau \leq \|x_\alpha^\gamma\|$, and

$$\mu(\|x_\alpha^\gamma\|) \geq \mu(\tau) > 0,$$

provided that a gauge function $\mu(t)$ of J^μ is increasing. Consequently, the estimate

$$\|x_\alpha^\gamma\| \leq \|x^*\| + \frac{M}{\mu(\tau)}\frac{\delta + \sigma + \beta(h,\sigma)}{\alpha} \quad \forall x^* \in N$$

follows from (4.3.16). Owing to (4.3.5), we now deduce that

$$\limsup_{\alpha \to 0} \|x_\alpha^\gamma\| \leq \|x^*\| \quad \forall x^* \in N. \qquad (4.3.18)$$

Combining (4.3.17) and (4.3.18) gives

$$\|\bar{x}\| \leq \liminf_{\alpha \to 0} \|x_\alpha^\gamma\| \leq \limsup_{\alpha \to 0} \|x_\alpha^\gamma\| \leq \|x^*\| \quad \forall x^* \in N. \qquad (4.3.19)$$

This shows that $\bar{x} = \bar{x}^*$. If we put $x^* = \bar{x}^*$ in (4.3.19), then $\|x_\alpha^\gamma\| \to \|\bar{x}^*\|$. The theorem is proved in an E-space X. ∎

Remark 4.3.2 *Let A be a bounded operator on Ω and*

$$\|\zeta\|_* \leq \varphi(\|x\|) \quad \forall x \in \Omega, \quad \forall \zeta \in Ax,$$

where $\varphi(t)$ is a non-negative and non-decreasing function. If $g_1(t)$ is also non-decreasing, then the inequality (4.3.3) can be rewritten as

$$d_*(\zeta, A^h z_\sigma) \leq g_2(\|x\|)\beta(h,\sigma) \quad \forall \zeta \in Ax, \quad \forall x \in \Omega, \qquad (4.3.20)$$

with $g_2(t) = g_1(\varphi(t))$. Theorem 4.3.1 remains still valid if (4.3.3) is replaced by (4.3.20).

4.4 Examples of Variational Inequalities

We present examples of variational inequalities on exactly and approximately given sets.

Example 4.4.1 Consider the contact problem of deformable bodies with an ideally smooth boundary. Let there be a beam of length l with firmly fixed edges. Direct the axis OX on the axis of the beam and OY up through the beam left end. Let $\mathcal{E}(x)$ be an elastic modulus, $\mathcal{I}(x)$ the moment of inertia of a section and $k(x) = \mathcal{E}(x)\mathcal{I}(x)$, $x \in [0, l]$. The beam bends under the influence of the force generated by a fixed rigid body pressing down it, with absolutely smooth contact surface defined by the equation $y = \chi(x)$; points lying beyond this surface satisfy the inequality $y \leq \chi(x)$. The following differential equation defines the equilibrium state of the beam:

$$\frac{d^2}{dx^2}\left[k(x)\frac{d^2w(x)}{dx^2}\right] = -q(x), \tag{4.4.1}$$

where $w(x)$ is a beam deformation and $q(x)$ is a rigid body reaction, with the boundary conditions

$$w(0) = w(l) = w'(0) = w'(l) = 0. \tag{4.4.2}$$

Besides, we have:
a) the one-sided restriction

$$w(x) \leq \chi(x) \tag{4.4.3}$$

characterizing the non-penetration of beam points into the rigid body;
b) the condition

$$q(x) \geq 0 \tag{4.4.4}$$

determining the direction of the rigid body reaction;
c) the equation

$$\Big(w(x) - \chi(x)\Big)q(x) = 0, \quad x \in (0, l), \tag{4.4.5}$$

implying that either (4.4.3) or (4.4.4) turns into a strict equality at any point $x \in (0, l)$.
Introduce the space

$$X = \{v \mid v = v(x),\ x \in [0, l],\ v \in W_2^2[0, l],\ v(0) = v(l) = v'(0) = v'(l) = 0\}$$

and define

$$\|v\| = \left(\int_0^l |v''(x)|^2 dx\right)^{1/2}.$$

Next, we assume that any solution $w(x)$ of the given problem has derivatives up to the fourth order. By (4.4.1),

$$\int_0^l \frac{d^2}{dx^2}\left[k(x)\frac{d^2w(x)}{dx^2}\right]v(x)dx = -\int_0^l q(x)v(x)dx \quad \forall v \in X,\quad w \in X. \tag{4.4.6}$$

That the obtained equality is equivalent to (4.4.1) results from the basic lemma of the calculus of variations ([84], Lemma 1, p.9). Apply twice the formula of integration by parts to the left-hand side of (4.4.6). Taking into account the boundary conditions (4.4.2) we deduce

$$\int_0^l k(x)w''(x)v''(x)dx = -\int_0^l q(x)v(x)dx \quad \forall v \in X, \quad w \in X. \tag{4.4.7}$$

The function $w(x)$ satisfying (4.4.7) is called a weak solution of the problem (4.4.1) - (4.4.5). Suppose that in (4.4.7) $v(x) = u(x) - w(x)$, where $u \in X$, and analyze the sign of the function

$$\xi(x) = q(x)\Big(u(x) - w(x)\Big).$$

If there is no contact between the beam and rigid body at a point $x \in (0, l)$, then $q(x) = 0$ and $\xi(x) = 0$. If there is contact at that point then we assert that $u(x) \le w(x)$ and, hence, $\xi(x) \le 0$. Consequently, solution $w(x)$ of the problem (4.4.1) - (4.4.5) satisfies the variational inequality

$$\int_0^l k(x)w''(x)\Big(u''(x) - w''(x)\Big)dx \ge 0 \quad \forall u \in \Omega, \quad w \in \Omega, \tag{4.4.8}$$

where

$$\Omega = \{v \mid v = v(x) \in X, \ v(x) \le \chi(x) \ \text{for almost all} \ x \in [0, l]\}$$

is a convex closed subset in X.

Let $w(x)$ now be a solution of (4.4.8). If there is no contact between the beam and rigid body at a point $x \in (0, l)$, then difference $u(x) - w(x)$ may take both positive and negative values. Integrating the left-hand side of (4.4.8) by parts twice and taking into account again the boundary conditions (4.4.2), we come to the inequality

$$\int_0^l \frac{d^2}{dx^2}\left[k(x)\frac{d^2w(x)}{dx^2}\right]\Big(u(x) - w(x)\Big)dx \ge 0 \quad \forall u \in \Omega, \quad w \in \Omega. \tag{4.4.9}$$

Then by repeating the reasoning of the proof of the basic lemma of the calculus of variations, we establish that

$$\frac{d^2}{dx^2}\left[k(x)\frac{d^2w(x)}{dx^2}\right] = 0. \tag{4.4.10}$$

Let $x \in (0, l)$ be a contact point of the beam and rigid body. Then $u(x) - w(x) \le 0$. Using (4.4.9) and the proof by contradiction, we deduce the inequality

$$\frac{d^2}{dx^2}\left[k(x)\frac{d^2w(x)}{dx^2}\right] = -q(x) \le 0. \tag{4.4.11}$$

By virtue of (4.4.10) and (4.4.11), $w(x)$ satisfies the equation (4.4.1) with condition (4.4.4). Thus, the equivalence of the problem (4.4.1) - (4.4.5) and variational inequality (4.4.8) has been established.

We introduce the operator $A : X \to X^*$ by the equality

$$\langle Aw, v \rangle = \int_0^l k(x)w''(x)v''(x)dx \quad \forall w, v \in X.$$

One can verify that A is a linear, continuous, monotone and potential map [83, 120, 128]. Using the natural assumption: There exists a constant $C > 0$ such that

$$|k(x)| \leq C \quad \forall x \in [0, l],$$

we are able to prove the boundedness of A. By the definition of A, inequality (4.4.8) can be rewritten in the form

$$\langle Aw, u - w \rangle \geq 0 \quad \forall u \in \Omega, \quad w \in \Omega. \tag{4.4.12}$$

We assume that, in place of $k(x)$, a perturbed function $k^h(x)$ is given, which depends on the positive parameter h, such that

$$|k(x) - k^h(x)| \leq h \quad \forall x \in [0, l]. \tag{4.4.13}$$

The function $k(x)$ induces the perturbed operator $A^h : X \to X^*$ as follows:

$$\langle A^h w, v \rangle = \int_0^l k^h(x) w''(x) v''(x) dx \quad \forall w, v \in X.$$

Then

$$\|Av - A^h v\| \leq h\|v\| \quad \forall v \in X,$$

that is, in the condition (4.2.29) $g(t) \equiv t$. Moreover, the operators A^h and A have the same properties.

The duality mapping J in X can be defined as follows:

$$\langle Jw, v \rangle = \int_0^l w''(x) v''(x) dx \quad \forall w, v \in X.$$

Let $w_\alpha^h \in \Omega$ be a unique solution of the regularized variational inequality

$$\int_0^l \left(k^h(x) w''(x) + \alpha w''(x) \right) \left(u''(x) - w''(x) \right) dx \geq 0 \quad \forall u \in \Omega, \quad w \in \Omega.$$

Then

$$\langle A^h w_\alpha^h + \alpha J w_\alpha^h, u - w_\alpha^h \rangle \geq 0 \quad \forall u \in \Omega.$$

Suppose that the variational inequality (4.4.12) is solvable and parameter h in (4.4.13) is such that

$$\lim_{\alpha \to 0} \frac{h}{\alpha} = 0.$$

Then Theorem 4.1.1 implies the strong convergence of $\{w_\alpha^h\}$ to $\bar{w}^*$ as $\alpha \to 0$, where $\bar{w}^*$ is the minimal norm solution of (4.4.12). Observe that the equation of the residual principle (4.1.44) accepts the form:

$$\rho(\bar{\alpha}) = \bar{\alpha}\|w_\alpha^h\| = \left(M + \|w_\alpha^h\| \right) h^s, \quad M > 1, \quad s \in (0, 1].$$

Example 4.4.2   We investigate the membrane bend problem, whose deflection is restricted by the hard fixed obstacle given by function $z = \chi(x,y)$, where $(x,y) \in D$, D is a bounded domain in the plane XOY with boundary $\partial D = \Gamma$. Let the membrane be jammed at the contour Γ, charged with pressure P, perpendicular to a median plane. Let deflection $w(x,y)$ of the membrane occur in the direction of the axis OZ and function $Q(x,y)$ be the obstacle reaction. Considering the problem to be nonlinear we come to the following equation [120]:

$$-\frac{\partial}{\partial x}a_1(x,y,w'_x) - \frac{\partial}{\partial y}a_2(x,y,w'_y) = -P(x,y,w) - Q(x,y), \qquad (4.4.14)$$

with

$$w(x,y)\mid_\Gamma = 0, \quad w(x,y) \le \chi(x,y), \qquad (4.4.15)$$

$$Q(x,y) \ge 0, \quad Q(x,y)\Big(w(x,y) - \chi(x,y)\Big) = 0. \qquad (4.4.16)$$

Assume that the nonlinear functions in (4.4.14) satisfy the following conditions:
a) $a_i(x,y,\xi)$, $i = 1,2$, and $P(x,y,\xi)$ are measurable on D for all $\xi \in R^1$, continuous and non-decreasing with respect to ξ for almost all $x,y \in D$;
b) there exist $c_i > 0$ and $k_i(x,y) \in L^q(D)$, $q = \dfrac{p}{p-1}$, $p > 2$, $i = 1,2,3$, such that

$$|a_i(x,y,\xi)| \le c_i\Big(k_i(x,y) + |\xi|^{p-1}\Big), \quad i = 1,2,$$

and

$$|P(x,y,\xi)| \le c_3\Big(k_3(x,y) + |\xi|^{p-1}\Big).$$

Introduce the space

$$X = \{w \mid w = w(x,y) \in W_1^p(D), \ (x,y) \in D, \ w(x,y)|_\Gamma = 0\}$$

and define

$$\|w\| = \Big(\int\int_D \Big[|w'_x(x,y)|^p + |w'_y(x,y)|^p\Big]dxdy\Big)^{1/p}.$$

Construct the operator A as follows: For all $v,w \in X$

$$\langle Aw, v\rangle = \int\int_D \Big(a_1(x,y,w'_x)v'_x + a_2(x,y,w'_y)v'_y + P(x,y,w)v\Big)dxdy. \qquad (4.4.17)$$

It follows from the conditions b) that the operator A acts from X in X^* [128, 220, 221], while the properties of the functions $a_1(x,y,\xi)$, $a_2(x,y,\xi)$ and $P(x,y,\xi)$ enable us to assert that A is monotone (see Section 1.3, Examples 5, 6). Moreover, it is possible to prove that A is bounded and potential [83, 143]. Repeating the arguments of the previous example, we make sure that the problem (4.4.14) - (4.4.16) can be stated as variational inequality of the type (4.4.12) on the set

$$\Omega = \{w \mid w = w(x,y) \in X, \ w(x,y) \le \chi(x,y), \quad \text{for almost all } (x,y) \in D\}.$$

Suppose that an operator $A^h : X \to X^*$ is defined by the equality (4.4.17) with perturbed functions $a_1^h(x, y, \xi)$, $a_2^h(x, y, \xi)$ and $P^h(x, y, \xi)$, instead of $a_1(x, y, \xi)$, $a_2(x, y, \xi)$, $P(x, y, \xi)$, which also satisfy a) and b). In addition, for any element $u \in X$, let the following inequalities hold:

$$\|a_1^h(x, y, u_x') - a_1(x, y, u_x')\|_{L^q} \le h\bar{g}(\|u\|),$$

$$\|a_2^h(x, y, u_y') - a_2(x, y, u_y')\|_{L^q} \le h\bar{g}(\|u\|),$$

$$\|P^h(x, y, u) - P(x, y, u)\|_{L^q} \le h\bar{g}(\|u\|),$$

where $\bar{g}(t)$ is a non-negative, continuous and increasing function for $t \ge 0$. Then it is not difficult to be sure that there exists a constant $C > 0$ such that

$$\|Au - A^h u\|_* \le Ch\bar{g}(\|u\|) \quad \forall u \in X. \tag{4.4.18}$$

The duality mapping J is defined by the following expression:

$$\langle Ju, v \rangle = \|u\|^{2-p} \int \int_D \left(|u_x'|^{p-2} u_x' v_x' + |u_y'|^{p-2} u_y' v_y' \right) dx dy$$

$$\forall u, v \in X, \ u \ne \theta_X, \ J\theta_X = \theta_{X^*}.$$

Now the regularized variational inequality and the equation of the residual principle are written similarly to Example 4.4.1.

Example 4.4.3 Consider again Example 4.4.1 but now on approximately given sets Ω_σ, which is defined by the inexact function $\chi_\sigma(x)$ in place of $\chi(x)$. Denote

$$\Omega_\sigma = \{v \mid v = v(x) \in X, \ v(x) \le \chi_\sigma(x) \text{ for almost all } x \in [0, l]\},$$

assuming that

$$\|\chi_\sigma(x) - \chi(x)\| \le \sigma. \tag{4.4.19}$$

We need to get the estimate (4.3.2) for $\|z - z_\sigma\|$, where $z \in \Omega$, $z_\sigma \in \Omega_\sigma$. However, (4.4.19) is not enough for this aim. It is possible to apply the inequality [83]

$$\int \int_G \left| \frac{\partial^k v(x, y)}{\partial x^{k_1} \partial y^{k_2}} \right|^p dx dy \le C \sum_m \int \int_G \left| \frac{\partial^m v(x, y)}{\partial x^{m_1} \partial y^{m_2}} \right|^p dx dy,$$

where

$$p > 1, \ k = k_1 + k_2, \ m = m_1 + m_2, \ 0 \le k \le m,$$

and positive C depends on p, k, m and G. As a result, the required estimate is obtained only in the space $L^2[0, l]$.

Observe that the function $w''(x)$, the modulus of which coincides with a curvature of the curve $y = w(x)$ in the framework of the analyzed model, is the main characteristic of the problem we are considering [90, 120]. This fact enables us to solve the problem changing its statement and preserving the basic requirements (4.4.1) - (4.4.5).

For variational inequality (4.4.12), we introduce the set

$$\Omega' = \{v \mid v = v(x) \in \Omega, \ v''(x) \le \chi''(x) \ \text{for almost all} \ x \in [0, l]\},$$

and define perturbed sets as follows:

$$\Omega'_\sigma = \{v \mid v = v(x) \in \Omega_\sigma, \ v''(x) \le \chi''_\sigma(x) \ \text{for almost all} \ x \in [0, l]\}.$$

Moreover, we assume that

$$\|\chi''_\sigma(x) - \chi''(x)\|_2 \le \sigma \quad \forall x \in [0, l]. \tag{4.4.20}$$

It is then obvious that for every element $z \in \Omega'$ ($z \in \Omega'_\sigma$) there exists an element $z_\sigma \in \Omega'_\sigma$ ($z_\sigma \in \Omega'$) such that $\|z - z_\sigma\| \le \sigma$, that is, conditions (4.3.2) and (4.3.4) of Theorem 4.3.1 are satisfied with $a_1(t) = a_2(t) = 1$ for $t \ge 0$. Since $\|Av - A^h v\| \le h\|v\|$ and A^h are bounded, we conclude that there exists a constant $c > 0$ such that

$$\|Az - A^h z_\sigma\| \ \le \ \|Az - A^h z\| + \|A^h z - A^h z_\sigma\|$$

$$\le \ h\|z\| + c\sigma \le (\|z\| + c)(h + \sigma).$$

Hence, the requirement (4.3.3) of Theorem 4.3.1 holds.

Observe that the replacement of Ω and Ω_σ by Ω' and Ω'_σ, respectively, does not disturb the structure and qualitative characteristics of a desired solution; however, it allows us to describe perturbations of the set Ω' satisfying the conditions of Theorem 4.3.1. This replacement may be treated as a construction of the set on which the regularizing operator is formed or as accounting for a priori information about unknown solutions.

Assume that the variational inequality

$$\langle Aw, v - w \rangle \ge 0 \quad \forall v \in \Omega', \quad w \in \Omega', \tag{4.4.21}$$

is solvable, and in (4.4.13) and (4.4.20)

$$\frac{h + \sigma}{\alpha} \to 0 \quad \text{as} \quad \alpha \to 0.$$

Next we find a solution sequence $\{w^\gamma_\alpha\}$ of the regularized variational inequality:

$$\langle A^h w^\gamma_\alpha + \alpha J^\mu w^\gamma_\alpha, v - w^\gamma_\alpha \rangle \ge 0 \quad \forall v \in \Omega'_\sigma, \quad w^\gamma_\alpha \in \Omega'_\sigma,$$

where $\gamma = (h, \sigma)$. Then the strong convergence of $\{w^\gamma_\alpha\}$ to the minimal norm solution of (4.4.21) as $\alpha \to 0$ is proved by Theorem 4.3.1. It is known [120] that every solution of (4.4.21) is a solution of problem (4.4.1) - (4.4.5). Thus, we have constructed the sequence $\{w^\gamma_\alpha\}$ which converges in the norm of the space X to some solution of the problem (4.4.1) - (4.4.5).

Remark 4.4.4 *In the definitions of Ω' and Ω'_σ, we can change $\chi''(x)$ and $\chi''_\sigma(x)$ by the functions $\psi(x)$ and $\psi_\sigma(x)$ such that*

$$\chi''(x) \le \psi(x), \quad \chi''_\sigma(x) \le \psi_\sigma(x)$$

and

$$\|\psi(x) - \psi_\sigma(x)\|_2 \le \sigma \quad \forall x \in [0, l].$$

Example 4.4.5 Consider Example 4.4.2 again. By the same reasoning as in the previous example, define sets

$$\Omega' = \{v \mid v = v(x, y) \in \Omega, \ v'_x(x, y) \le \chi'_x(x, y), \ v'_y(x, y) \le \chi'_y(x, y)$$

$$\text{for almost all } (x, y) \in D\}$$

and

$$\Omega'_\sigma = \{v \mid v = v(x, y) \in \Omega_\sigma, \ v'_x(x, y) \le (\chi_\sigma)'_x(x, y), \ v'_y(x, y) \le (\chi_\sigma)'_y(x, y)$$

$$\text{for almost all } x, y \in D\}.$$

Let, for all $x, y \in D$ and for $\tau = \dfrac{1}{p}$, the following inequalities hold:

$$\|\chi'_x(x, y) - (\chi_\sigma)'_x(x, y)\|_{L^p} \le \frac{\sigma}{2^\tau}$$

and

$$\|\chi'_y(x, y) - (\chi_\sigma)'_y(x, y)\|_{L^p} \le \frac{\sigma}{2^\tau}.$$

Under these conditions, it is not difficult to verify that $d(u, \Omega'_\sigma) \le \sigma$ for all $u \in \Omega'$ and $d(u, \Omega') \le \sigma$ for all $u \in \Omega'_\sigma$. In order to evaluate from above the norm $\|Aw - A^h w_\sigma\|_*$, where $w \in \Omega'$, $w_\sigma \in \Omega'_\sigma$ and $\|w - w_\sigma\| \le \sigma$, we assume that [220]:

$$|(a_i^h)'_\xi(x, y, \xi)| \le c'_i\Big(d_i(x, y) + |\xi|^{p-2}\Big), \quad i = 1, 2, \tag{4.4.22}$$

and

$$|(P^h)'_\xi(x, y, \xi)| \le c'_3\Big(d_3(x, y) + |\xi|^{p-2}\Big). \tag{4.4.23}$$

Here $c'_i > 0$, $d_i(x, y) \in L^{p'}(D)$ are non-negative functions for i=1,2,3, $p' = \dfrac{p}{p-2}$. Applying the Lagrange formula, the Hölder and Minkovsky integral inequalities and taking into account the conditions (4.4.22) and (4.4.23), write down the chain of relations

$$\int\int_D \Big(a_1^h(x, y, w'_x) - a_1^h(x, y, u'_x)\Big)v'_x dx dy$$

$$= \int\int_D (a_1^h)'_\xi\Big(x, y, w'_x + \theta(u'_x - w'_x)\Big)(w'_x - u'_x)v'_x dx dy$$

$$\leq \quad \|v\| \left[\int \int_D \left|(a_1^h)'_\xi \big(x,y,w'_x + \theta(u'_x - w'_x)\big)\right|^{\frac{p}{p-1}} |w'_x - u'_x|^{\frac{p}{p-1}} dxdy \right]^{\frac{p-1}{p}}$$

$$\leq \quad c'_1 \|v\| \|w - u\| \left[\int \int_D \big(d_1(x,y) + |w'_x + \theta(u'_x - w'_x)|^{p-2}\big)^{\frac{p}{p-2}} dxdy \right]^{\frac{p-2}{p}}$$

$$\leq \quad c'_1 \|v\| \|w - u\| \left[\left(\int \int_D (d_1(x,y))^{\frac{p}{p-2}} dxdy \right)^{\frac{p-2}{p}} \right.$$

$$+ \quad \left. \left(\int \int_D |w'_x|^p dxdy + \int \int_D |u'_x|^p dxdy \right)^{\frac{p-2}{p}} \right]$$

$$\leq \quad c'_1 \|v\| \|w - u\| (\|d_1(x,y)\|_{L^{p'}} + 2R^{p-2}),$$

where $0 < \theta < 1$, $R = max\,\{\|w\|, \|u\|\}$.

Evaluating analogously two other terms of the dual product $\langle A^h w - A^h u, v \rangle$, namely,

$$\int \int_D \big(a_2^h(x,y,w'_y) - a_2^h(x,y,u'_y)\big)v'_y dxdy$$

and

$$\int \int_D \big(P^h(x,y,w) - P^h(x,y,u)\big)vdxdy,$$

we come to the following estimate:

$$\|A^h w - A^h u\|_* \leq c'(m + 2R^{p-2})\|w - u\|, \tag{4.4.24}$$

where

$$c' = \sum_{i=1}^{3} c'_i \quad \text{and} \quad m = \sum_{i=1}^{3} \|d_i(x,y)\|_q.$$

Consequently, we established that an operator A^h is Lipschitz-continuous on each bounded set. By (4.4.18) and (4.4.24), if $\|w - w_\sigma\| \leq \sigma$ then

$$\|Aw - A^h w_\sigma\|_* \quad \leq \quad \|Aw - A^h w\|_* + \|A^h w - A^h w_\sigma\|_*$$

$$\leq \quad \Big(C\bar{g}(\|w\|) + c'\big(m + 2(\|w\| + \bar{\sigma})^{p-2}\big)\Big)(h + \sigma), \quad \sigma \in (0, \bar{\sigma}).$$

Thus, the conditions (4.3.2) - (4.3.4) of Theorem 4.3.1 are also satisfied for problem (4.4.14) - (4.4.16).

Example 4.4.6 Suppose that in a bounded set $G \subset R^3$, the filtration equation

$$-div(g(x, |\nabla u|^2)\nabla u) = f(x), \quad x \in G, \quad \partial G = \Gamma = \Gamma^0 + \Gamma^1 + \Gamma^2, \tag{4.4.25}$$

is solved (see Example 8 in Section 1.3) with the following boundary conditions:

$$u(x)\,|_{\Gamma^0} = 0, \quad \bar{v} \cdot \bar{n}\,|_{\Gamma^1} = 0, \tag{4.4.26}$$

$$\bar{v} \cdot \bar{n} \mid_{\Gamma^2} = 0 \quad \text{as} \quad u(x) \geq 0 \quad \text{and} \quad \bar{v} \cdot \bar{n} \mid_{\Gamma^2} \leq 0 \quad \text{as} \quad u(x) = 0, \qquad (4.4.27)$$

where $\bar{v}(x) = g(x, |\nabla u|^2)\nabla u$, $\bar{n}$ is a unit vector of exterior normal, $\bar{v} \cdot \bar{n}$ is the scalar product of the vectors $\bar{v}$ and $\bar{n}$. Assume that in place of G, its approximations

$$G_\sigma \subset R^3 \quad \text{with} \quad \partial G_\sigma = \Gamma^0 + \Gamma^1_\sigma + \Gamma^2_\sigma$$

are known. Let $G^\triangle_\sigma$ be a symmetric difference of sets G and G_σ, that is,

$$G^\triangle_\sigma = (G \cup G_\sigma) \backslash (G \cap G_\sigma),$$

and let measure $\mu(G^\triangle_\sigma) \leq \sigma$. Introduce the set $D = \bigcup_{\sigma \geq 0} G_\sigma$, where $G_0 = G$, and define the space

$$X = \{u \mid u = u(x) \in W^p_1(D), \ u \mid_{\Gamma^0} = 0\}$$

with the norm

$$\|u\| = \left(\int_D |\nabla u|^p dx \right)^{1/p}, \quad x \in R^3.$$

Construct sets Ω_σ, $\sigma \geq 0$, as follows: $\Omega_0 = \Omega$ and if $\sigma \neq 0$ then

$$\Omega_\sigma = \{u \mid u = u(x) \in X, \ u(x) \geq 0 \ \text{for almost all} \ x \in \Gamma^2_\sigma, \ u(x) = 0 \ \text{as} \ x \in D \backslash G_\sigma\}.$$

It is well known (see, for instance, [123]) that if Γ and $w(x)$ are sufficiently smooth, then the problem (4.4.25) - (4.4.27) is equivalent to the variational inequality

$$\langle Aw, w - v \rangle = \int_G \Big(g(x, |\nabla w|^2)\nabla w \cdot \nabla(w - v) - f(x)(w - v) \Big) dx \leq 0 \quad \forall v \in \Omega, \quad w \in \Omega,$$

with the monotone operator $A : X \to X^*$.

Let $u(x) \in \Omega$. Consider the function $u_\sigma(x) \in \Omega_\sigma$, which coincides with $u(x)$ on $G_\sigma \cap G$, and assume that $u(x) = 0$ at any point $x \in G_\sigma \backslash G$. Then

$$\|u - u_\sigma\| \leq \left(\int_{G \backslash G_\sigma} |\nabla u|^p dx \right)^{1/p}.$$

Since the Lebesgue integral is absolutely continuous [117], we deduce the inequality

$$\|u - u_\sigma\| \leq a(u, \sigma),$$

where $a(u, \sigma) \to 0$ as $\sigma \to 0$ for fixed $u(x)$. However, we cannot obtain the function $a(u, \sigma)$ in the analytical form and prove the convergence theorems for the regularization methods.

4.5 Variational Inequalities with Unbounded Operators

As in Section 4.2, let $\Re = (0, \delta^*] \times (0, h^*] \times (0, \sigma^*]$, let X be a reflexive strictly convex Banach space, X^* be also strictly convex, $A : X \to 2^{X^*}$ and $A^h : X \to 2^{X^*}$ be maximal monotone operators for all $h \in (0, h^*]$, $D(A) = D(A^h)$, sets $\Omega \subseteq int\ D(A)$ and $\Omega_\sigma \subseteq int\ D(A^h)$ as $\sigma \in (0, \sigma^*]$, Ω_σ uniformly approximate Ω with the estimate

$$\mathcal{H}_X(\Omega, \Omega_\sigma) \leq \sigma. \tag{4.5.1}$$

We also assume that there is no inequality of the kind (4.2.2) which imposes the growth condition upon the operators A^h. At the same time, the conditions (4.1.1), (4.1.5), (4.1.6) and (4.2.1) are fulfilled for all $x \in \Omega \cup \Omega_\sigma$, where, in addition, a function $g(t)$ is bounded on bounded sets.

Let $x_\alpha^\gamma \in \Omega_\sigma$ with $\gamma = (\delta, h, \sigma)$ be a solution of the variational inequality

$$\langle A^h x + \alpha J x - f^\delta, z - x \rangle \geq 0 \quad \forall z \in \Omega_\sigma. \tag{4.5.2}$$

Theorem 4.5.1 *There exists a constant $r_0 > 0$ such that $x_\alpha^\gamma \to \bar{x}^*$ if $\sigma < min\ \{r_0, \sigma^*\}$ and (4.2.7) is satisfied.*

Proof. First of all, prove that $\{x_\alpha^\gamma\}$ is bounded as $\alpha \to 0$. To this end, consider the auxiliary variational inequality problem: To find $x \in \Omega_\sigma$ such that

$$\langle Ax + \alpha J x - f, z - x \rangle \geq 0 \quad \forall z \in \Omega_\sigma. \tag{4.5.3}$$

Denote by x_α^σ its unique solution for fixed $\alpha > 0$ and $\sigma \in \Re$. Show that $\{x_\alpha^\sigma\}$ is bounded as $\alpha \to 0$. By Lemma 1.5.14, there exist constants $r_0 > 0$ and $c_0 > 0$ such that

$$D_1 = \langle y_\alpha^\sigma - f, x_\alpha^\sigma - \bar{x}^* \rangle \geq r_0 \|y_\alpha^\sigma - f\|_* - c_0(\|x_\alpha^\sigma - \bar{x}^*\| + r_0), \tag{4.5.4}$$

where

$$c_0 = sup\ \{\|y - f\|_* \mid y \in Ax, \ x \in B(\bar{x}^*, r_0)\} < \infty.$$

Take $y_\alpha^\sigma \in Ax_\alpha^\sigma$ satisfying the inequality

$$\langle y_\alpha^\sigma + \alpha J x_\alpha^\sigma - f, z - x_\alpha^\sigma \rangle \geq 0 \quad \forall z \in \Omega_\sigma. \tag{4.5.5}$$

Evaluate D_1 from above. Since (4.5.1) holds for $\sigma \in \Re$, there exists $z_\sigma \in \Omega_\sigma$ such that $\|\bar{x}^* - z_\sigma\| \leq \sigma$ and at that

$$\langle y_\alpha^\sigma + \alpha J x_\alpha^\sigma - f, z_\sigma - x_\alpha^\sigma \rangle \geq 0.$$

Therefore, by the relations

$$\langle J x_\alpha^\sigma, x_\alpha^\sigma - \bar{x}^* \rangle = \|x_\alpha^\sigma\|^2 - \langle J x_\alpha^\sigma, \bar{x}^* \rangle \geq -\|x_\alpha^\sigma\| \|\bar{x}^*\|,$$

one has

$$
\begin{aligned}
D_1 &= \langle y_\alpha^\sigma + \alpha J x_\alpha^\sigma - f, x_\alpha^\sigma - \bar{x}^* \rangle - \alpha \langle J x_\alpha^\sigma, x_\alpha^\sigma - \bar{x}^* \rangle \\[6pt]
&= \langle y_\alpha^\sigma + \alpha J x_\alpha^\sigma - f, x_\alpha^\sigma - z_\sigma \rangle + \langle y_\alpha^\sigma + \alpha J x_\alpha^\sigma - f, z_\sigma - \bar{x}^* \rangle \\[6pt]
&\quad - \alpha \langle J x_\alpha^\sigma, x_\alpha^\sigma - \bar{x}^* \rangle \leq \sigma \|y_\alpha^\sigma + \alpha J x_\alpha^\sigma - f\|_* + \alpha \|x_\alpha^\sigma\| \|\bar{x}^*\| \\[6pt]
&\leq \alpha \|x_\alpha^\sigma\|(\|\bar{x}^*\| + \sigma) + \sigma \|y_\alpha^\sigma - f\|_*. \tag{4.5.6}
\end{aligned}
$$

Using (4.5.4) and (4.5.6) we obtain

$$\alpha\|x_\alpha^\sigma\|(\|\bar{x}^*\| + \sigma) + \sigma\|y_\alpha^\sigma - f\|_* \geq r_0\|y_\alpha^\sigma - f\|_* - c_0(\|x_\alpha^\sigma\| + \|\bar{x}^*\| + r_0).$$

Hence,

$$(r_0 - \sigma)\|y_\alpha^\sigma - f\|_* \leq (\alpha\|\bar{x}^*\| + \alpha\sigma + c_0)\|x_\alpha^\sigma\| + c_0(\|\bar{x}^*\| + r_0). \tag{4.5.7}$$

Since $\sigma < r_0$, we come to the following estimate:

$$\|y_\alpha^\sigma - f\|_* \leq M_1\|x_\alpha^\sigma\| + M_2, \tag{4.5.8}$$

where

$$M_1 \geq \frac{\alpha\|\bar{x}^*\| + \alpha\sigma + c_0}{r_0 - \sigma}$$

and

$$M_2 \geq \frac{c_0(\|\bar{x}^*\| + r_0)}{r_0 - \sigma}.$$

Thus, A has not more than linear growth on solutions x_α^σ of the variational inequality (4.5.3). We emphasize that this fact is proved only by the condition that the semi-deviation $\beta(\Omega, \Omega_\sigma) \leq \sigma$.

Let now $x_\alpha \in \Omega$ be a solution of the variational inequality

$$\langle Ax + \alpha Jx - f, z - x \rangle \geq 0 \quad \forall z \in \Omega. \tag{4.5.9}$$

By virtue of the monotonicity of A and by the properties of J, we get

$$D_2 = \langle y_\alpha^\sigma + \alpha Jx_\alpha^\sigma - y_\alpha - \alpha Jx_\alpha, x_\alpha^\sigma - x_\alpha \rangle \geq \alpha(\|x_\alpha^\sigma\| - \|x_\alpha\|)^2,$$

where $y_\alpha \in Ax_\alpha$ satisfies the inequality

$$\langle y_\alpha + \alpha Jx_\alpha - f, z - x_\alpha \rangle \geq 0 \quad \forall z \in \Omega.$$

On the other hand, there exist elements $z_\alpha^\sigma \in \Omega_\sigma$ and $u_\alpha^\sigma \in \Omega$ such that $\|z_\alpha^\sigma - x_\alpha\| \leq \sigma$ and $\|u_\alpha^\sigma - x_\alpha^\sigma\| \leq \sigma$. Then firstly

$$\langle y_\alpha^\sigma + \alpha Jx_\alpha^\sigma - f, x_\alpha^\sigma - z_\alpha^\sigma \rangle \leq 0,$$

and secondly

$$\langle y_\alpha + \alpha Jx_\alpha - f, x_\alpha - u_\alpha^\sigma \rangle \leq 0.$$

We come to the upper estimate of D_2, namely,

$$D_2 \leq \sigma(\|y_\alpha^\sigma + \alpha Jx_\alpha^\sigma - f\|_* + \|y_\alpha + \alpha Jx_\alpha - f\|_*).$$

Convergence of the sequence $\{x_\alpha\}$ to $\bar{x}^*$ as $\alpha \to 0$ follows from the proof of Theorem 4.1.1, while (4.5.8) implies the boundedness of $\{y_\alpha\}$ if we put there $\sigma = 0$. Hence, there exists a constant $C_1 > 0$ such that

$$\|y_\alpha + \alpha Jx_\alpha - f\|_* \leq C_1.$$

Then
$$D_2 \leq \sigma \left(\|y_\alpha^\sigma - f\|_* + \alpha\|x_\alpha^\sigma\| + C_1 \right) \leq \sigma \left[(M_1 + \alpha)\|x_\alpha^\sigma\| + M_2 + C_1 \right],$$
and we obtain the following quadratic inequality with respect to $\|x_\alpha^\sigma\|$:

$$\|x_\alpha^\sigma\|^2 - 2\|x_\alpha^\sigma\|\left(\|x_\alpha\| + \frac{\sigma}{2\alpha}M_1 + \frac{\sigma}{2} \right) + \|x_\alpha\|^2 - \frac{\sigma}{\alpha}(C_1 + M_2) \leq 0. \qquad (4.5.10)$$

It is easy to calculate that

$$
\begin{aligned}
\|x_\alpha^\sigma\| \ \leq \ & \|x_\alpha\| + \frac{\sigma}{2\alpha}M_1 + \frac{\sigma}{2} \\
& + \left[\left(\frac{\sigma}{2\alpha}M_1 + \frac{\sigma}{2} \right)^2 + 2\|x_\alpha\|\left(\frac{\sigma}{2\alpha}M_1 + \frac{\sigma}{2} \right) + \frac{\sigma}{\alpha}(C_1 + M_2) \right]^{1/2} \leq C_2,
\end{aligned}
$$

where $C_2 > 0$. Owing to (4.2.7) one can consider, without loss of generality, that $\frac{\sigma}{\alpha} \leq C$, $C > 0$. Then the sequence $\{x_\alpha^\sigma\}$ is uniformly bounded because $\sigma < r_0$.

Now we are going to show that the solutions x_α^γ are bounded. To this end, evaluate from above the following expression:

$$D_3 = \langle y_\alpha^\gamma + \alpha J x_\alpha^\gamma - f^\delta - y_\alpha^\sigma - \alpha J x_\alpha^\sigma + f, x_\alpha^\gamma - x_\alpha^\sigma \rangle,$$

where $y_\alpha^\gamma \in A^h x_\alpha^\gamma$ satisfies the inequality

$$\langle y_\alpha^\gamma + \alpha J x_\alpha^\gamma - f^\delta, z - x_\alpha^\gamma \rangle \geq 0 \quad \forall z \in \Omega_\sigma. \qquad (4.5.11)$$

Since operators A^h are monotone, we have

$$D_3 \geq \alpha \left(\|x_\alpha^\gamma\| - \|x_\alpha^\sigma\| \right)^2 - \delta\|x_\alpha^\gamma - x_\alpha^\sigma\| - hg(\|x_\alpha^\sigma\|)\|x_\alpha^\gamma - x_\alpha^\sigma\|.$$

In view of (4.5.5) and (4.5.11), it is not difficult to see that $D_3 \leq 0$. Hence,

$$
\begin{aligned}
\alpha\|x_\alpha^\gamma\|^2 \ - \ & 2\alpha\|x_\alpha^\gamma\|\|x_\alpha^\sigma\| + \alpha\|x_\alpha^\sigma\|^2 - \delta\|x_\alpha^\gamma\| - \delta\|x_\alpha^\sigma\| \\
- \ & hg(\|x_\alpha^\sigma\|)\|x_\alpha^\gamma\| - hg(\|x_\alpha^\sigma\|)\|x_\alpha^\sigma\| \leq 0.
\end{aligned}
$$

Consequently, we obtain the quadratic inequality again but now with respect to $\|x_\alpha^\gamma\|$:

$$
\begin{aligned}
\|x_\alpha^\gamma\|^2 \ - \ & 2\|x_\alpha^\gamma\|\left(\|x_\alpha^\sigma\| + \frac{\delta + hg(\|x_\alpha^\sigma\|)}{2\alpha} \right) + \|x_\alpha^\sigma\|^2 \\
- \ & \frac{\delta}{\alpha}\|x_\alpha^\sigma\| - \frac{h}{\alpha}g(\|x_\alpha^\sigma\|)\|x_\alpha^\sigma\| \leq 0,
\end{aligned}
$$

which yields the estimate

$$
\begin{aligned}
\|x_\alpha^\gamma\| \ \leq \ & \|x_\alpha^\sigma\| + \frac{\delta}{2\alpha} + \frac{h}{2\alpha}g(\|x_\alpha^\sigma\|) \\
& + \left[\left(\frac{\delta}{2\alpha} + \frac{hg(\|x_\alpha^\sigma\|)}{2\alpha} \right)^2 + 2\|x_\alpha^\sigma\|\frac{\delta + hg(\|x_\alpha^\sigma\|)}{\alpha} \right]^{1/2}.
\end{aligned}
$$

Since the sequences $\{x_\alpha^\sigma\}$ and $\left\{\dfrac{\delta+h}{\alpha}\right\}$ are bounded as $\alpha \to 0$, there exists a constant $C_3 > 0$ such that $\|x_\alpha^\gamma\| \leq C_3$.

Prove the boundedness of $\|y_\alpha^\gamma\|$ as $\alpha \to 0$. For this, establish first the boundedness of elements $v_\alpha^\gamma \in Ax_\alpha^\gamma$ such that

$$\|y_\alpha^\gamma - v_\alpha^\gamma\| \leq g(\|x_\alpha^\gamma\|)h. \tag{4.5.12}$$

Observe that $\{v_\alpha^\gamma\}$ exists according to (4.1.6). By Lemma 1.5.14,

$$D_4 = \langle v_\alpha^\gamma - f, x_\alpha^\gamma - \bar{x}^* \rangle \geq r_0 \|v_\alpha^\gamma - f\|_* - c_0(r_0 + \|x_\alpha^\gamma - \bar{x}^*\|). \tag{4.5.13}$$

Evaluate D_4 from above. We have

$$\begin{aligned}
D_4 &= \langle v_\alpha^\gamma - y_\alpha^\gamma + y_\alpha^\gamma - f^\delta + f^\delta - f, x_\alpha^\gamma - \bar{x}^* \rangle \\[2mm]
&\leq hg(\|x_\alpha^\gamma\|)\|x_\alpha^\gamma - \bar{x}^*\| + \delta\|x_\alpha^\gamma - \bar{x}^*\| \\[2mm]
&\quad - \alpha\langle Jx_\alpha^\gamma, x_\alpha^\gamma - \bar{x}^* \rangle + \langle y_\alpha^\gamma + \alpha Jx_\alpha^\gamma - f^\delta, x_\alpha^\gamma - \bar{x}^* \rangle.
\end{aligned}$$

We present the last term in the equivalent form

$$\langle y_\alpha^\gamma + \alpha Jx_\alpha^\gamma - f^\delta, x_\alpha^\gamma - \bar{x}^* \rangle = \langle y_\alpha^\gamma + \alpha Jx_\alpha^\gamma - f^\delta, x_\alpha^\gamma - z_\sigma + z_\sigma - \bar{x}^* \rangle,$$

where $z_\sigma \in \Omega_\sigma$ such that

$$\|z_\sigma - \bar{x}^*\| \leq \sigma.$$

Taking into account the inequality

$$\langle y_\alpha^\gamma + \alpha Jx_\alpha^\gamma - f^\delta, z_\sigma - x_\alpha^\gamma \rangle \geq 0,$$

we deduce

$$\langle y_\alpha^\gamma + \alpha Jx_\alpha^\gamma - f^\delta, x_\alpha^\gamma - \bar{x}^* \rangle \leq \langle y_\alpha^\gamma + \alpha Jx_\alpha^\gamma - f^\delta, z_\sigma - \bar{x}^* \rangle \leq \sigma\|y_\alpha^\gamma + \alpha Jx_\alpha^\gamma - f^\delta\|_*.$$

Then

$$\begin{aligned}
D_4 &\leq hg(\|x_\alpha^\gamma\|)\|x_\alpha^\gamma - \bar{x}^*\| + \delta\|x_\alpha^\gamma - \bar{x}^*\| \\[2mm]
&\quad + \alpha\|x_\alpha^\gamma\|\|x_\alpha^\gamma - \bar{x}^*\| + \sigma\|y_\alpha^\gamma + \alpha Jx_\alpha^\gamma - f^\delta\|_*.
\end{aligned}$$

By the fact that

$$\|y_\alpha^\gamma + \alpha Jx_\alpha^\gamma - f^\delta\|_* \leq \|y_\alpha^\gamma - v_\alpha^\gamma\|_* + \|v_\alpha^\gamma - f^\delta\|_* + \alpha\|x_\alpha^\gamma\|,$$

we obtain the final estimate

$$\begin{aligned}
D_4 &\leq hg(\|x_\alpha^\gamma\|)\|x_\alpha^\gamma - \bar{x}^*\| + \delta\|x_\alpha^\gamma - \bar{x}^*\| + \alpha\|x_\alpha^\gamma\|\|x_\alpha^\gamma - \bar{x}^*\| \\[2mm]
&\quad + \sigma hg(\|x_\alpha^\gamma\|) + \sigma\|v_\alpha^\gamma - f^\delta\|_* + \alpha\sigma\|x_\alpha^\gamma\|. \tag{4.5.14}
\end{aligned}$$

It follows from (4.5.13) and (4.5.14) that if $\sigma < r_0$ then

$$\|v_\alpha^\gamma - f\|_* \leq \frac{1}{r_0 - \sigma}\Big((c_0 + hg(\|x_\alpha^\gamma\|) + \delta + \alpha\|x_\alpha^\gamma\|)\,\|x_\alpha^\gamma - \bar{x}^*\|$$

$$+ \quad c_0 r_0 + \sigma\,(hg(\|x_\alpha^\gamma\|) + \alpha\|x_\alpha^\gamma\|)\Big),$$

which implies the boundedness of $\{v_\alpha^\gamma\}$. Owing to (4.5.12), this allows us to conclude that $\{y_\alpha^\gamma\}$ is also bounded.

As usual, from the sequence $\{x_\alpha^\gamma\}$, one can choose a subsequence which weakly converges to some $\bar{x} \in X$ as $\alpha \to 0$. Then, similarly to Theorem 4.2.5, we establish the inclusion $\bar{x} \in N$. The proof of the last assertions that $\bar{x} = \bar{x}^*$ and $x_\alpha^\gamma \to \bar{x}^*$ as $\alpha \to 0$ is done by the previous scheme (see Sections 4.1 and 4.2). ∎

Further we use the weaker approximation criterion of the exact set Ω by perturbed sets Ω_σ. Assume that there exist bounded functionals φ_1 and φ_2 with $dom\ \varphi_1$ and $dom\ \varphi_2$, respectively, such that $\Omega \subset dom\ \varphi_1$, $\Omega_\sigma \subset dom\ \varphi_2$,

$$d(x, \Omega_\sigma) \leq \sigma_1\varphi_1(\|x\|) \quad \forall x \in \Omega, \tag{4.5.15}$$

$$d(x, \Omega) \leq \sigma_2\varphi_2(\|x\|) \quad \forall x \in \Omega_\sigma \tag{4.5.16}$$

and

$$\limsup_{\|x\| \to \infty} \Big\{ \frac{\varphi_2(\|x\|)}{\|x\|^2} \mid x \in D(\varphi_2) \Big\} = 0. \tag{4.5.17}$$

Let $\sigma = max\{\sigma_1, \sigma_2\}$. We note the changes in the proof of Theorem 4.5.1 that are brought about by these conditions. Instead of (4.5.6), we have the following estimate:

$$D_1 \leq \alpha\|x_\alpha^\sigma\|\Big(\|\bar{x}^*\| + \sigma_1\varphi_1(\|\bar{x}^*\|)\Big) + \sigma_1\|y_\alpha^\sigma - f\|_*\varphi_1(\|\bar{x}^*\|).$$

Therefore, if $\sigma < \dfrac{r_0}{\varphi_1(\|\bar{x}^*\|)}$ then

$$\|y_\alpha^\sigma - f\|_* \leq M_1\|x_\alpha^\sigma\| + M_2,$$

where

$$M_1 \geq \frac{\alpha\|\bar{x}^*\| + \alpha\sigma_1\varphi_1(\|\bar{x}^*\|) + c_0}{r_0 - \sigma_1\varphi_1(\|\bar{x}^*\|)}$$

and

$$M_2 \geq \frac{c_0(\|\bar{x}^*\| + r_0)}{r_0 - \sigma_1\varphi_1(\|\bar{x}^*\|)}.$$

By (4.5.15) and (4.5.16), there exist $z_\alpha^\sigma \in \Omega_\sigma$ and $u_\alpha^\sigma \in \Omega$ such that

$$\|z_\alpha^\sigma - x_\alpha\| \leq \sigma_1\varphi_1(\|x_\alpha\|)$$

and

$$\|u_\alpha^\sigma - x_\alpha^\sigma\| \leq \sigma_2\varphi_2(\|x_\alpha^\sigma\|).$$

Therefore the quadratic inequality (4.5.10) for $\|x_\alpha^\sigma\|$ is established in the following form:

$$
\begin{aligned}
\|x_\alpha^\sigma\|^2 \quad &- \quad 2\|x_\alpha^\sigma\|\Big(\|x_\alpha\| + \frac{\sigma_1}{2\alpha}\varphi_1(\|x_\alpha\|)M_1 + \frac{\sigma_1}{2}\varphi_1(\|x_\alpha\|)\Big)\\
&+ \quad \|x_\alpha\|^2 - \frac{\sigma_2}{\alpha}\varphi_2(\|x_\alpha^\sigma\|)C_1 - \frac{\sigma_1}{\alpha}\varphi_1(\|x_\alpha\|)M_2 \leq 0.
\end{aligned}
$$

Taking into account the condition (4.5.17), we ascertain the boundedness of $\|x_\alpha^\sigma\|$ as $\alpha \to 0$. Now it is not difficult to complete the proof of Theorem 4.5.1 by making use of assumptions (4.5.15) - (4.5.17). Note that if the condition (4.5.17) is not fulfilled then, in place of J in the regularized inequality (4.5.2), we should apply the duality mapping J^μ with a gauge function $\mu(t)$ (see Section 4.1, 4.2). Furthermore, Lemma 1.5.14 shows that $\bar{x}^*$ in (4.5.4) may be replaced by arbitrary fixed element $x \in \Omega$.

4.6   Variational Inequalities with Non-Monotone Perturbations

Let $\Re = (0, \delta^*] \times (0, h^*] \times (0, \sigma^*]$, X be a reflexive strictly convex Banach space together with its dual space X^*, A and A^h be demicontinuous operators, $A : X \to X^*$ be monotone, Ω and Ω_σ be convex and closed sets for all $\sigma \in (0, \sigma^*]$. In the sequel, we do not consider that $A^h : X \to X^*$ are necessarily monotone. However, we assume that

$$\|Ax - A^h x\| \leq g(\|x\|)h \quad \forall x \in \Omega \cup \Omega_\sigma, \quad \forall h \in \Re, \qquad (4.6.1)$$

where $g(t)$ is a non-negative and increasing function for all $t \geq 0$. Let the conditions (4.1.5), (4.2.29) and (4.2.40) be fulfilled. In this case, it is unknown if the regularized variational inequality (4.2.41) is solvable. Therefore, define approximations to a solution of (7.1.1) by the variational inequality with small offset:

$$\langle A^h x + \alpha J^\mu x - f^\delta, z - x \rangle \geq -\epsilon g(\|x\|)\|z - x\| \quad \forall z \in \Omega_\sigma, \quad x \in \Omega_\sigma, \qquad (4.6.2)$$

where $\epsilon \geq h$ and there exists $t_0 > 0$ such that a gauge function $\mu(t) \geq max\ \{g(t), a(t), \kappa(t)\}$ for all $t \geq t_0$.

Lemma 4.6.1 *A variational inequality (4.6.2) has at least one solution.*

   Proof. Using the condition (4.6.1), it is enough to make certain that a solution of (4.2.41) with exact operator A satisfies the variational inequality (4.6.2). Observe that the solvability of this inequality follows from Theorem 1.11.9 (see also Lemma 2.6.3). ∎

   Let $\gamma = (\delta, h, \sigma) \in \Re$ and x_α^γ be one of solutions of (4.6.2), that is,

$$\langle A^h x_\alpha^\gamma + \alpha J^\mu x_\alpha^\gamma - f^\delta, z - x_\alpha^\gamma \rangle \geq -\epsilon g(\|x_\alpha^\gamma\|)\|z - x_\alpha^\gamma\| \quad \forall z \in \Omega_\sigma. \qquad (4.6.3)$$

Definition 4.6.2 *It is said that an operator* $A : X \to X^*$ *has S-property if the weak convergence* $x_n \rightharpoonup x$ *and convergence* $\langle Ax_n - Ax, x_n - x \rangle \to 0$ *imply the strong convergence* $x_n \to x$ *as* $n \to \infty$.

Theorem 4.6.3 *Assume that the conditions of the present section are satisfied and, in addition, operator A has the S-property, (4.2.1) and (4.6.1) hold and a solution set N of (7.1.1) is not empty. Let*

$$\lim_{\alpha \to 0} \frac{\delta + \epsilon + \sigma}{\alpha} = 0.$$

Then $\{x_\alpha^\gamma\}$ converges strongly to the minimal norm solution $\bar{x}^* \in N$.

Proof. As in the proof of Theorem 4.2.1, by (4.1.3) and (4.6.3), we obtain

$$\langle A^h x_\alpha^\gamma + \alpha J^\mu x_\alpha^\gamma - f^\delta, v^\sigma - x_\alpha^\gamma \rangle \quad + \quad \langle A x^* - f, u_\alpha^\gamma - x^* \rangle$$

$$\geq \quad -\epsilon g(\|x_\alpha^\gamma\|)\|x_\alpha^\gamma - v^\sigma\|, \tag{4.6.4}$$

where $x^* \in N$, $v^\sigma \in \Omega_\sigma$, $u_\alpha^\gamma \in \Omega$,

$$\|x^* - v^\sigma\| \leq a(\|x^*\|)\sigma \tag{4.6.5}$$

and

$$\|u_\alpha^\gamma - x_\alpha^\gamma\| \leq a(\|x_\alpha^\gamma\|)\sigma. \tag{4.6.6}$$

The monotonicity property of A and inequalities (4.1.5), (4.6.1) and (4.6.4) yield the relation

$$\alpha \langle J^\mu x_\alpha^\gamma, x_\alpha^\gamma - v^\sigma \rangle \quad \leq \quad \Big((h + \epsilon)g(\|x_\alpha^\gamma\|) + \delta\Big)\|x_\alpha^\gamma - x^*\|$$

$$+ \quad \langle A x^* - f, u_\alpha^\gamma - x_\alpha^\gamma \rangle + \langle A^h x_\alpha^\gamma - f^\delta, v^\sigma - x^* \rangle$$

$$+ \quad \epsilon g(\|x_\alpha^\gamma\|)\|v^\sigma - x^*\| \quad \forall x^* \in N. \tag{4.6.7}$$

By (4.6.5), (4.6.6) and (4.2.40), one has

$$\mu(\|x_\alpha^\gamma\|)\|x_\alpha^\gamma\| \quad \leq \quad \Big(\frac{h + \epsilon}{\alpha}g(\|x_\alpha^\gamma\|) + \frac{\delta}{\alpha}\Big)\|x_\alpha^\gamma - x^*\| + \mu(\|x_\alpha^\gamma\|)\Big(\|x^*\| + a(\|x^*\|)\sigma\Big)$$

$$+ \quad \frac{\sigma}{\alpha}\Big(\kappa(\|x_\alpha^\gamma\|)a(\|x^*\|) + \kappa(\|x^*\|)a(\|x_\alpha^\gamma\|) + \epsilon g(\|x_\alpha^\gamma\|)a(\|x^*\|)\Big).$$

Since $\dfrac{\epsilon}{\alpha} \to 0$ as $\alpha \to 0$ (and consequently, $\dfrac{h}{\alpha} \to 0$) and since a gauge function $\mu(t) \geq max\ \{g(t), a(t), \kappa(t)\}$, we conclude from the last inequality that solutions x_α^γ are bounded. After this, the weak convergence $x_\alpha^\gamma \rightharpoonup \bar{x} \in \Omega$ is established as in Theorem 4.2.1.

We prove the strong convergence of $\{x_\alpha^\gamma\}$ to $\bar{x}$. The monotonicity of A and J^μ implies

$$0 \quad \leq \quad \langle A x_\alpha^\gamma - A\bar{x}, x_\alpha^\gamma - \bar{x} \rangle \leq \langle A x_\alpha^\gamma + \alpha J^\mu x_\alpha^\gamma - A\bar{x} - \alpha J^\mu \bar{x}, x_\alpha^\gamma - \bar{x} \rangle$$

$$= \quad \langle A x_\alpha^\gamma + \alpha J^\mu x_\alpha^\gamma, x_\alpha^\gamma - \bar{x} \rangle - \langle A\bar{x} + \alpha J^\mu \bar{x}, x_\alpha^\gamma - \bar{x} \rangle. \tag{4.6.8}$$

In view of the weak convergence of $\{x_\alpha^\gamma\}$ to $\bar{x}$, we have

$$\lim_{\alpha \to 0} \langle A\bar{x} + \alpha J^\mu \bar{x}, x_\alpha^\gamma - \bar{x} \rangle = 0.$$

By virtue of (4.6.1), the next to last term in (4.6.8) admits the following estimate:

$$\langle Ax_\alpha^\gamma + \alpha J^\mu x_\alpha^\gamma, x_\alpha^\gamma - \bar{x} \rangle \le \langle A^h x_\alpha^\gamma + \alpha J^\mu x_\alpha^\gamma, x_\alpha^\gamma - \bar{x} \rangle + hg(\|x_\alpha^\gamma\|)\|x_\alpha^\gamma - \bar{x}\|. \qquad (4.6.9)$$

Using further (4.6.3) for all $\bar{x}_\sigma \in \Omega_\sigma$ such that $\|\bar{x}_\sigma - \bar{x}\| \to 0$ as $\sigma \to 0$, we deduce

$$\langle A^h x_\alpha^\gamma \ + \ \alpha J^\mu x_\alpha^\gamma, x_\alpha^\gamma - \bar{x} \rangle = \langle A^h x_\alpha^\gamma + \alpha J^\mu x_\alpha^\gamma - f^\delta, x_\alpha^\gamma - \bar{x}_\sigma \rangle$$

$$+ \ \langle f^\delta, x_\alpha^\gamma - \bar{x}_\sigma \rangle + \langle A^h x_\alpha^\gamma + \alpha J^\mu x_\alpha^\gamma, \bar{x}_\sigma - \bar{x} \rangle$$

$$\le \ \epsilon g(\|x_\alpha^\gamma\|)\|x_\alpha^\gamma - \bar{x}_\sigma\| + \langle f^\delta, x_\alpha^\gamma - \bar{x} \rangle$$

$$+ \ \langle A^h x_\alpha^\gamma + \alpha J^\mu x_\alpha^\gamma, \bar{x}_\sigma - \bar{x} \rangle. \qquad (4.6.10)$$

In view of (4.2.40), it results that $\{Ax_\alpha^\gamma\}$ is bounded together with bounded $\{x_\alpha^\gamma\}$. Since $\bar{x}_\sigma \to \bar{x}$ and $x_\alpha^\gamma \rightharpoonup \bar{x}$, we have from (4.6.10) the limit inequality

$$\limsup_{\alpha \to 0} \langle A^h x_\alpha^\gamma + \alpha J^\mu x_\alpha^\gamma, x_\alpha^\gamma - \bar{x} \rangle \le 0.$$

Now we can conclude, by (4.6.8) and (4.6.9), that

$$\lim_{\alpha \to 0} \langle Ax_\alpha^\gamma - A\bar{x}, x_\alpha^\gamma - \bar{x} \rangle = 0.$$

Finally, the S-property of A implies the strong convergence of $\{x_\alpha^\gamma\}$ to $\bar{x} \in \Omega$.

Show that $\bar{x} \in N$. If $z \in \Omega$ then there exists a sequence $\{z_\sigma\} \in \Omega_\sigma$ such that $\{z_\sigma\} \to z$ as $\sigma \to 0$. Put in (4.6.3) $z = z_\sigma$ and take into account (4.6.1). We obtain

$$\langle Ax_\alpha^\gamma + \alpha J^\mu x_\alpha^\gamma - f^\delta, z_\sigma - x_\alpha^\gamma \rangle \ge -(\epsilon + h)g(\|x_\alpha^\gamma\|)\|z_\sigma - x_\alpha^\gamma\|.$$

Passing to the limit as $\alpha \to 0$, provided that A is demicontinuous, one gets

$$\langle A\bar{x} - f, z - \bar{x} \rangle \ge 0 \quad \forall z \in \Omega, \quad \bar{x} \in \Omega.$$

This means that $\bar{x} \in N$.

Prove that $\bar{x} = \bar{x}^*$. Applying the monotonicity property of J^μ, we rewrite (4.6.7) as

$$\langle J^\mu x^*, x_\alpha^\gamma - x^* \rangle \ \le \ \left(\frac{h+\epsilon}{\alpha} g(\|x_\alpha^\gamma\|) + \frac{\delta}{\alpha} \right)\|x_\alpha^\gamma - x^*\|$$

$$+ \ \frac{\sigma}{\alpha}\Big(\kappa(\|x_\alpha^\gamma\|)a(\|x^*\|) + \kappa(\|x^*\|)a(\|x_\alpha^\gamma\|) + \epsilon g(\|x_\alpha^\gamma\|)a(\|x^*\|) \Big)$$

$$+ \ \mu(\|x_\alpha^\gamma\|)a(\|x^*\|)\sigma \quad \forall x^* \in N.$$

If $\alpha \to 0$ then

$$\langle J^\mu x^*, \bar{x} - x^* \rangle \le 0 \quad \forall x^* \in N.$$

This implies the equality $\bar{x} = \bar{x}^*$. ∎

Remark 4.6.4 *If operators $A : X \to 2^{X^*}$ and $A^h : X \to 2^{X^*}$ are not single-valued but at the same time A is maximal monotone, the value set of A^h at every point of $D(A^h)$ is convex and closed (it occurs, for instance, if A^h is either semimonotone or pseudomonotone), then the conclusion of Theorem 4.6.3 is valid provided that the inequality (4.6.1) is replaced by (4.1.6).*

Remark 4.6.5 *If the sequence*

$$\left\{ \frac{\delta + \epsilon + \sigma}{\alpha} \right\}$$

is bounded as $\alpha \to 0$, say by C, and $\mu(t) \geq C \max \{g(t),\ \alpha(t),\ \kappa(t)\}$ for $t \geq t_0$, then Theorem 4.6.3 asserts that every strong limit of $\{x_\alpha^\gamma\}$ belongs to N.

4.7 Variational Inequalities with Mosco-Approximation of the Constraint Sets

We study in this section the regularization method for variational inequalities with s-w-demiclosed operators and Mosco-approximation of the constraint sets. Let X be a reflexive strictly convex Banach space together with its dual space X^*.

Definition 4.7.1 *An operator $A : X \to 2^{X^*}$ is called strongly-weakly demiclosed (s-w-demiclosed for short) on a set $\Omega \subseteq D(A)$ if for any sequences $\{z_n\}$ and $\{\xi_n\}$ such that $z_n \in \Omega$, $\xi_n \in Az_n$, $z_n \to z$, $\xi_n \rightharpoonup \xi$, it follows that $\xi \in Az$.*

Definition 4.7.2 *An operator $A : X \to 2^{X^*}$ is called weakly-strongly demiclosed (w-s-demiclosed for short) on a set $\Omega \subseteq D(A)$ if for any sequences $\{z_n\}$ and $\{\xi_n\}$ such that $z_n \in \Omega$, $\xi_n \in Az_n$, $z_n \rightharpoonup z$, $\xi_n \to \xi$, it follows that $\xi \in Az$.*

Definition 4.7.3 *An operator $A : X \to 2^{X^*}$ is called demiclosed on a set $\Omega \subseteq D(A)$ if it is s-w-demiclosed and w-s-demiclosed at the same time.*

By Lemma 1.4.5, any maximal monotone operator is demiclosed.

The following proposition is proved similarly to Theorem 1.4.7.

Proposition 4.7.4 *Let $A : X \to 2^{X^*}$ be a monotone s-w-demiclosed operator in $D(A)$. Suppose that $\Omega \subseteq int\ D(A)$ and that image Az is a nonempty closed convex subset of X^* at each point $z \in \Omega$. Let $x_0 \in \Omega$. If the inequality*

$$\langle \xi - f, z - x_0 \rangle \geq 0 \quad \forall z \in \Omega, \quad \forall \xi \in Az, \tag{4.7.1}$$

holds then $f \in Ax_0$.

Proof. Suppose, by contradiction, that $f \notin Ax_0$. Then according to the strong separation theorem, there exists $y \in X$ such that

$$\langle f, y \rangle > sup \{ \langle g, y \rangle \mid \forall g \in Ax_0 \} \qquad (4.7.2)$$

because Ax_0 is a convex and closed set. Since $x_0 \in int\ \Omega$, there exists $\bar{t} > 0$ such that

$$y_t = x_0 + ty \in \Omega \quad \forall t \in [0, \bar{t}].$$

Obviously, the sets Ay_t are not empty for each $t \in [0, \bar{t}]$, therefore, there exist a vector $g_t \in Ay_t$. If $t \to 0$ then $y_t \to x_0$. By virtue of the local boundedness of A at any interior point of $D(A)$, we conclude that $g_t \rightharpoonup \bar{g} \in X^*$, where $\bar{g}$ depends on y. Since A is demiclosed, $\bar{g} \in Ax^0$. This fact and (4.7.2) imply

$$\langle f, y \rangle > sup \{ \langle g, y \rangle \mid \forall g \in Ax_0 \} \geq \langle \bar{g}, y \rangle. \qquad (4.7.3)$$

On the other hand, by (4.7.1),

$$\langle g_t - f, y \rangle \geq 0 \quad \forall g_t \in Ay_t.$$

Setting $t \to 0$, one gets

$$\langle \bar{g} - f, y \rangle \geq 0,$$

that is,

$$\langle \bar{g}, y \rangle \geq \langle f, y \rangle$$

which contradicts (4.7.3). ∎

In the sequel, operator A will be called convex-valued on Ω if its image Az is a nonempty convex subset of X^* at each point $z \in \Omega$. Assume that $\Omega \subseteq D(A)$ is a convex closed subset and $f \in X^*$. Consider the variational inequality problem with s-w-demiclosed operator A : To find $x \in \Omega$ such that

$$\langle Ax - f, z - x \rangle \geq 0 \quad \forall z \in \Omega. \qquad (4.7.4)$$

A solution x^0 of (4.7.4) is understood in the sense of Definition 1.11.1, that is, there exists $\zeta^0 \in Ax^0$ such that

$$\langle \zeta^0 - f, z - x^0 \rangle \geq 0 \quad \forall z \in \Omega. \qquad (4.7.5)$$

Next we present the Minty–Browder type lemma for variational inequalities with s-w-demiclosed operators.

Lemma 4.7.5 *Assume that $A : X \to 2^{X^*}$ is a monotone operator, $\Omega \subseteq D(A)$ is a nonempty convex and closed set and $x^0 \in \Omega$.*
(i) If x^0 is a solution of the variational inequality (4.7.4) defined by (4.7.5), then the inequality

$$\langle \zeta - f, z - x^0 \rangle \geq 0 \quad \forall z \in \Omega, \quad \forall \zeta \in Az, \qquad (4.7.6)$$

holds.
(ii) If $\Omega \subseteq int\ D(A)$ and if the operator A is s-w-demiclosed and convex valued, then the converse implication is also true.

Proof. The claim (i) has been established in Lemma 1.11.3. We prove (ii). First of all, we show that for any maximal monotone extension $\bar{A}$ of A, one has $\bar{A}x = Ax$ for any $x \in \Omega$. To this end, let $x \in \Omega$ and observe that, since the operator A is monotone and $x \in int\, D(A)$, A is locally bounded at x. Hence, if $\{x_n\}$ is a sequence in X such that $\lim_{n \to \infty} x_n = x$, and if $\{\xi^n\}$ is a sequence such that $\xi^n \in Ax_n$ for all $n \geq 1$, then $\{\xi^n\}$ is bounded. It has weak accumulation points because X^* is reflexive. From s-w-demiclosedness of A it follows that any such point belongs to Ax.

Denote by Rx the closed convex hull of the set of weak accumulation points of all sequences $\{\xi^n\}$ as described above. The set Ax is convex, therefore, it is closed by the reason of the demiclosedness of A on Ω. It results from this that $Rx \subseteq Ax$. Obviously, $Ax \subseteq \bar{A}x$. Thus, $Rx \subseteq \bar{A}x$.

We claim that the inclusion $Rx \supseteq \bar{A}x$ holds too. Suppose, by contradiction, that this is not the case. Then there exists $\eta \in \bar{A}x$ such that $\eta \notin Rx$. According to the strong separation theorem, there is a vector $w \in X$ satisfying the inequality

$$\langle \xi - \eta, w \rangle < 0 \quad \forall \xi \in Rx. \tag{4.7.7}$$

Similarly to the proof of Proposition 4.7.4, we can show that the condition $x \in int\, D(A)$ and monotonicity property of $\bar{A}$ imply existence of the weak accumulation point $\bar{\xi} \in Ax$ of the sequence $\{\zeta_n\}$, $\zeta_n \in Ax_n$, $x_n \to x$, such that

$$\langle \bar{\xi} - \eta, w \rangle \geq 0.$$

Clearly, $\bar{\xi} \in Rx$ and this contradicts (4.7.7). Hence, $Ax = Rx = \bar{A}x$.

So, $Ax = \bar{A}x$ for any $x \in \Omega$. Therefore, variational inequality (4.7.4) and variational inequality

$$\langle \bar{A}x - f, z - x \rangle \geq 0 \quad \forall z \in \Omega, \quad x \in \Omega \tag{4.7.8}$$

have the same solution set. The maximal monotone operator $\bar{A}$ also satisfies the requirements of Lemma 1.11.4. Consequently, x^0 is a solution of (4.7.8). Therefore, it is a solution of (4.7.4) too. The lemma is proved. ∎

Definition 4.7.6 *Let in (4.7.4) A be a monotone operator and let $\bar{A}$ be a maximal monotone extension of A. A solution of the variational inequality*

$$\langle \bar{A}x - f, z - x \rangle \geq 0 \quad \forall z \in \Omega, \quad x \in \Omega \tag{4.7.9}$$

(in the sense of Definition 1.11.1) is called a generalized solution of (4.7.4).

Lemma 4.7.7 *If the operator $A : X \to 2^{X^*}$ is monotone, if $\Omega \subseteq int\, D(A)$ is a nonempty convex and closed set, and if A is s-w-demiclosed and convex-valued on Ω, then any generalized solution of (4.7.4) is its solution in the sense of (4.7.5), and vice versa.*

Proof. Suppose that x^0 is a generalized solution of (4.7.4). Then for some maximal monotone extension $\bar{A}$ of A and for some $\bar{\zeta}^0 \in \bar{A}x^0$ we have

$$\langle \bar{\zeta}^0 - f, z - x^0 \rangle \geq 0 \quad \forall z \in \Omega.$$

According to Lemma 1.11.3 applied to $\bar{A}$, it results that for each $z \in \Omega$ and for any $\zeta \in \bar{A}z$,

$$\langle \zeta - f, z - x^0 \rangle \geq 0.$$

In particular, the last holds for each $z \in \Omega$ and for any $\zeta \in Az$ because $Az \subseteq \bar{A}z$. Lemma 4.7.5 asserts now that x^0 is a solution of (4.7.4).

The inverse proposition follows from Definition 1.11.1. Namely, if there exists $\zeta^0 \in Ax^0$ such that

$$\langle \zeta^0 - f, z - x^0 \rangle \geq 0 \quad \forall z \in \Omega,$$

then the same inequality holds with $\zeta^0 = \bar{\zeta}^0 \in \bar{A}x^0$. ∎

Corollary 4.7.8 *Let A be a monotone s-w-demiclosed and convex-valued on Ω operator and $\Omega \subseteq int\, D(A)$. If the generalized solution set N of the variational inequality (4.7.4) is not empty, then it is convex and closed.*

Consider the regularized variational inequality

$$\langle Ax + \alpha Jx - f, z - x \rangle \geq 0 \quad \forall z \in \Omega, \quad x \in \Omega. \tag{4.7.10}$$

Lemma 4.7.9 *If $A : X \to 2^{X^*}$ is a monotone s-w-demiclosed and convex-valued operator on convex closed set $\Omega \subseteq int\, D(A)$, then (4.7.10) has a unique solution for any $\alpha > 0$.*

Proof. The operator $A + \alpha J$ is s-w-demiclosed on Ω and monotone because A and J are so. Let $\bar{A}$ be some maximal monotone extension of A. Then $\bar{A} + \alpha J$ is a maximal monotone extension of $A + \alpha J$. Due to Lemma 4.7.7, the variational inequality (4.7.10) and the variational inequality

$$\langle \bar{A}x + \alpha Jx - f, z - x \rangle \geq 0 \quad \forall z \in \Omega, \quad x \in \Omega \tag{4.7.11}$$

have the same solution sets. But we know that (4.7.11) has a unique solution because of Theorem 1.11.11. ∎

Definition 4.7.10 *Let $\{\alpha_n\}$ be a sequence of positive real numbers converging to zero. The sequence of sets $\{\Omega_n\}$ is said to be fast Mosco-convergent to Ω (fast M-convergent for short) relative to the sequence $\{\alpha_n\}$ if the following conditions are satisfied:*
(j) if $y \in \Omega$ then $\theta_X \in s - \liminf_{n \to \infty} \alpha_{i_n}^{-1}(y - \Omega_n)$;
(jj) if $\{z^n\}$ is a weakly convergent sequence in X such that, for some subsequence $\{\Omega_{i_n}\}$ of $\{\Omega_n\}$, we have $z^n \in \Omega_{i_n}$ for all $n > 0$, then $\theta_X \in w - \limsup_{n \to \infty} \alpha_{i_n}^{-1}(z^n - \Omega)$.

Remark 4.7.11 *A reasoning similar to that involved in the proof of Lemma 1.9 in [154] shows that if $\{\Omega_n\}$ is fast M-convergent to Ω relative to some sequence of positive real numbers $\{\alpha_n\}$ with $\lim_{n \to \infty} \alpha_n = 0$, then the sequence $\{\Omega_n\}$ is M-convergent to Ω.*

Assume that in the variational inequality (4.7.4), instead of A, f and Ω, we have the sequences $\{A^n\}$, $\{f^n\}$ and $\{\Omega_n\}$, $n = 1, 2, ...$, satisfying the following conditions:
1) $A^n : X \to 2^{X^*}$ are monotone s-w-demiclosed and convex-valued on Ω_n operators, and

$$\mathcal{H}_{X^*}(A^n x, Ax) \le h_n g(\|x\|) \quad \forall x \in \Omega_n,$$

where $h_n \ge 0$, $g(t)$ is the non-negative bounded function for all $t \ge 0$;
2) $\|f - f^n\|_* \le \delta_n$, $\quad \delta_n \ge 0$;
3) Ω_n is a convex closed set, $\Omega_n \subseteq int\, D(A) \cap int\, D(A^n)$, a sequence $\{\Omega_n\}$ is M-convergent to Ω.

We introduce the following regularized variational inequality:

$$\langle A^n x + \alpha_n Jx - f^n, z - x \rangle \ge 0 \quad \forall z \in \Omega_n, \quad x \in \Omega_n, \quad \alpha_n > 0. \tag{4.7.12}$$

Under our assumptions, according to Lemma 4.7.9, the inequality (4.7.12) has a unique solution x^n in the sense of (4.7.5). More precisely, there is some element $u^n \in A^n x^n$ such that

$$\langle u^n + \alpha_n Jx^n - f^n, z - x^n \rangle \ge 0 \quad \forall z \in \Omega_n. \tag{4.7.13}$$

Show that the sequence $\{x^n\}$ is bounded. To this end, along with (4.7.12), consider the variational inequality

$$\langle Ax + \alpha_n Jx - f, z - x \rangle \ge 0 \quad \forall z \in \Omega_n, \quad x \in \Omega_n, \tag{4.7.14}$$

and denote its (unique) solution by y^n. This means that there exists $v^n \in Ay^n$ such that

$$\langle v^n + \alpha_n Jy^n - f, z - y^n \rangle \ge 0 \quad \forall z \in \Omega_n. \tag{4.7.15}$$

Lemma 4.7.12 *Let $A : X \to 2^{X^*}$ be a monotone convex-valued and s-w-demiclosed operator on closed convex set $\Omega \subseteq int\, D(A)$. Assume that the conditions 1) - 3) are fulfilled, $\alpha_n > 0$, $n = 1, 2, ...$, $\lim\limits_{n \to \infty} \alpha_n = 0$ and the sequence*

$$\left\{ \frac{\delta_n + h_n}{\alpha_n} \right\} \tag{4.7.16}$$

is bounded as $n \to \infty$. Then the sequence $\{x^n\}$ of solutions to the variational inequality (4.7.12) is bounded if one of the following requirements is satisfied:
a) For each real number $\beta > 0$, the set

$$M_\beta = \{ x \in X \mid \|\xi\|_* \le \beta \|x\| \quad \forall \xi \in Ax \}$$

is bounded;
b) There exists a bounded sequence $\{w^n\}$ such that $w^n \in \Omega_n$ and $\{\alpha_n^{-1} d_*(f^n, Aw^n)\}$ is bounded. Here $d_*(u, G)$ is a distance between an element $u \in X^*$ and set $G \subset X^*$;
c) There exists an element $\bar{x} \in \bigcap_{n=1}^{\infty} \Omega_n$, at which the operator A is coercive with respect to $\bar{x}$, that is, for any $\{\xi^n\}$ and $\{z^n\}$, where $\xi^n \in Az^n$ and $z^n \in D(A)$, we have

$$\lim_{\|z^n\| \to \infty} \frac{\langle \xi^n, z^n - \bar{x} \rangle}{\|z^n\|} = \infty.$$

Proof. First of all, show that if $\{y^n\}$ is bounded then $\{x^n\}$ is also. Really, due to the condition 1), for any $v^n \in Ay^n$, it is possible to find $\tilde{v}^n \in A^n y^n$ such that

$$\|v^n - \tilde{v}^n\|_* \leq h_n g(\|y^n\|). \tag{4.7.17}$$

Presuming $z = y^n$ in (4.7.13) and $z = x^n$ in (4.7.15), and adding the obtained inequalities, we get

$$\langle u^n + \alpha_n J x^n - f^n - v^n - \alpha_n J y^n + f, x^n - y^n \rangle \leq 0.$$

Rewrite it in the equivalent form:

$$\alpha_n \langle J x^n - J y^n, x^n - y^n \rangle \quad + \quad \langle f - f^n, x^n - y^n \rangle + \langle u^n - \tilde{v}^n, x^n - y^n \rangle$$

$$+ \quad \langle \tilde{v}^n - v^n, x^n - y^n \rangle \leq 0. \tag{4.7.18}$$

Since $u^n \in A^n x^n$, $\tilde{v}^n \in A^n y^n$ and A^n is monotone, the last term in the left-hand side of (4.7.18) is non-negative and, therefore, it can be omitted. Taking into account the assumption 2) and applying inequalities (1.5.3) and (4.7.17), we deduce from (4.7.18) the relation

$$(\|x^n\| - \|y^n\|)^2 \leq \frac{\delta_n + h_n g(\|y^n\|)}{\alpha_n}(\|x^n\| + \|y^n\|).$$

It is clear that if $\{y^n\}$ and $\left\{\dfrac{\delta_n + h_n}{\alpha_n}\right\}$ are bounded, then $\{x^n\}$ is bounded too.

Suppose that the condition a) holds. Prove that $\{y^n\}$ is bounded in this case. Indeed, take $z \in \Omega \subseteq int\, D(A)$. By Lemma 1.5.14, there exist $r_0 > 0$ and $c_0 > 0$ such that for any $y \in int\, D(A)$ and for any $v \in Ay$, we have

$$\langle v - f, y - z \rangle \geq r_0 \|v - f\|_* - c_0(\|y - z\| + r_0). \tag{4.7.19}$$

According to the condition 3), the set $\Omega_n \subseteq int\, D(A)$. Therefore, in (4.7.19) we can put $y = y^n \in \Omega_n$ and $v = v^n \in Ay^n$. Then

$$\langle v^n - f, y^n - z \rangle \geq r_0 \|v^n - f\|_* - c_0(\|y^n - z\| + r_0). \tag{4.7.20}$$

Fix any $\bar{z} \in \Omega$. By virtue of M-convergence of $\{\Omega_n\}$ to Ω, we are able to construct a sequence $\{z^n\}$ such that $z^n \in \Omega_n$ and $z^n \to \bar{z}$. Rewrite (4.7.15) with $z = z^n$ as

$$\langle v^n + \alpha_n J y^n - f, z^n - y^n \rangle \geq 0.$$

Then

$$\langle v^n - f, y^n - \bar{z} \rangle = \langle v^n + \alpha_n J y^n - f, y^n - z^n \rangle + \langle v^n + \alpha_n J x^n - f, z^n - \bar{z} \rangle$$

$$- \alpha_n \langle J y^n, y^n - \bar{z} \rangle \leq \langle v^n + \alpha_n J y^n - f, z^n - \bar{z} \rangle + \alpha_n \|y^n\| \|\bar{z}\|$$

$$\leq \|v^n - f\|_* \|z^n - \bar{z}\| + \alpha_n \|y^n\|(\|z^n - \bar{z}\| + \|\bar{z}\|). \tag{4.7.21}$$

Now (4.7.20) with $z = \bar{z}$ and (4.7.21) yield the inequality

$$(r_0 - \|z^n - \bar{z}\|)\|v^n - f\|_* \leq \|y^n\|\Big(\alpha_n(\|z^n - \bar{z}\| + \|\bar{z}\|) + c_0\Big) + c_0(r_0 + \|\bar{z}\|). \qquad (4.7.22)$$

Since $z^n \to \bar{z}$, the estimate $r_0 - \|z^n - \bar{z}\| > \dfrac{r_0}{2}$ holds for a sufficiently large $n > 0$. By (4.7.22), we conclude that there are $c_1 > 0$ and $c_2 > 0$ such that

$$\|v^n\|_* \leq c_1\|y^n\| + c_2. \qquad (4.7.23)$$

Suppose that $\{y^n\}$ is unbounded. Then (4.7.23) implies that

$$\|v^n\|_* \leq \beta\|y^n\|$$

for some $\beta > 0$. This means that y^n belongs to M_β which is bounded according to the condition a). This fact leads to a contradiction. Thus, $\{y^n\}$ is bounded together with $\{x^n\}$.

Assume that the condition b) holds. It is obvious that the sequence $\{w^n\}$ within satisfies the inclusion $A^n w^n \subseteq \bar{A}^n w^n$, where $\bar{A}^n$ is a maximal monotone extension of A^n. We know that the set $\bar{A}^n w^n$ is convex and closed as well. Therefore, there exists $\zeta^n \in \bar{A}^n w^n$ such that

$$\|\zeta^n - f^n\|_* = d_*(f^n, \bar{A}^n w^n) \leq d_*(f^n, A^n w^n). \qquad (4.7.24)$$

If in (4.7.13) we put $z = w^n$, then one gets

$$\langle u^n - \zeta^n, x^n - w^n \rangle + \langle \zeta^n - f^n, x^n - w^n \rangle + \alpha_n \langle Jx^n, x^n - w^n \rangle \leq 0. \qquad (4.7.25)$$

Recall that $\bar{A}^n$ is monotone. Hence, the first term in the left-hand side of (4.7.25) is non-negative. Therefore, by (4.7.24) and (4.7.25), we come to the estimate

$$\|x^n\|^2 \leq \|x^n\|\|w^n\| + \alpha_n^{-1} d_*(f^n, A^n w^n)\left(\|x^n\| + \|w^n\|\right),$$

which gives the boundedness of $\{x^n\}$.

Suppose that the condition c) holds. Since $\bar{x} \in \cap_{n=1}^\infty \Omega_n$, putting $z = \bar{x}$ in (4.7.15), we have

$$\langle v^n - f, y^n - \bar{x} \rangle + \alpha_n\|y^n\|^2 \leq \alpha_n\|y^n\|\|\bar{x}\|.$$

Moreover,

$$\langle v^n - f, y^n - \bar{x} \rangle \leq \alpha_n\|y^n\|\|\bar{x}\|.$$

From this, it is easy to see that

$$\frac{\langle v^n, y^n - \bar{x} \rangle}{\|y^n\|} \leq \|f\|_* + \alpha_n\|\bar{x}\| + \frac{\|f\|_*\|\bar{x}\|}{\|y^n\|}.$$

Furthermore, if we assume that $\{y^n\}$ is unbounded then

$$\lim_{n \to \infty} \frac{\langle v^n, y^n - \bar{x} \rangle}{\|y^n\|} \leq \|f\|_*$$

which contradicts the condition c). Thus, $\{y^n\}$ is bounded together with $\{x^n\}$. It accomplishes the proof of the lemma. ∎

Theorem 4.7.13 *Let X be an E-space, X^* be strictly convex. Assume that an operator $A : X \to 2^{X^*}$ is monotone s-w-demiclosed and convex-valued on a closed convex set $\Omega \subseteq int\ D(A)$, the variational inequality (4.7.4) has a nonempty solution set N, $\{\alpha_n\}$ is a sequence of positive numbers such that $\lim_{n \to \infty} \alpha_n = 0$. Let the conditions 1) - 3) be fulfilled, $\{\Omega_n\}$ be fast M-convergent to Ω relative to the sequence $\{\alpha_n\}$ and*

$$\lim_{n \to \infty} \frac{\delta_n + h_n}{\alpha_n} = 0. \tag{4.7.26}$$

Suppose that one of the conditions a) - c) of Lemma 4.7.12 holds. Then a sequence $\{x^n\}$ generated by (4.7.12) strongly converges in X to the minimal norm solution $\bar{x}^* \in N$ as $n \to \infty$.

Proof. First of all, observe that due to Lemma 4.7.12 $\{x^n\}$ is bounded. Then there exists a weak accumulation point of $\{x^n\}$. Let $x^n \rightharpoonup x^*$ (in reality, some subsequence of $\{x^n\}$ weakly converges to x^*, however, as before, we do not change its denotation). Since $x^n \in \Omega_n$, the fast M-convergence of $\{\Omega_n\}$ to Ω implies inclusion $x^* \in \Omega$ (see Remark 4.7.11). Show that $x^* \in N$. Take an arbitrary element $z \in \Omega$ and fix it. For any $n > 0$, construct $z^n \in \Omega_n$ such that $z^n \to z$ as $n \to \infty$. Presuming z^n in (4.7.13), in place of z, we get for some $u^n \in A^n x^n$

$$\langle u^n + \alpha_n J x^n - f^n, x^n - z^n \rangle \leq 0. \tag{4.7.27}$$

According to the assumption 1), for $u^n \in A^n x^n$ one can find $\tilde{u}^n \in A x^n$ satisfying the estimate

$$\|u^n - \tilde{u}^n\|_* \leq h_n g(\|x^n\|). \tag{4.7.28}$$

Further, take any $u \in Az$. It follows from (4.7.27) that

$$\langle u - f, x^n - z \rangle \quad + \quad \langle u^n + \alpha_n J x^n - f^n, z - z^n \rangle + \langle f - f^n, x^n - z \rangle$$

$$+ \quad \alpha_n \langle J x^n, x^n - z \rangle + \langle u^n - \tilde{u}^n, x^n - z \rangle + \langle \tilde{u}^n - u, x^n - z \rangle \leq 0.$$

By the monotonicity of A, the last term in the left-hand side of the previous inequality can be omitted. Then, owing to the assumption 2) and estimate (4.7.28), we come to the following relation:

$$\langle u - f, x^n - z \rangle \leq \|u^n + \alpha_n J x^n - f^n\|_* \|z^n - z\|$$

$$+ \left(\delta_n + \alpha_n \|x^n\| + h_n g(\|x^n\|) \right) \|x^n - z\|. \tag{4.7.29}$$

Using the boundedness of $\{x^n\}$ and (4.7.19), prove that $\{u^n\}$ is bounded. Indeed, we can write down for some constants $r_0 > 0$ and $c_0 > 0$ the following inequality:

$$\langle \tilde{u}^n - f, x^n - z \rangle \geq r_0 \|\tilde{u}^n - f\|_* - c_0(\|x^n - z\| + r_0), \quad \tilde{u}^n \in A x^n,$$

because of $z \in int\ D(A)$. It is clear that there exists $c_1 > 0$ such that

$$\langle \tilde{u}^n - f, x^n - z \rangle \geq r_0 \|\tilde{u}^n - f\|_* - c_1. \tag{4.7.30}$$

At the same time,

$$\langle \tilde{u}^n - f, x^n - z \rangle \;=\; \langle \tilde{u}^n - u^n, x^n - z \rangle + \langle u^n + \alpha_n J x^n - f^n, x^n - z^n \rangle$$

$$+ \;\; \langle u^n + \alpha_n J x^n - f^n, z^n - z \rangle + \langle f^n - f, x^n - z \rangle - \alpha_n \langle J x^n, x^n - z \rangle$$

$$\leq \;\; \Big(h_n g(\|x^n\|) + \delta_n + \alpha_n \|x^n\| \Big) \|x^n - z\|$$

$$+ \;\; \Big(\|\tilde{u}^n - f\|_* + h_n g(\|x^n\|) + \delta_n \Big) \|z^n - z\|.$$

Therefore, there exists a constant $c_2 > 0$ such that

$$\langle \tilde{u}^n - f, x^n - z \rangle \leq \|\tilde{u}^n - f\|_* \|z^n - z\| + c_2. \tag{4.7.31}$$

Combining (4.7.30) with (4.7.31), one gets

$$r_0 \|\tilde{u}^n - f\|_* \leq \|\tilde{u}^n - f\|_* \|z^n - z\| + c_1 + c_2.$$

By virtue of the strong convergence of $\{z^n\}$ to z, we conclude that there exists $\bar{n} > 0$ such that for any $n \geq \bar{n}$,

$$\|\tilde{u}^n\|_* \leq \|f\|_* + \frac{c_1 + c_2}{r_0 - \|z^n - z\|}.$$

This means that $\{\tilde{u}^n\}$ is bounded. Furthermore, in view of (4.7.28), $\{u^n\}$ is also bounded. Then the boundedness of $\{x^n\}$, its weak convergence to $x^* \in \Omega$ and (4.7.29) imply the inequality

$$\langle u - f, x^* - z \rangle \leq 0 \quad \forall z \in \Omega, \quad \forall u \in Az,$$

because $\alpha_n \to 0$, $\delta_n \to 0$, $h_n \to 0$, $f^n \to f$ and $z^n \to z$ as $n \to \infty$. Since A is s-w-demiclosed and convex-valued, Lemma 4.7.5 allows us to establish that $x^* \in \Omega$ is a solution of (4.7.4).

Under the hypothesis of the theorem, solution set N is convex closed and nonempty and, therefore, it contains the minimal norm element $\bar{x}^*$. We are going to show that the only weak accumulation point of $\{x^n\}$ is $\bar{x}^*$. To this end, we apply the fast Mosco-convergence of $\{\Omega_n\}$ to Ω relative to the sequence $\{\alpha_n\}$. Due to Definition 4.7.10, there exist $\{q^n\}$ and $\{\tilde{q}^n\}$ such that $q^n \in \Omega_n$, $\tilde{q}^n \in \Omega$ and if $n \to \infty$ then

$$\alpha_n^{-1}(q^n - \bar{x}^*) \to 0 \tag{4.7.32}$$

and

$$\left\{ \alpha_n^{-1}(x^n - \tilde{q}^n) \right\} \rightharpoonup 0. \tag{4.7.33}$$

Since $\bar{x}^*$ is the solution of (4.7.4), there is $u^* \in A\bar{x}^*$ such that

$$\langle u^* - f, z - \bar{x}^* \rangle \geq 0 \quad \forall z \in \Omega. \tag{4.7.34}$$

Putting $z = q^n$ in (4.7.13) and $z = \tilde{q}^n$ in (4.7.34) and adding the obtained inequalities, by simple algebra, we have

$$
\begin{aligned}
\|x^n\|^2 \leq\ & \|x^n\|\|\bar{x}^*\| + \alpha_n^{-1}(\|\bar{x}^* - q^n\|\|u^n + \alpha_n Jx^n - f^n\|_* \\
& + \langle u^* - f, \tilde{q}^n - x^n \rangle + \langle \tilde{u}^n - u^n, x^n - \bar{x}^* \rangle \\
& + \langle u^* - \tilde{u}^n, x^n - \bar{x}^* \rangle + \langle f^n - f, x^n - \bar{x}^* \rangle).
\end{aligned}
$$

Further we use the condition 2), monotonicity property of A and (4.7.28) to get

$$
\begin{aligned}
\|x^n\|^2 \leq\ & \|x^n\|\|\bar{x}^*\| + \alpha_n^{-1}\|\bar{x}^* - q^n\|\|u^n + \alpha_n Jx^n - f^n\|_* \\
& + \alpha_n^{-1}\langle u^* - f, \tilde{q}^n - x^n \rangle + \frac{\delta_n + h_n g(\|x^n\|)}{\alpha_n}\|x^n - \bar{x}^*\|.
\end{aligned} \tag{4.7.35}
$$

If $\theta_X \in N$ then $\bar{x}^* = \theta_X$ and it results from (4.7.35), (4.7.26), (4.7.32) and (4.7.33) that $\lim_{n \to \infty} \|x^n\| = 0$. Then the theorem is proved. Let now $\theta_X \notin N$. Then the weak convergence of $\{x^n\}$ to $x^* \in N$ implies

$$
0 < \|x^*\| \leq \liminf_{n \to \infty} \|x^n\|. \tag{4.7.36}
$$

Therefore, one can consider that $\|x^n\| > \mu > 0$. Dividing (4.7.35) on $\|x^n\|$ and passing to the limit as $n \to \infty$, we conclude, according to (4.7.26), (4.7.32) and (4.7.33), that

$$
\limsup_{n \to \infty} \|x^n\| \leq \|\bar{x}^*\|. \tag{4.7.37}
$$

Finally, by (4.7.36) and (4.7.37), we establish the equality $x^* = \bar{x}^*$ and limit relation $\|x^n\| \to \|\bar{x}^*\|$. Hence, $\{x^n\}$ has a unique weak accumulation point $\bar{x}^*$. Since X is an E-space, the whole sequence $\{x^n\}$ strongly converges to $\bar{x}^*$. The proof is completed. ∎

4.8 Variational Inequalities with Hypomonotone Approximations

Let X be a reflexive strictly convex and smooth Banach space, Ω be a closed and convex subset of X, $A : X \to 2^{X^*}$ be a monotone s-w-demiclosed and convex-valued operator on Ω, $\Omega \subseteq int\, D(A)$, $f \in X^*$.

We study the variational inequality (4.7.4) and suppose that it has a nonempty solution set N. As in the previous section, a solution of (4.7.4) is understood in the sense of Definition 1.11.1. We have shown there that in these circumstances a solution of (4.7.4) satisfies the inequality (4.7.6) and vice versa. In addition, each solution of (4.7.4) coincides with a solution of the variational inequality (4.7.9) with maximal monotone extension $\bar{A}$ of A. By Corollary 4.7.8, the set N is convex and closed.

Definition 4.8.1 An operator $A : X \to 2^{X^*}$ is called hypomonotone in $D(A)$ if there exists a constant $c > 0$ such that

$$
\langle u - v, x - y \rangle \geq -c\|x - y\|^2 \quad \forall x,\, y \in D(A), \quad \forall u \in Ax, \quad \forall v \in Ay.
$$

Definition 4.8.2 *An operator* $A : X \to 2^{X^*}$ *is called strongly hypomonotone in* $D(A)$ *if there exists a constant* $c > 0$ *such that*

$$\langle u - v, x - y \rangle \geq -c(\|x\| - \|y\|)^2 \quad \forall x, \, y \in D(A), \quad \forall u \in Ax, \quad \forall v \in Ay.$$

It is clear that the strong hypomonotonicity implies hypomonotonicity.

Introduce $\Re = (0, \delta^*] \times (0, h^*] \times (0, \sigma^*]$ for some positive δ^*, h^* and σ^*. Assume that the perturbed date f^δ, Ω_σ and A^h for the variational inequality (4.7.4) satisfy the following conditions:
1) $\|f - f^\delta\|_* \leq \delta \quad \forall \delta \in (0, \delta^*]$;
2) Ω_σ is a convex and closed set for any $\sigma \in (0, \sigma^*]$;
3) $A^h : X \to 2^{X^*}$ is a s-w-demiclosed operator for any $h \in (0, h^*]$ and convex-valued for any $x \in \Omega_\sigma \subseteq int \, D(A^h)$, there exists $\eta(h) > 0$ such that $\eta(h) \to 0$ as $h \to 0$ and for any $x, y \in \Omega_\sigma$,

$$\langle u^h - v^h, x - y \rangle \geq -\eta(h)(\|x\|| - \|y\|)^2 \quad \forall u^h \in A^h x, \quad \forall v^h \in A^h y; \quad (4.8.1)$$

4) For any $x \in \Omega$, there exist $x_\sigma \in \Omega_\sigma$, $a(t)$ and $g(t)$ such that

$$\|x - x_\sigma\| \leq a(\|x\|)\sigma,$$

and

$$d_*(\zeta, A^h x_\sigma) \leq g(\|\zeta\|_*)\xi(h, \sigma) \quad \forall \zeta \in Ax;$$

5) For any $z_\sigma \in \Omega_\sigma$, there exist $z \in \Omega$ and $b(t)$ such that

$$\|z - z_\sigma\| \leq b(\|z_\sigma\|)\sigma.$$

In 4) - 5), the functions $a(t)$, $b(t)$, $g(t)$ are non-negative and bounded on bounded sets for all $t \geq 0$, and $\xi(h, \sigma) \to 0$ as h, $\sigma \to 0$. Observe that s-w-demiclosedness of A on Ω and A^h on Ω_σ imply the closedness property of Ax and $A^h x$ at each point $x \in \Omega$ and at each point $x \in \Omega_\sigma$, respectively.

Lemma 4.8.3 *For all* $\alpha > \eta(h)$, *the operator* $B = A^h + \alpha J$ *is strictly monotone on* Ω_σ *and coercive relative to any element of* Ω_σ. *In addition,* $Bx = \bar{B}x$ *for all* $x \in \Omega_\sigma$, *where* $\bar{B}$ *is an arbitrary maximal monotone extension of* B.

Proof. Let arbitrary $x, y \in \Omega_\sigma$ and $u^h \in A^h x$, $v^h \in A^h y$. Then by (4.8.1) and by the property (1.5.3) of normalized duality mapping J,

$$\langle u^h + \alpha Jx - v^h - \alpha Jy, x - y \rangle \geq (\alpha - \eta(h))(\|x\|| - \|y\|)^2 \geq 0. \quad (4.8.2)$$

Thus, B is the monotone operator.

Let $x \neq y$. By virtue of (4.8.1), if $\|x\| = \|y\|$ then we have

$$\langle u^h - v^h, x - y \rangle \geq 0.$$

Consequently,

$$\langle u^h + \alpha Jx - v^h - \alpha Jy, x - y \rangle \geq \alpha \langle Jx - Jy, x - y \rangle > 0$$

because J is strictly monotone. Taking into account (4.8.2) we conclude that B is strictly monotone on Ω_σ.

Fix $x_\sigma \in \Omega_\sigma$ and $u^\lambda \in A^h x_\sigma$, where $\lambda = (h, \sigma)$. Let $v \in A^h x$, where $x \in \Omega_\sigma$. Then (4.8.2) allows us to write down the inequality

$$
\begin{aligned}
\langle v + \alpha Jx, x - x_\sigma \rangle \; &= \; \langle v + \alpha Jx - u^\lambda - \alpha Jx_\sigma, x - x_\sigma \rangle \\
&+ \; \langle u^\lambda + \alpha Jx_\sigma, x - x_\sigma \rangle \\
&\geq \; (\alpha - \eta(h))(\|x\| - \|x_\sigma\|)^2 \\
&- \; \|u^\lambda + \alpha Jx_\sigma\|_*(\|x\| + \|x_\sigma\|).
\end{aligned}
$$

It results from this that B is coercive relative to $x_\sigma \in \Omega_\sigma$.

Let $\bar{B}$ be a maximal monotone extension of B. Take $x \in \Omega_\sigma$. Since $\Omega_\sigma \subseteq int\, D(A^h)$, we assert that the monotone operator B is locally bounded at the point x. Therefore, if $x_n \to x$ then $u_n \rightharpoonup u$, where $u_n \in Bx_n$. Further, B is s-w-demiclosed because A^h is s-w-demiclosed and J is demicontinuous. Consequently, $u \in Bx$. The rest of the proof follows the pattern of Lemma 4.7.5. ∎

Consider the problem: To find $x_\alpha^\gamma \in \Omega_\sigma$, where $\gamma = (\delta\,, h,\; \sigma) \in \Re$, satisfying the regularized variational inequality (4.5.2) under the conditions 1) - 5) of this section.

Lemma 4.8.4 *The variational inequality (4.5.2) has a unique solution x_α^γ for any fixed $\gamma \in \Re$.*

Proof. We have shown above that maximal monotone extension $\bar{B}$ of the operator $B = A^h + \alpha J$ is coercive on Ω_σ relative to some point $\tilde{x} \in \Omega_\sigma$. According to Corollary 1.11.13, the variational inequality

$$\langle \bar{B}x - f^\delta, z - x \rangle \geq 0 \quad \forall z \in \Omega_\sigma, \quad x \in \Omega_\sigma,$$

has at least one solution. However, this inequality coincides with (4.5.2) because $\bar{B}x = Bx$ for all $x \in \Omega_\sigma$. Uniqueness of the solution of (4.5.2) results from the strict monotonicity of B proved in Lemma 4.8.3. ∎

Theorem 4.8.5 *Suppose that X is an E-space, X^* is strictly convex, $A : X \to 2^{X^*}$ is a monotone s-w-demiclosed operator and convex-valued on $\Omega \subseteq int\, D(A)$, Ω is a convex closed set, variational inequality (4.7.4) has a nonempty solution set N, approximation data f^δ, A^h and Ω_σ satisfy the conditions 1) - 5) and, in addition,*

$$\text{a)} \quad \lim_{\alpha \to 0} \frac{\delta + \eta(h) + \sigma + \xi(h, \sigma)}{\alpha} = 0,$$

$$\text{b)} \quad \limsup_{t \to \infty} \frac{b(t)}{t^2} \leq Q, \quad 0 < Q < \infty.$$

Then a solution sequence $\{x_\alpha^\gamma\}$ of the variational inequality (4.5.2) strongly converges to the minimal norm solution $\bar{x}^*$ of (4.7.4) as $\alpha \to 0$.

Proof. If $x^* \in N$ then there is $\zeta^* \in Ax^*$ such that

$$\langle \zeta^* - f, z - x^* \rangle \geq 0 \quad \forall z \in \Omega. \tag{4.8.3}$$

Since $x_\alpha^\gamma \in \Omega_\sigma$ is a solution of (4.5.2), there exists $\zeta_\alpha^\gamma \in A^h x_\alpha^\gamma$ satisfying the inequality

$$\langle \zeta_\alpha^\gamma + \alpha J x_\alpha^\gamma - f^\delta, z - x_\alpha^\gamma \rangle \geq 0 \quad \forall z \in \Omega_\sigma. \tag{4.8.4}$$

By virtue of the condition 4) for $x = x^* \in N$, we are able to find $x_\sigma \in \Omega_\sigma$ and $\zeta^\lambda \in A^h x_\sigma$ such that for any $\lambda = (h, \sigma)$,

$$\|x^* - x_\sigma\| \leq a(\|x^*\|)\sigma \tag{4.8.5}$$

and

$$\|\zeta^* - \zeta^\lambda\|_* \leq g(\|\zeta^*\|_*)\xi(h, \sigma). \tag{4.8.6}$$

In its turn, by virtue of the condition 5), for $x_\alpha^\gamma \in \Omega_\sigma$ there exists $y_\alpha^\gamma \in \Omega$ such that

$$\|x_\alpha^\gamma - y_\alpha^\gamma\| \leq b(\|x_\alpha^\gamma\|)\sigma. \tag{4.8.7}$$

Add (4.8.3) and (4.8.4) putting there $z = y_\alpha^\gamma$ and in $z = x_\sigma$, respectively. Then we obtain

$$\langle \zeta_\alpha^\gamma + \alpha J x_\alpha^\gamma - f^\delta, x_\alpha^\gamma - x_\sigma \rangle + \langle \zeta^* - f, x^* - y_\alpha^\gamma \rangle \leq 0.$$

We rewrite this inequality in the following equivalent form:

$$\begin{aligned} 0 \;\geq\;& \alpha \langle J x_\alpha^\gamma, x_\alpha^\gamma - x_\sigma \rangle + \langle \zeta_\alpha^\gamma - \zeta^\lambda, x_\alpha^\gamma - x_\sigma \rangle + \langle \zeta^\lambda - \zeta^*, x_\alpha^\gamma - x_\sigma \rangle \\[2mm] &+ \;\langle \zeta^* - f, x^* - y_\alpha^\gamma + x_\alpha^\gamma - x_\sigma \rangle + \langle f - f^\delta, x_\alpha^\gamma - x_\sigma \rangle. \end{aligned}$$

Taking into account the hypomonotonicity property (4.8.1) and the estimates (4.8.5) - (4.8.7), we deduce

$$\begin{aligned} \|x_\alpha^\gamma\|^2 \;\leq\;& \|x_\alpha^\gamma\| \Big(\|x^*\| + a(\|x^*\|)\sigma \Big) \\[2mm] &+ \; \Big(\frac{\delta}{\alpha} + \frac{\eta(h)}{\alpha} + \frac{\xi(h,\sigma)}{\alpha} g(\|\zeta^*\|_*) \Big) \Big(\|x_\alpha^\gamma\| + \|x^*\| + a(\|x^*\|)\sigma \Big) \\[2mm] &+ \; \frac{\sigma}{\alpha} \|\zeta^* - f\|_* \Big(a(\|x^*\|) + b(\|x_\alpha^\gamma\|) \Big). \end{aligned}$$

The conditions a) and b) of the theorem imply the boundedness of $\{x_\alpha^\gamma\}$ as $\alpha \to 0$. The rest of the proof follows the pattern of Theorem 4.3.1. ∎

If the function $b(t)$ does not possess the condition b), then in place of normalized duality mapping J, it should be used in (4.5.2) duality mapping J^μ with a suitable gauge function $\mu(t)$.

4.9   Variational Inequalities with Pseudomonotone Operators

Let X be a reflexive strictly convex Banach space together with its dual space X^*, $A : X \to X^*$ be a pseudomonotone operator, Ω be a convex closed set in $D(A)$. We study in X the variational inequality (4.7.4) with such operator A. If A is coercive and $\theta_X \in \Omega$ then, in accordance with Theorem 1.13.10, (4.7.4) has a nonempty solution set N.

Lemma 4.9.1 N *is a weakly closed set.*

Proof. Let $x_n \in N$ and $x_n \rightharpoonup x$. Due to the Mazur theorem, $x \in \Omega$. Show that $x \in N$. Since x_n is a solution of (4.7.4), we have

$$\langle Ax_n - f, x_n - z \rangle \leq 0 \quad \forall z \in \Omega. \tag{4.9.1}$$

Assuming in (4.9.1) $z = x$, one gets

$$\langle Ax_n - f, x_n - x \rangle \leq 0.$$

Consequently,

$$\limsup_{n \to \infty} \langle Ax_n - f, x_n - x \rangle \leq 0.$$

The last implies

$$\limsup_{n \to \infty} \langle Ax_n, x_n - x \rangle \leq 0$$

because

$$\langle f, x_n - x \rangle \to 0 \quad \text{as} \quad n \to \infty.$$

The pseudomonotonicity of A allows us to write down the following relation:

$$\liminf_{n \to \infty} \langle Ax_n, x_n - z \rangle \geq \langle Ax, x - z \rangle \quad \forall z \in \Omega.$$

By the equality

$$\lim_{n \to \infty} \langle f, x_n - z \rangle = \langle f, x - z \rangle \quad \forall z \in \Omega,$$

we obtain

$$\liminf_{n \to \infty} \langle Ax_n - f, z - x_n \rangle \leq \langle Ax - f, z - x \rangle \quad \forall z \in \Omega. \tag{4.9.2}$$

In addition, (4.9.1) yields the limit estimate

$$\liminf_{n \to \infty} \langle Ax_n - f, z - x_n \rangle \geq 0 \quad \forall z \in \Omega,$$

which together with (4.9.2) imply (4.7.4). Thus, we have proved that $x \in N$. ∎

Let f, Ω and A be given approximately, that is, in reality, perturbed data f^δ, Ω_σ and A^h are known such that for any $\gamma = (\delta, h, \sigma) \in \Re$, where $\Re = (0, \delta^*] \times (0, h^*] \times (0, \sigma^*]$ with some positive δ^*, h^* and σ^*, there hold:
1) $\|f - f^\delta\|_* \leq \delta$;
2) $\Omega_\sigma \subset D(A^h)$ are convex and closed sets and $\theta_X \in \Omega_\sigma$;

3) $\{A^h\}$ is the sequence of pseudomonotone operators acting from X to X^*, single-valued on Ω_σ and satisfying the condition

$$\|A^h x - A x\|_* \leq h g(\|x\|) \quad \forall x \in \Omega_\sigma, \tag{4.9.3}$$

where the function $g(t)$ is non-negative, non-decreasing and bounded on each bounded interval of $t \geq 0$;

4) $\{\Omega_\sigma\}$ Mosco-approximates Ω. In other words, every point $x \in \Omega$ is a strong limit of some sequence $\{x_\sigma\}$, $x_\sigma \in \Omega_\sigma$, as $\sigma \to 0$, and every weak limit point $\tilde{x}$ of any subsequence $\{\tilde{u}_\sigma\} \subset \Omega_\sigma$, belongs to Ω.

We will find approximation solutions $x_\alpha^\gamma \in \Omega_\sigma$ from the regularized variational inequality (4.5.2) again. By Lemma 1.13.6, the mapping $B = A^h + \alpha J$ is pseudomonotone as the sum of two pseudomonotone operators A^h and αJ. Observe further that, under our hypothesis, Theorem 1.13.10 guarantees solvability of (4.5.2) if B is coercive on Ω_σ. The following lemma forms the sufficient condition for B to be coercive. Denote

$$G = \bigcup_{\sigma \in (0.\sigma^*]} \Omega_\sigma \subset D(A). \tag{4.9.4}$$

Lemma 4.9.2 *If operator A is coercive on each set Ω_σ, and there exists a function $c(t) \to \infty$ as $t \to \infty$ such that*

$$\frac{\langle A x, x \rangle}{\|x\|} \geq c(\|x\|) \quad \forall x \in G, \tag{4.9.5}$$

and if

$$\lim_{t \to \infty} \frac{c(t)}{g(t)} \geq c_0 > 0, \tag{4.9.6}$$

then for all $\alpha > 0$ the operator $B = A^h + \alpha J$ is coercive on Ω_σ.

Proof. Indeed, using (4.9.3) and (4.9.5), we deduce for any $x \in \Omega_\sigma$,

$$
\begin{aligned}
\liminf_{\|x\| \to \infty} \frac{\langle A^h x, x \rangle}{\|x\|} &= \liminf_{\|x\| \to \infty} \left[\frac{\langle A^h x - A x, x \rangle}{\|x\|} + \frac{\langle A x, x \rangle}{\|x\|} \right] \\
&\geq \liminf_{\|x\| \to \infty} \Big(c(\|x\|) - h g(\|x\|) \Big) \\
&= \liminf_{\|x\| \to \infty} g(\|x\|) \Big(\frac{c(\|x\|)}{g(\|x\|)} - h \Big).
\end{aligned}
$$

Consequently, in view of (4.9.6), there exists $\epsilon > 0$ such that

$$\liminf_{\|x\| \to \infty} \frac{\langle A^h x, x \rangle}{\|x\|} \geq (c_0 - h - \epsilon) \lim_{t \to \infty} g(t) \geq c_h,$$

where $c_h < \infty$ when $h > 0$ is small enough. In this case, for all $\alpha > 0$,

$$\frac{\langle A^h x + \alpha J x, x \rangle}{\|x\|} = \frac{\langle A^h x, x \rangle}{\|x\|} + \alpha \|x\| \geq c_h - \epsilon + \alpha \|x\| \to \infty$$

as $\|x\| \to \infty$. The lemma is proved. ∎

We present now the main result of this section.

Theorem 4.9.3 *Assume that the conditions 1) - 4) of this section are fulfilled. Let $A : X \to X^*$ be a pseudomonotone bounded and coercive operator on each set Ω_σ and let (4.9.5) and (4.9.6) hold. If A has the S-property,*

$$\limsup_{t \to \infty} \frac{g(t)}{t} \leq Q < \infty \tag{4.9.7}$$

and

$$\limsup_{\alpha, h \to 0} \frac{Qh}{\alpha} < 1, \tag{4.9.8}$$

then solution sets of (4.7.4) and (4.5.2) are nonempty. A set N_0 of all strong limit points of the sequence $\{x_\alpha^\gamma\}$, as $\Delta = (\gamma, \alpha) \to 0$, is also nonempty and belongs to N, no matter how solutions x_α^γ are chosen.

Proof. The first assertion of the theorem follows from Lemma 4.9.2 and Theorem 1.13.10. Construct the sequence $\{x_\alpha^\gamma\}$ using any solutions of the variational inequality (4.5.2) for given $\gamma \in \Re$ and $\alpha > 0$. Prove that $\{x_\alpha^\gamma\}$ is bounded as $\Delta \to 0$. Since $\theta_X \in \Omega_\sigma$ for all $\sigma \in (0.\sigma^*]$, assume $z = \theta_X$ in (4.5.2). Then we have

$$\langle A^h x_\alpha^\gamma + \alpha J x_\alpha^\gamma - f^\delta, x_\alpha^\gamma \rangle \leq 0,$$

which is equivalent to the inequality

$$\langle A^h x_\alpha^\gamma - A x_\alpha^\gamma, x_\alpha^\gamma \rangle + \langle A x_\alpha^\gamma, x_\alpha^\gamma \rangle - \langle f^\delta, x_\alpha^\gamma \rangle + \alpha \|x_\alpha^\gamma\|^2 \leq 0.$$

By (4.9.3), we deduce

$$\langle A x_\alpha^\gamma, x_\alpha^\gamma \rangle + \alpha \|x_\alpha^\gamma\|^2 \leq \left(hg(\|x_\alpha^\gamma\|) + \|f^\delta\|_* \right) \|x_\alpha^\gamma\|. \tag{4.9.9}$$

Suppose that $\{x_\alpha^\gamma\}$ is unbounded, i.e., $\|x_\alpha^\gamma\| \to \infty$ as $\Delta \to 0$. Then (4.9.5) implies the estimate

$$\langle A x_\alpha^\gamma, x_\alpha^\gamma \rangle \geq c(\|x_\alpha^\gamma\|)\|x_\alpha^\gamma\|.$$

Now (4.9.9) gives the following relation:

$$c(\|x_\alpha^\gamma\|)\|x_\alpha^\gamma\| + \alpha \|x_\alpha^\gamma\|^2 \leq \left(hg(\|x_\alpha^\gamma\|) + \|f^\delta\|_* \right) \|x_\alpha^\gamma\|.$$

If α is small enough then the last can be transformed as

$$1 \leq \frac{h}{\alpha} \frac{g(\|x_\alpha^\gamma\|)}{\|x_\alpha^\gamma\|} + \frac{\|f^\delta\|_* - c(\|x_\alpha^\gamma\|)}{\alpha \|x_\alpha^\gamma\|} \leq \frac{h}{\alpha} \frac{g(\|x_\alpha^\gamma\|)}{\|x_\alpha^\gamma\|}.$$

Here we used the boundedness of $\|f^\delta\|_*$, the condition 1) and the fact that $c(t) \to \infty$ as $t \to \infty$. Passing to the limit in the latter inequality as $\Delta \to 0$ and taking into account

(4.9.7) and (4.9.8), we come to a contradiction. Thus, the sequence $\{x_\alpha^\gamma\}$ is bounded as $\Delta \to 0$.

Let $x_\alpha^\gamma \rightharpoonup \bar{x}$. Since $x_\alpha^\gamma \in \Omega_\sigma$, we have that $\bar{x} \in \Omega$ by reason of the Mosco-approximation properties. Applying these properties once more for $\bar{x}$, one can construct a sequence $\{\bar{x}_\sigma\}$, $\bar{x}_\sigma \in \Omega_\sigma$, such that $\bar{x}_\sigma \to \bar{x}$ as $\sigma \to 0$. Assuming in (4.5.2) $z = \bar{x}_\sigma$ we obtain

$$\langle A^h x_\alpha^\gamma + \alpha J x_\alpha^\gamma - f^\delta, \bar{x}_\sigma - x_\alpha^\gamma \rangle \geq 0.$$

Then (4.9.3) and the boundedness of $\{x_\alpha^\gamma\}$ and $\{\bar{x}_\sigma\}$ yield the inequality

$$\limsup_{\Delta \to 0} \langle A x_\alpha^\gamma + \alpha J x_\alpha^\gamma - f^\delta, x_\alpha^\gamma - \bar{x}_\sigma \rangle \leq \limsup_{\Delta \to 0} hg(\|x_\alpha^\gamma\|)\|x_\alpha^\gamma - \bar{x}_\sigma\| = 0. \tag{4.9.10}$$

Since $x_\alpha^\gamma \rightharpoonup \bar{x}$, we conclude that

$$\alpha \langle J x_\alpha^\gamma, x_\alpha^\gamma - \bar{x}_\sigma \rangle \to 0$$

and

$$\langle f^\delta, x_\alpha^\gamma - \bar{x}_\sigma \rangle \to 0$$

as $\Delta \to 0$. Then it results from (4.9.10) that

$$\limsup_{\Delta \to 0} \langle A x_\alpha^\gamma, x_\alpha^\gamma - \bar{x}_\sigma \rangle \leq 0. \tag{4.9.11}$$

The sequence $\{A x_\alpha^\gamma\}$ is bounded because the operator A and $\{x_\alpha^\gamma\}$ are bounded. Thus,

$$\langle A x_\alpha^\gamma, \bar{x}_\sigma - \bar{x} \rangle \to 0 \quad \text{as} \quad \Delta \to 0.$$

Therefore, (4.9.11) implies

$$\limsup_{\Delta \to 0} \langle A x_\alpha^\gamma, x_\alpha^\gamma - \bar{x} \rangle \leq 0. \tag{4.9.12}$$

By the definition of pseudomonotone operators, one gets

$$\liminf_{\Delta \to 0} \langle A x_\alpha^\gamma, x_\alpha^\gamma - x \rangle \geq \langle A\bar{x}, \bar{x} - x \rangle \quad \forall x \in \Omega. \tag{4.9.13}$$

In view of (4.9.3), we are able now to write down the following relations:

$$\langle A^h x_\alpha^\gamma + \alpha J x_\alpha^\gamma \quad - \quad f^\delta, x_\alpha^\gamma - x \rangle = \langle A^h x_\alpha^\gamma - A x_\alpha^\gamma + \alpha J x_\alpha^\gamma - f^\delta, x_\alpha^\gamma - x \rangle$$

$$+ \quad \langle A x_\alpha^\gamma, x_\alpha^\gamma - x \rangle$$

$$\geq \quad -hg(\|x_\alpha^\gamma\|)\|x_\alpha^\gamma - x\| + \alpha \langle J x_\alpha^\gamma, x_\alpha^\gamma - x \rangle$$

$$- \quad \langle f^\delta, x_\alpha^\gamma - x \rangle + \langle A x_\alpha^\gamma, x_\alpha^\gamma - x \rangle \quad \forall x \in \Omega. \tag{4.9.14}$$

It is obvious that if $\Delta \to 0$ then

$$hg(\|x_\alpha^\gamma\|)\|x_\alpha^\gamma - x\| \to 0,$$

$$\alpha\langle Jx_\alpha^\gamma, x_\alpha^\gamma - x\rangle \to 0$$

and

$$\langle f^\delta, x_\alpha^\gamma - x\rangle \to \langle f, \bar{x} - x\rangle.$$

Together with (4.9.13), this allows us to deduce from (4.9.14) the inequality

$$\liminf_{\Delta\to 0} \langle A^h x_\alpha^\gamma + \alpha Jx_\alpha^\gamma - f^\delta, x_\alpha^\gamma - x\rangle \geq \langle A\bar{x} - f, \bar{x} - x\rangle \quad \forall x \in \Omega, \quad \bar{x} \in \Omega. \qquad (4.9.15)$$

Fix $x \in \Omega$. According to the Mosco-approximation of Ω by Ω_σ, construct the sequence $\{x_\sigma\}$, $x_\sigma \in \Omega_\sigma$, such that $x_\sigma \to x$ as $\sigma \to 0$. Let

$$u_\alpha^\gamma = A^h x_\alpha^\gamma + \alpha Jx_\alpha^\gamma - f^\delta. \qquad (4.9.16)$$

Then

$$\langle u_\alpha^\gamma, x_\alpha^\gamma - x\rangle = \langle u_\alpha^\gamma, x_\alpha^\gamma - x_\sigma\rangle + \langle u_\alpha^\gamma, x_\sigma - x\rangle.$$

Since $\{x_\alpha^\gamma\}$ and $\{Ax_\alpha^\gamma\}$ are bounded, (4.9.3) implies the boundedness of $\{A^h x_\alpha^\gamma\}$. Consequently, by (4.9.16), $\{u_\alpha^\gamma\}$ is also bounded as $\Delta \to 0$. In addition, we have $\alpha Jx_\alpha^\gamma \to \theta_{X^*}$ and $f^\delta \to f$. Owing to the strong convergence of $\{x_\sigma\}$ to x, we conclude that

$$\langle u_\alpha^\gamma, x_\sigma - x\rangle \to 0 \quad \text{as} \quad \Delta \to 0.$$

Put in (4.5.2) $z = x_\sigma$. Then

$$\langle u_\alpha^\gamma, x_\alpha^\gamma - x_\sigma\rangle \leq 0.$$

Now (4.9.15) gives

$$\langle A\bar{x} - f, \bar{x} - x\rangle \leq 0 \quad \forall x \in \Omega, \quad \bar{x} \in \Omega.$$

This means that $\bar{x} \in N$.

Further, asuming in (4.9.13) $x = \bar{x}$, we obtain

$$\liminf_{\Delta\to 0} \langle Ax_\alpha^\gamma, x_\alpha^\gamma - \bar{x}\rangle \geq 0.$$

Then (4.9.12) gives the limit equality

$$\lim_{\Delta\to 0} \langle Ax_\alpha^\gamma, x_\alpha^\gamma - \bar{x}\rangle = 0.$$

Since $x_\alpha^\gamma \rightharpoonup \bar{x}$, the last implies

$$\lim_{\Delta\to 0} \langle Ax_\alpha^\gamma - A\bar{x}, x_\alpha^\gamma - \bar{x}\rangle = 0.$$

Finally, the S-property of A guarantees strong convergence $\{x_\alpha^\gamma\}$ to $\bar{x} \in N_0 \subseteq N$ as $\Delta \to 0$. The proof is accomplished. ∎

Note that normalized duality mapping J in the inequality (4.5.2) can be replaced by the duality mapping J^μ with a gauge function $\mu(t)$. Theorem 4.9.3 remains still valid if, instead of (4.9.7), it used the inequality

$$\lim_{t\to+\infty} \frac{g(t)}{\mu(t)} \leq Q < \infty.$$

4.10   Variational Inequalities of Mixed Type

Let X be an E-space, X^* be a strongly convex space, $A : X \to X^*$ be a monotone bounded hemicontinuous operator with $D(A) = X$ and $\varphi : X \to R^1$ be a properly convex lower semicontinuous functional.

Consider the problem of solving the mixed variational inequality: To find $x \in X$ such that

$$\langle Ax - f, x - y \rangle + \varphi(x) - \varphi(y) \leq 0 \quad \forall y \in X. \tag{4.10.1}$$

As usual, we start to study the existence of a solution x for (4.10.1). To this end, first of all, we need the following theorem [128]:

Theorem 4.10.1 *If there exists $x_0 \in \operatorname{dom} \varphi$ satisfying the limit relation*

$$\lim_{\|x\| \to \infty} \frac{\langle Ax, x - x_0 \rangle + \varphi(x)}{\|x\|} = \infty, \tag{4.10.2}$$

then (4.10.1) has at least one solution.

We first prove two auxiliary lemmas.

Lemma 4.10.2 *The mixed variational inequality (4.10.1) is equivalent to*

$$\langle Ay - f, x - y \rangle + \varphi(x) - \varphi(y) \leq 0 \quad \forall y \in X, \quad x \in X. \tag{4.10.3}$$

Proof. Since A is monotone, we have from (4.10.1) that

$$\langle Ay - f, x - y \rangle \leq \langle Ax - f, x - y \rangle \leq \varphi(y) - \varphi(x) \quad \forall y \in X, \quad x \in X.$$

This means that (4.10.3) is valid.

Let now (4.10.3) be assumed. If $y = y_t = tx + (1 - t)z$, $t \in [0, 1]$, where z is any fixed element of X, then (4.10.3) accepts the form:

$$(1 - t)\langle Ay_t - f, x - z \rangle + \varphi(x) - \varphi(tx + (1 - t)z) \leq 0 \quad \forall z \in X.$$

Taking into account the convexity of φ we obtain

$$(1 - t)\langle Ay_t - f, x - z \rangle + \varphi(x) - t\varphi(x) - (1 - t)\varphi(z) \leq 0 \quad \forall z \in X.$$

Consequently,

$$\langle Ay_t - f, x - z \rangle + \varphi(x) - \varphi(z) \leq 0 \quad \forall z \in X.$$

Letting $t \to 1$ one gets (4.10.1). ∎

Lemma 4.10.3 *The solution set N of (4.10.1) is closed and convex if it is not empty.*

Proof. Let x_1 and x_2 be two different elements of N. Then, by Lemma 4.10.2,

$$\langle Ay - f, x_1 - y \rangle + \varphi(x_1) - \varphi(y) \leq 0 \quad \forall y \in X$$

and

$$\langle Ay - f, x_2 - y \rangle + \varphi(x_2) - \varphi(y) \leq 0 \quad \forall y \in X.$$

Multiplying these inequalities, respectively, on t and $1 - t$ and adding them, we get

$$\langle Ay - f, z - y \rangle + t\varphi(x_1) + (1 - t)\varphi(x_2) - \varphi(y) \leq 0 \quad \forall y \in X, \tag{4.10.4}$$

where $z = tx_1 + (1 - t)x_2$. Since φ is convex,

$$\varphi(z) \leq t\varphi(x_1) + (1 - t)\varphi(x_2).$$

Hence,

$$\langle Ay - f, z - y \rangle + \varphi(z) - \varphi(y) \leq 0 \quad \forall y \in X$$

because of (4.10.4). Thus, $z \in N$ and convexity of N is proved.

Show that N is closed. Let $x_n \to x$, where $x_n \in N$. Then

$$\langle Ax_n - f, x_n - y \rangle + \varphi(x_n) - \varphi(y) \leq 0 \quad \forall y \in X.$$

Since a hemicontinuous monotone operator on X is demicontinuous and since φ is a lower semicontinuous functional, the last inequality implies (4.10.1) as $n \to \infty$. ∎

Let $\Re^* = (0, \delta^*] \times (0, h^*] \times (0, \epsilon^*]$ with some positive δ^*, h^* and ϵ^* and let $\gamma = (\delta, h, \epsilon) \in \Re^*$. Assume that the solution set N of the inequality (4.10.1) is nonempty, and its data A, f, φ are given with approximations A^h, f^δ, φ_ϵ satisfying the conditions:
1) $\|f - f^\delta\|_* \leq \delta$;
2) $A^h : X \to X^*$ are monotone hemicontinuous operators, $D(A^h) = D(A) = X$, and

$$\|A^h x - Ax\|_* \leq hg(\|x\|) \quad \forall x \in X, \tag{4.10.5}$$

where $g(s)$ is a non-negative bounded function for $s \geq 0$;
3) $\varphi_\epsilon : X \to R^1$ are properly convex lower semicontinuous functionals and there exist positive numbers c_ϵ and R_ϵ such that

$$\varphi_\epsilon(x) \geq -c_\epsilon\|x\| \quad \text{as} \quad \|x\| > R_\epsilon$$

and

$$|\varphi_\epsilon(x) - \varphi(x)| \leq \epsilon q(\|x\|) \quad \forall x \in X, \tag{4.10.6}$$

where $q(s)$ has the same properties as $g(s)$.

The regularization method for the mixed variational inequality (4.10.1) is written as

$$\langle A^h x + \alpha Jx - f^\delta, x - y \rangle + \varphi_\epsilon(x) - \varphi_\epsilon(y) \leq 0 \quad \forall y \in X, \quad x \in X. \tag{4.10.7}$$

Let $x_\epsilon \in dom\ \varphi_\epsilon$. The monotonicity of A^h and assumption 3) imply for $\|x\| > R_\epsilon$ the following inequality:

$$\frac{\langle A^h x + \alpha Jx, x - x_\epsilon \rangle + \varphi_\epsilon(x)}{\|x\|} \geq \alpha(\|x\| - \|x_\epsilon\|) - \|A^h x_\epsilon\| \left(1 + \frac{\|x_\epsilon\|}{\|x\|}\right) - c_\epsilon.$$

Consequently, (4.10.2) is fulfilled for the operator $A^h + \alpha J$ and for the functional φ_ϵ. Thus, a solution of (4.10.7) exists.

Lemma 4.10.4 *The inequality (4.10.7) has a unique solution.*

Proof. Let x_1 and x_2 be two different solutions of (4.10.7). Then

$$\langle A^h x_1 + \alpha Jx_1 - f^\delta, x_1 - y \rangle + \varphi_\epsilon(x_1) - \varphi_\epsilon(y) \leq 0 \quad \forall y \in X, \tag{4.10.8}$$

and

$$\langle A^h x_2 + \alpha Jx_2 - f^\delta, x_2 - y \rangle + \varphi_\epsilon(x_2) - \varphi_\epsilon(y) \leq 0 \quad \forall y \in X. \tag{4.10.9}$$

Put $y = x_2$ in (4.10.8) and $y = x_1$ in (4.10.9) and add the obtained inequalities. We obtain

$$\langle A^h x_1 - A^h x_2, x_1 - x_2 \rangle + \alpha \langle Jx_1 - Jx_2, x_1 - x_2 \rangle \leq 0.$$

Due to the monotonicity of A^h and strict monotonicity of J, the letter occurs only if $x_1 = x_2$. Thus, the lemma holds. ∎

Denote the unique solution of (4.10.7) by x_α^γ. Then

$$\langle A^h x_\alpha^\gamma + \alpha Jx_\alpha^\gamma - f^\delta, x_\alpha^\gamma - y \rangle + \varphi_\epsilon(x_\alpha^\gamma) - \varphi_\epsilon(y) \leq 0 \quad \forall y \in X. \tag{4.10.10}$$

Theorem 4.10.5 *Let X be an E-space, X^* be a strictly convex space, $A : X \to X^*$ be a monotone hemicontinuous and bounded operator, $D(A) = X$, $\varphi : X \to R^1$ be a properly convex lower semicontinuous functional, a solution set N of the mixed variational inequality (4.10.1) be nonempty. Assume that the conditions 1) - 3) are fulfilled and, in addition,*

$$\limsup_{s \to \infty} \frac{q(s)}{s^2} \leq Q, \quad 0 < Q < \infty, \tag{4.10.11}$$

and

$$\lim_{\alpha \to 0} \frac{\delta + h + \epsilon}{\alpha} = 0. \tag{4.10.12}$$

Then the sequence $\{x_\alpha^\gamma\}$ converges strongly to the minimal norm solution $\bar{x}^*$ of (4.10.1).

Proof. Let $x^* \in N$. Put $x = x^*$, $y = x_\alpha^\gamma$ in (4.10.1) and $y = x^*$ in (4.10.10). Adding the obtained results one gets

$$\begin{aligned}
\langle Ax^* - A^h x^*, x^* - x_\alpha^\gamma \rangle \quad &+ \quad \langle A^h x^* - A^h x_\alpha^\gamma, x^* - x_\alpha^\gamma \rangle \\
&+ \quad \alpha \langle Jx_\alpha^\gamma, x_\alpha^\gamma - x^* \rangle + \langle f - f^\delta, x_\alpha^\gamma - x^* \rangle \\
&+ \quad \varphi(x^*) - \varphi_\epsilon(x^*) + \varphi_\epsilon(x_\alpha^\gamma) - \varphi(x_\alpha^\gamma) \leq 0. \tag{4.10.13}
\end{aligned}$$

Using the conditions 1) - 3) we deduce

$$\|x_\alpha^\gamma\|^2 \;\leq\; \|x_\alpha^\gamma\|\|x^*\| + \left(\frac{\delta}{\alpha} + \frac{h}{\alpha}g(\|x^*\|)\right)\left(\|x_\alpha^\gamma\| + \|x^*\|\right)$$

$$+ \;\frac{\epsilon}{\alpha}\Big(q(\|x^*\|) + q(\|x_\alpha^\gamma\|)\Big) \quad \forall x^* \in N.$$

Now (4.10.11) and (4.10.12) guarantee that $\{x_\alpha^\gamma\}$ is bounded as $\alpha \to 0$. Hence, $x_\alpha^\gamma \to \bar{x} \in X$. Due to Lemma 4.10.2, (4.10.10) yields the inequality

$$\langle A^h y + \alpha J y - f^\delta, x_\alpha^\gamma - y\rangle + \varphi_\epsilon(x_\alpha^\gamma) - \varphi_\epsilon(y) \leq 0 \quad \forall y \in X. \tag{4.10.14}$$

According to Theorem 1.1.13, the functional φ is weakly lower semicontinuous. Therefore,

$$\varphi(\bar{x}) \leq \liminf_{\alpha\to 0} \varphi(x_\alpha^\gamma). \tag{4.10.15}$$

Since $\{x_\alpha^\gamma\}$ and $q(s)$ are bounded, by (4.10.6), we have

$$\varphi(x_\alpha^\gamma) \leq \varphi_\epsilon(x_\alpha^\gamma) + c_2\epsilon, \tag{4.10.16}$$

where $c_2 > 0$. Obviously, $\epsilon \to 0$ as $\alpha \to 0$ because of (4.10.12). Then (4.10.5), (4.10.6), (4.10.14), (4.10.15), (4.10.16) and the condition 1) imply

$$\langle Ay - f, \bar{x} - y\rangle + \varphi(\bar{x}) - \varphi(y) \leq 0 \quad \forall y \in X.$$

It is not difficult to be now sure that the inclusion $\bar{x} \in N$ follows from Lemma 4.10.2.

Further, by (4.10.13), there exists a constant $c_3 > 0$ such that for all $x^* \in N$

$$\Big(\|x_\alpha^\gamma\| - \|x^*\|\Big)^2 \;\leq\; \langle Jx_\alpha^\gamma - Jx^*, x_\alpha^\gamma - x^*\rangle$$

$$\leq\; \langle Jx^*, x^* - x_\alpha^\gamma\rangle + c_3\frac{\delta + h + \epsilon}{\alpha}.$$

According to Lemma 4.10.3, the set N is closed and convex. Therefore, there exists a unique $\bar{x}^*$. Then we obtain that in E-space X the sequence $x_\alpha^\gamma \to \bar{x}^*$ (see, for example, Section 2.2). ∎

Observe that if the function $q(s)$ does not satisfy (4.10.11) then in (4.10.7) we may use the duality mapping J^μ with a suitable gauge function $\mu(t)$ in place of J.

Bibliographical Notes and Remarks

The results of Section 4.1 are due to [35, 189, 201]. The idea to apply the generalized residual in oder to state the residual principle for variational inequalities belongs to Ryazantseva. In connection with this, we need to emphasize that, generally speaking, the classical residual of the variational inequality (4.1.7) in the form $\rho(\alpha) = \|A^h x_\alpha^\gamma - f^\delta\|_*$ on solutions x_α^γ of the regularized inequality (4.1.14) does not make sense even in the case of a continuous operator A^h. At the same time, it turned out that the generalized residual principle is a very effective tool for choosing the regularization parameter α as it is done, for instance, in Theorems 4.1.8 - 4.1.11.

The operator regularization method for solving variational inequalities on approximately given sets was studied by Ryazantseva in [190, 201]. Theorem 4.3.1 was proved by Alber, Butnariu and Ryazantseva and can be found in [18]. The examples of variational inequalities in Section 4.4 have been adopted from [120]. The convergence of the operator regularization method for variational inequalities with unbounded monotone operators on inexact sets was established by Alber and Notik in [27]. It was also shown in [27] that the assertion similar to Theorem 2.2.5 holds for variational inequalities in Banach spaces.

The so-called variational inequalities with small offset were introduced and investigated by Liskovets in [133, 138]. The regularization method for solving variational inequalities with Mosco-approximation of constraint sets was studied in [19]. The similar version for monotone inclusions with Mosco-approximation of operators was considered in [17]. Mosco perturbations for variational inequalities with inverse-strongly-monotone operators in a Hilbert space has been developed in [140, 155], where the estimates of convergence rate are also presented. The statement and convergence analysis of regularization algorithms for solving variational inequalities with hypomonotone mappings follows the paper [20]. The results of Section 4.9 have appeared in [137]. Theorem 4.10.5 for the mixed variational inequalities with monotone hemicontinuous operators A and A^h was proved in [139].

Chapter 5

APPLICATIONS OF THE REGULARIZATION METHODS

5.1 Computation of Unbounded Monotone Operators

It has been already marked that monotone operators in a Banach space are not bounded in general. Hence, the problem of the value computation

$$y_0 \in Ax_0 \tag{5.1.1}$$

of an unbounded monotone operator $A : X \rightarrow 2^{X^*}$ at a point $x_0 \in X$ belongs to the class of unstable problems [99, 130, 131]. For solving the problem (5.1.1), as before, the regularization methods in operator form are used in this chapter.

In the sequel, we assume that $y_0 \in R(A)$ but, in place of x_0, a sequence $\{x_\delta\} \subset X$ is given such that $\|x_\delta - x_0\| \leq \delta$, where $\delta \in (0, \delta^*]$. Note that elements x_δ do not necessarily belong to $D(A)$. Using x_δ, we should construct in X^* an approximation sequence which converges weakly or strongly as $\delta \rightarrow 0$ to a certain value of the operator A at a point x_0. Consider separately the cases of the Hilbert and Banach spaces.

1. Let $X = H$ be a Hilbert space, $A : H \rightarrow H$ be a maximal monotone and hemicontinuous (i.e., single-valued) operator. Then an element $y_0 \in H$ satisfying (5.1.1) is unique. Put into correspondence with (5.1.1) the following regularized equation:

$$x_\alpha^\delta + \alpha Ax_\alpha^\delta = x_\delta, \quad \alpha > 0, \tag{5.1.2}$$

and establish the connection $\alpha = \alpha(\delta)$, which guarantees the limit relation $Ax_\alpha^\delta = y_\alpha^\delta \rightarrow y_0$, when $\alpha(\delta) \rightarrow 0$ as $\delta \rightarrow 0$. By virtue of Theorem 1.7.4, equation (5.1.2) with any right-hand side x_δ has a unique solution. Introduce the intermediate equation corresponding to the exact value x_0 as

$$x_\alpha^0 + \alpha Ax_\alpha^0 = x_0, \quad \alpha > 0. \tag{5.1.3}$$

By the monotonicity of A,

$$(Ax - Ax_0, x - x_0) \geq 0 \quad \forall x \in D(A). \tag{5.1.4}$$

Assume that in (5.1.4) $x = x_\alpha^0$. Then, by (5.1.3),

$$(Ax_\alpha^0 - Ax_0, x_\alpha^0 - x_0) = -\alpha(Ax_\alpha^0 - Ax_0, Ax_\alpha^0) \geq 0, \qquad (5.1.5)$$

and the estimate

$$\|Ax_\alpha^0\| \leq \|y_0\| \qquad (5.1.6)$$

appears from (5.1.5). It allows us to prove by the standard arguments that Ax_α^0 weakly converges to some element $\bar{y} \in H$ as $\alpha \to 0$. On the other hand, (5.1.3) implies

$$\|x_\alpha^0 - x_0\| = \alpha\|Ax_\alpha^0\| \to 0 \quad \text{as} \quad \alpha \to 0.$$

Show that $\bar{y} = Ax_0$. Let $x = x_0 + tv$, where v is an arbitrary element of H and $t \geq 0$. Due to Theorem 1.7.19, $D(A)$ is open because A is a maximal monotone and hemicontinuous operator. Therefore, $x \in D(A)$ if t is sufficiently small. Since A is monotone, we have

$$(Ax_\alpha^0 - A(x_0 + tv), x_\alpha^0 - x_0 - tv) \geq 0.$$

Let first $\alpha \to 0$. By the strong convergence of x_α^0 to x_0 and by weak convergence of Ax_α^0 to $\bar{y}$, we deduce the inequality

$$-t(\bar{y} - A(x_0 + tv), v) \geq 0.$$

Hence,

$$(A(x_0 + tv) - \bar{y}, v) \geq 0.$$

Let now $t \to 0$. In view of the hemicontinuity of A,

$$(Ax_0 - \bar{y}, v) \geq 0 \quad \forall v \in H.$$

This means that $Ax_0 = \bar{y}$. Since the operator A is single-valued, the equality $\bar{y} = y_0$ holds. Observe that the same results can be established by Lemma 1.4.5. Thus, we have proved a weak convergence

$$Ax_\alpha^0 \rightharpoonup y_0 \quad \text{as} \quad \alpha \to 0. \qquad (5.1.7)$$

It is easy to see that (5.1.5) involves the inequality

$$(Ax_\alpha^0, Ax_\alpha^0 - y_0) \leq 0.$$

Then by the obvious equality

$$(Ax_\alpha^0, Ax_\alpha^0 - y_0) - (y_0, Ax_\alpha^0 - y_0) = \|Ax_\alpha^0 - y_0\|^2,$$

we obtain

$$\|Ax_\alpha^0 - y_0\|^2 \leq -(y_0, Ax_\alpha^0 - y_0).$$

Consequently, the convergence $Ax_\alpha^0 \to y_0$ results from (5.1.7). It is clear that

$$\|Ax_\alpha^\delta - y_0\| \leq \|Ax_\alpha^\delta - Ax_\alpha^0\| + \|Ax_\alpha^0 - y_0\|. \qquad (5.1.8)$$

Then (5.1.2), (5.1.3) and the monotonicity A imply the following estimates:

$$\|Ax_\alpha^\delta - Ax_\alpha^0\| \leq \alpha^{-1}(\|x_\delta - x_0\| + \|x_\alpha^\delta - x_\alpha^0\|)$$

and

$$\|x_\alpha^\delta - x_\alpha^0\| \le \|x_\delta - x_0\| \le \delta.$$

Thus, by (5.1.6),

$$\|Ax_\alpha^\delta\| \le \frac{2\delta}{\alpha} + \|y_0\|. \tag{5.1.9}$$

At the same time, the equality

$$x_\alpha^\delta - x_0 + \alpha Ax_\alpha^\delta - \alpha Ax_0 + \alpha Ax_0 = x_\delta - x_0$$

gives

$$\|x_\alpha^\delta - x_0\| \le \delta + \alpha\|y_0\|. \tag{5.1.10}$$

Hence, $x_\alpha^\delta \to x_0$, when $\alpha \to 0$ and $\delta \to 0$. The same condition ensures the weak convergence of Ax_α^δ to y_0. Finally,

$$\|Ax_\alpha^\delta - Ax_\alpha^0\| \le \frac{2\delta}{\alpha}.$$

If $\dfrac{\delta}{\alpha} \to 0$ then $Ax_\alpha^\delta \to Ax_0$. It follows from (5.1.8) that $Ax_\alpha^\delta \to Ax_\alpha^0$ as $\alpha \to 0$. Thus, we have proved

Theorem 5.1.1 *The sequence $\{Ax_\alpha^\delta\}$, where x_α^δ are solutions of the equation (5.1.2), converges strongly (weakly) to $y_0 = Ax_0$ if $\dfrac{\delta}{\alpha} \to 0$ ($\delta \to 0$) as $\alpha \to 0$, and the estimate (5.1.9) holds.*

Corollary 5.1.2 *Let x_α^δ be solutions of the equation (5.1.2) and $Ax_\alpha^\delta \to y_0$ when $\delta \to 0$, $\alpha \to 0$ and $\|x_\delta - x_0\| \le \delta$. Then $y_0 = Ax_0$, that is, $y_0 \in R(A)$.*

These assertions may be combined by means of the following theorem.

Theorem 5.1.3 *Let x_α^δ be solutions of the equation (5.1.2). The sequence $\{Ax_\alpha^\delta\}$ converges to an element $y_0 \in H$ as $\alpha \to 0$, $\dfrac{\delta}{\alpha} \to 0$ and $\|x_\delta - x_0\| \le \delta$ if and only if $y_0 = Ax_0$.*

2. Next we discuss behavior of the functions $\rho(\alpha) = \|x_\alpha^\delta - x_\delta\|$ and $\sigma(\alpha) = \|Ax_\alpha^\delta\|$ on a semi-infinite interval $[\alpha_0, \infty)$ with fixed $\delta \in (0, \delta^*]$ and $\alpha_0 > 0$, where x_α^δ is a solution of the equation (5.1.2). Denote $y_\alpha^\delta = Ax_\alpha^\delta$. The function $\rho(\alpha)$ is single-valued because the equation (5.1.2) is uniquely solvable. The function $\sigma(\alpha)$ is single-valued and bounded if $\dfrac{\delta}{\alpha}$ is bounded (it holds for $\alpha \ge \alpha_0 > 0$). A continuity of $\sigma(\alpha)$ is shown by analogy with the case of Section 3.1 (see Lemma 3.1.1). Without loss of generality, take α_1 and α_2 such that $\alpha_2 > \alpha_1$. It is obvious that

$$0 \le (y_{\alpha_1}^\delta - y_{\alpha_2}^\delta, x_{\alpha_1}^\delta - x_{\alpha_2}^\delta) = (y_{\alpha_1}^\delta - y_{\alpha_2}^\delta, \alpha_2 y_{\alpha_2}^\delta - \alpha_1 y_{\alpha_1}^\delta).$$

From this, we have

$$\alpha_1\|y_{\alpha_1}^\delta - y_{\alpha_2}^\delta\|^2 \le (\alpha_2 - \alpha_1)(y_{\alpha_1}^\delta - y_{\alpha_2}^\delta, y_{\alpha_2}^\delta). \tag{5.1.11}$$

Then we find that, firstly,

$$\left| \|y_{\alpha_1}^\delta\| - \|y_{\alpha_2}^\delta\| \right| \le \frac{\alpha_2 - \alpha_1}{\alpha_1} \|y_{\alpha_2}^\delta\|, \tag{5.1.12}$$

and, secondly,

$$(y_{\alpha_1}^\delta - y_{\alpha_2}^\delta, y_{\alpha_2}^\delta) \ge 0.$$

We come to the relation

$$\|y_{\alpha_2}^\delta\| \le \|y_{\alpha_1}^\delta\| \quad \text{as} \quad \alpha_2 > \alpha_1. \tag{5.1.13}$$

Since $\|y_\alpha^\delta\|$ is bounded, (5.1.12) guarantees that $\sigma(\alpha)$ and, consequently, $\rho(\alpha)$ are continuous because $\rho(\alpha) = \alpha\sigma(\alpha)$. The result (5.1.13) shows that $\sigma(\alpha)$ does not increase. Hence, $\sigma(\alpha)$ approaches a finite limit $\bar\sigma \ge 0$ as $\alpha \to \infty$. We prove below that $\bar\sigma = 0$.

It is well known that if operator A is monotone then inverse map A^{-1} is also monotone. Suppose that A^{-1} is defined at some point $u^0 \in H$, that is, $u^0 \in R(A)$. We have

$$(Ax_\alpha^\delta - u_0, x_\alpha^\delta - v_0) \ge 0 \quad \forall v_0 \in A^{-1}u_0.$$

Let $\theta_H \in R(A)$ and $x^0 \in A^{-1}(\theta_H)$, i.e., $Ax^0 = \theta_H$. Assuming in the last inequality that $u_0 = \theta_H$ and $v_0 = x^0$ we obtain

$$(Ax_\alpha^\delta, x_\alpha^\delta - x^0) \ge 0.$$

Then it results from the equation (5.1.2) that

$$(x_\delta - x_\alpha^\delta, x_\alpha^\delta - x^0) \ge 0.$$

The simple calculations allow us to derive for $x^0 \in A^{-1}(\theta_H)$ the following estimate:

$$\|x_\alpha^\delta - x_\delta\| \le \|x^0 - x_\delta\| \le \|x^0 - x_0\| + \delta. \tag{5.1.14}$$

Thus, if $\theta_H \in R(A)$ then the sequence $\{x_\alpha^\delta\}$ is bounded for all $\alpha > 0$ and a bound does not depend on α and δ. Therefore, $x_\alpha^\delta \rightharpoonup \bar x \in H$ and

$$\sigma(\alpha) = \|Ax_\alpha^\delta\| = \alpha^{-1}\|x_\delta - x_\alpha^\delta\| \to 0 \quad \text{as} \quad \alpha \to \infty.$$

In addition, $Ax_\alpha^\delta \to \theta_H$.

Due to Lemma 1.4.5, the graph of A is demiclosed and this implies that $A\bar x = \theta_H$. Consequently, $\bar x \in Q = \{x \mid Ax = 0\}$. After that, taking into account the weak convergence of x_α^δ to $\bar x$ and the inequality (5.1.14), we establish, as in Lemma 3.1.4, strong convergence of x_α^δ to x_*^0, as $\alpha \to \infty$, where $x_*^0 \in Q$ and

$$\|x_*^0 - x_\delta\| = min\{\|x^0 - x_\delta\| \mid x^0 \in A^{-1}(\theta_H)\}. \tag{5.1.15}$$

This leads to the limit equality

$$\lim_{\alpha \to \infty} \rho(\alpha) = \|x_*^0 - x_\delta\|. \tag{5.1.16}$$

Let $p \in (0, 1]$. By reason of (5.1.10), we can write

$$\|x_\alpha^\delta - x_\delta\| \le 2\delta + \alpha\|y_0\|.$$

It is clear that there exists $\alpha \ge \alpha_0 > 0$ such that $\alpha\|y_0\| < \delta^p$. Then

$$\rho(\alpha) < 2\delta + \delta^p \le \delta^p(1 + 2\delta^{*1-p}) = k\delta^p, \quad k = 1 + 2\delta^{*1-p}.$$

Assume that δ satisfies the inequality

$$k\delta^p < \|x_*^0 - x_\delta\|, \tag{5.1.17}$$

where x_*^0 is defined by (5.1.15). Then

$$k\delta^p < \|x_*^0 - x_0\| + \delta.$$

If we suppose now that $y_0 = \theta_H$, that is, $x_*^0 = x_0$, then we will come to a contradiction. Hence, (5.1.17) really implies that $y_0 = Ax_0 \ne \theta_H$.

Since $\rho(\alpha)$ is continuous and (5.1.16) is true, there exists at least one $\bar{\alpha}$ such that

$$\rho(\bar{\alpha}) = \|x_{\bar{\alpha}}^\delta - x_\delta\| = k\delta^p, \tag{5.1.18}$$

where $x_{\bar{\alpha}}^\delta$ is a solution of (5.1.2) with $\alpha = \bar{\alpha}$. Based upon that, $\bar{\alpha} > \dfrac{\delta^p}{\|y_0\|}$ because otherwise $\rho(\bar{\alpha}) < k\delta^p$ which contradicts (5.1.18). Hence,

$$\frac{\delta}{\bar{\alpha}} \le \delta^{1-p}\|y_0\|$$

and

$$\|y_{\bar{\alpha}}^\delta\| = \|Ax_{\bar{\alpha}}^\delta\| = \frac{k\delta^p}{\bar{\alpha}} \le k\|y_0\|.$$

In other words, the sequence $\{y_{\bar{\alpha}}^\delta\}$ is bounded if $x_{\bar{\alpha}}^\delta$ satisfy (5.1.18) for every $\delta \in (0, \delta^*]$. Therefore, $\{y_{\bar{\alpha}}^\delta\}$ is weakly compact in H. Then we assert that there exists $\bar{y} \in H$ such that $y_{\bar{\alpha}}^\delta \rightharpoonup \bar{y}$ as $\delta \to 0$. On the other hand,

$$\|x_{\bar{\alpha}}^\delta - x_0\| \le \|x_{\bar{\alpha}}^\delta - x_\delta\| + \|x_\delta - x_0\| \le k\delta^p + \delta \to 0, \ \delta \to 0,$$

i.e., $x_{\bar{\alpha}}^\delta \to x_0$. According to Lemma 1.4.5, the graph of operator A is demiclosed, therefore, $Ax_0 = \bar{y} = y_0$. We obtain

$$
\begin{aligned}
\|Ax_{\bar{\alpha}}^\delta - Ax_0\|^2 &= \bar{\alpha}^{-1}(\bar{\alpha}Ax_{\bar{\alpha}}^\delta, Ax_{\bar{\alpha}}^\delta - Ax_0) - (Ax_0, Ax_{\bar{\alpha}}^\delta - Ax_0) \\
&= -\bar{\alpha}^{-1}(x_{\bar{\alpha}}^\delta - x_\delta, Ax_{\bar{\alpha}}^\delta - Ax_0) - (Ax_0, Ax_{\bar{\alpha}}^\delta - Ax_0) \\
&= -\bar{\alpha}^{-1}(x_{\bar{\alpha}}^\delta - x_0, Ax_{\bar{\alpha}}^\delta - Ax_0) - (Ax_0, Ax_{\bar{\alpha}}^\delta - Ax_0) \\
&\quad + \bar{\alpha}^{-1}(x_\delta - x_0, Ax_{\bar{\alpha}}^\delta - Ax_0) \\
&\le \frac{\delta}{\bar{\alpha}}\|Ax_{\bar{\alpha}}^\delta - Ax_0\| - (Ax_0, Ax_{\bar{\alpha}}^\delta - Ax_0).
\end{aligned}
$$

If $p \in (0,1)$ then $\dfrac{\delta}{\bar{\alpha}} \to 0$ as $\delta \to 0$. Hence, $Ax_{\bar{\alpha}}^{\delta} \to Ax_0$. If $p = 1$ then $\dfrac{\delta}{\bar{\alpha}} \leq \|y_0\|$. In this case we may assert only the weak convergence of $Ax_{\bar{\alpha}}^{\delta}$ to Ax_0.

Finally, it is necessary to recall that $Ax_0 \neq \theta_H$ as $\delta \in (0, \delta^*]$, therefore, there exists a constant $\mu > 0$ such that $\|Ax_{\bar{\alpha}}^{\delta}\| \geq \mu$ for sufficiently small $\delta > 0$. Since

$$\rho(\bar{\alpha}) = \bar{\alpha}\|Ax_{\bar{\alpha}}^{\delta}\| = k\delta^p,$$

we establish that $\bar{\alpha} \to 0$ as $\delta \to 0$. Thus, on the basis of Theorem 5.1.1, the residual principle in the form (5.1.18) satisfies the general conditions for the operator regularization methods to be convergent. However, we do not state the obtained result as a separate theorem because we will give it below in the more general form.

3. Observe further in what directions these results can be generalized.

a) We can assume that $\theta_H \notin R(A)$, $z^0 \in R(A)$, $z^0 \neq y_0$, and consider the equation

$$x_{\alpha}^{\delta} + \alpha(Ax_{\alpha}^{\delta} - z^0) = x_{\delta}. \tag{5.1.19}$$

The boundedness of $\{x_{\alpha}^{\delta}\}$ can be proved making use of monotonicity of A and following implications:

$$(Ax_{\alpha}^{\delta} - z^0, x_{\alpha}^{\delta} - u^0) \geq 0 \quad \forall u^0 \in A^{-1}z^0$$

$$\Longrightarrow \quad (x_{\delta} - x_{\alpha}^{\delta}, x_{\alpha}^{\delta} - u^0) \geq 0$$

$$\Longrightarrow \quad -\|x_{\alpha}^{\delta} - x_{\delta}\|^2 + (x_{\delta} - x_{\alpha}^{\delta}, x_{\delta} - u^0) \geq 0$$

$$\Longrightarrow \quad \|x_{\alpha}^{\delta} - x_{\delta}\| \leq \|x_{\delta} - u^0\| \leq \|x_{\delta} - x_0\| + \|x_0 - u^0\|$$

$$\Longrightarrow \quad Ax_{\alpha}^{\delta} \to z^0 \quad \text{as} \quad \alpha \to \infty.$$

By means of the scheme applying in Lemma 3.1.9, we can prove that

$$\lim_{\alpha \to \infty} \rho(\alpha) = \|x^* - x_{\delta}\|,$$

where $x^* \in A^{-1}z^0$ satisfies the equality

$$\|x^* - x_{\delta}\| = min\{\|x - x_{\delta}\| \mid x \in A^{-1}z^0\}. \tag{5.1.20}$$

Now in the proof of the residual principle, it suffices to replace $x_*^0 \in A^{-1}(\theta_H)$ by $x^* \in A^{-1}z^0$.

b) The hemicontinuity property of A can be omitted. Recall that A is maximal monotone and, as well known, its inverse operator A^{-1} is maximal monotone too, so, they are multiple-valued, in general. By Theorem 1.4.9, Ax_0 is a convex closed set for all $x_0 \in D(A)$. Since H is a reflexive strictly convex space, there exists a unique element y_0^*, satisfying the condition

$$\|y_0^* - z^0\| = min\{\|y - z^0\| \mid y \in Ax_0\}. \tag{5.1.21}$$

Presuming $z^0 \notin Ax_0$, we can prove again all the assertions of this section.

4. Next we deal with the more general case. Suppose that X is a strictly convex Banach space, X^* is an E-space, $A : X \to 2^{X^*}$ is a maximal monotone operator, and, as before, $J^* : X^* \to X$ is a duality mapping in X^*. Consider the equation

$$x + \alpha J^*(Ax - z^0) = x_\delta, \qquad (5.1.22)$$

where $z^0 \notin Ax_0$. Let x_α^δ be a solution of (5.1.22). Then there exists $y_\alpha^\delta \in Ax_\alpha^\delta$ such that

$$x_\alpha^\delta + \alpha J^*(y_\alpha^\delta - z^0) = x_\delta. \qquad (5.1.23)$$

Applying now Theorem 1.7.4 to the equation

$$A^{-1}y + \alpha J^*(y - z^0) = x_\delta, \quad \alpha > 0, \qquad (5.1.24)$$

with maximal monotone operator $A^{-1} : X^* \to 2^X$, we conclude that there exists a unique element $y = y_\alpha^\delta \in Ax_\alpha^\delta$ satisfying the equation (5.1.24). Hence, x_α^δ in (5.1.23) is also unique.

By the monotonicity of A,

$$\langle y_\alpha^\delta - y_0, x_\alpha^\delta - x_0 \rangle = \langle y_\alpha^\delta - y_0, x_\alpha^\delta - x_\delta \rangle + \langle y_\alpha^\delta - y_0, x_\delta - x_0 \rangle \geq 0. \qquad (5.1.25)$$

It is easy to see that

$$\langle y_\alpha^\delta - y_0, x_\alpha^\delta - x_\delta \rangle = -\alpha \langle y_\alpha^\delta - z^0, J^*(y_\alpha^\delta - z^0) \rangle - \alpha \langle z^0 - y_0, J^*(y_\alpha^\delta - z^0) \rangle$$

because of (5.1.23). Present the last term in the left-hand side of the inequality (5.1.25) as

$$\langle y_\alpha^\delta - y_0, x_\delta - x_0 \rangle = \langle y_\alpha^\delta - z^0, x_\delta - x_0 \rangle + \langle z^0 - y_0, x_\delta - x_0 \rangle.$$

Since

$$\langle y_\alpha^\delta - z^0, J^*(y_\alpha^\delta - z^0) \rangle = \|y_\alpha^\delta - z^0\|_*^2$$

and

$$\langle z^0 - y_0, J^*(y_\alpha^\delta - z^0) \rangle \geq -\|z^0 - y_0\|_* \|y_\alpha^\delta - z^0\|_*,$$

we have

$$\alpha \|y_\alpha^\delta - z^0\|_*^2 - \alpha \|z^0 - y_0\|_* \|y_\alpha^\delta - z^0\|_* - (\|y_\alpha^\delta - z_0\|_* + \|z^0 - y_0\|_*) \|x_\delta - x_0\| \leq 0.$$

One can be sure that the estimate $\|x_\delta - x_0\| \leq \delta$ leads to the quadratic inequality for norms of regularized solutions similar to those that we have obtained in Sections 2.1 and 2.2:

$$\|y_\alpha^\delta - z^0\|_*^2 - \left(\|z^0 - y_0\|_* + \frac{\delta}{\alpha} \right) \|y_\alpha^\delta - z^0\|_* - \frac{\delta}{\alpha} \|z^0 - y_0\|_* \leq 0. \qquad (5.1.26)$$

There are two possibilities to evaluate $\|y_\alpha^\delta - z^0\|_*$ from (5.1.26):

$$\|y_\alpha^\delta - z^0\|_* \leq 2\frac{\delta}{\alpha} + \|z^0 - y_0\|_* \qquad (5.1.27)$$

and

$$\|y_\alpha^\delta - z^0\|_* \leq \frac{\delta}{\alpha} + 2\|z^0 - y_0\|_*. \tag{5.1.28}$$

They show that if $\dfrac{\delta}{\alpha}$ is bounded as $\alpha \to 0$ then $\|y_\alpha^\delta - z^0\|_*$ is bounded as well. Hence,

$$\alpha\|y_\alpha^\delta - z^0\|_* = \|x_\alpha^\delta - x_\delta\| \to 0 \quad \text{as} \quad \alpha \to 0.$$

By analogy with Theorem 5.1.1 we prove that $y_\alpha^\delta \rightharpoonup \bar{y}_0 \in X^*$ as $\alpha \to 0$, and then $\bar{y}_0 \in Ax_0$. Return again to the equation (5.1.23) which yields the equality

$$\alpha \langle y_\alpha^\delta - \bar{y}_0, J^*(y_\alpha^\delta - z^0) - J^*(\bar{y}_0 - z^0)\rangle$$

$$+ \langle y_\alpha^\delta - \bar{y}_0, x_\alpha^\delta - x_0\rangle + \alpha\langle y_\alpha^\delta - \bar{y}_0, J^*(\bar{y}_0 - z^0)\rangle$$

$$= \langle y_\alpha^\delta - \bar{y}_0, x_\delta - x_0\rangle. \tag{5.1.29}$$

The second term in the left-hand side of (5.1.29) is non-negative because A is monotone. Therefore, in view of the monotonicity of J^*,

$$0 \;\leq\; \langle y_\alpha^\delta - \bar{y}_0, J^*(y_\alpha^\delta - z^0) - J^*(\bar{y}_0 - z^0)\rangle$$

$$\leq\; -\langle y_\alpha^\delta - \bar{y}_0, J^*(\bar{y}_0 - z^0)\rangle + \alpha^{-1}\delta\|y_\alpha^\delta - \bar{y}_0\|_*.$$

Since y_α^δ weakly converges to $\bar{y}_0$ as $\alpha \to 0$ and $\|y_\alpha^\delta - \bar{y}_0\|_*$ is bounded, we have

$$\langle y_\alpha^\delta - \bar{y}_0, J^*(y_\alpha^\delta - z^0) - J^*(\bar{y}_0 - z^0)\rangle \to 0,$$

provided that $\dfrac{\delta}{\alpha} \to 0$. On the other hand, by (1.5.3),

$$\langle y_\alpha^\delta - \bar{y}_0, J^*(y_\alpha^\delta - z^0) - J^*(\bar{y}_0 - z^0)\rangle \geq (\|y_\alpha^\delta - z^0\|_* - \|\bar{y}_0 - z^0\|_*)^2.$$

Consequently,

$$\|y_\alpha^\delta - z^0\|_* \to \|\bar{y}_0 - z^0\|_* \quad \text{as} \quad \alpha \to 0, \quad \frac{\delta}{\alpha} \to 0.$$

We know that the weak convergence and convergence of norms imply in the E-space X^* strong convergence. Thus, $y_\alpha^\delta \to \bar{y}_0 \in Ax_0$. Choose and fix an element $y \in Ax_0$. Replace in (5.1.29) $\bar{y}_0$ by y. In view of the inequality

$$\alpha\langle y_\alpha^\delta - y, J^*(y_\alpha^\delta - z^0) - J^*(y - z^0)\rangle + \langle y_\alpha^\delta - y, x_\alpha^\delta - x_0\rangle \geq 0,$$

we derive the estimate

$$\langle y_\alpha^\delta - y, J^*(y - z^0)\rangle \leq \frac{\delta}{\alpha}\|y_\alpha^\delta - y\|_*.$$

Letting $\alpha \to 0$, one gets

$$\langle \bar{y}_0 - y, J^*(y - z^0)\rangle \leq 0 \quad \forall y \in Ax_0, \quad \bar{y}_0 \in Ax_0,$$

which proves that $\bar{y}_0 = y_0^*$ (see (5.1.21)). Since y_0^* is unique, the whole sequence $\{y_\alpha^\delta\}$ converges to y_0^*.

5. Consider further the equation (5.1.23) separately with $\alpha = \alpha_1$ and $\alpha = \alpha_2$ and with fixed $\delta > 0$. It is not difficult to be sure that

$$\alpha_1 \langle y_{\alpha_1}^\delta - y_{\alpha_2}^\delta, J^*(y_{\alpha_1}^\delta - z^0) - J^*(y_{\alpha_2}^\delta - z^0) \rangle \le (\alpha_2 - \alpha_1) \langle y_{\alpha_1}^\delta - y_{\alpha_2}^\delta, J^*(y_{\alpha_2}^\delta - z^0) \rangle. \quad (5.1.30)$$

We find from (5.1.30) that

$$(\|y_{\alpha_1}^\delta - z^0\|_* - \|y_{\alpha_2}^\delta - z^0\|_*)^2 \le \frac{|\alpha_2 - \alpha_1|}{\alpha_1} \|y_{\alpha_2}^\delta - z^0\|_* \|y_{\alpha_1}^\delta - y_{\alpha_2}^\delta\|_*.$$

This inequality guarantees the continuity of the functions $\sigma(\alpha) = \|y_\alpha^\delta - z^0\|_*$ and $\rho(\alpha) = \alpha\sigma(\alpha)$ with respect to $\alpha \ge \alpha_0 > 0$. By (5.1.30), if $\alpha_2 \ge \alpha_1$ then

$$\langle y_{\alpha_1}^\delta - y_{\alpha_2}^\delta, J^*(y_{\alpha_2}^\delta - z^0) \rangle \ge 0.$$

This implies the following result:

$$\|y_{\alpha_2}^\delta - z^0\|_* \le \|y_{\alpha_1}^\delta - z^0\|_*.$$

As in the case of Hilbert space, one can show that $\|x_\alpha^\delta\|$ is bounded and

$$\lim_{\alpha \to \infty} \|y_\alpha^\delta - z^0\|_* = 0.$$

In this case, the equality

$$y_\alpha^\delta - z^0 = \alpha^{-1}(J^*)^{-1}(x_\delta - x_\alpha^\delta) = \alpha^{-1}J(x_\delta - x_\alpha^\delta)$$

is used. It is valid because $J^*J = I_X$ and $JJ^* = I_{X^*}$ in our assumption about X and X^*. Thus, $y_\alpha^\delta \to z^0$ as $\alpha \to \infty$. Assuming $z^0 \in R(A)$ we conclude that

$$\lim_{\alpha \to \infty} \rho(\alpha) = \|x^* - x_\delta\|,$$

where $x^* \in A^{-1}z^0$, that is, $z^0 \in Ax^*$. Moreover, (5.1.20) holds. Then (5.1.27) leads to the estimate

$$\rho(\alpha) = \alpha\|y_\alpha^\delta - z^0\|_* \le 2\delta + \alpha\|y_0 - z^0\|_*.$$

The rest of the proof (including the assertion that $\sigma(\alpha)$ and $\rho(\alpha)$ are single-valued and continuous functions) follows the pattern of the Hilbert space case. Let us state the final result.

Theorem 5.1.4 (the generalized residual principle). *Let X be a strictly convex Banach space, X^* be an E-space, $A : X \to 2^{X^*}$ be an unbounded maximal monotone (possibly multiple-valued) operator and $z^0 \in R(A)$. Assume that for any $\delta \in (0, \delta^*]$ there holds the condition*

$$\|x^* - x_\delta\| > k\delta^p, \quad 0 < p \le 1, \quad k > 1 + 2\delta^{*1-p}, \quad (5.1.31)$$

where $x^* \in A^{-1}z^0$ satisfies (5.1.20). Then there exists at least one $\alpha = \bar{\alpha}$ such that the generalized residual

$$\rho(\bar{\alpha}) = \bar{\alpha}\|y_{\bar{\alpha}}^\delta - z^0\|_* = k\delta^p,$$

where $y_{\bar{\alpha}}^\delta \in Ax_{\bar{\alpha}}^\delta$ and $x_{\bar{\alpha}}^\delta$ is a solution of the equation (5.1.22) with $\alpha = \bar{\alpha}$, i.e.,

$$x_{\bar{\alpha}}^\delta + \bar{\alpha}J^*(y_{\bar{\alpha}}^\delta - z^0) = x_\delta.$$

Furthermore, (i) if $p \in (0, 1)$ and $\delta \to 0$, then $y_{\bar{\alpha}}^\delta \to y_0^*$, $\dfrac{\delta}{\bar{\alpha}} \to 0$ and $\bar{\alpha} \to 0$; (ii) if $p = 1$, $\delta \to 0$ and operator A is single-valued at a point x_0, then $y_{\bar{\alpha}}^\delta \rightharpoonup y_0 = Ax_0$, $\dfrac{\delta}{\bar{\alpha}} \leq const$ and $\bar{\alpha} \to 0$.

Remark 5.1.5 *Instead of (5.1.22), it is possible to make use of the equation*

$$\alpha Ax + J(x - x_\delta) = 0.$$

Remark 5.1.6 *The problems of the value computation of approximately given monotone operators are considered in a similar way.*

Remark 5.1.7 *If we choose $z^0 \notin R(A)$, then requirement (5.1.31) in the last theorem should be omitted (see Section 3.1).*

5.2   Computation of Unbounded Semimonotone Operators

We study the operator regularization method for problems of the value computation of unbounded semimonotone maps. Let X be a strictly convex Banach space, X^* be an E-space, $A : X \to 2^{X^*}$ be a semimonotone unbounded operator with $D(A) = X$. Assume that T is a monotone operator such that $T = A + C$, where an operator $C : X \to X^*$ is strongly continuous.

If $\bar{T}$ is a maximal monotone extension of T then $\bar{A} = \bar{T} - C$. Denote

$$\bar{R}(x_0) = \{y \in X^* \mid \langle v - y, x - x_0 \rangle \geq 0 \quad \forall v \in \bar{T}x, \quad \forall x \in X\}.$$

According to Theorem 1.4.9, the set $\bar{R}(x_0)$ is convex and closed.

Definition 5.2.1 *A set*

$$R(x_0) = \{v \mid v = y - Cx_0, \ y \in \bar{R}(x_0)\}$$

is called the generalized value set of semimonotone operator A at a point x_0.

The set $R(x_0)$ is also convex and closed in X^*. Therefore, under our conditions, there exists a unique element y^* satisfying the equality

$$\|y^*\|_* = min\ \{\|y\|_* \mid y \in R(x_0)\}.$$

Let a sequence $\{x_\delta\}$ be given in place of x_0 such that $\|x_0 - x_\delta\| \leq \delta$, $\delta \in (0, \delta^*]$. Our aim is to construct a sequence $\{y^\delta\} \subset X^*$, which strongly converges to $y^* \in R(x_0)$ as $\delta \to 0$.

In the beginning, we solve the value computation problem of T at a point x_0, when operator C is known. Due to the results of Section 5.1, we are able to define the sequence $\{y_\alpha^\delta\}$, $y_\alpha^\delta \in X^*$, which strongly converges in X^* to the minimal norm element $\bar{y}_0 \in \bar{R}(x_0)$ as $\dfrac{\delta}{\alpha} \to 0$, $\alpha \to 0$. Then

$$u_\alpha^\delta = y_\alpha^\delta - Cx_\alpha^\delta \to u \in R(x_0),$$

where x_α^δ is a solution of the equation

$$\alpha Tx + J(x - x_\delta) = 0, \quad \alpha > 0,$$

in the sense of Definition 1.9.3. Note that continuity of C is enough in these arguments.

However, there are problems of this type (see, e.g., [78]) in which it is impossible to give explicit representation of A by the operator C. We present the method for solving such problems, as well.

Suppose that the space X has the M-property and there exists $r_1 > 0$ such that the inequality

$$\langle y, x - x_0 \rangle \geq 0 \quad \forall y \in \bar{A}x, \quad x \in X, \tag{5.2.1}$$

is satisfied provided that $\|x - x_0\| \geq r_1$. Consider in X the equation

$$\alpha Ax + J(x - x_\delta) = 0. \tag{5.2.2}$$

Let $\dfrac{\delta}{\alpha} \to 0$ as $\alpha \to 0$. Then we have for all $y \in \bar{A}x$, $x \in X$ and $\|x - x_0\| \geq r_1$ that

$$\begin{aligned}
\langle \alpha y + J(x - x_\delta), x - x_0 \rangle &= \alpha \langle y, x - x_0 \rangle + \|x - x_\delta\|^2 \\
&+ \langle J(x - x_\delta), x_\delta - x_0 \rangle \geq \|x - x_\delta\|\Big(\|x - x_\delta\| - \delta\Big).
\end{aligned}$$

It follows from this that there is a constant $r_2 \geq r_1$ such that if $\|x - x_0\| \geq r_2$ then the inequality

$$\langle \alpha y + J(x - x_\delta), x - x_0 \rangle \geq 0 \quad \forall y \in \bar{A}x$$

holds. By Theorem 1.10.6, we conclude that there exists at least one solution x_α^δ of the equation (5.2.2) in the sense of Definition 1.10.3 with the estimate $\|x_\alpha^\delta - x_0\| \leq r_2$.

Define the sequence $\{y_\alpha^\delta\}$ as

$$y_\alpha^\delta = -\alpha^{-1} J(x_\alpha^\delta - x_\delta). \tag{5.2.3}$$

Theorem 5.2.2 *Let X be strictly convex and have the M-property, X^* be an E-space, $A : X \to 2^{X^*}$ be a semimonotone operator with $D(A) = X$, $\|x_\delta - x_0\| \leq \delta$, $\dfrac{\delta}{\alpha} \to 0$ as $\alpha \to 0$, and the condition (5.2.1) hold. Then the sequence $\{y_\alpha^\delta\}$ defined by (5.2.3) converges strongly in X^* to the minimal norm element $y^* \in R(x_0)$.*

Proof. Rewrite (5.2.3) as

$$\alpha y_\alpha^\delta = -J(x_\alpha^\delta - x_\delta).$$

Since $J^* = J^{-1}$, where J^* is a duality mapping in X^*, there exists $y_\alpha^\delta \in \bar{A}x_\alpha^\delta$ such that

$$\alpha J^* y_\alpha^\delta + x_\alpha^\delta - x_\delta = \theta_X. \tag{5.2.4}$$

For all $x \in X$ and $y \in \bar{A}x$, from the inequality

$$
\begin{aligned}
\langle y, \alpha J^* y + x - x_\delta \rangle &= \alpha \|y\|_*^2 + \langle y, x - x_0 \rangle + \langle y, x_0 - x_\delta \rangle \\
&\geq \alpha \|y\|_* \left(\|y\|_* - \frac{\delta}{\alpha} \right) + \langle y, x - x_0 \rangle,
\end{aligned}
$$

it results that $\{y_\alpha^\delta\}$ is bounded in X^*. Hence, $y_\alpha^\delta \rightharpoonup \bar{y} \in X^*$. Then, by (5.2.4), one gets

$$\|x_\alpha^\delta - x_0\| \leq \|x_\alpha^\delta - x_\delta\| + \|x_\delta - x_0\| \leq \alpha \|y_\alpha^\delta\|_* + \delta.$$

Consequently, $x_\alpha^\delta \to x_0$ as $\alpha \to 0$.

We show that $\bar{y} \in R(x_0)$. Indeed, the monotonicity of $\bar{T}$ implies

$$\langle v - y_\alpha^\delta - Cx_\alpha^\delta, x - x_\alpha^\delta \rangle \geq 0 \quad \forall v \in \bar{T}x, \quad \forall x \in X. \tag{5.2.5}$$

Since the operator $C : X \to X^*$ is strongly continuous, we have that $Cx_\alpha^\delta \to Cx_0$ as $\alpha \to 0$. Letting $\alpha \to 0$ in (5.2.5), we derive the inequality

$$\langle v - \bar{y} - Cx_0, x - x_0 \rangle \geq 0 \quad \forall v \in \bar{T}x, \quad \forall x \in X.$$

This means that $\bar{y} + Cx_0 \in \bar{T}x_0$ or $\bar{y} \in R(x_0)$. Rewrite now (5.2.4) in the equivalent form:

$$\alpha J^* y_\alpha^\delta - \alpha J^* u + x_\alpha^\delta - x_0 = x_\delta - x_0 - \alpha J^* u \quad \forall u \in R(x_0).$$

Then one has

$$
\begin{aligned}
\alpha \langle z_\alpha^\delta - v, J^* y_\alpha^\delta - J^* u \rangle + \langle z_\alpha^\delta - v, x_\alpha^\delta - x_0 \rangle \\
= \langle z_\alpha^\delta - v, x_\delta - x_0 \rangle - \alpha \langle z_\alpha^\delta - v, J^* u \rangle,
\end{aligned}
\tag{5.2.6}
$$

where $u \in R(x_0)$ and

$$z_\alpha^\delta = y_\alpha^\delta + Cx_\alpha^\delta, v = u + Cx_0.$$

Using further the properties of J^* we conclude that

$$\langle z_\alpha^\delta - v, J^* y_\alpha^\delta - J^* u \rangle \geq (\|y_\alpha^\delta\|_* - \|u\|_*)^2 + \langle Cx_\alpha^\delta - Cx_0, J^* y_\alpha^\delta - J^* u \rangle. \tag{5.2.7}$$

Since $z_\alpha^\delta \in \bar{T}x_\alpha^\delta$ and $v \in \bar{T}x_0$, then the monotonicity of $\bar{T}$ gives

$$\langle z_\alpha^\delta - v, x_\alpha^\delta - x_0 \rangle \geq 0.$$

Combining (5.2.6) and (5.2.7) with $u = \bar{y}$, we come to the relation

$$(\|y_\alpha^\delta\|_* - \|\bar{y}\|_*)^2 \quad + \quad \langle Cx_\alpha^\delta - Cx_0, J^* y_\alpha^\delta - J^* \bar{y} \rangle$$

$$\leq \quad \frac{\delta}{\alpha} \|z_\alpha^\delta - \bar{v}\|_* - \langle z_\alpha^\delta - \bar{v}, J^* \bar{y} \rangle, \tag{5.2.8}$$

where $\bar{v} = \bar{y} + Cx_0$. Hence, $\|y_\alpha^\delta\|_* \to \|\bar{y}\|_*$ as $\alpha \to 0$ and $\dfrac{\delta}{\alpha} \to 0$. Taking into account the fact that X^* is a E-space, in which weak convergence and convergence of norms imply strong convergence, one gets that $y_\alpha^\delta \to \bar{y}$. Next we use the proof scheme of Section 2.2. Presuming $\bar{y} = y$ and $\bar{v} = y + Cx_0$ in (5.2.8), where y is an arbitrary element of $R(x_0)$, and passing to the limit as $\alpha \to 0$ we obtain that $\bar{y} = y^*$. ∎

In place of the equation (5.2.2), we may consider the following more general regularized equation:

$$\alpha(Ax - z^0) + J(x - x_\delta) = 0, \tag{5.2.9}$$

where z^0 is a fixed element of X^*. In this case, the sequence $\{y_\alpha^\delta\}$ defined by the equality

$$y_\alpha^\delta = z^0 - \alpha^{-1} J(x_\alpha^\delta - x_\delta),$$

where x_α^δ is a solution of (5.2.9), converges by norm of X^* as $\dfrac{\delta}{\alpha} \to 0$, $\alpha \to 0$ to the element $y_0^* \in R(x_0)$ satisfying the condition:

$$\|y_0^* - z^0\| = min\{\|y - y^0\| \mid y \in R(x_0)\}.$$

Remark 5.2.3 *It is also possible to prove the strong convergence of the regularization method in the case when, instead of the operator A, a sequence A^h of the semimonotone maps is given with $D(A^h) = X$,*

$$\mathcal{H}_{X^*}(R(x), R^h(x)) \leq g(\|x\|)h \quad \forall x \in X.$$

Here $R^h(x)$ is the generalized value set of the operator A^h at a point x.

5.3 Computation of Unbounded Accretive Operators

Let X be a reflexive Banach space, X^* be strongly convex, $A : X \to X$ be an accretive operator with domain $D(A)$, $x_0 \in D(A)$ and

$$y_0 = Ax_0. \tag{5.3.1}$$

Suppose that, instead of x_0, its perturbed values x_δ are given such that $\|x_\delta - x_0\| \leq \delta$, where $\delta \in (0, \delta^*]$. Using elements x_δ we will construct approximations to $y_0 \in X$. Below we present the computation value theorem for unbounded accretive operators in a more precise setting.

1. We sudy first the case of hemicontinuous operators.

Theorem 5.3.1 *Assume that X is an E-space and it possesses an approximation, X^* is strongly convex, $A : X \to X$ is an accretive hemicontinuous operator, $D(A) = X$, duality mapping $J : X \to X^*$ is continuous and weak-to-weak continuous, $\dfrac{\delta}{\alpha} \to 0$ as $\alpha \to 0$. Then the sequence $\{Ax_\alpha^\delta\}$, where x_α^δ are solutions of the regularized equation*

$$x + \alpha Ax = x_\delta, \quad \alpha > 0, \tag{5.3.2}$$

converges by the norm of X to y_0 as $\alpha \to 0$.

   Proof. First of all, observe that, due to Theorem 1.15.23, the equation (5.3.2) is uniquely solvable. Then

$$x_\alpha^\delta + \alpha Ax_\alpha^\delta = x_\delta. \tag{5.3.3}$$

Hence, Ax_α^δ is well defined. Apply further the proof scheme of Theorem 5.1.1 considering the following auxiliary equation:

$$x_\alpha^0 + \alpha Ax_\alpha^0 = x_0. \tag{5.3.4}$$

Since X is smooth, it is obvious that

$$J(x_\alpha^0 - x_0) = -\alpha J(Ax_\alpha^0),$$

and, by the accretiveness condition of A,

$$-\alpha \langle J(Ax_\alpha^0), Ax_\alpha^0 - Ax_0 \rangle = \langle J(x_\alpha^0 - x_0), Ax_\alpha^0 - Ax_0 \rangle \geq 0.$$

This inequality yields the estimate

$$\|Ax_\alpha^0\| \leq \|Ax_0\| = \|y_0\|, \tag{5.3.5}$$

that is, $\{Ax_\alpha^0\}$ is bounded. Then by (5.3.4), we easily obtain the limit result:

$$x_\alpha^0 \to x_0 \quad \text{as} \quad \alpha \to 0.$$

From the boundedness of $\{Ax_\alpha^0\}$, it also follows that $Ax_\alpha^0 \rightharpoonup \bar{y} \in X$ as $\alpha \to 0$. Then, by Lemma 1.15.12 and by Theorem 1.15.14, we deduce that $\bar{y} = Ax_0 = y_0$. Moreover, taking into account (5.3.5) and the weak convergence of $\{Ax_\alpha^0\}$ to $\bar{y}$ in E-space, it is proved, as in Theorem 5.1.1, that $Ax_\alpha^0 \to Ax_0$.
   Show now the strong convergence of $\{Ax_\alpha^\delta\}$ to $\bar{y}$. Firstly, the equations (5.3.3) and (5.3.4) together give

$$\alpha \langle J(x_\alpha^\delta - x_\alpha^0), Ax_\alpha^\delta - Ax_\alpha^0 \rangle + \|x_\alpha^\delta - x_\alpha^0\|^2 = \langle J(x_\alpha^\delta - x_\alpha^0), x_\delta - x_0 \rangle.$$

Using again the accretiveness of A we have

$$\|x_\alpha^\delta - x_\alpha^0\|^2 \leq \langle J(x_\alpha^\delta - x_\alpha^0), x_\delta - x_0 \rangle \leq \|x_\alpha^\delta - x_\alpha^0\| \|x_\delta - x_0\|.$$

Thus,

$$\|x_\alpha^\delta - x_\alpha^0\| \leq \|x_\delta - x_0\| \leq \delta. \tag{5.3.6}$$

Secondly, the same equations (5.3.3) and (5.3.4) allow us to derive the following relations:

$$\|Ax_\alpha^\delta - Ax_\alpha^0\| = \frac{1}{\alpha}\|x_\delta - x_\alpha^\delta - x_0 + x_\alpha^0\| \le \frac{1}{\alpha}\left(\|x_\delta - x_0\| + \|x_\alpha^\delta - x_\alpha^0\|\right). \tag{5.3.7}$$

We see that the estimate

$$\|Ax_\alpha^\delta - Ax_\alpha^0\| \le \frac{2\delta}{\alpha} \tag{5.3.8}$$

arises from (5.3.7) and (5.3.6). Hence, the conditions of the theorem give

$$\|Ax_\alpha^\delta - Ax_\alpha^0\| \to 0 \quad \text{as} \quad \alpha \to 0.$$

Then the assertion of the theorem follows from the evident inequality

$$\|Ax_\alpha^\delta - Ax_0\| \le \|Ax_\alpha^\delta - Ax_\alpha^0\| + \|Ax_\alpha^0 - Ax_0\|.$$

The proof is accomplished. ∎

Corollary 5.3.2 *Under the condition of Theorem 5.3.1, if $\{Ax_\alpha^\delta\}$ is bounded as $\dfrac{\delta}{\alpha} \to 0$, $\alpha \to 0$, then any weak accumulation point of $\{Ax_\alpha^\delta\}$ belongs to $R(A)$.*

Proof. It results from (5.3.3) that $x_\alpha^\delta \to x_0$. On the other hand, the boundedness of $\{Ax_\alpha^\delta\}$ guarantees that there exist some $\bar{y}$ such that

$$Ax_\alpha^\delta \rightharpoonup \bar{y} \in X.$$

Due to Theorem 1.15.14, maximal accretive operators are weak closed. Then $\bar{y} = Ax_0$, that is, $Ax_0 = y_0 \in R(A)$. ∎

2. From here, we consider the case of an m-accretive (possibly, multiple-valued) operator $A : X \to 2^X$. Let $Ax_0 \ne \emptyset$. As it already earlier observed (see Lemmas 1.15.13 and 1.15.20), Ax_0 is a convex and closed set. We are going to find an element

$$y_0 \in Ax_0. \tag{5.3.9}$$

Theorem 5.3.3 *Let X be an E-space, X^* be strictly convex, J be a continuous and weak-to-weak continuous duality mapping in X, $A : X \to 2^X$ be an m-accretive operator, $\dfrac{\delta}{\alpha} \to 0$ as $\alpha \to 0$. Then sequence $\{y_\alpha^\delta\}$, where $y_\alpha^\delta \in Ax_\alpha^\delta$, x_α^δ is a solution of the regularized equation (5.3.2) and*

$$x_\alpha^\delta + \alpha y_\alpha^\delta = x_\delta, \tag{5.3.10}$$

converges strongly to the minimal norm element $y_0^* \in Ax_0$ as $\alpha \to 0$.

Proof. Since $B = \alpha A + I$ is strictly accretive, the equation (5.3.2) has a unique solution x_α^δ for all $\alpha > 0$ and for all $x_\delta \in X$. Hence, there exists a unique element $y_\alpha^\delta \in Ax_\alpha^\delta$ such that (5.3.10) holds. We study the properties of $\{y_\alpha^\delta\}$. Let x_α^0 be a solution of (5.3.2) with the right-hand side x_0, that is, there exists $y_\alpha^0 \in Ax_\alpha^0$ such that

$$x_\alpha^0 + \alpha y_\alpha^0 = x_0. \tag{5.3.11}$$

As in the proof of the previous theorem, it is established that

$$\|y_\alpha^0\| \le \|y\| \quad \forall y \in Ax_0.$$

Therefore,

$$\|y_\alpha^0\| \le \|y_0^*\|, \tag{5.3.12}$$

where

$$\|y_0^*\| = min\{\|y\| \mid y \in Ax_0\}. \tag{5.3.13}$$

Note that y_0^* satisfying (5.3.13) exists and it is unique in E-space X. Since the sequence $\{y_\alpha^0\}$ is bounded, there is $\bar{y} \in X$ such that

$$y_\alpha^0 \rightharpoonup \bar{y} \quad \text{as} \quad \alpha \to 0.$$

Moreover, the equality (5.3.11) guarantees strong convergence of x_α^0 to x_0. Recall that the operator A is maximal accretive (as m-accretive), and duality mapping J is continuous. Then we have, by Lemma 1.15.12, that $\bar{y} \in Ax_0$. Now the weak convergence of y_α^0 to $\bar{y}$ and (5.3.12) allow us to conclude that $\bar{y} = y_0^*$ and

$$\|y_\alpha^0\| \to \|y_0^*\| \quad \text{as} \quad \alpha \to 0.$$

This means in E-space X the strong convergence of $y_\alpha^0 \to y_0^*$. Using (5.3.1) and (5.3.10), as in Theorem 5.3.1, one can prove that

$$\|y_\alpha^\delta - y_\alpha^0\| \to 0$$

because $\dfrac{\delta}{\alpha} \to 0$ when $\alpha \to 0$. Finally, from the inequality

$$\|y_\alpha^\delta - y_0^*\| \le \|y_\alpha^\delta - y_\alpha^0\| + \|y_\alpha^0 - y_0^*\|,$$

it follows that

$$y_\alpha^\delta \to y_0^* \quad \text{as} \quad \alpha \to 0. \tag{5.3.14}$$

The theorem is proved. ∎

Remark 5.3.4 *We are able to study the regularized equation (5.3.2) in a more general form*

$$x + \alpha(Ax - z^0) = x_\delta,$$

where $z^0 \in X$ is a fixed element. The assertions of Theorems 5.3.1 and 5.3.3 remain still valid; further, the element $y_0^* \in Ax_0$ is defined as

$$\|y_0^* - z^0\| = min\{\|y - z^0\| \mid y \in Ax_0\}.$$

3. We next investigate a choice possibility of the regularization parameter α from the residual principle in the value computation problem of unbounded accretive operators. Suppose that the conditions of Theorem 5.3.3 are satisfied. As usual, we first study the properties of the functions $\sigma(\alpha) = \|y_\alpha^\delta\|$ and $\rho(\alpha) = \|x_\alpha^\delta - x_0\| = \alpha\|y_\alpha^\delta\|$, where x_α^δ and y_α^δ are from (5.3.10) and $\delta \in (0, \delta^*]$ is fixed. It is easily established that $\sigma(\alpha)$ and $\rho(\alpha)$ are single-valued because the equation (5.3.2) has a unique solution x_α^δ for each $\alpha > 0$. Write down (5.3.10) with $\alpha = \beta$ and $y_\beta^\delta \in Ax_\beta^\delta$ as

$$x_\beta^\delta + \beta y_\beta^\delta = x_\delta. \qquad (5.3.15)$$

By (5.3.10) and (5.3.15), we find

$$J(x_\alpha^\delta - x_\beta^\delta) = J(\beta y_\beta^\delta - \alpha y_\alpha^\delta). \qquad (5.3.16)$$

Since A is accretive, one has

$$\langle J(x_\alpha^\delta - x_\beta^\delta), y_\alpha^\delta - y_\beta^\delta \rangle \geq 0.$$

Taking into consideration (5.3.16) we obtain

$$\langle J(\beta y_\beta^\delta - \alpha y_\alpha^\delta), y_\alpha^\delta - y_\beta^\delta \rangle \geq 0.$$

This is equivalent to the inequality

$$\langle J(\alpha y_\alpha^\delta - \beta y_\beta^\delta), \alpha y_\alpha^\delta - \beta y_\beta^\delta + (\beta - \alpha)y_\beta^\delta \rangle \leq 0.$$

Then the properties of duality mapping J yield

$$\|\alpha y_\alpha^\delta - \beta y_\beta^\delta\| \leq |\beta - \alpha| \, \|y_\beta^\delta\|.$$

Therefore,

$$\left|\alpha\|y_\alpha^\delta\| - \beta\|y_\beta^\delta\|\right| \leq \|\alpha y_\alpha^\delta - \beta y_\beta^\delta\| \leq |\beta - \alpha| \, \|y_\beta^\delta\|. \qquad (5.3.17)$$

As in Theorem 5.3.1, the inequality

$$\|y_\alpha^\delta - y_\alpha^0\| \leq \frac{2\delta}{\alpha}$$

together with (5.3.12) give the following estimate:

$$\|y_\alpha^\delta\| \leq \|y_0^*\| + \frac{2\delta}{\alpha}. \qquad (5.3.18)$$

This shows that $\{y_\alpha^\delta\}$ is bounded if $\alpha \geq \alpha_0 > 0$. Then the continuity of $\rho(\alpha) = \alpha\|y_\alpha^\delta\|$ follows from (5.3.17).

We study the behavior of $\rho(\alpha)$ as $\alpha \to \infty$. Let $\theta_X \in R(A)$, $N_0 = \{x \mid \theta_X \in Ax\}$. Then the accretiveness condition of A implies

$$\langle J(x_\alpha^\delta - x), y_\alpha^\delta \rangle \geq 0 \quad \forall x \in N_0. \qquad (5.3.19)$$

The equation (5.3.10) yields the equality

$$y_\alpha^\delta = \frac{x_\delta - x_\alpha^\delta}{\alpha},\tag{5.3.20}$$

and (5.3.19) can be rewritten in the form

$$\langle J(x_\alpha^\delta - x), x_\delta - x_\alpha^\delta \rangle \geq 0 \quad \forall x \in N_0.\tag{5.3.21}$$

This leads to the estimate

$$\|x_\alpha^\delta - x\|^2 + \langle J(x - x_\alpha^\delta), x_\delta - x \rangle \leq 0 \quad \forall x \in N_0.\tag{5.3.22}$$

Thus,

$$\|x_\alpha^\delta - x\| \leq \|x_\delta - x\| \quad \forall x \in N_0,\tag{5.3.23}$$

that is, $\{x_\alpha^\delta\}$ is bounded and, by (5.3.20), $y_\alpha^\delta \to \theta_X$. In addition, the weak convergence of x_α^δ to $\bar{x} \in X$ as $\alpha \to \infty$ is established as before. Since J is weak-to-weak continuous, we deduce by help of Lemma 1.15.12 that $\theta_X \in A\bar{x}$ which means that $\bar{x} \in N_0$.

Pay attention to the inequality (5.3.22). Assuming there $x = \bar{x}$ and passing to the limit as $\alpha \to \infty$, one gets

$$\lim_{\alpha \to \infty} x_\alpha^\delta = \bar{x}.$$

At the same time, the inequality (5.3.21) gives, as $\alpha \to \infty$,

$$\langle J(\bar{x} - x), x_\delta - \bar{x} \rangle \geq 0 \quad \forall x \in N_0.\tag{5.3.24}$$

Show that $\bar{x}$ defined by (5.3.24) is unique. Let $\alpha \to \infty$ and $\{x_\alpha^\delta\}$ have two accumulation points denoted by $\bar{x}_1$ and $\bar{x}_2$. Then for them (5.3.24) holds, that is,

$$\langle J(\bar{x}_1 - x), x_\delta - \bar{x}_1 \rangle \geq 0 \quad \forall x \in N_0$$

and

$$\langle J(\bar{x}_2 - x), x_\delta - \bar{x}_2 \rangle \geq 0 \quad \forall x \in N_0.$$

Assuming in the first and in the second inequalities $x = \bar{x}_2$ and $x = \bar{x}_1$, respectively, and summing them, we thus obtain

$$\langle J(\bar{x}_1 - \bar{x}_2), \bar{x}_1 - \bar{x}_2 \rangle \leq 0,$$

from which the equality $\bar{x}_1 = \bar{x}_2$ follows. Hence, the whole sequence $\{x_\alpha^\delta\}$ converges as $\alpha \to \infty$ to $\bar{x} \in N_0$, that satisfies inequality (5.3.24). In that case

$$\rho(\alpha) = \alpha\|y_\alpha^\delta\| = \|x_\alpha^\delta - x_\delta\| \to \|\bar{x} - x_\delta\| \quad \text{as} \quad \alpha \to \infty.$$

The properties proved above of the generalized residual enable us to establish a generalized residual principle for the value computation problem of accretive operators.

Theorem 5.3.5 *Assume that X is an E-space, a duality mapping J in X is continuous and weak-to-weak continuous, $A : X \to 2^X$ in the problem (5.3.9) is an unbounded m-accretive operator. $N_0 = \{x \mid \theta_X \in Ax\} \neq \emptyset$, $\|x_0 - x_\delta\| \leq \delta$, $\delta \in (0, \delta^*]$,*

$$\|\bar{x} - x_\delta\| > k\delta^p, \quad k = 1 + 2\delta^{*\,1-p}, \quad p \in (0, 1], \tag{5.3.25}$$

where $\bar{x} \in N_0$ and satisfies the inequality (5.3.24). Then there exists at least one $\alpha = \bar{\alpha}$ such that

$$\rho(\bar{\alpha}) = \|x_{\bar{\alpha}}^\delta - x_\delta\| = \bar{\alpha}\|y_{\bar{\alpha}}^\delta\| = k\delta^p, \tag{5.3.26}$$

where $x_{\bar{\alpha}}^\delta$ is a solution of the equation (5.3.2) with $\alpha = \bar{\alpha}$. Moreover, $\bar{\alpha} \to 0$ as $\delta \to 0$, and (i) if $p \in (0, 1)$ then $y_{\bar{\alpha}}^\delta \to y_0^*$, where y_0^* is defined by (5.3.13), and $\dfrac{\delta}{\bar{\alpha}} \to 0$ as $\delta \to 0$; (ii) if $p = 1$ and A is single-valued at x_0 then $y_{\bar{\alpha}}^\delta \rightharpoonup y_0 = Ax_0$ and there exists a constant $C > 0$ such that $\dfrac{\delta}{\bar{\alpha}} \leq C$ as $\delta \to 0$.

Proof. Choosing $\alpha < \dfrac{\delta^p}{\|y_0^*\|}$ we deduce by (5.3.18) that

$$\rho(\alpha) < k\delta^p.$$

Since $\rho(\alpha)$ is continuous and

$$\lim_{\alpha \to \infty} \rho(\alpha) = \|\bar{x} - x_\delta\| > k\delta^p,$$

we have (5.3.26) with $\bar{\alpha} \geq \dfrac{\delta^p}{\|y_0^*\|}$. Then

$$\|y_{\bar{\alpha}}^\delta\| = \frac{k\delta^p}{\bar{\alpha}} \leq k\|y_0^*\|.$$

Observe that the dependence of $\left\{\dfrac{\delta}{\bar{\alpha}}\right\}$ on p is proved as in Theorem 5.1.4. This allows us to consider that $y_{\bar{\alpha}}^\delta \rightharpoonup \bar{y} \in X$ as $\delta \to 0$. By (5.3.26), we conclude that $x_{\bar{\alpha}}^\delta \to x_0$ as $\delta \to 0$. In its turn, demiclosedness of A yields the inclusion $\bar{y} \in Ax_0$. Obviously, (5.3.18) holds for $y_\alpha^\delta = y_{\bar{\alpha}}^\delta$ and $y_0^* \in Ax_0$. Now we are able to show, similarly to Theorem 5.1.4, that $\bar{y} = y_0^*$ and $\|y_{\bar{\alpha}}^\delta\| \to \|y_0^*\|$. Since X is the E-space, the strong convergence of $y_{\bar{\alpha}}^\delta$ to y_0^* is thus established for $p \in (0, 1)$. If $p = 1$ and A is a single-valued operator, then $\left\{\dfrac{\delta}{\bar{\alpha}}\right\}$ is bounded and the sequence $\{y_{\bar{\alpha}}^\delta\}$ weakly converges to $y_0 = Ax_0$ as $\delta \to 0$. In order to be sure that $\lim_{\delta \to 0} \bar{\alpha} = 0$, it is necessary to repeat the corresponding arguments of the proof of Theorem 5.1.4. ∎

Remark 5.3.6 *If $\theta_X \notin R(A)$ then requirement (5.3.25) of Theorem 5.3.5 may be omitted.*

Remark 5.3.7 *If $\theta_X \notin D(A)$ then we choose $z^0 \in R(A)$ and find regularized solutions from the following equation:*

$$x_\alpha^\delta + \alpha(y_\alpha^\delta - z^0) = x_\delta, \quad y_\alpha^\delta \in Ax_\alpha^\delta.$$

Taking $\rho(\alpha) = \alpha\|y_\alpha^\delta - z^0\|$ and $N_0 = \{x \mid z^0 \in Ax\}$, one can get all the conclusions of Theorem 5.3.5. With this, y_0^* is defined as the minimization problem:

$$\|y_0^* - z^0\| = \min\,\{\|y - z^0\| \mid y \in Ax_0\}.$$

5.4   Hammerstein Type Operator Equations

Let X be an E-space, X^* be strictly convex, $A : X \rightarrow 2^{X^*}$ and $B : X^* \rightarrow 2^X$ be maximal monotone operators with domains $D(A)$ and $D(B)$, respectively, $f \in X$. An equation

$$x + BAx = f \tag{5.4.1}$$

is said to be the Hammerstein type equation.

1. First of all, study the solvability of the problem (5.4.1). Let $C = A^{-1} : X^* \rightarrow 2^X$ and $\phi \in Ax$. Then $x \in A^{-1}\phi = C\phi$, and (5.4.1) may be rewritten as

$$C\phi + B\phi = f. \tag{5.4.2}$$

Recall that C is a maximal monotone operator because A is so. In order to apply to (5.4.2) the known existence theorems for equations with maximal monotone operators, it is necessary to impose such conditions on A and B which guarantee that the sum $C + B$ is also maximal monotone. By Theorem 1.8.3, these conditions are the following:

$$int \ D(C) \cap D(B) \neq \emptyset \quad \text{or} \quad D(C) \cap int \ D(B) \neq \emptyset.$$

Turning to operators A and B of the equation (5.4.1) we see that the operator $C + B$ is maximal monotone if

$$int \ R(A) \cap D(B) \neq \emptyset \quad \text{or} \quad R(A) \cap int \ D(B) \neq \emptyset. \tag{5.4.3}$$

Joining (5.4.3) to Theorem 1.7.5 we get

Theorem 5.4.1 *If $A : X \rightarrow 2^{X^*}$ and $B : X^* \rightarrow 2^X$ are maximal monotone operators, operator $A^{-1} + B$ is coercive and the condition (5.4.3) holds, then equation (5.4.1) with any right-hand side $f \in X$ has at least one solution.*

We present some assertions necessary to construct the operator regularization method for the equation (5.4.2). Consider the space product (see [206])

$$\mathcal{Z} = X \times X^* = \{\zeta = [x, \phi] \mid x \in X, \ \phi \in X^*\}$$

with the natural linear operation

$$\alpha\zeta_1 + \beta\zeta_2 = [\alpha x_1 + \beta x_2, \alpha\phi_1 + \beta\phi_2],$$

where α and β are real numbers, $\zeta_1 = [x_1, \phi_1]$ and $\zeta_2 = [x_2, \phi_2]$ are elements from $\mathcal{Z}$. If the norm of an element $\zeta = [x, \phi]$ is defined by the formula

$$\|\zeta\|_{\mathcal{Z}} = \left(\|x\|^2 + \|\phi\|_*^2\right)^{1/2}$$

then $\mathcal{Z}$ is the Banach space and $\mathcal{Z}^* = X^* \times X$ is its dual space. A pairing of the spaces $\mathcal{Z}$ and $\mathcal{Z}^*$ is given by the dual product of elements $\zeta = [x, \phi] \in \mathcal{Z}$ and $\eta^* = [\psi, y] \in \mathcal{Z}^*$ with respect to the equality:

$$\langle \eta^*, \zeta \rangle = \langle \psi, x \rangle + \langle \phi, y \rangle.$$

Lemma 5.4.2 *Let $\{\zeta_n\}$, $n = 1, 2, ...$, be a sequence of $\mathcal{Z}$ such that $\zeta_n = [x_n, \phi_n]$, and let $\zeta_0 = [x_0, \phi_0] \in \mathcal{Z}$. Then the limit relations*

$$\zeta_n \rightharpoonup \zeta_0, \quad \|\zeta_n\|_{\mathcal{Z}} \to \|\zeta_0\|_{\mathcal{Z}} \tag{5.4.4}$$

and

$$x_n \rightharpoonup x_0, \quad \phi_n \rightharpoonup \phi_0, \quad \|x_n\| \to \|x_0\|, \quad \|\phi_n\|_* \to \|\phi_0\|_* \tag{5.4.5}$$

are equivalent as $n \to \infty$.

Proof. The implication $(5.4.5) \implies (5.4.4)$ is obvious. Let $(5.4.4)$ be valid and assume that $\eta^* = [\psi, y]$ is an element of $\mathcal{Z}^*$. Then the limit $\langle \eta^*, \zeta_n \rangle \to \langle \eta^*, \zeta_0 \rangle$ as $n \to \infty$ implies

$$\langle \psi, x_n \rangle + \langle \phi_n, y \rangle \to \langle \psi, x_0 \rangle + \langle \phi_0, y \rangle. \tag{5.4.6}$$

If we put $y = \theta_X$ in $(5.4.6)$ then

$$\langle \psi, x_n \rangle \to \langle \psi, x_0 \rangle \quad \forall \psi \in X^*,$$

that is, $x_n \rightharpoonup x_0$ as $n \to \infty$. If $\psi = \theta_{X^*}$ in $(5.4.6)$ then $\phi_n \rightharpoonup \phi_0$, that leads to the estimates

$$\|x_0\| \leq \liminf_{n \to \infty} \|x_n\| \quad \text{and} \quad \|\phi_0\|_* \leq \liminf_{n \to \infty} \|\phi_n\|_*. \tag{5.4.7}$$

Since

$$\|\zeta_n\|_{\mathcal{Z}} \to \|\zeta_0\|_{\mathcal{Z}} = \left(\|x_0\|^2 + \|\phi_0\|_*^2 \right)^{1/2},$$

one can consider that $\|x_{n_k}\| \to a$ and $\|\phi_{n_k}\|_* \to b$, where $\{x_{n_k}\}$ and $\{\phi_{n_k}\}$ are subsequences of $\{x_n\}$ and $\{\phi_n\}$, respectively. Further,

$$a^2 + b^2 = \|x_0\|^2 + \|\phi_0\|_*^2.$$

Now it follows from $(5.4.7)$ that $\|x_0\| = a$ and $\|\phi_0\|_* = b$. Thus, $\|x_n\| \to \|x_0\|$ and $\|\phi_n\|_* \to \|\phi_0\|_*$. The lemma is proved. ∎

Lemma 5.4.3 *If $J : X \to X^*$ and $J^* : X^* \to X$ are normalized duality mappings in X and X^*, respectively, then the operator $J_{\mathcal{Z}} : \mathcal{Z} \to \mathcal{Z}^*$, defined by the formula*

$$J_{\mathcal{Z}}\zeta = [Jx, J^*\phi] \quad \forall \zeta = [x, \phi] \in \mathcal{Z}, \tag{5.4.8}$$

is a normalized duality mapping in $\mathcal{Z}$. And conversely: every normalized duality mapping in $\mathcal{Z}$ has the form $(5.4.8)$.

Proof. We verify Definition 1.5.1 with $\mu(t) = t$ for $J_{\mathcal{Z}}$. Toward this end, choose an element $\zeta = [x, \phi] \in \mathcal{Z}$ and, using the properties of J and J^*, find

$$\langle J_{\mathcal{Z}}\zeta, \zeta \rangle = \langle Jx, x \rangle + \langle \phi, J^*\phi \rangle = \|x\|^2 + \|\phi\|_*^2 = \|\zeta\|_{\mathcal{Z}}^2$$

and

$$\|J_{\mathcal{Z}}\zeta\|_{\mathcal{Z}^*} = \left(\|Jx\|_*^2 + \|J^*\phi\|^2 \right)^{1/2} = \left(\|x\|^2 + \|\phi\|_*^2 \right)^{1/2} = \|\zeta\|_{\mathcal{Z}}.$$

Thus , $J_{\mathcal{Z}} : \mathcal{Z} \to \mathcal{Z}^*$ is a duality mapping in $\mathcal{Z}$. Let then operator $\bar{J}_{\mathcal{Z}} : \mathcal{Z} \to \mathcal{Z}^*$ be such that

$$\langle \bar{J}_{\mathcal{Z}} \zeta, \zeta \rangle = \|\zeta\|_{\mathcal{Z}}^2 \quad \text{and} \quad \|\bar{J}_{\mathcal{Z}} \zeta\|_{\mathcal{Z}^*} = \|\zeta\|_{\mathcal{Z}}. \tag{5.4.9}$$

Assuming that $\bar{J}_{\mathcal{Z}} \zeta = \eta^* = [\psi, y]$, $\psi \in X^*$, $y \in X$, write down the following equalities:

$$\langle \bar{J}_{\mathcal{Z}} \zeta, \zeta \rangle = \langle \eta^*, \zeta \rangle = \langle \psi, x \rangle + \langle \phi, y \rangle,$$

which, in view of (5.4.9), gives

$$\|x\|^2 + \|\phi\|_*^2 = \langle \psi, x \rangle + \langle \phi, y \rangle = \|\psi\|_*^2 + \|y\|^2. \tag{5.4.10}$$

Show that $\psi = Jx$, $y = J^*\phi$. For that, it is necessary to establish that equalities $\langle \psi, x \rangle = \|\psi\|_* \|x\|$ and $\langle \phi, y \rangle = \|\phi\|_* \|y\|$ are true. Indeed, if it is not the case then, by (5.4.10), we come to a contradiction because

$$\|x\|^2 + \|\phi\|_*^2 \quad < \quad \|\psi\|_* \|x\| + \|y\| \|\phi\|_*$$

$$\leq \quad 2^{-1} \left(\|\psi\|_*^2 + \|x\|^2 \right) + 2^{-1} \left(\|y\|^2 + \|\phi\|_*^2 \right)$$

$$= \quad \|x\|^2 + \|\phi\|_*^2.$$

Hence, (5.4.10) may be rewritten in the form

$$\|x\|^2 + \|\phi\|_*^2 = \|\psi\|_* \|x\| + \|y\| \|\phi\|_* = \|\psi\|_*^2 + \|y\|^2.$$

Then

$$(\|x\| - \|\psi\|_*)^2 + (\|\phi\|_* - \|y\|)^2 = 0,$$

i.e., $\|x\| = \|\psi\|_*$ and $\|\phi\|_* = \|y\|$. Therefore, $\psi = Jx$ and $y = J^*\phi$. ∎

Lemma 5.4.4 *A space $\mathcal{Z}$ is strictly convex if and only if X and X^* are strictly convex.*

Proof. A space $\mathcal{Z}$ is strictly convex if and only if $J_{\mathcal{Z}}$ is strictly monotone. In its turn, in view of $J_{\mathcal{Z}} \zeta = [Jx, J^*\phi]$ for all $\zeta = [x, \phi] \in \mathcal{Z}$, operator $J_{\mathcal{Z}}$ is strictly monotone if and only if J and J^* are strictly monotone too. ∎

2. Pass from equation (5.4.1) to the system

$$\begin{cases} Ax - \phi = \theta_{X^*}, \\ x + B\phi = f. \end{cases}$$

It is equivalent to the following operator equation:

$$T\zeta = \hbar, \tag{5.4.11}$$

where $\zeta = [x, \phi] \in \mathcal{Z}$, $\hbar = [\theta_{X^*}, f] \in \mathcal{Z}^*$ and

$$T : \mathcal{Z} \to 2^{\mathcal{Z}^*}, \quad T\zeta = [Ax - \phi, x + B\phi] \in \mathcal{Z}^*.$$

We study now properties of the operator T. First of all, we assert that T is monotone. Indeed, for $\zeta_1 = [x_1, \phi_1] \in \mathcal{Z}$ and $\zeta_2 = [x_2, \phi_2] \in \mathcal{Z}$, the equality

$$\langle \vartheta_1 - \vartheta_2, \zeta_1 - \zeta_2 \rangle = \langle \xi_1 - \xi_2, x_1 - x_2 \rangle + \langle \phi_1 - \phi_2, w_1 - w_2 \rangle, \qquad (5.4.12)$$

holds for all $\xi_i \in Ax_i$, for all $w_i \in B\phi_i$ and for $\vartheta_i = [\xi_i - \phi_i, w_i - x_i]$, $i = 1, 2$. We note that the inclusion $\phi_i \in Ax_i$ implies $x_i \in A^{-1}\phi_i$. Therefore, (5.4.12) may be rewritten as

$$\langle \vartheta_1 - \vartheta_2, \zeta_1 - \zeta_2 \rangle = \langle \phi_1 - \phi_2, x_1 - x_2 \rangle + \langle \phi_1 - \phi_2, w_1 - w_2 \rangle,$$

where $\phi_i \in D(B)$ and $\phi_i \in D(A^{-1})$, $i = 1, 2$. From this equality follows the fact that if condition (5.4.3) is satisfied then the operator T is maximal monotone in $D(T)$.

Assume further that operators A and B are maximal monotone, in addition, A is hemicontinuous, $D(A) = X$, and one of the conditions (5.4.3) is satisfied.

Lemma 5.4.5 *If the solution set N of the equation (5.4.1) is nonempty, then it is convex and closed.*

Proof. Let M be a solution set of the equation (5.4.11). By Corollary 1.4.10, M is convex and closed in $\mathcal{Z}$. Since $M = \{[x, \phi] \mid x \in N, \ \phi \in Ax\}$, we conclude, in view of (5.4.11), that $M = N \times A(N)$. Then obviously N is convex. Show that N is closed. Let $x_n \in N$ and $x_n \to x_0$ as $n \to \infty$. Due to Theorem 1.3.20, under our conditions, operator A is demicontinuous. By the Mazur theorem, the set M is weakly closed, therefore, we get for $\zeta_n = [x_n, Ax_n] \in M$ the following limit relation:

$$\zeta_n \rightharpoonup [x_0, Ax_0] \in M.$$

Hence, $x_0 \in N$. ∎

Construct the regularized operator equation in the form

$$x + (B + \alpha J^*)(A + \alpha J)x = f^\delta, \qquad (5.4.13)$$

where $\alpha > 0$, f^δ is a δ-approximation of f such that $\|f - f^\delta\| \leq \delta$.

Lemma 5.4.6 *Equation (5.4.13) is uniquely solvable for any $f^\delta \in X$.*

Proof. Let $B^\alpha = B + \alpha J^*$ and $A^\alpha = A + \alpha J$. We introduce an operator

$$T^\alpha \zeta = [A^\alpha x - \phi, x + B^\alpha \phi] = T\zeta + \alpha J_{\mathcal{Z}}\zeta,$$

where $\zeta = [x, \phi]$ and $J_{\mathcal{Z}}\zeta = [Jx, J^*\phi]$. It is clear that solvability of (5.4.13) is equivalent to solvability of the equation

$$T^\alpha \zeta = \hbar^\delta$$

with the maximal monotone operator $T^\alpha = T + \alpha J_{\mathcal{Z}}$ and right-hand side $\hbar^\delta = [\theta_{X^*}, f^\delta]$. Then the conclusion of the lemma follows from Theorem 1.7.4. ∎

Theorem 5.4.7 *Let a solution set N of the equation (5.4.1) be not empty, $A : X \to X^*$ be a hemicontinuous monotone operator with $D(A) = X$. Assume that $B : X^* \to 2^X$ is maximal monotone, $f^\delta \in X$ are such that $\|f - f^\delta\| \leq \delta$ and one of the conditions (5.4.3) holds. If $\dfrac{\delta}{\alpha} \to 0$ as $\alpha \to 0$, then the sequence $\{x_\alpha^\delta\}$ of solutions of the regularized equation (5.4.13) converges strongly in X to the solution $\bar{x}^* \in N$ which is defined as*

$$\|\bar{x}^*\|^2 + \|A\bar{x}^*\|^2 = min\{\|x^*\|^2 + \|Ax^*\|^2 \mid x^* \in N\}.$$

Proof. Validity of this theorem follows from the results of Section 2.2 and from Lemmas 5.4.5 and 5.4.6. ∎

Under the conditions of Theorem 5.4.7 (not assuming solvability of (5.4.1)), if solutions of the regularized equation (5.4.13) converge strongly as $\alpha \to 0$ then the equation (5.4.1) is solvable. Indeed, since $\hbar^\delta \in T^\alpha \zeta_\alpha^\delta$, where $\hbar^\delta = [\theta_{X^*}, f^\delta] \in \mathcal{Z}^*$ and $\zeta_\alpha^\delta = [x_\alpha^\delta, Ax_\alpha^\delta] \in \mathcal{Z}$, the monotonicity condition of T^α can be written as

$$\langle A^\alpha x - \phi, x - x_\alpha^\delta \rangle + \langle \phi - A^\alpha x_\alpha^\delta, x + w^\alpha - f^\delta \rangle \geq 0, \quad [x, \phi] \in D(T^\alpha), \quad w^\alpha \in B^\alpha \phi.$$

By the assumptions, $x_\alpha^\delta \to \bar{x}$, $Ax_\alpha^\delta \rightharpoonup A\bar{x}$. Therefore, passing to the limit in the last inequality as $\alpha \to 0$, we come to the inequality

$$\langle Ax - \phi, x - \bar{x} \rangle + \langle \phi - A\bar{x}, x + w - f \rangle \geq 0, \quad w \in B\phi,$$

which means that $\bar{x}$ is a solution of (5.4.1). In addition, if A is weak-to-weak continuous, then for solvability of (5.4.1) it is necessary to assume that the operator method of regularization (5.4.13) converges weakly.

As in Chapters 2 and 3, we are able to consider for the equation (5.4.1) the case of approximately given operators and also state the corresponding residual principle. For instance, under the conditions of Theorem 5.4.7, the residual should be defined by the formula

$$\rho(\alpha) = \alpha \left(\|x_\alpha^\delta\|^2 + \|Ax_\alpha^\delta\|^2 \right)^{1/2}. \tag{5.4.14}$$

In addition, in all assertions above, the hemicontinuity of A can be replaced by its maximal monotonicity; and then in the proofs the local boundedness of A at any point $x \in X$ is used. At the same time, (5.4.14) is rewritten as

$$\rho(\alpha) = \alpha \left(\|x_\alpha^\delta\|^2 + \|y_\alpha^\delta\|^2 \right)^{1/2},$$

where $y_\alpha^\delta \in Ax_\alpha^\delta$ and

$$f^\delta - x_\alpha^\delta \in (B + \alpha J^*)(y_\alpha^\delta + \alpha x_\alpha^\delta).$$

5.5   Pseudo-Solutions of Monotone Equations

Let X be a Banach space, $A : X \to 2^{X^*}$ be a maximal monotone operator, $f \in X^*$. Suppose that the range $R(A)$ is a closed set. Due to Corollary 1.7.18, it is convex. Consider an equation

$$Ax = f. \tag{5.5.1}$$

Assuming that $f \notin R(A)$ we do not require in the sequel solvability of (5.5.1) in the classical sense. Our aim in this section is to study pseudo-solutions of (5.5.1).

Definition 5.5.1 *An element $z \in D(A)$ is called a pseudo-solution of the equation (5.5.1) if*

$$\|g - f\|_* = min\{\|y - f\|_* \mid y \in Ax, \ x \in D(A)\}, \quad g \in Az.$$

Under the conditions above, there exists a unique element $g \in R(A)$ such that

$$\mu = min\{\|y - f\|_* \mid y \in R(A)\} = \|g - f\|_* > 0. \tag{5.5.2}$$

Since $g \in R(A)$, nonempty pseudo-solution set N of (5.5.1) coincides with the classical solution set of the equation

$$Au = g. \tag{5.5.3}$$

Hence, N is convex and closed. It is known that in this case g belongs to the boundary of $R(A)$. If $int\ R(A) \neq \emptyset$ then N is unbounded. Indeed, by Theorem 1.7.19 applied to the operator A^{-1}, we conclude that N contains at least one semi-line.

1. Assume first that X is a Hilbert space H, $A : H \to 2^H$ is a maximal monotone operator and $f \in H$. Let $\bar{x}^*$ be a minimal norm pseudo-solution of (5.5.1), i.e., $\|\bar{x}^*\| = min\{\|x^*\| \mid x^* \in N\}$. Assume that, instead of exact right-hand side f, a sequence $\{f^\delta\}$ is given such that $\|f - f^\delta\| \leq \delta$, where $0\delta \in (0, \delta^*]$. Construct in H approximations strongly converging to $\bar{x}^* \in N$. To this end, consider the regularized operator equation

$$Ax + \alpha x = f^\delta, \quad \alpha > 0. \tag{5.5.4}$$

Denote by x_α^δ its unique solution and let

$$\mu^\delta = min\ \{\|y - f^\delta\| \mid y \in R(A)\}. \tag{5.5.5}$$

It is clear that there exists $g^\delta \in R(A)$ such that

$$\mu^\delta = \|g^\delta - f^\delta\|. \tag{5.5.6}$$

Lemma 5.5.2 *The following estimate holds:*

$$\|g - f^\delta\| \leq \mu^\delta + \delta. \tag{5.5.7}$$

Proof. By (1.3.5), the problems (5.5.2) and (5.5.5) are equivalent, respectively, to inequalities

$$(g - f, g - y) \leq 0 \quad \forall y \in R(A), \tag{5.5.8}$$

and

$$(g^\delta - f^\delta, g^\delta - y) \leq 0 \quad \forall y \in R(A). \tag{5.5.9}$$

Substituting $y = g^\delta$ for the first inequality and $y = g$ for the second one, and summing the obtained inequalities we obtain after simple algebra the estimate

$$\|g - g^\delta\| \leq \|f - f^\delta\| \leq \delta. \tag{5.5.10}$$

Hence,

$$\|g - f^\delta\| \leq \|g^\delta - f^\delta\| + \|g^\delta - g\| \leq \mu^\delta + \delta. \quad \blacksquare$$

Since the equation (5.5.1) is unsolvable in the classical sense, we conclude by Theorem 2.1.4 that $\{x_\alpha^\delta\}$ becomes unbounded as $\alpha \to 0$ and $\dfrac{\delta}{\alpha} \to 0$. Thus, the strong convergence of $\{x_\alpha^\delta\}$ to $\bar{x}^*$ may be established only if the regularization parameter α approaches some $\alpha_0 > 0$, which is unknown as well. Hence, the main problem is to point the way of choosing α in (5.5.4) ensuring convergence of $\{x_\alpha^\delta\}$ to $\bar{x}^*$ as $\delta \to 0$.

Theorem 5.5.3 *Let $A : H \to 2^H$ be a maximal monotone operator, $R(A)$ be closed and for all $x \in D(A)$ such that $\|x\| \geq r > 0$, the inequality*

$$(y - g, x) \geq 0 \quad \forall y \in Ax \tag{5.5.11}$$

be fulfilled, where g solves the problem (5.5.2). If $\theta_H \in D(A)$, then assume in addition that

$$\|y^0 - f^\delta\| > \mu^\delta + k\delta, \quad k > 1, \tag{5.5.12}$$

where

$$\|y^0 - f^\delta\| = min\,\{\|y - f^\delta\| \mid y \in A(\theta_H)\}. \tag{5.5.13}$$

Under these conditions, there exists at least one $\bar{\alpha}$ satisfying the equation

$$\rho(\bar{\alpha}) = \mu^\delta + k\delta. \tag{5.5.14}$$

If $\delta \to 0$, then every strong accumulation point of the sequence $\{x_{\bar{\alpha}}^\delta\}$, where $x_{\bar{\alpha}}^\delta$ is a solution of (5.5.4) with $\alpha = \bar{\alpha}$, belongs to the pseudo-solutions set N of (5.5.1).

Proof. Choose any fixed $x^* \in N$. There exists $y_\alpha^\delta \in Ax_\alpha^\delta$ such that

$$y_\alpha^\delta + \alpha x_\alpha^\delta = f^\delta. \tag{5.5.15}$$

Subtract $g \in Ax^*$ from both sides of the equality (5.5.15) and consider the scalar product of the obtained difference and the element $x_\alpha^\delta - x$. Then, similarly to (3.3.5), one gets

$$\|x_\alpha^\delta\| \leq 2\|x^*\| + \frac{\mu^\delta + \delta}{\alpha}. \tag{5.5.16}$$

We have proved in Lemma 3.1.4 that the function $\rho(\alpha) = \alpha \|x_\alpha^\delta\|$ with fixed δ is continuous as $\alpha \geq \alpha_0 > 0$. If $\theta_H \in D(A)$ then

$$\lim_{\alpha \to \infty} \rho(\alpha) = \|y^0 - f^\delta\|.$$

Otherwise,

$$\lim_{\alpha \to \infty} \rho(\alpha) = +\infty.$$

By (5.5.16), if $\alpha < C$, where

$$C = \frac{(k-1)\delta}{2\|x^*\|}, \quad x^* \in N,$$

then the estimate $\rho(\alpha) < \mu^\delta + k\delta$ is satisfied. Using (5.5.12) and continuity of $\rho(\alpha)$ we establish (5.5.14) with $\bar{\alpha} > C$.

Calculate the limit of μ^δ as $\delta \to 0$. The obvious inequalities

$$\mu^\delta = \|f^\delta - g^\delta\| \leq \|f - g\| + \|f - f^\delta\| + \|g - g^\delta\|$$

and

$$\mu = \|f - g\| \leq \|f^\delta - g^\delta\| + \|f - f^\delta\| + \|g - g^\delta\|$$

together with (5.5.10) imply

$$\mu - 2\delta \leq \mu^\delta \leq \mu + 2\delta.$$

Therefore,

$$\lim_{\delta \to 0} \mu^\delta = \mu. \tag{5.5.17}$$

We investigate the behavior of $\{x_{\bar{\alpha}}^\delta\}$ as $\delta \to 0$. By virtue of (5.5.17) and (5.5.14), if $y_{\bar{\alpha}}^\delta \in Ax_{\bar{\alpha}}^\delta$ and

$$y_{\bar{\alpha}}^\delta + \bar{\alpha} x_{\bar{\alpha}}^\delta = f^\delta, \tag{5.5.18}$$

then

$$\|y_{\bar{\alpha}}^\delta - f^\delta\| \to \|g - f\| = \mu.$$

Hence, the sequence $\{y_{\bar{\alpha}}^\delta\}$ is bounded and there exists $u \in R(A)$ such that $y_{\bar{\alpha}}^\delta \rightharpoonup u$. Then (5.5.14) yields the following relations:

$$\|u - f\| \leq \liminf_{\delta \to 0} \|y_{\bar{\alpha}}^\delta - f^\delta\| \leq \limsup_{\delta \to 0} \|y_{\bar{\alpha}}^\delta - f^\delta\|$$

$$= \lim_{\delta \to 0}(\mu^\delta + k\delta) = \mu = \|g - f\|.$$

Since g satisfying (5.5.2) is unique, we conclude that $u = g$. Now, due to the weak convergence of $\{y_{\bar{\alpha}}^\delta\}$ to g, strong convergence of $y_{\bar{\alpha}}^\delta$ to g follows as $\delta \to 0$.

We show by a contradiction that $\{x_{\bar{\alpha}}^\delta\}$ is bounded as $\delta \to 0$. Assume that $\|x_{\bar{\alpha}}^\delta\| \to \infty$ as $\delta \to 0$. Then for a sufficiently small $0 < \delta \leq \delta^*$, making use of (5.5.11), (5.5.14) and (5.5.18), we have

$$0 = (y_{\bar{\alpha}}^\delta + \bar{\alpha} x_{\bar{\alpha}}^\delta - f^\delta, x_{\bar{\alpha}}^\delta) = (y_{\bar{\alpha}}^\delta - g, x_{\bar{\alpha}}^\delta) + (g - f^\delta, x_{\bar{\alpha}}^\delta) + \bar{\alpha}\|x_{\bar{\alpha}}^\delta\|^2$$

$$\geq \|x_{\bar{\alpha}}^\delta\|(\mu^\delta + k\delta - \|g - f^\delta\|). \tag{5.5.19}$$

If $\theta_H \in D(A)$ and $x_{\bar{\alpha}}^\delta = \theta_H$ then (5.5.18) gives the equality $y_{\bar{\alpha}}^\delta = f^\delta$, where $y_{\bar{\alpha}}^\delta \in A(\theta_H)$. Then we obtain from (5.5.12) and (5.5.13) that

$$0 < \mu^\delta + k\delta < \|y^0 - f^\delta\| \le \|y_{\bar{\alpha}}^\delta - f^\delta\| = 0.$$

But it is impossible. Hence, $\|x_{\bar{\alpha}}^\delta\| \ne 0$. In its turn, (5.5.7) and (5.5.19) imply

$$0 \ge \|x_{\bar{\alpha}}^\delta\|(k-1)\delta > 0,$$

which contradicts the assumption that $\|x_{\bar{\alpha}}^\delta\| \to \infty$. Therefore $\{x_{\bar{\alpha}}^\delta\}$ is bounded and there exists a subsequence $\{x_{\bar{\beta}}^\delta\} \subset \{x_{\bar{\alpha}}^\delta\}$ such that $x_{\bar{\beta}}^\delta \rightharpoonup \bar{x} \in H$. Recall that the strong convergence of $y_{\bar{\alpha}}^\delta$ to $g \in H$ has been already proved. Hence, $\bar{x} \in N$.

Replacing in (5.5.18) $\bar{\alpha}$ by $\bar{\beta}$ we find that

$$(y_{\bar{\beta}}^\delta - f^\delta, x_{\bar{\beta}}^\delta) + \bar{\beta}\|x_{\bar{\beta}}^\delta\|^2 = 0.$$

Then by (5.5.14),

$$(y_{\bar{\beta}}^\delta - f^\delta, x_{\bar{\beta}}^\delta) + (\mu^\delta + k\delta)\|x_{\bar{\beta}}^\delta\| = 0.$$

Consequently,

$$\|x_{\bar{\beta}}^\delta\| = \frac{(f^\delta - y_{\bar{\beta}}^\delta, x_{\bar{\beta}}^\delta)}{\mu^\delta + k\delta}.$$

This shows that $\|x_{\bar{\beta}}^\delta\|$ has a limit as $\delta \to 0$. Moreover,

$$\lim_{\delta \to 0} \|x_{\bar{\beta}}^\delta\| \le \|\bar{x}\|.$$

Due to the weak convergence of $x_{\bar{\beta}}^\delta$ to $\bar{x}$, we get

$$\|\bar{x}\| \le \lim_{\delta \to 0} \|x_{\bar{\beta}}^\delta\| \le \|\bar{x}\|.$$

This means that $\|x_{\bar{\beta}}^\delta\| \to \|\bar{x}\|$, which implies in a Hilbert space the strong convergence of $\{x_{\bar{\beta}}^\delta\}$ to $\bar{x} \in N$ as $\delta \to 0$. The proof is accomplished. ∎

Remark 5.5.4 *According to Theorem 1.7.9, inequality (5.5.11) is one of the sufficient conditions guaranteing the inclusion $g \in R(A)$.*

Theorem 5.5.5 *Let H be a Hilbert space, $A : H \to 2^H$ be a maximal monotone operator, $R(A)$ be a closed set in H, $0 < \delta < 1$. If $\theta_H \in D(A)$, then it is additionally assumed that*

$$\|y^0 - f^\delta\| > \mu^\delta + k\delta^p, \quad k > 1, \quad p \in (0,1),$$

where y^0 is defined by (5.5.13). Then there exists at least one value $\bar{\alpha}$ of the regularization parameter α such that

$$\bar{\alpha}\|x_{\bar{\alpha}}^\delta\| = k\delta^p.$$

Moreover, $x_{\bar{\alpha}}^\delta \to \bar{x}^*$, $\dfrac{\delta}{\bar{\alpha}} \to 0$ and $\bar{\alpha} \to 0$ as $\delta \to 0$, where $x_{\bar{\alpha}}^\delta$ is a solution of the equation

$$Ax + \alpha x = g^\delta \tag{5.5.20}$$

with $\alpha = \bar{\alpha}$.

Proof. By (5.5.6), the following inequalities obviously hold:

$$\|y^0 - g^\delta\| \geq \|y^0 - f^\delta\| - \|g^\delta - f^\delta\| > k\delta^p.$$

Observe that the problem (5.5.2) is well-posed. Applying Theorem 3.1.12 and Remark 3.1.14 to equations (5.5.20) and (5.5.3), and taking into account the estimate (5.5.10), we obtain the required conclusion. ∎

2. We study the more general situation in a Banach space X. Namely, suppose now that $A : X \to 2^{X^*}$, a sequence of maximal monotone operators $\{A^h\}$ is given in place of A, $D(A) = D(A^h)$, $h \in (0, h^*]$, and

$$\mathcal{H}_{X^*}(Ax, A^h x) \leq \eta(\|x\|)h \quad \forall x \in D(A), \tag{5.5.21}$$

where $\eta(t)$ is a continuous non-negative increasing function for $t \geq 0$. We do not assume that the sets $R(A)$ and $R(A^h)$ are closed, however, presume that there exists $g \in R(A)$ satisfying (5.5.2). The incompatibility measure of approximate equations $A^h x = f^\delta$ is defined by the quantity

$$\mu^\gamma = \|g^\gamma - f^\delta\|_* = min\{\|y - f^\delta\|_* \mid y \in \overline{R(A^h)}\}, \; \gamma = (\delta, h). \tag{5.5.22}$$

Here

$$\|f^\delta - f\|_* \leq \delta, \quad \delta \in (0, \delta^*], \tag{5.5.23}$$

$\overline{R(A^h)}$ means the closure of $R(A^h)$ and $g^\gamma \in \overline{R(A^h)}$. By Corollary 1.7.18, $\overline{R(A^h)}$ is convex.

First of all, note that under our conditions the operator regularization method does not always converge. Let us give an example.

Example 5.5.6 Let $A : R^1 \to 2^{R^1}$ be given by the formula

$$Ax = \begin{cases} 1, & x < 0, \\ [1,3], & x = 0, \\ 3, & x > 0; \end{cases}$$

$f = f^\delta = 0$, $A^h x = Ax + hx$, i.e., in (5.5.21) $\eta(t) \equiv t$, $t \geq 0$. Then $g = 1$, $\bar{x}^* = 0$, $g^\gamma = 0$ for all $h > 0$, $\gamma = (0, h)$, $x_\alpha^\gamma = -(h + \alpha)^{-1}$. Here x_α^δ is a solution of the equality $A^h x + \alpha x = g^\gamma$. Hence $x_\alpha^\gamma \to -\infty$ if α and h tend to zero in any way.

This situation is explained by the fact that it is not always possible to get estimates of the type $\|g - g^\gamma\| \leq \tilde{\delta}(\gamma)$, which guarantee a realization of the sufficient convergence condition of the operator regularization method:

$$\frac{h + \tilde{\delta}(\gamma)}{\alpha} \to 0 \quad \text{as} \quad \alpha \to 0$$

(see Theorem 2.2.1). Our investigations show, in order to obtain such estimates, it is necessary along with (5.5.21) to use additional proximity properties of $R(A)$ and $R(A^h)$. Since $\overline{R(A)}$ and $\overline{R(A^h)}$ are convex, we assume that

$$\tau(r, \overline{R(A)}, \overline{R(A^h)}) \leq ha(r), \tag{5.5.24}$$

where $a(r)$ is a non-negative non-decreasing and continuous function for $r \geq 0$, $a(r) \to \infty$ as $r \to \infty$ (see the definition of $\tau(r, \overline{R(A)}, R(A^h))$ in Section 4.2).

Observe that the requirement (5.5.21) is not a consequence of (5.5.24) and vice versa. For the operator A, given in Example 5.5.6, the condition (5.5.21) is satisfied with $g(t) = t$ for all $t \geq 0$. The one gets

$$s(r, R(A), R(A^h)) = 0$$

and

$$s(r, R(A^h), R(A)) = r + 1, \ r > 1.$$

Thus, (5.5.24) does not hold.

Example 5.5.7 Consider the operator $A : R^1 \to 2^{R^1}$ given in Example 5.5.6, and let

$$A^h x = \begin{cases} \frac{x}{\sqrt{x^2+h^2}} + 2, & |x| \leq \tilde{x}, \\ 3 - h, & x > \tilde{x}, \\ 1 + h, & x < -\tilde{x}, \end{cases}$$

where

$$\tilde{x} = \frac{(1-h)\sqrt{h}}{\sqrt{2-h}}, \quad 0 < h < 1.$$

It is obvious that $\tau(r, R(A), R(A^h)) = h$ for all $r > 1$, but $\mathcal{H}_{R^1}(A(0), A^h(0)) = 1$. Hence, the property (5.5.24) is fulfilled, however, (5.5.21) does not hold. Thus, (5.5.21) and (5.5.24) are independent, in general.

The quantity $\|g - g^\gamma\|_*$ can be evaluated by making use of geometrical characteristics of the space X^*, in other words, by the properties of duality mapping $(J^s)^* : X^* \to X$ with the gauge function $\mu(t) = t^{s-1}$, $s \geq 2$. Assume that

$$\|(J^s)^* z_1 - (J^s)^* z_2\| \leq d(R)\|z_1 - z_2\|_*^\sigma \quad \forall z_1, z_2 \in X^*, \tag{5.5.25}$$

where $0 < \sigma \leq 1$, $R = max\{\|z_1\|_*, \|z_2\|_*\}$, $d(R)$ is a non-negative non-decreasing function for $R \geq 0$. Besides,

$$\langle z_1 - z_2, (J^s)^* z_1 - (J^s)^* z_2 \rangle \geq m\|z_1 - z_2\|_*^s \quad \forall z_1, z_2 \in X^*, \quad m > 0. \tag{5.5.26}$$

The problems of finding elements g and g^γ are equivalent to the following variational inequalities:

$$\langle g - y, (J^s)^* (g - f) \rangle \leq 0 \quad \forall y \in \overline{R(A)}, \quad g \in R(A), \tag{5.5.27}$$

and

$$\langle g^\gamma - y, (J^s)^* (g^\gamma - f^\delta) \rangle \leq 0 \quad \forall y \in \overline{R(A^h)}, \quad g^\gamma \in \overline{R(A^h)}. \tag{5.5.28}$$

By virtue of condition (5.5.24) for $g \in R(A)$, there exists $y^h \in \overline{R(A^h)}$ such that

$$\|g - y^h\|_* \leq a(\|g\|_*)h. \tag{5.5.29}$$

For $g^\gamma \in \overline{R(A^h)}$, there is $u_\gamma \in \overline{R(A)}$ satisfying the inequality

$$\|g^\gamma - u_\gamma\|_* \leq a(\|g^\gamma\|_*)h. \tag{5.5.30}$$

Presuming $y = u_\gamma$ in (5.5.27) and $y = y^h$ in (5.5.28), and then summing the corresponding inequalities side by side, we obtain the following result:

$$\langle g - g^\gamma, (J^s)^*(g - f) \rangle \quad + \quad \langle g^\gamma - u_\gamma, (J^s)^*(g - f) \rangle + \langle g^\gamma - g, (J^s)^*(g^\gamma - f) \rangle$$

$$+ \quad \langle g^\gamma - g, (J^s)^*(g^\gamma - f^\delta) - (J^s)^*(g^\gamma - f) \rangle$$

$$+ \quad \langle g - y^h, (J^s)^*(g^\gamma - f^\delta) \rangle \leq 0. \tag{5.5.31}$$

It follows from (5.5.25) and (5.5.26) that, respectively,

$$|\langle g^\gamma - g, (J^s)^*(g^\gamma - f^\delta) - (J^s)^*(g^\gamma - f) \rangle| \quad \leq \quad d(R_\gamma)\|g^\gamma - g\|_* \|f - f^\delta\|_*^\sigma$$

$$\leq \quad d(R_\gamma)\delta^\sigma \|g^\gamma - g\|_*,$$

where $R_\gamma = max\{\mu^\gamma, \|g^\gamma - f\|_*\}$, and

$$\langle g - g^\gamma, (J^s)^*(g - f) - (J^s)^*(g^\gamma - f) \rangle \geq m\|g^\gamma - g\|^s.$$

Further, (5.5.2), (5.5.22), (5.5.29), (5.5.30) and the definition of $(J^s)^*$ yield the estimates

$$|\langle g^\gamma - u_\gamma, (J^s)^*(g - f) \rangle| \leq \mu^{s-1}a(\|g^\gamma\|_*)h$$

and

$$|\langle g - y^h, (J^s)^*(g^\gamma - f^\delta) \rangle| \leq (\mu^\gamma)^{s-1}a(\|g\|_*)h.$$

Thus, (5.5.31) leads to the following inequality:

$$m\|g^\gamma - g\|_*^s \quad \leq \quad \mu^{s-1}a(\|g^\gamma\|_*)h + (\mu^\gamma)^{s-1}a(\|g\|_*)h$$

$$+ \quad d(R_\gamma)\delta^\sigma \|g^\gamma - g\|_*. \tag{5.5.32}$$

Now from the relations

$$\mu^\gamma = \|g^\gamma - f^\delta\|_* \leq \|y_0 - f\|_* + \delta,$$

where y_0 is a fixed element of $\overline{R(A^h)}$, we conclude that there exist constants $c_1 > 0$ and $c_2 > 0$ such that $\mu^\gamma \leq c_1$ and $\|g^\gamma\|_* \leq c_2$. Therefore, taking $R_\gamma = max\{c_1, c_2 + \|f\|_*\} = c_3$, (5.5.32) can be rewritten as

$$m\|g^\gamma - g\|_*^s \quad \leq \quad \left(\mu^{s-1}a(c_2) + c_1^{s-1}a(\|g\|_*) \right) h$$

$$+ \quad d(c_3)\delta^\sigma \|g^\gamma - g\|_*.$$

Thus,

$$\|g^\gamma - g\|_*^s \le \beta_1 h + \beta_2 \delta^\sigma \|g^\gamma - g\|_*, \tag{5.5.33}$$

where

$$\beta_1 \ge \frac{\mu^{s-1} a(c_2) + c_1^{s-1} a(\|g\|_*)}{m} \quad \text{and} \quad \beta_2 \ge \frac{d(c_3)}{m}.$$

Consider the function $\varphi(t) = t^s - \bar{a}t - \bar{b}$, where $\bar{a} > 0$, $\bar{b} > 0$, $s \ge 2$, $t \ge 0$. It is easy to see that $\varphi(0) = -\bar{b} < 0$ and $\varphi'(t_0) = 0$ for

$$t_0 = \left(\frac{\bar{a}}{s}\right)^\nu \le \bar{a}^\nu, \quad \text{and} \quad \nu = \frac{1}{s-1}.$$

Thus, $\varphi(t)$ has a minimum at the point $t = t_0$. Choose $t = \bar{t}$ such that $\bar{t}^{s-1} = \bar{a} + \bar{b}^r$, where $r = (s-1)s^{-1}$. Then

$$\varphi(\bar{t}) = \bar{t}(\bar{t}^{s-1} - \bar{a}) - \bar{b} = \bar{b}^r(\bar{a} + \bar{b}^r)^\nu - \bar{b} = \bar{b}(1 + \bar{a}\bar{b}^{-r})^\nu - \bar{b} > 0.$$

Moreover, $\bar{t} > t_0$. Hence, the function $\varphi(t) \le 0$ on some interval $[0, t_2]$ with $t_2 \le \bar{t}$. Therefore, by (5.5.33), we conclude that

$$\|g^\gamma - g\|_* \le \left(\beta_1^r h^r + \beta_2 \delta^\sigma\right)^\nu. \tag{5.5.34}$$

Thus, $g^\gamma \to g$ as $\gamma \to 0$. This allows us to apply the operator regularization method for solving equation $Ax = g$ in the form

$$A^h x + \alpha J^p x = g^\gamma, \quad \alpha > 0, \tag{5.5.35}$$

where $J^p : X \to X^*$ is a duality mapping with the gauge function $\mu(t) = t^{p-1}$, $p \ge 2$. Note that parameter p is chosen independently of s. Finally, using Theorems 2.2.1 and (2.2.10) we obtain the following result:

Theorem 5.5.8 *Let $A : X \to 2^{X^*}$ be a maximal monotone operator, $\{A^h\}$ be a sequence of maximal monotone operators acting from X to X^*, $D(A) = D(A^h)$ for all $h \in (0, h^*]$, an element g satisfy (5.5.2) and $g \in R(A)$. Assume that the conditions (5.5.21), (5.5.23), (5.5.24) are fulfilled and duality mapping $(J^\mu)^*$ in X^* with the gauge function $\mu(t) = t^{s-1}$, $s \ge 2$, has the properties (5.5.25) and (5.5.26). Let*

$$\frac{\delta^\sigma}{\alpha^{s-1}} \to 0 \quad \text{and} \quad \frac{h^{1/s}}{\alpha} \to 0 \quad \text{as} \quad \alpha \to 0.$$

Then the sequence $\{x_\alpha^\gamma\}$, where x_α^γ is a solution of (5.5.35), strongly converges to the minimal norm pseudo-solution of the equation (5.5.1).

Remark 5.5.9 *In the spaces l^p, L^p and W_m^p with $p > 1$, $m > 0$, the estimates (5.5.25) and (5.5.26) for the duality mapping $(J^s)^*$ are actually fulfilled with some $s \ge 2$ (see (1.6.52), (1.6.53) and (1.6.56) - (1.6.59)).*

Remark 5.5.10 *If X is a Hilbert space then $\sigma = 1$, $s = 2$, and, by (5.5.34), there exist constants $R > 0$ and $C(R) > 0$ such that*

$$max\{\|g - g^\gamma\| \mid \|f\| \leq R\} \leq C(R)\sqrt{h} \tag{5.5.36}$$

for $\delta = 0$. The estimate (5.5.36) can not be improved relative to the order of h (see the example in [73]).

Remark 5.5.11 *It is possible to construct the pseudo-residual principle for the regularization method (5.5.35) if β_1 and β_2 are known. For this aim, it is proposed to use the quantity $\triangle = (\beta_1^r h^r + \beta_2 \delta^\sigma)^\nu$ in place of δ (see Section 3.1).*

Remark 5.5.12 *If $A : X \to 2^X$ is an m-accretive operator and the conditions of Theorem 1.15.35 hold, then the set $\overline{R(A)}$ is convex. Therefore, the pseudo-solution concept can be extended to equations with such operators and Theorem 5.5.8 is proved again.*

Instead of (5.5.1), consider a variational inequality

$$\langle Ax - f, z - x \rangle \geq 0 \quad \forall z \in \Omega, \quad x \in \Omega, \tag{5.5.37}$$

where $\Omega \subset D(A)$ is a convex closed set and either $int\ \Omega \neq \emptyset$ or $int\ D(A) \cap \Omega \neq \emptyset$. By Lemma 1.11.7, inequality (5.5.37) is equivalent to the equation

$$Ax + \partial I_\Omega x = f, \tag{5.5.38}$$

where ∂I_Ω is a subdifferential of the indicator function I_Ω of Ω. If we define a pseudo-solution of (5.5.37) as a pseudo-solution of (5.5.38), then all the results of this section can be stated for variational inequalities.

3. The procedure of finding g^γ in the minimization problem (5.5.22) meets essential difficulties connected with the specific character of the minimization set $\overline{R(A^h)}$. By reason of that, to construct the element g^γ with a fixed γ, it is possible to use the following lemma:

Lemma 5.5.13 *Let a functional $\|\psi\|_*^s$, where $s \geq 2$, be uniformly convex on the space X^* with a modulus of convexity ct^s, $c > 0$, $s \geq 2$, and let an element x_λ be a solution of the equation*

$$A^h x + J^p(\lambda x) = f^\delta, \quad \lambda > 0, \quad p \geq 2. \tag{5.5.39}$$

Then $f^\delta - J^p(\lambda x_\lambda) \to g^\gamma$ as $\lambda \to 0$, where $g^\gamma \in \overline{R(A^h)}$ and defined by (5.5.22).

Proof. Take $\epsilon > 0$ and $g_\epsilon \in R(A^h)$ such that $\|g_\epsilon - g^\gamma\|_* \leq \epsilon$. Then there exists $u_\epsilon \in D(A^h)$, for which $g_\epsilon \in A^h(u_\epsilon)$. Since x_λ is a solution of (5.5.39), there is an element $y_\lambda \in A^h x_\lambda$ satisfying the equality

$$y_\lambda + J^p(v_\lambda) = f^\delta,$$

where $v_\lambda = \lambda x_\lambda$. Then we have

$$\langle y_\lambda - g_\epsilon, x_\lambda - u_\epsilon \rangle + \langle J^p(v_\lambda), x_\lambda - u_\epsilon \rangle = \langle f^\delta - g_\epsilon, x_\lambda - u_\epsilon \rangle. \tag{5.5.40}$$

Due to the monotonicity of A^h, the first term in the left-hand side of (5.5.40) is non-negative. Therefore,

$$\langle J^p(v_\lambda), v_\lambda - \lambda u_\epsilon \rangle \leq \langle f^\delta - g_\epsilon, v_\lambda - \lambda u_\epsilon \rangle. \tag{5.5.41}$$

By the Cauchy−Schwartz inequality and definition of J^p, (5.5.41) implies

$$\|v_\lambda\|^p \leq \|\lambda u_\epsilon\| \|v_\lambda\|^{p-1} + (\epsilon + \|w_\gamma\|_*)(\|\lambda u_\epsilon\| + \|v_\lambda\|),$$

where $w_\gamma = g^\gamma - f^\delta$. In accordance with (2.2.10), there exists a constant $K_1 > 1$ such that the estimate

$$\|v_\lambda\| \leq (\epsilon + \|w_\gamma\|_*)^{1/(p-1)} + K_1 \|\lambda u_\epsilon\| \tag{5.5.42}$$

holds if additionally $K_1(p - 1) \geq 2$. Taking into account the inequality

$$a^r - b^r \leq ra^{r-1}(a - b), \quad a > 0, \quad b > 0, \quad r \geq 1, \tag{5.5.43}$$

from (5.5.42) conclude that

$$\|J^p(v_\lambda)\|_* \quad - \quad (\|w_\gamma\|_* + \epsilon) = \|v_\lambda\|^{p-1} - (\|w_\gamma\|_* + \epsilon)$$

$$\leq \quad (p-1)\|v_\lambda\|^{p-2}\Big(\|v_\lambda\| - (\|w_\gamma\|_* + \epsilon)^{1/(p-1)}\Big)$$

$$\leq \quad (p-1)K_1\|v_\lambda\|^{p-2}\|\lambda u_\epsilon\|. \tag{5.5.44}$$

It follows from (5.5.42) that $\|v_\lambda\|$ is bounded for a sufficiently small λ. Therefore, in view of (5.5.44), we can write

$$\|J^p(v_\lambda)\|_* - \|w_\gamma\|_* \leq K_2\lambda + \epsilon, \quad K_2 > 0.$$

Applying the inequality (5.5.43) once more, we deduce that

$$\|J^p(v_\lambda)\|^s - \|w_\gamma\|^s \leq K_3(\lambda + \epsilon), \quad K_3 > 0.$$

Obviously, the inclusion $-J^p(v_\lambda) \in R(A^h) - f^\delta$ holds and

$$\|w_\gamma\|_* = min\{\|y\|_* \mid y \in \overline{R(A^h)} - f^\delta\}.$$

It arises from Theorem 1.1.24 for a uniformly convex functional $\varphi(u)$ with modulus of convexity $\chi(t)$ and convex closed set $\Omega \subseteq dom\ (\varphi)$ that

$$\chi(\|u - u_*\|) \leq \varphi(u) - \varphi(u_*) \quad \forall u \in \Omega, \quad \varphi(u_*) = min\{\varphi(u) \mid u \in \Omega\}. \tag{5.5.45}$$

Then we derive the inequality

$$\|J^p(v_\lambda) + w_\gamma\|_*^s \leq c^{-1} (\|J^p(v_\lambda)\|^s - \|w_\gamma\|^s) \leq c^{-1}K_3(\lambda + \epsilon), \tag{5.5.46}$$

where w_γ is a minimal point of $\|\psi\|_*^s$ on the set

$$S = \{y - f^\delta \mid y \in \overline{R(A^h)}\}.$$

Since ϵ is arbitrary, we come to the conclusion of the lemma. ∎

Remark 5.5.14 If $g^\gamma \in R(A^h)$ then in the proof of Lemma 5.5.13, one can put $g_\epsilon = g^\gamma$, $g^\gamma \in A^h(u_\gamma)$, $u_\epsilon = u_\gamma$ and $\epsilon = 0$. In this case, (5.5.46) gives

$$\|J^p(v_\lambda) + w_\gamma\|_*^s \le c^{-1} K_3 \lambda. \tag{5.5.47}$$

In a Hilbert space, $s = p = 2$, and then there exists $K_4 > 0$ such that

$$\|v_\lambda + w_\gamma\| \le K_4 \sqrt{\lambda}.$$

In conclusion, we present the following example. Suppose that $A : R^2 \to R^2$ is defined by matrix

$$A = \begin{pmatrix} 1 & 2 \\ 2 & 4 \end{pmatrix}.$$

It is easy to see that A is positive and $det\, A = 0$. Analyze A on a set $R_+^2 = \{(x,y) \in R^2 \mid y \ge 0\}$. Observe that if A is defined on the boundary of R_+^2, as in Theorem 1.7.19, then we obtain nonlinear maximal monotone operator $\bar A : R^2 \to 2^{R^2}$. By analogy, consider the perturbed matrix

$$A^h = \begin{pmatrix} (2+h)^2(4+h)^{-1} & 2+h \\ 2+h & 4+h \end{pmatrix}$$

with $h > 0$ and construct maximal monotone operator $\bar A^h : R^2 \to 2^{R^2}$ with $D(\bar A^h) = R_+^2$. Then Theorem 1.7.19 and the Kronecker–Capelli Theorem [122] allow us to get the following representations:

$$R(\bar A) = \{(x,y) \in R^2 \mid y \le 2x\}, \quad R(\bar A^h) = \{(x,y) \in R^2 \mid y \le (2+h)^{-1}(4+h)x\}.$$

Find the estimates for $\bar A$ and $\bar A^h$ like (5.5.24). It is obvious that, in our circumstances, the quantities $s(r, R(\bar A), R(\bar A^h))$ and $s(r, R(\bar A^h), R(\bar A))$, defined by (4.2.28) coincide, and the simple calculations give

$$\tau(r, R(\bar A), R(\bar A^h)) = \frac{rh}{\sqrt{10(h^2 + 6h + 10)}}.$$

Hence, we may assume in (5.5.24) $a(r) = \dfrac{r}{10}$.

If $det\, A^h \ne 0$ then $R(\bar A^h) = R^2$ and like (5.5.24) is not true. Observe, for instance, that if $f = (1;5)^T$ then the inequality $\|g^\gamma - g\| \le ch$ holds with some $c > 0$ and solutions of the regularized equation

$$A^h x + \alpha x = g^\gamma$$

strongly converge to

$$x^* = \left(\frac{11}{25}; \frac{22}{25}\right)$$

as $\alpha \to 0$ and $h \to 0$. For general monotone nonlinear equations, necessary and sufficient conditions for existence of solutions are not obtained. Therefore, in this case, it is not possible to describe $R(A)$ and $R(A^h)$.

5.6   Minimization Problems

In this section, we study the minimization problem of a proper lower semicontinuous non-differentiable, in general, functional $\varphi(x) : X \to R^1$: To find a point $x^* \in X$ such that

$$\varphi(x^*) = min\{\varphi(x) \mid x \in X\}. \qquad (5.6.1)$$

Note that, by these hypotheses, $\varphi(x)$ has no conditions of strong or uniform convexity. Therefore, the problem (5.6.1) is ill-posed and, according to the ill-posedness concept, we assume that it has a solution. Denote by N a solution set of (5.6.1) and by $\partial \varphi = A : X \to 2^{X^*}$ a subdifferential of $\varphi(x)$ with $D(A) = X$. Suppose that X is an E-space with strictly convex dual space X^*. Then, in view of Lemma 1.2.5 and Theorem 1.7.15, the problem (5.6.1) is equivalent to the equation

$$Ax = 0 \qquad (5.6.2)$$

with the maximal monotone operator A.

In the sequel, we assume that, instead of φ, there is a sequence $\{\varphi^h\}$ of bounded from below proper convex lower semicontinuous functionals $\varphi^h : X \to R^1$, depending on the parameter $h > 0$ and satisfying the inequality

$$|\varphi(x) - \varphi^h(x)| \le g(\|x\|)h \quad \forall x \in X, \qquad (5.6.3)$$

where $g(t)$ is a non-negative continuous and non-decreasing function for all $t \ge 0$. Thus, in reality, the problems (5.6.1) and (5.6.2) are replaced, respectively, by

$$\varphi^h(x) \to min \quad s.t. \quad x \in X \qquad (5.6.4)$$

and

$$A^h x = 0, \qquad (5.6.5)$$

where $\partial \varphi^h = A^h : X \to 2^{X^*}$ is the maximal monotone operator again. According to Theorem 2.2.1, in order to obtain strong approximations to $x^* \in N$ by means of operator A^h in (5.6.5) it is possible to use the regularized equation

$$A^h x + \alpha s J^s x = 0 \qquad (5.6.6)$$

where $J^s : X \to X^*$ is a duality mapping with the gauge function $\mu(t) = t^{s-1}$, $s \ge 2$:

$$J^s x = \frac{1}{s} grad \, \|x\|^s,$$

and apply the theory and methods of Chapters 2 and 3. At the same time, it is clear that (5.6.6) is equivalent to minimization problem

$$\Phi_h^\alpha(x) \to min \quad s.t. \quad x \in X \qquad (5.6.7)$$

for the regularized functional

$$\Phi_h^\alpha(x) = \varphi^h(x) + \alpha\|x\|^s, \quad \alpha > 0, \quad s \ge 2, \qquad (5.6.8)$$

which is strictly convex and also non-differentiable, in general. To solve this problem one can apply the well developed methods for approximate minimization of nonsmooth functionals. This approach is used, for instance, in [85].

We present further the functional criterion of choosing a regularization parameter for the problem (5.6.7), (5.6.8). Let x_α^h be its unique minimum point of (5.6.8),

$$m_h(\alpha) = \Phi_h^\alpha(x_\alpha^h),$$

$$m^h = inf\ \{\varphi^h(x) \mid x \in X\}$$

and

$$m = inf\ \{\varphi(x) \mid x \in X\}.$$

Next we state a known proposition.

Lemma 5.6.1 (cf. Lemma 3.6.1). *Any function $m_h(\alpha)$ is continuous and non-decreasing as $\alpha \geq 0$. Moreover,*

$$\lim_{\alpha \to 0+} m_h(\alpha) = m^h \tag{5.6.9}$$

and

$$\lim_{\alpha \to \infty} m_h(\alpha) = \varphi^h(\theta_X).$$

In addition, if $m^h < \varphi(\theta_X$, then $m_h(\alpha)$ increases on the interval $(0, \alpha_h^*)$, where

$$\alpha_h^* = sup\ \{\alpha \mid m_h(\alpha) < \varphi(\theta_X)\}.$$

We introduce the following denotations:

$$\Gamma(\alpha) = \varphi^h(x_\alpha^h) = m_h(\alpha) - \alpha\|x_\alpha^h\|^s;$$

$$\xi(\alpha) = \Gamma(\alpha) - hg(\|x_\alpha^h\|) - \tau^h;$$

$$\tau^h = max\{m, m^h\}.$$

We claim that the function $\Gamma(\alpha)$ is continuous. Indeed, by Theorem 1.2.8 and Lemma 1.2.5, the minimization problem of the functional (5.6.8) on X leads to the equation

$$\partial\varphi^h(x) + \alpha s J^s x = 0,$$

where $\partial\varphi^h(x)$ is a subdifferential of $\varphi^h(x)$ at a point x. Continuity of the function $\sigma(\alpha) = \|x_\alpha^h\|$ is established as in Lemma 3.1.1. Then, by making use of Lemma 5.6.1, we obtain the claim.

It additionally follows from Lemma 3.1.1 that $\sigma(\alpha)$ is non-increasing for all $\alpha \geq \alpha_0 > 0$ and

$$\lim_{\alpha \to \infty} \sigma(\alpha) = 0.$$

Taking into account these properties, it is not difficult to verify the following assertions:
a) $\xi(\alpha)$ is continuous as $\alpha \geq \alpha_0 > 0$;

b) $\lim_{\alpha \to \infty} \xi(\alpha) = \varphi^h(\theta_X) - hg(0) - \tau^h$;

c) the equality

$$\xi(\alpha) = m_h(\alpha) - \tau^h - \alpha\|x_\alpha^h\|^s - hg(\|x_\alpha^h\|)$$

implies the limit relation

$$\lim_{\alpha \to 0+} \xi(\alpha) < 0.$$

We state the obvious proposition.

Lemma 5.6.2 *If*

$$\varphi^h(\theta_X) - hg(0) - \tau^h > 0, \tag{5.6.10}$$

then there exists at least one solution $\bar{\alpha} = \bar{\alpha}(h)$ *of the equation*

$$\xi(\bar{\alpha}) = 0. \tag{5.6.11}$$

Remark 5.6.3 *If* $\theta_X \in N$ *then the inequality*

$$\varphi^h(\theta_X) \leq \tau^h + hg(0)$$

is satisfied. Consequently, the condition (5.6.10) of Lemma 5.6.2 implies the fact that $\theta_X \notin N$.

Theorem 5.6.4 *Let* $\{x_{\bar{\alpha}}^h\}$ *be a minimal point sequence of the smoothing functional (5.6.8) with* $\alpha = \bar{\alpha}$ *defined by (5.6.11). If the condition (5.6.10) is filfilled, then*

$$\lim_{h \to 0} x_{\bar{\alpha}}^h = \bar{x}^*,$$

where $\bar{x}^* \in N$ *is a minimal norm solution of the problem (5.6.1).*

Proof. Put in (5.6.8) $\alpha = \bar{\alpha}$. Construct the sequence $\{x_{\bar{\alpha}}^h\}$ and study its behaviour when $h \to 0$. First of all, the inclusion $x^* \in N$ yields the inequality

$$\varphi^h(x_{\bar{\alpha}}^h) + \bar{\alpha}\|x_{\bar{\alpha}}^h\|^s \leq \varphi^h(x^*) + \bar{\alpha}\|x^*\|^s \quad \forall x^* \in N.$$

By (5.6.11), one gets

$$hg(\|x_{\bar{\alpha}}^h\|) \quad + \quad \bar{\alpha}\|x_{\bar{\alpha}}^h\|^s = \varphi^h(x_{\bar{\alpha}}^h) + \bar{\alpha}\|x_{\bar{\alpha}}^h\|^s - \tau^h$$

$$\leq \quad hg(\|x^*\|) + \bar{\alpha}\|x^*\|^s + m - \tau^h.$$

Since $m - \tau^h \leq 0$, we have

$$hg(\|x_{\bar{\alpha}}^h\|) + \bar{\alpha}\|x_{\bar{\alpha}}^h\|^s \leq hg(\|x^*\|) + \bar{\alpha}\|x^*\|^s \quad \forall x^* \in N.$$

Recalling that the function $g(t)$ is non-decreasing we obtain the estimate

$$\|x_{\bar{\alpha}}^h\| \leq \|x^*\| \quad \forall x^* \in N. \tag{5.6.12}$$

Thus, $x_{\bar{\alpha}}^h \rightharpoonup \bar{x} \in X$ as $h \to 0$. Further, taking into account (5.6.12) and weak lower semicontinuity of the norm in X, we can write

$$\|\bar{x}\| \leq \liminf_{h \to 0} \|x_{\bar{\alpha}}^h\| \leq \limsup_{h \to 0} \|x_{\bar{\alpha}}^h\| \leq \|x^*\|. \qquad (5.6.13)$$

Show that $\bar{x} \in N$. Indeed, using (5.6.11) and (5.6.12), it is not difficult to verify that the following inequalities are satisfied:

$$
\begin{aligned}
0 &\leq \varphi(x_{\bar{\alpha}}^h) - \varphi(x^*) = \varphi(x_{\bar{\alpha}}^h) - \varphi^h(x_{\bar{\alpha}}^h) + \varphi^h(x_{\bar{\alpha}}^h) - \varphi(x^*) \\
&\leq 2hg(\|x_{\bar{\alpha}}^h\|) + \tau^h - m \leq 2hg(\|x^*\|) + \tau^h - m.
\end{aligned} \qquad (5.6.14)
$$

Besides,

$$m^h = \inf\{\varphi^h(x) \mid x \in X\} \leq \varphi^h(x^*) \leq hg(\|x^*\|) + m, \qquad (5.6.15)$$

that is,

$$\limsup_{h \to 0} m^h \leq m.$$

Therefore,

$$\limsup_{h \to 0} \tau^h \leq m.$$

By definition of τ^h,

$$\liminf_{h \to 0} \tau^h \geq m.$$

Hence,

$$\lim_{h \to 0} \tau^h = m. \qquad (5.6.16)$$

Then from (5.6.14) we deduce that

$$\lim_{h \to 0} \varphi(x_{\bar{\alpha}}^h) = \varphi(x^*). \qquad (5.6.17)$$

By (5.6.17) and by weak lower semicontinuity of the functional φ (see Theorem 1.1.13), we have

$$\varphi(\bar{x}) \leq \lim_{h \to 0} \varphi(x_{\bar{\alpha}}^h) = \varphi(x^*) \quad \forall x^* \in N.$$

Thus, $\bar{x} \in N$. Finally, by (5.6.13), we conclude that $\bar{x} = \bar{x}^*$ and $\|x_{\bar{\alpha}}^h\| \to \|\bar{x}^*\|$. Thus, the theorem is proved. ∎

Theorem 5.6.5 *Under the conditions of Lemma 5.6.2, if a functional φ is Gâteaux-differentiable at the point $\bar{x}^*$, then $\bar{\alpha}(h) \to 0$ as $h \to 0$.*

Proof. Let $\bar{\alpha} = \bar{\alpha}(h) \to \hat{\alpha}$ as $h \to 0$. First of all, observe that $\hat{\alpha}$ is finite. Indeed,

$$\bar{\alpha} = \frac{m_h(\bar{\alpha}) - \varphi^h(x_{\bar{\alpha}}^h)}{\|x_{\bar{\alpha}}^h\|^s}$$

$$= \frac{m_h(\bar{\alpha}) - hg(\|x_{\bar{\alpha}}^h\|) - \tau^h}{\|x_{\bar{\alpha}}^h\|^s}$$

$$\leq \frac{\varphi^h(\theta_X) - hg(\|x_{\bar{\alpha}}^h\|) - \tau^h}{\|x_{\bar{\alpha}}^h\|^s}$$

$$\leq \frac{\varphi^h(\theta_X) - hg(0) - \tau^h}{c},$$

where a constant $c > 0$ satisfies the inequality $\|x_{\bar{\alpha}}^h\|^s \geq c$ with sufficiently small h. The existence of c follows from Theorem 5.6.4 and from the condition $\theta_X \notin N$ because of Remark 5.6.3.

Prove now the theorem by contradiction. Let $\hat{\alpha} > 0$. Then the properties of the sequence $\{x_{\bar{\alpha}}^h\}$ imply

$$\lim_{h \to 0} \Phi_h^{\bar{\alpha}}(x_{\bar{\alpha}}^h) = \Phi_0^{\hat{\alpha}}(\bar{x}^*) = \varphi(\bar{x}^*) + \hat{\alpha}\|\bar{x}^*\|^s. \tag{5.6.18}$$

Since $x_{\bar{\alpha}}^h$ is a minimum point of the functional $\Phi_h^{\bar{\alpha}}$, we have

$$\Phi_h^{\bar{\alpha}}(x_{\bar{\alpha}}^h) \leq \varphi^h(x) + \bar{\alpha}\|x\|^s \quad \forall x \in X.$$

Passing to the limit as $h \to 0$ and using (5.6.18), we deduce that $\bar{x}^*$ is a minimum point of the functionals $\varphi(x)$ and $\varphi_+(x) = \varphi(x) + \hat{\alpha}\|x\|^s$ at the same time. In view of Lemmas 1.2.4 and 1.2.5, the latter means that

$$\varphi'(\bar{x}^*) = \theta_{X^*} \quad \text{and} \quad \varphi_+'(\bar{x}^*) = \varphi'(\bar{x}^*) + \hat{\alpha} \, grad \, \|\bar{x}^*\|^s = \theta_{X^*},$$

where $\varphi'(x)$ is the Gâteaux derivative of $\varphi(x)$. Thus,

$$\hat{\alpha} \, grad \, \|\bar{x}^*\|^s = \theta_{X^*}, \quad \hat{\alpha} \neq 0.$$

Therefore, $grad \, \|\bar{x}^*\|^s = \theta_{X^*}$ and $\bar{x}^* = \theta_X$. This contradicts the assumption (5.6.10) (see Remark 5.6.3). ∎

The next example shows that the conclusion of Theorem 5.6.5 may be wrong if the smoothness demand of the functional φ at the point $\bar{x}^*$ is disturbed.

Example 5.6.6 Let $\varphi : R^1 \to R^1$ and $\varphi^h : R^1 \to R^1$ be expressed by the formulas

$$\varphi(x) = |x + 1| + 1$$

and

$$\varphi^h(x) = |x + 1| + 1 + h, \quad h > 0.$$

It is easy to see that $N = \{-1\}$ and $g(t) = 1$ in (5.6.3). Take $s = 2$ in (5.6.8). The condition (5.6.10) of Lemma 5.6.2 are satisfied if $h < 1$. One can verify that $x_{\bar{\alpha}}^h = h - 1$, where $\bar{\alpha} = \dfrac{1}{2 - 2h}$. Hence, $x_{\bar{\alpha}}^h \to -1 = \bar{x}^*$ and $\bar{\alpha} \to \dfrac{1}{2}$ as $h \to 0$. Note that if $\alpha = \hat{\alpha} = \dfrac{1}{2}$ then functionals $\varphi(x)$ and $\varphi(x) + \hat{\alpha}x^2$ have the same minimum point $\bar{x}^* = -1$. Moreover, $\bar{x}^* = -1$ is a unique minimum point of the functional $\varphi(x) + \hat{\alpha}x^2$ for all $\alpha \in (0, \dfrac{1}{2})$.

Remark 5.6.7 *By (5.6.17), we have*

$$\lim_{h \to 0} \varphi(x_{\tilde{\alpha}}^h) = m,$$

that is, the sequence $\{x_{\tilde{\alpha}}^h\}$ converges to $\bar{x}^*$ also with respect to functional φ.

Remark 5.6.8 *The relations (5.6.11) and (5.6.15) imply*

$$\varphi^h(x_{\tilde{\alpha}}^h) - m \le h\Big(g(\|x_{\tilde{\alpha}}^h\|) + g(\|x^*\|)\Big).$$

In other words, there exists a constant $C > 0$ such that

$$\frac{\varphi^h(x_{\tilde{\alpha}}^h) - m}{h} \le C.$$

Remark 5.6.9 *Instead of m^h, the quantity $m^h + \triangle(h)$ can be used, where $\triangle(h) \to 0$ as $h \to 0$.*

In the filtration problem (see Sections 1.3 and 2.2), operators A and A^h are potential. Moreover, their potentials are defined by the following expressions:

$$\varphi(u) = \int_\Omega \int_0^{|\nabla u|} g(x, \xi^2)\xi d\xi dx$$

and

$$\varphi^h(u) = \int_\Omega \int_0^{|\nabla u|} g^h(x, \xi^2)\xi d\xi dx.$$

If $\varphi^h(\theta_X) - h - \tau^h > 0$ then (5.6.11) is written by the equation

$$\varphi^h(u_{\tilde{\alpha}}^h) - h(c\|u_{\tilde{\alpha}}^h\|_{1,p}^{p-1} + 1) - \tau^h = 0.$$

5.7   Optimal Control Problems

In this section we are interested in the following optimal control problem:

$$min \ \{\Phi(u) = \varphi(u) + \psi(z(u)) \mid u \in \Omega\}, \tag{5.7.1}$$

where the system state $z = z(u) \in \mathcal{D}$ is connected with a control $u \in \Omega$ by the variational inequality

$$(A(u, z) - f, v - z) \ge 0 \quad \forall v \in \mathcal{D}. \tag{5.7.2}$$

Here $\Omega \subset H$ is a convex closed bounded set, $\mathcal{D} \subset H_1$ is a convex closed set, $f \in H_1$ is fixed, H and H_1 are Hilbert spaces, $\varphi : \Omega \to R^1$ is a bounded from below and weakly lower semicontinuous functional with the $\mathcal{H}$-property, which intends that the relations $u_n \rightharpoonup u$ in H and $\varphi(u_n) \to \varphi(u)$ imply strong convergence $u_n \to u$. We assume in the sequel that $\psi : H_1 \to R^1$ is a strongly convex and Fréchet differentiable functional.

Under the solution of the problem (5.7.1), (5.7.2), we understand an element $u_0 \in \Omega$ such that for a certain solution $z = z(u_0)$ of the inequality (5.7.2) there hold the equalities

$$\Phi(u_0) = \varphi(u_0) + \psi(z(u_0)) = min\ \{\Phi(u)|u \in \Omega\},$$

where a minimum is taken for all $u \in \Omega$ and for all $z(u)$ satisfying (5.7.2).

Suppose that the operator $A : \Omega \times H_1 \rightarrow H_1$ has the following properties:
1) $A(u, z)$ is strongly continuous with respect to u and demicontinuous with respect to z;
2) $A(u, z)$ is monotone with respect to z for all $u \in \Omega$.

Let the problem (5.7.1), (5.7.2) possess a nonempty solution set N. As usual, we assume further that in place of f its δ-approximations f^δ are known such that

$$\|f - f^\delta\|_{H_1} \leq \delta,$$

where $\delta \in (0, \delta^*]$. The problem is posed: For given $\{f^\delta\}$, to construct a sequence $\{u_\delta\}$ which converges strongly to some element of the set N. In our assumptions, the variational inequality (5.7.2) and the minimization problem (5.7.1), (5.7.2) are ill-posed. Indeed, for given control u, either solvability of (5.7.2) or uniqueness of its solutions and stability with respect to data perturbations can not be guaranteed. Furthermore, in general, the functional Φ is not uniformly convex in the control space. Therefore, the minimization problem (5.7.1) can be also unstable. By virtue of the mentioned aspects, strong approximations to solutions of (5.7.1) and (5.7.2) can be constructed only by making use of some regularization procedure.

Denote by $Z(u)$ a solution set of the variational inequality (5.7.2) with any fixed $u \in \Omega$. It follows from the properties of A that if $Z(u)$ is nonempty then it is convex and closed. List several properties of the functional ψ and operator $B = grad\ \psi$. Since ψ is strongly convex, there exists a constant $c > 0$ such that

$$\psi(y + t(x - y)) - \psi(y) \leq t[\psi(x) - \psi(y)] - t(1 - t)c\|x - y\|^2_{H_1} \quad \forall x, y \in H_1,\ t \in (0, 1). \quad (5.7.3)$$

Dividing (5.7.3) by t and then passing to the limit as $t \rightarrow 0$ one gets

$$(By, x - y) \leq \psi(x) - \psi(y) - c\|x - y\|^2_{H_1}. \quad (5.7.4)$$

Interchanging x and y in (5.7.4) and adding the obtained inequality to (5.7.4), we establish the property of strong monotonicity of the operator B :

$$(Bx - By, x - y) \geq 2c\|x - y\|^2_{H_1} \quad \forall x, y \in H_1. \quad (5.7.5)$$

We emphasize that we proved above the inequality (1.1.8) and (1.1.11) for ψ and $grad\ \psi$, respectively. If $y = \theta_{H_1}$ then it is not difficult to see from (5.7.5) that

$$(Bx, x) \geq 2c\|x\|^2_{H_1} - \|B(\theta_{H_1})\|\|x\|_{H_1}.$$

This means that the operator $B : H_1 \rightarrow H_1$ is coercive. Then, due to Corollary 1.7.7, the equation $Bx = \theta_{H_1}$ with strongly monotone operator has a unique solution. In other words, the functional $\psi(z)$ has on H_1 a unique minimizer (see Theorem 1.1.21). Hence, ψ is bounded on H_1 from below.

Next, it follows from (5.7.4) that for $y = \theta_{H_1}$,

$$\psi(x) \geq \psi(\theta_{H_1}) + c\|x\|_{H_1}^2 - \|B(\theta_{H_1})\|\|x\|_{H_1}.$$

Then it is clear that

$$\lim_{\|x\| \to \infty} \psi(x) = \infty. \tag{5.7.6}$$

Let $z_n \rightharpoonup z$ in H_1 as $n \to \infty$ and $\psi(z_n) \to \psi(z)$. Presuming $x = z_n$ and $y = z$ in (5.7.4) we obtain

$$(Bz, z_n - z) \leq \psi(z_n) - \psi(z) - c\|z_n - z\|_{H_1}^2,$$

from which we deduce the strong convergence of $\{z_n\}$ to z as $n \to \infty$. Therefore, the functional ψ has $\mathcal{H}$-property. Moreover, according to Theorem 1.1.16, ψ is weakly lower semicontinuous.

So, in order to solve the variational inequality (5.7.2), we have to apply some regularization method. To this end, we introduce the family of operators $\{R^\delta(u, \cdot)\}$, $R^\delta(u, \cdot) : H_1 \to H_1$, $\delta \in (0, \delta^*]$, such that for every $u \in \Omega$ provided that $Z(u) \neq \emptyset$, there holds the convergence

$$R^\delta(u, f^\delta) \to z \in Z(u) \quad \text{as} \quad \delta \to 0. \tag{5.7.7}$$

We study the problem: To find an element $u_\delta \in \Omega$ such that

$$\Phi_*^\delta = \inf \{\Phi^\delta(u) \mid u \in \Omega\} \leq \Phi^\delta(u_\delta) \leq \Phi_*^\delta + \epsilon(\delta), \tag{5.7.8}$$

where $\epsilon(\delta) > 0$, $\lim_{\delta \to 0} \epsilon(\delta) = 0$ and

$$\Phi^\delta(u) = \varphi(u) + \psi(z_\delta(u)), \quad z_\delta(u) = R^\delta(u, f^\delta). \tag{5.7.9}$$

Since the functionals φ and ψ are bounded from below, respectively, on Ω and on H_1, there exists an element u_δ satisfying (5.7.8) and (5.7.9). We accept it as an approximate solution of the problem (5.7.1), (5.7.2). Denote

$$\Phi_0 = \min \{\Phi(u) \mid u \in \Omega\}$$

and

$$Z_0 = \{z(u_0) \mid \varphi(u_0) + \psi(z(u_0)) = \Phi_0, \ u_0 \in N\}.$$

Theorem 5.7.1 *Let the assumptions of this section be held and the regularizing algorithm R^δ satisfy the following conditions:*
(i) the element $z \in Z(u)$ in (5.7.7) is a minimizer of $\psi(v)$ on $Z(u)$, that is,

$$\psi(z) = \min\{\psi(v) \mid v \in Z(u)\}; \tag{5.7.10}$$

(ii) for any sequence $\{u_\beta\}$, the limit relations $u_\beta \rightharpoonup u$, $R^\delta(u_\beta, f^\delta) \to z$ as $\beta \to 0$ and $\delta \to 0$ imply $z \in Z(u)$.
Then sets of strong limit points of the sequences $\{u_\delta\}$ and $\{z_\delta(u_\delta)\}$ are nonempty and belong, respectively, to N and Z_0.

Proof. First of all, note that uniqueness of the minimizer $z \in Z(u)$ satisfying condition (5.7.10) arises from Theorem 1.1.23 because $Z(u)$ is convex and closed, ψ is strongly convex and (5.7.6) holds. For the problem (5.7.1), (5.7.2), take the minimizing control $u_0 \in N$ and corresponding state $z_0 = z(u_0) \in Z(u_0)$ such that $\Phi(u_0) = \Phi_0$. In view of (5.7.1), only one z_0 is defined by the equality

$$\psi(z_0) = min \; \{\psi(v) \mid v \in Z(u_0)\}. \tag{5.7.11}$$

Then (5.7.8) implies the inequality

$$\varphi(u_\delta) + \psi(z_\delta(u_\delta)) \leq \varphi(u_0) + \psi(z_\delta(u_0)) + \epsilon(\delta). \tag{5.7.12}$$

By making use of the condition (i), we conclude that $z_\delta(u_0) \to z_0$ as $\delta \to 0$. Taking into account continuity of ψ on H_1 and equality $\Phi_0 = \varphi(u_0) + \psi(z_0)$, one has

$$\limsup_{\delta \to 0} \left(\varphi(u_\delta) + \psi(z_\delta(u_\delta)) \right) \leq \limsup_{\delta \to 0} \left(\varphi(u_0) + \psi(z_\delta(u_0)) + \epsilon(\delta) \right) = \Phi_0. \tag{5.7.13}$$

Since the functional φ is bounded from below and the property (5.7.6) holds, we obtain the boundedness of the sequence $\{z_\delta(u_\delta)\}$ as $\delta \to 0$. In its turn, the sequence $\{u_\delta\}$ is also bounded because Ω is so. Thus, the limit results

$$u_\delta \rightharpoonup \bar{u} \in \Omega \quad \text{and} \quad z_\delta(u_\delta) \rightharpoonup \bar{z} \tag{5.7.14}$$

are established. Now the condition (ii) yields the inclusion $\bar{z} \in Z(\bar{u})$. Therefore,

$$\Phi_0 \leq \varphi(\bar{u}) + \psi(\bar{z}) \leq \liminf_{\delta \to 0} \left(\varphi(u_\delta) + \psi(z_\delta(u_\delta)) \right), \tag{5.7.15}$$

and combination of (5.7.13) with (5.7.15) forms the following relations:

$$\Phi_0 \;\; \leq \;\; \varphi(\bar{u}) + \psi(\bar{z}) \leq \liminf_{\delta \to 0} \left(\varphi(u_\delta) + \psi(z_\delta(u_\delta)) \right)$$

$$\leq \;\; \limsup_{\delta \to 0} \left(\varphi(u_\delta) + \psi(z_\delta(u_\delta)) \right) \leq \Phi_0.$$

Consequently,

$$\lim_{\delta \to 0} \left(\varphi(u_\delta) + \psi(z_\delta(u_\delta)) \right) = \varphi(\bar{u}) + \psi(\bar{z}) = \Phi_0.$$

Besides, $\bar{u} \in N$ and $\bar{z} \in Z_0$. Then weak lower semicontinuity of the functionals φ and ψ gives

$$\lim_{\delta \to 0} \varphi(u_\delta) = \varphi(\bar{u}) \quad \text{and} \quad \lim_{\delta \to 0} \psi(z_\delta(u_\delta)) = \psi(\bar{z}). \tag{5.7.16}$$

In conclusion, it just remains to recall that φ and ψ have the $\mathcal{H}$-property. Then the strong convergence result

$$u_\delta \to \bar{u}, \quad z_\delta(u_\delta) \to \bar{z}$$

follows from (5.7.14) and from (5.7.16) as $\delta \to 0$. The theorem is proved. ∎

We present the regularization algorithm satisfying the conditions of Theorem 5.7.1. Fix $\delta \in (0, \delta^*]$ and $u \in \Omega$. Consider a mapping $R^\delta(u, \cdot) : H_1 \to H_1$ which assigns to each element $f^\delta \in H_1$ a solution $z_\delta(u) \in \mathcal{D}$ of the regularized variational inequality

$$(A(u, z) + \alpha(\delta)Bz - f^\delta, v - z) \geq 0 \quad \forall v \in \mathcal{D}, \tag{5.7.17}$$

where

$$\alpha(\delta) > 0, \quad \alpha(\delta) \to 0, \quad \frac{\delta}{\alpha(\delta)} \to 0 \quad \text{as} \quad \delta \to 0.$$

The condition (i) for the chosen family $\{R^\delta\}$ is established on the basis of the property (5.7.5) to the potential operator B (see the deduction of (2.2.10), the results of which are easily transferred on variational inequalities). Let $u_\beta \rightharpoonup u$ and $z_\delta(u_\beta) \rightharpoonup z$ as $\beta, \delta \to 0$. Due to Lemma 1.11.4, we are able to proceed from (5.7.17) to the equivalent variational inequality

$$(A(u_\beta, v) + \alpha(\delta)Bv - f^\delta, v - z_\delta(u_\beta)) \geq 0 \quad \forall v \in \mathcal{D}.$$

Passing there to the limit as $\beta \to 0$ and $\delta \to 0$ and making use of condition 1), one gets

$$(A(u, v) - f, v - z) \geq 0 \quad \forall v \in \mathcal{D}.$$

The same Lemma 1.11.4 allows us to assert that $z \in Z(u)$. Thus, the property (ii) holds.

Remark 5.7.2 *Theorem 5.7.1 remains still valid if the set Ω is unbounded and the functional φ has the property*

$$\lim_{\|u\| \to \infty} \varphi(u) = \infty, \quad u \in \Omega.$$

If $\psi(z) = 2^{-1}\|z\|^2$ then $B = I$ and (5.7.17) takes the form of the regularization method studied in Chapter 4.

5.8   Fixed Point Problems

We are going to study fixed point problems with nonexpansive mapping $T : \Omega \to \Omega$, where $\Omega \subseteq X$ is a convex closed set and X is a uniformly smooth Banach space. The problem is to find a fixed point x^* of T, in other words, to find a solution x^* of the equation

$$x = Tx. \tag{5.8.1}$$

It is clear that (5.8.1) is equivalent to the equation

$$Ax = 0 \tag{5.8.2}$$

with the accretive operator $A = I - T : X \to X$ (see Lemma 1.15.10), that is,

$$\langle J(x - y), Ax - Ay \rangle \geq 0 \quad \forall x, y \in \Omega. \tag{5.8.3}$$

If x^* is a solution of (5.8.1) then $Ax^* = \theta_X$. In the sequel, we assume that the fixed point set $F(T)$ of T is not empty. Then it is closed and convex.

Introduce parameter ω such that $0 < \omega < 1$ and $\omega \to 1$. Obviously, if x^* is a solution of (5.8.2) then it is solution of the equation

$$\omega A x = 0 \tag{5.8.4}$$

for any fixed $\omega > 0$. Using Corollary 2.7.4 with $\alpha = 1 - \omega \to 0$, consider for (5.8.4) the operator regularization method

$$\omega A x + (1 - \omega)(x - z_0) = 0, \tag{5.8.5}$$

where some $z_0 \in \Omega$. It is easy to see that (5.8.5) is equivalent to the equation

$$x = (1 - \omega)z_0 + \omega T x. \tag{5.8.6}$$

Denote

$$T_\omega x = (1 - \omega)z_0 + \omega T x.$$

Since Ω is convex and closed, we have that $T_\omega : \Omega \to \Omega$, and (5.8.6) can be rewritten as

$$x = T_\omega x. \tag{5.8.7}$$

Theorem 5.8.1 *Let a Banach space X possess an approximation, Ω be a closed convex subset of X, $T : \Omega \to \Omega$ be a nonexpansive mapping, $z_0 \in \Omega$. Then for each $0 < \omega < 1$, operator T_ω is a strong contraction of Ω into Ω with the estimate*

$$\|T_\omega x - T_\omega y\| \leq \omega \|x - y\|.$$

Hence, T_ω has a unique fixed point $x_\omega \in \Omega$.

Proof. Since T is a nonexpansive mapping, we have

$$\|T_\omega x - T_\omega y\| \leq \omega \|T x - T y \leq \omega \|x - y\|.$$

Then the assertion results from the Banach principle for strong contractive maps. ∎

Due to Theorem 5.8.1, the equation (5.8.6) has a unique solution x_ω and the successive approximation method

$$x_{n+1} = (1 - \omega)z_0 + \omega T x_n$$

converges strongly to x_ω. Let $\omega_k \to 1$ as $k \to \infty$. Consider the equation

$$\omega_k A x + (1 - \omega_k)(x - z_0) = 0 \tag{5.8.8}$$

with fixed k and denote by x_k its unique solution. Let J be weak-to-weak continuous. Then Corollary 2.7.4 implies that $x_k \to \bar{x}^* \in F(T)$ as $k \to \infty$. Moreover,

$$\langle J(\bar{x}^* - x^*), \bar{x}^* - z_0 \rangle \geq 0; \quad \forall x^* \in F(T).$$

The goal of this section is to prove strong convergence of the generalized successive approximation method for the regularized equation (5.8.6) in the following form:

$$y_{n+1} = (1 - \omega_n)z_0 + \omega_n T y_n, \quad n = 0, 1, 2, \ldots, \tag{5.8.9}$$

where

$$\lim_{n \to \infty} \omega_n = 1 \quad \text{and} \quad \sum_{n=0}^{\infty}(1 - \omega_n) = \infty. \tag{5.8.10}$$

Theorem 5.8.2 *Let Ω be a closed convex subset of a uniformly smooth Banach space X, $T : \Omega \to \Omega$ be a nonexpansive mapping with a fixed point set $F(T) \neq \emptyset$, $z_0 \in \Omega$, and $\{\omega_n\}$ be an increasing in $(0,1)$ sequence satisfying (5.8.10). If X has weak-to-weak duality mapping then the sequence $\{y_n\}$ generated by (5.8.9) converges strongly to the fixed point $\bar{x}^* = Q_{\Omega}z_0$ of T, where $Q_{\Omega} : \Omega \to F(T)$ is a unique sunny nonexpansive retraction.*

Proof. Let $x^* \in F(T)$. It follows from (5.8.9) that

$$\|y_{n+1} - x^*\| \leq (1 - \omega_n)\|z_0 - x^*\| + \omega_n\|T y_n - T x^*\|$$

$$\leq (1 - \omega_n)\|z_0 - x^*\| + \omega_n\|y_n - x^*\|.$$

Denoting $\lambda_n = \|y_n - x^*\|$ we have

$$\lambda_{n+1} \leq \omega_n \lambda_n + (1 - \omega_n)\|z_0 - x^*\|.$$

Let $q_n = 1 - \omega_n$. Then $\displaystyle\sum_{n=1}^{\infty} q_n = \infty$, and the previous inequality is rewritten as

$$\lambda_{n+1} \leq \lambda_n - q_n \lambda_n + q_n\|z_0 - x^*\|.$$

According to Lemma 7.1.1, the sequence $\{\lambda_n\}$ is bounded, namely,

$$\|y_n - x^*\| \leq max\{2\|z_0 - x^*\|, \|y_0 - x^*\|\} = M_1.$$

Consequently, $\{y_n\}$ is also bounded.

It is not difficult to calculate the following difference:

$$y_{n+1} - y_n = (q_n - q_{n-1})(z_0 - x^*) + (1 - q_n)(T y_n - T y_{n-1}) + (q_{n-1} - q_n)(T y_{n-1} - T x^*).$$

We have

$$\|(q_n - q_{n-1})(z_0 - x^*) + (q_{n-1} - q_n)(T y_{n-1} - T x^*)\|$$

$$\leq |q_n - q_{n-1}|(\|z_0 - x^*\| + \|y_{n-1} - x^*\|).$$

Therefore, there exists a constant $M_2 > 0$ such that

$$\|y_{n+1} - y_n\| \leq (1 - q_n)\|y_n - y_{n-1}\| + M_2|q_n - q_{n-1}|.$$

Denoting $\lambda_n = \|y_n - y_{n-1}\|$ one gets

$$\lambda_{n+1} \leq \lambda_n - q_n \lambda_n + M_2|q_n - q_{n-1}|.$$

Lemma 7.1.2 implies now that $\lim_{n\to\infty} \|y_n - y_{n-1}\| = 0$ because of

$$\lim_{n\to\infty} \frac{|q_n - q_{n-1}|}{q_n} = 0.$$

Since

$$y_{n+1} - Ty_{n+1} = q_n(z_0 - x^*) + (Ty_n - Ty_{n+1}) - q_n(Ty_n - Tx^*),$$

we deduce

$$\|y_{n+1} - Ty_{n+1}\| \leq q_n(\|z_0 - x^*\| + M_1) + \|y_n - y_{n+1}\|.$$

Therefore,

$$\lim_{n\to\infty}(y_n - Ty_n) = \theta_X. \tag{5.8.11}$$

Let $J : X \to X^*$ be a weak-to-weak continuous normalized duality mapping and $\Phi(t) = 2^{-1}t^2$. Then by Lemmas 1.5.7 and 1.2.7, we have $\Phi'(\|x\|) = Jx$ and

$$\Phi(\|x + y\|) - \Phi(\|x\|) = \int_0^1 \langle J(x + ty), y\rangle dt$$

$$= \langle Jx, y\rangle + \int_0^1 \langle J(x + ty) - Jx, y\rangle dt. \tag{5.8.12}$$

Evaluate the last term in the previous equalities. By virtue of $(1.6.5)$, for all $x, y \in X$ such that $\|x\| \leq M$ and $\|y\| \leq M$ we have

$$\int_0^1 \langle J(x + ty) - Jx, y\rangle dt \leq \int_0^1 t^{-1}\Big(8\|ty\|^2 + c_1\rho_X(\|ty\|)\Big)dt,$$

where $c_1 = 8\max\{L, M\}$. Since $\rho_X(\tau)$ is convex, $\rho_X(\|ty\|) \leq t\rho_X(\|y\|)$. Therefore,

$$\int_0^1 \langle J(x + ty) - Jx, y\rangle dt \leq 8\int_0^1 \|y\|^2 t\,dt + c_1\int_0^1 \rho_X(\|y\|)dt = 4\|y\|^2 + c_1\rho_X(\|y\|).$$

Thus,

$$\Phi(\|x + y\|) - \Phi(\|x\|) \leq \langle Jx, y\rangle + 4\|y\|^2 + c_1\rho_X(\|y\|).$$

It is easy to verify the equality

$$y_{n+1} - Q_\Omega z_0 = (1 - q_n)(Ty_n - Q_\Omega z_0) + q_n(z_0 - Q_\Omega z_0).$$

It now follows that

$$\Phi(\|y_{n+1} - Q_\Omega z_0\|) \leq \Phi\Big((1 - q_n)(\|Ty_n - Q_\Omega z_0\|)\Big)$$

$$+ q_n(1 - q_n)\Big\langle J(Ty_n - Q_\Omega z_0), z_0 - Q_\Omega z_0 \Big\rangle \tag{5.8.13}$$

$$+ 4q_n^2\|z_0 - Q_\Omega z_0\|^2 + \bar{c}_1\rho_X(q_n\|z_0 - Q_\Omega z_0\|),$$

where $\bar{c}_1 = 8\max\{L, M_3, \|z_0 - Q_\Omega z_0\|\}$ and M_3 satisfies the inequality

$$\|Ty_n - Q_\Omega z_0\| = \|Ty_n - Tx^* + Tx^* - Q_\Omega z_0\|$$

$$\leq \|y_n - x^*\| + \|x^* - Q_\Omega z_0\| \leq M_1 + \|x^* - Q_\Omega z_0\| = M_3.$$

Denote

$$\zeta_n = \left\langle J(Ty_n - Q_\Omega z_0), z_0 - Q_\Omega z_0 \right\rangle. \qquad (5.8.14)$$

We want further to show that $\lim_{n\to\infty} \zeta_n \leq 0$. If this is not the case then there would exist a subsequence $\{y_{n_k}\}$ of $\{y_n\}$ and $\epsilon > 0$ such that

$$\lim_{n\to\infty} \left\langle J(Ty_{n_k} - Q_\Omega z_0), z_0 - Q_\Omega z_0 \right\rangle > \epsilon. \qquad (5.8.15)$$

The space X is reflexive, therefore, by the Mazur theorem, Ω is weakly closed. Then we can assume that $\{y_{n_k}\}$ converges weakly to a point $\bar{y} \in \Omega$. Since, in view of Lemma 1.5.13, X also satisfies the Opial condition, (5.8.11) implies that this weak accumulation point $\bar{y}$ belongs to $F(T)$. Actually, if it is not true then one has

$$\liminf_{k\to\infty} \|y_{n_k} - \bar{y}\| < \liminf_{k\to\infty} \|y_{n_k} - T\bar{y}\|$$

$$\leq \liminf_{k\to\infty} (\|y_{n_k} - Ty_{n_k}\| + \|y_{n_k} - \bar{y}\|) = \liminf_{k\to\infty} \|y_{n_k} - \bar{y}\|.$$

This is a contradiction and, therefore, $\bar{y} \in F(T)$. Then (5.8.15) yields

$$\langle J(\bar{y} - Q_\Omega z_0), z_0 - Q_\Omega z_0 \rangle > \epsilon$$

which, in its turn, contradicts Proposition 1.5.20 with the corresponding result

$$\langle J(\bar{y} - Q_\Omega x), x - Q_\Omega x \rangle \leq 0 \quad \forall x \in \Omega, \quad \forall \bar{y} \in F(T).$$

Hence, (5.8.15) is not true and $\lim_{n\to\infty} \zeta_n \leq 0$.

We prove that in reality $\lim_{n\to\infty} \zeta_n = 0$. Indeed, since $\Phi(\|x\|)$ is convex and increasing and T is nonexpansive, we have

$$\Phi\Big((1 - q_n)(\|Ty_n - Q_\Omega z_0\|)\Big) \leq (1 - q_n)\Phi(\|Ty_n - TQ_\Omega z_0\|)$$
$$\leq (1 - q_n)\Phi(\|y_n - Q_\Omega z_0\|).$$

Consequently,

$$\Phi(\|y_{n+1} - Q_\Omega z_0\|) \leq \Phi(\|y_n - Q_\Omega z_0\|) - q_n\Phi(\|y_n - Q_\Omega z_0\|) + \gamma_n, \qquad (5.8.16)$$

where

$$\gamma_n = q_n(1 - q_n)\zeta_n + \mu_n$$

and

$$\mu_n = 4q_n^2\|z_0 - Q_\Omega z_0\|^2 + \bar{c}_1\rho_X(\|z_0 - Q_\Omega z_0\|q_n).$$

Recall that $q_n \to 0$ and $\sum_{n=0}^{\infty} q_n = \infty$. Since X is a uniformly smooth Banach space, we deduce

$$\lim_{n\to\infty} \frac{\mu_n}{q_n} = \lim_{n\to\infty} \left(4\|z_0 - Q_\Omega m z_0\|^2 q_n + \bar{c}_1 \frac{\rho_X(\|z_0 - Q_\Omega z_0\|q_n)}{q_n}\right) = 0.$$

Then

$$\lim_{n\to\infty} \frac{\gamma_n}{q_n} = \lim_{n\to\infty}\left((1-q_n)\zeta_n + \frac{\mu_n}{q_n}\right) = \lim_{n\to\infty}\zeta_n.$$

Rewrite now (5.8.16) in the following form:

$$\lambda_{n+1} \le \lambda_n - q_n\lambda_n + \gamma_n, \qquad (5.8.17)$$

where

$$\lambda_n = \Phi(\|y_n - Q_\Omega z_0\|).$$

There may be only one alternative for any $n \ge 0$:

$$(\mathrm{H}_1): \quad \lambda_n \le \frac{1}{\displaystyle\sum_{i=0}^{n} q_i} + \frac{\gamma_n}{q_n},$$

or

$$(\mathrm{H}_2): \quad \lambda_n > \frac{1}{\displaystyle\sum_{i=0}^{n} q_i} + \frac{\gamma_n}{q_n}.$$

If we assume that $\displaystyle\lim_{n\to\infty}\zeta_n < 0$ then also

$$\lim_{n\to\infty}\frac{\gamma_n}{q_n} < 0.$$

We show that this is wrong. First of all, we claim that (H_1) happens infinitely many times. If this is not the case, there exists $\bar{n} > 1$ such that the hypotheses (H_2) holds for all $n \ge \bar{n}$. Then

$$\gamma_n \le q_n\lambda_n - \frac{q_n}{\displaystyle\sum_{i=0}^{n} q_i}$$

and (5.8.17) yields

$$\lambda_{n+1} \le \lambda_n - q_n\lambda_n + q_n\lambda_n - \frac{q_n}{\displaystyle\sum_{i=0}^{n} q_i} = \lambda_n - \frac{q_n}{\displaystyle\sum_{0}^{n} q_i}.$$

Hence

$$\lambda_{n+1} \le \lambda_{\bar{n}} - \sum_{j=\bar{n}}^{n} \frac{q_j}{\displaystyle\sum_{i=0}^{j} q_i},$$

which is a contradiction because $\lambda_n \ge 0$ for any $n \ge 1$ and

$$\sum_{j=\bar{n}}^{n} \frac{q_j}{\displaystyle\sum_{i=0}^{j} q_i} \to \infty \quad \text{as} \quad n \to \infty.$$

Consequently, our claim is true. But then we come to a contradiction with the assumption that $\lim_{n \to \infty} \zeta_n < 0$ because from the unbounded hypothesis (H_1) we obtain that there exists $\tilde{n} > \bar{n}$ such that $\lambda_{\tilde{n}} < 0$, which is not possible. That means that $\lim_{n \to \infty} \frac{\gamma_n}{q_n} = 0$.

Due to Lemma 7.1.2 for recurrent inequality (5.8.17),

$$\Phi(\|y_n - Q_\Omega z_0\|) \to 0 \quad \text{as} \quad n \to \infty.$$

Hence, $\{y_n\}$ converges strongly to Qz_0. The proof is complete. ∎

Assume that $J^\mu : X \to X^*$ is the duality mapping with a gauge function $\mu : R^1_+ \to R^1_+$, and

$$\Phi(t) = \int_0^t \mu(\tau)d\tau.$$

Denote J^μ by J^p if $\mu(t) = t^{p-1}$ with $1 < p < \infty$. Then $\Phi(t) = p^{-1}t^p$ and $\Phi'(\|x\|) = J^p x$. Similarly to (5.8.12) one gets

$$\Phi(\|x + y\|) - \Phi(\|x\|) = \int_0^1 \langle J^p(x + ty), y\rangle dt$$

$$= \langle J^p x, y\rangle + \int_0^1 \langle J^p(x + ty) - J^p x, y\rangle dt.$$

Using (1.6.59) and (1.6.4) we obtain that if $\|x\| \leq M$ and $\|y\| \leq M$, then there exist constants $K_1 > 0$ and $K_2 > 0$ such that

$$\Phi(\|x + y\|) - \Phi(\|x\|) \leq \langle J^p x, y\rangle + K_1 M^p \rho_X(K_2 M^{-1}\|y\|).$$

As before, we prove that $\{y_n\}$ converges strongly to $Q_\Omega z_0$ if J^p is a weak-to-weak continuous duality mapping.

Bibliographical Notes and Remarks

The regularization methods (5.1.2) and (5.1.22) for the value computation of an unbounded monotone operator were studied by Alber in [8]. The results of Sections 5.2 and 5.3 are stated in [186] and [11], respectively. Observe that Theorems 5.1.1, 5.1.3, 5.1.4 can be proved with inessential changes for discontinuous monotone operators and Theorem 5.3.1 for discontinuous accretive operators. In these cases, Rx_0 is understood as a generalized value set of A at a point x_0, namely, $R(x_0) = \{y \mid y \in \bar{A}x_0\}$, where $\bar{A}$ is a maximal monotone and maximal accretive extension of A, respectively.

The regularized Hammerstein equation was constructed and investigated in [157, 158]. Other results are discussed in [50, 66, 136]. The pseudo-solutions of monotone equations described in Section 5.5 are due to Ryazantseva [196]. Lemma 5.5.13 uses the approach of [114] and [115]. Section 5.6 follows the paper [199]. Note that in [208], the authors propose to examine a smoothing functional in the form $\phi(\|A^h x - f^\delta\|) + \alpha\|x\|^s$, where A^h is weakly continuous and $\omega(x) = \phi(\|A^h x - f^\delta\|)$ is convex. However, no direct means for construction of ϕ are given.

The ill-posed optimal control problem (5.7.1), (5.7.2) was investigated in [116]. The regularized method of the successive approximations for finding fixed points of nonexpansive mappings was studied by Browder [57], Halpern [92], Reich [172], Takahashi and Kim [214] and others. Section 5.8, mainly, follows the paper [172]. General fixed point theory is well described in [87, 88].

Chapter 6

SPECIAL TOPICS ON REGULARIZATION METHODS

6.1 Quasi-Solution Method

In this section, we study the quasi-solution method for monotone equations and establish its connection with the operator regularization.

Let X be a reflexive strictly convex space together with its dual space X^*, $A : X \to 2^{X^*}$ be a maximal strictly monotone operator, $\mathcal{M} \subset D(A)$ be a closed convex compact set and $int\ \mathcal{M} \neq \emptyset$, $f \in X^*$.

Definition 6.1.1 *An element $x_0 \in \mathcal{M}$ is said to be a v-quasi-solution on $\mathcal{M}$ of the equation*

$$Ax = f \tag{6.1.1}$$

if it satisfies the inequality

$$\langle Ax - f, x_0 - x \rangle \leq 0 \quad \forall x \in \mathcal{M}. \tag{6.1.2}$$

Observe that a solution of (6.1.2) is understood in the sense of Definition 1.11.2.

Definition 6.1.1 considerably differs from the known definition of the classical quasi-solution which is given by the equality

$$\|Ax_0 - f\|_*^2 = min\{\|Ax - f\|_*^2 \mid x \in \mathcal{M}\}. \tag{6.1.3}$$

The fact is that, the functional $\|Ax - f\|_*^2$ is not necessarily convex on $\mathcal{M}$ in the case of nonlinear monotone operators A, therefore, there are no effective tools to investigate both theoretical and numerical aspects of the classical quasi-solutions. However, if A is a monotone and potential operator, i.e., there exists a convex function $\varphi : X \to R^1$ such that $A = \partial \varphi$, then by Lemma 1.11.4 and Theorem 1.11.14, we conclude that the variational inequality (6.1.2) is equivalent to the following minimization problem:

$$\varphi(x_0) - \langle f, x_0 \rangle = min\{\varphi(x) - \langle f, x \rangle \mid x \in \mathcal{M}\}.$$

Under these conditions the v-quasi-solution coincides with a quasi-solution defined in [130], Chapter 5.

Lemma 6.1.2 *A v-quasi-solution of the equation (6.1.1) on $\mathcal{M}$ exists, is unique and depends continuously on a right-hand side f.*

Proof. The compact set $\mathcal{M}$ is bounded in X. Hence, according to Theorem 1.11.9 and Remark 1.11.12, the inequality (6.1.2) is solvable on $\mathcal{M}$ for any $f \in X^*$. Moreover, its solution x_0 is unique, because A is a strictly monotone operator. Show that the v-quasi-solution of (6.1.1) continuously depends on f. Let f_n, $n = 1, 2, ...$, be given and $f_n \to f$ as $n \to \infty$. Denote by x_n a (unique) v-quasi-solution of the equation $Ax = f_n$ on $\mathcal{M}$ with fixed n. In other words, x_n satisfies the inequality

$$\langle y - f_n, x_n - x \rangle \leq 0 \quad \forall x \in \mathcal{M}, \quad \forall y \in Ax. \tag{6.1.4}$$

Since $x_n \in \mathcal{M}$ and $\mathcal{M}$ is compact, there exists $\bar{x} \in \mathcal{M}$ such that $x_n \to \bar{x}$ as $n \to \infty$. Passing in (6.1.4) to a limit, we obtain

$$\langle y - f, \bar{x} - x \rangle \leq 0 \quad \forall x \in \mathcal{M}, \quad \forall y \in Ax.$$

This means that $\bar{x}$ is a v-quasi-solution of (6.1.1). Now we conclude that $\bar{x} = x_0$ because $\bar{x}$ is unique. Hence, the whole sequence $x_n \to x_0$ as $n \to \infty$. ∎

Lemma 6.1.3 *If the equation (6.1.1) is solvable (in the sense of Definition 1.7.2) and its solution belongs to $\mathcal{M}$, then it coincides with v-quasi-solution x_0 of (6.1.1) on $\mathcal{M}$.*

Proof. Let x^* be a solution of (6.1.1), that is, $f \in Ax^*$. It is unique because A is a strictly monotone operator. Due to the fact that A is maximal monotone and $int\ \mathcal{M} \neq \emptyset$, we conclude, by Lemma 1.11.4, that a solution $x_0 \in \mathcal{M}$ of the inequality (6.1.2) satisfies also the inequality (1.11.1), i.e., there exists $\xi \in Ax_0$ such that

$$\langle \xi - f, x_0 - x \rangle \leq 0 \quad \forall x \in \mathcal{M}.$$

Since $f \in Ax^*$, we conclude that x^* is a v-quasi-solution of (6.1.1) on $\mathcal{M}$. Finally, uniqueness of v-quasi-solution following from Lemma 6.1.2 guarantees the equality $x_0 = x^*$. ∎

Theorem 6.1.4 *Assume that $A : X \to 2^{X^*}$ is a maximal strictly monotone operator, $f \in X^*$, the equation (6.1.1) has a solution x_0 belonging to $\mathcal{M}$, $f^\delta \in X^*$ approximate f such that (5.5.23) holds. Then the sequence $\{x^\delta\}$ of v-quasi-solutions on $\mathcal{M}$ of the equations $Ax = f^\delta$ strongly converges to x_0 as $\delta \to 0$.*

Proof. Let x^δ be a v-quasi-solution of equation $Ax = f^\delta$ on $\mathcal{M}$. Lemma 6.1.2 guarantees existence and uniqueness of $\{x^\delta\}$ and also its convergence to v-quasi-solution x^* of equation (6.1.1) on $\mathcal{M}$ as $\delta \to 0$. It remains to add that, due to Lemma 6.1.3, x^* coincides with the solution x_0 of (6.1.1) in the sense of Definition 1.7.2. ∎

Denote $\Re_1 = (0, \delta^*] \times (0, h^*]$ and $\Re_2 = (0, h^*] \times (0, \sigma^*]$. We present further the stability theorem of v-quasi-solutions.

Theorem 6.1.5 *Suppose that all the conditions of Theorem 6.1.4 are fulfilled, and operator A is also known with perturbations. namely, instead of A, a sequence $\{A^h\}$ of the maximal strictly monotone operators is given, such that the estimate*

$$\mathcal{H}_{X^*}(Ax, A^h x) \leq g(\|x\|)h \quad \forall x \in \mathcal{M} \tag{6.1.5}$$

holds. where $g(t)$ is a non-negative continuous function for $t \geq 0$. Let $\mathcal{M} \subset D(A^h)$ for all $h > 0$. Then a sequence of v-quasi-solutions $\{x^\gamma\}$, $\gamma = (\delta, h) \in \Re_1$, of the equation $A^h x = f^\delta$ on $\mathcal{M}$ converges strongly to a v-quasi-solution $\bar{x}$ of the equation (6.1.1) on $\mathcal{M}$ as $\gamma \to 0$.

Proof. An element $x^\gamma \in \mathcal{M}$ is defined by the inequality

$$\langle A^h x - f^\delta, x^\gamma - x \rangle \leq 0 \quad \forall x \in \mathcal{M}. \tag{6.1.6}$$

The existence and uniqueness of $x^\gamma \in \mathcal{M}$ result from Lemma 6.1.2. Then the relation

$$\langle \zeta^h - f^\delta, x^\gamma - x \rangle \leq 0 \quad \forall x \in \mathcal{M}. \tag{6.1.7}$$

holds all $\zeta^h \in A^h x$. Let $x \in \mathcal{M}$ be fixed. Take an arbitrary $\zeta \in Ax$. The condition (6.1.5) enables us to find $\zeta^h \in A^h x$ satisfying the estimate

$$\|\zeta^h - \zeta\|_* \leq g(\|x\|)h.$$

Thus, if $h \to 0$ then we construct a sequence $\{\zeta^h\}$ such that $\zeta^h \to \zeta$. Since $\mathcal{M}$ is a compact set, we conclude that $x^\gamma \to \bar{x} \in \mathcal{M}$. Passing in (6.1.7) to the limit as $\gamma \to 0$, one gets

$$\langle \zeta - f, \bar{x} - x \rangle \leq 0 \quad \forall \zeta \in Ax, \quad \forall x \in \mathcal{M}.$$

Hence, $\bar{x}$ is a v-quasi-solution of (6.1.1) on $\mathcal{M}$. The theorem is proved because $\bar{x}$ is unique. ∎

Remark 6.1.6 *If operators A^h, $h \in (0, h^*]$, are not strictly monotone, then a v-quasi-solution set on $\mathcal{M}$ of each equation $A^h x = f^\delta$ is convex and closed, and it is not singleton, in general. The assertion of Theorem 6.1.5 remains valid in this case, if x^γ in (6.1.6) is chosen arbitrarily.*

Assume that under the conditions of Theorem 6.1.5, in place of compact set $\mathcal{M}$, a sequence of convex and closed compact sets $\{\mathcal{M}_\sigma\}$ is known, where $\mathcal{M}_\sigma \subset D(A^h)$, int $\mathcal{M}_\sigma \neq \emptyset$ for all $(h, \sigma) \in \Re_2$. Suppose also that $\{\mathcal{M}_\sigma\}$ converges to the compact set $\mathcal{M}$ as $\sigma \to 0$ with respect to the Hausdorff metric. Then convergence of the v-quasi-solution sequence of equations $A^h x = f^\delta$ on $\mathcal{M}_\sigma$, as δ, h, $\sigma \to 0$ is established by similar arguments as in Sections 4.2 and 4.3. Note also that the method described above of constructing approximations to the solution of (6.1.1) does not require us to know the level of errors δ, h and σ in the initial data.

In what follows, we consider finite-dimensional approximations of v-quasi-solutions.

Definition 6.1.7 *We say that a sequence of finite-dimensional spaces $\{X_n\}$, $X_n \subset X$, is extremely dense in X if $P_n x \to x$ as $n \to \infty$ for all $x \in X$, where $\{P_n\}$ is a sequence of projectors $P_n : X \to X_n$.*

Theorem 6.1.8 *Assume that in addition to the conditions of Theorem 6.1.5, operators $A^h : X \to X^*$ are hemicontinuous, $\{X_n\}$, $n = 1, 2, ...$, are the ordered sequences of finite-dimensional subspaces of X which is extremely dense in X, $P_n : X \to X_n$ are linear operators and P_n^* are their conjugate, $f_n^\delta = P_n^* f^\delta$, $A_n^h = P_n^* A^h$, $\{\mathcal{M}_n\}$ is a sequence of convex closed compact sets, $\mathcal{M}_n \subset X_n$, $\mathcal{M}_n \subset D(A_n^h)$, $\mathcal{M}_n = P_n \mathcal{M}$ for all $n \geq 1$. Let $\gamma = (\delta, h) \in \Re_1$. Then a sequence $\{x_n^\gamma\}$ of v-quasi-solutions of equations $A_n^h x = f_n^\delta$ on $\mathcal{M}_n$ strongly converges to a v-quasi-solution x^γ of the equation $A^h x = f^\delta$ on $\mathcal{M}$ as $n \to \infty$.*

Proof. A solution $x_n^\gamma \in \mathcal{M}_n$ is uniquely defined by the inequality

$$\langle A_n^h x_n - f_n^\delta, x_n^\gamma - x_n \rangle \leq 0 \quad \forall x_n \in \mathcal{M}_n. \tag{6.1.8}$$

Rewrite (6.1.8) in the following form

$$\langle A^h x_n - f^\delta, x_n^\gamma - x_n \rangle \leq 0 \quad \forall x_n \in \mathcal{M}_n, \quad x_n^\gamma \in \mathcal{M}_n. \tag{6.1.9}$$

It is possible because $x_n \in X_n$ and $x_n^\gamma \in X_n$. Since $\mathcal{M}_n \subset \mathcal{M}$ for all $n \geq 1$, the sequence $\{x_n^\gamma\}$ is compact, hence, there exists $\bar{x}^\gamma \in \mathcal{M}$ such that $x_n^\gamma \to \bar{x}^\gamma$ as $n \to \infty$. Let in (6.1.8) and (6.1.9) $x_n = P_n x$, where $x \in \mathcal{M}$. Then $x_n \to x$ as $n \to \infty$, because a sequence $\{X_n\}$ is extremely dense in X. Letting in (6.1.9) $n \to \infty$ and taking into account the demicontinuity property of A^h we have

$$\langle A^h x - f^\delta, \bar{x}^\gamma - x \rangle \leq 0 \quad \forall x \in \mathcal{M}, \quad \bar{x}^\gamma \in \mathcal{M}.$$

From this inequality and from uniqueness of v-quasi-solution x^γ, the conclusion of the theorem follows. ∎

We establish a connection between the quasi-solution method and operator regularization method. Let $\mathcal{M}$ be a convex closed compact set defined by the formula

$$\mathcal{M} = \{x \in X \mid \psi(x) \leq 0\},$$

where $\psi : X \to R^1$ is a convex and continuous functional. We assume further that there exists at least one point $z_0 \in \mathcal{M}$ such that the Slater condition

$$\psi(z_0) < 0 \tag{6.1.10}$$

holds. Then the variational inequalities

$$\langle Ax - f, x_0 - x \rangle \leq 0 \quad \forall x \in \mathcal{M}, \quad x_0 \in \mathcal{M}, \tag{6.1.11}$$

and

$$\langle Ax + \alpha \partial \psi(x) - f, x_\alpha - x \rangle \leq 0 \quad \forall x \in \mathcal{M}, \quad x_\alpha \in \mathcal{M}, \quad \alpha > 0, \tag{6.1.12}$$

where $\partial \psi$ is a subdifferential of ψ, define, respectively, the quasi-solution method and operator regularization method. Due to the condition (6.1.10) and Theorem 1.11.18, the problems (6.1.11) and (6.1.12) are reduced mutually each to the other.

Remark 6.1.9 *If the variational inequality*

$$\langle Ax - f, x - z \rangle \leq 0 \quad \forall z \in \Omega, \quad x \in \Omega,$$

is being solved, where Ω is a convex closed set in $D(A)$ and $\mathcal{M} \subset \Omega$, then the definition of a v-quasi-solution on $\mathcal{M}$ given above and all the results of the present section remain still true.

6.2   Residual Method

Let X be an E-space, X^* be strictly convex, $A : X \to 2^{X^*}$ be a maximal monotone operator. Consider in X the equation (6.1.1). Let $N \neq \emptyset$ be its solution set, $f^\delta \in X^*$ be δ-approximations of $f \in X^*$, i.e., $\|f - f^\delta\|_* < \delta$, $\delta \in (0, \delta^*]$ with some positive δ^*.

1. The residual method for solving the problem (6.1.1) with linear operator A in a Hilbert space is reduced to the minimization problem

$$\varphi(x) = \|x\|^2 \to min \tag{6.2.1}$$

for a strongly convex functional φ on a convex closed set

$$\mathcal{M}^\delta = \{x \in X \mid \|Ax - f^\delta\| \leq \delta\}. \tag{6.2.2}$$

Observe that in the case of a nonlinear operator A, the set $\mathcal{M}^\delta$ is not convex, in general. In this case, the problem (6.2.1), (6.2.2) in the solving process meets with considerable difficulties. Therefore, we propose below another approach in which the constraint set $\mathcal{M}^\delta$ is defined by means of variational inequalities.

Let $G \subseteq D(A)$ be a convex closed bounded set in X. Then there exists a constant $C > 0$ such that $diam\, G \leq C$. Assume that $N_G = G \cap N$,

$$N \cap int\, G \neq \emptyset \quad \text{and} \quad \theta_X \notin N_G. \tag{6.2.3}$$

We introduce the following set:

$$\Omega^\delta = \{w \in G \mid \langle \xi - f^\delta, w - v \rangle \leq \delta C \quad \forall v \in G, \quad \forall \xi \in Av\}. \tag{6.2.4}$$

It is nonempty because $N_G \subset \Omega^\delta$ for all $\delta \in (0, \delta^*]$. Indeed, if $w_0 \in N_G$ then we have

$$\langle \zeta - f^\delta, w_0 - w \rangle = \langle \zeta - f, w_0 - w \rangle + \langle f - f^\delta, w_0 - w \rangle$$

$$\leq \delta \|w - w_0\| \leq \delta C \quad \forall w \in G, \quad \forall \zeta \in Aw. \tag{6.2.5}$$

Here we have taken into consideration that

$$\langle \zeta - f, w - w_0 \rangle \geq 0 \quad \forall w \in G, \quad \forall \zeta \in Aw, \tag{6.2.6}$$

which follows from the monotonicity of A. Moreover, Ω^δ is a convex and closed set.

Our aim is to find an element x^δ satisfying the condition

$$\|x^\delta\|^s = min\{\|x\|^s \mid x \in \Omega^\delta\}, \quad s \geq 2, \tag{6.2.7}$$

and prove that x^δ is the approximation to a solution of (6.1.1). Note that a number s should be chosen such that a functional $\|x\|^s$ has the best uniform convexity. For instance, in the Lebesgue spaces $L^p(G)$ and l^p with $p \geq 2$, the preferable choice is $s = p$ and if $p \in (1, 2]$ then $s = 2$. So, we assume further that the functional $\|x\|^s$ is either uniformly convex or strongly convex on X. Hence, the problem (6.2.4), (6.2.7) is uniquely solvable. Comparing it with (6.2.1), (6.2.2) above, it is natural to call (6.2.4), (6.2.7) the residual method.

We study the behavior of the sequence $\{x^\delta\}$ as $\delta \to 0$. Let $\bar{x}^*$ be a minimal norm element of N_G. Then $\|x^\delta\| \leq \|\bar{x}^*\|$ for all $\delta \in (0, \delta^*]$ because $N_G \subset \Omega^\delta$. Hence, $\{x^\delta\}$ is a bounded sequence and there exists $\bar{x} \in X$ such that $x^\delta \rightharpoonup \bar{x}$. Since $x^\delta \in G$ and G is convex and closed, the inclusion $\bar{x} \in G$ takes place. Then the definition of Ω^δ gives

$$\langle \zeta - f^\delta, x^\delta - v \rangle \leq \delta C \quad \forall v \in G, \quad \forall \zeta \in Av. \tag{6.2.8}$$

As $\delta \to 0$, we get

$$\langle \zeta - f, \bar{x} - v \rangle \leq 0 \quad \forall v \in G, \quad \forall \zeta \in Av. \tag{6.2.9}$$

Similarly to (6.2.4), introduce now a set

$$\Omega^0 = \{w \in G \mid \langle \zeta - f, w - v \rangle \leq 0 \quad \forall v \in G, \quad \forall \zeta \in Av\}.$$

As we noted above, $N_G \subset \Omega^0$. Let $w_0 \in \Omega^0$ and at the same time $w_0 \notin N_G$. If we assume that $w_0 \in int\ G$, then one can deduce from the inequality (6.2.6) and from Lemma 1.11.6 that $w_0 \in N_G$. Therefore, we conclude that $w_0 \in \partial G$. More precisely, $w_0 \in \partial \Omega^0$. But this is impossible because both the sets Ω^0 and N_G are convex and closed, and their intersections with $int\ G$ coincide. Hence, it results from (6.2.9) that $\bar{x} \in N_G$. Due to the weak convergence of x^δ to $\bar{x} \in G$, and since $\|x^\delta\| \leq \|x^*\|$ for all $x^* \in N_G$, we are able to write down the chain of inequalities

$$\|\bar{x}\| \leq \liminf_{\delta \to 0} \|x^\delta\| \leq \limsup_{\delta \to 0} \|x^\delta\| \leq \|x^*\| \quad \forall x^* \in N_G. \tag{6.2.10}$$

It results from this that $\bar{x} = \bar{x}^*$. Assuming in (6.2.10) $\bar{x} = \bar{x}^*$, we obtain that $\|x^\delta\| \to \|\bar{x}^*\|$ as $\delta \to 0$. Thus, the following theorem is proved:

Theorem 6.2.1 *Let X be an E-space, X^* be strictly convex, $A : X \to 2^{X^*}$ be a maximal monotone operator, G be a bounded convex closed set in $D(A)$, N be a nonempty solution set of (6.1.1) with the properties (6.2.3), a functional $\|x\|^s$ $(s \geq 2)$ be uniformly convex on X. Then the problem (6.2.7), (6.2.4) has a unique solution x^δ and $x^\delta \to \bar{x}^*$ as $\delta \to 0$, where $\bar{x}^*$ is the minimal norm vector of N_G.*

Corollary 6.2.2 *If in Theorem 6.2.1 an operator A is continuous, then $\|Ax^\delta - f\|_* \to 0$ as $\delta \to 0$. If it is maximal monotone and single-valued at the point $\bar{x}^*$, then $y^\delta \rightharpoonup f$ as $\delta \to 0$, where y^δ is any element of Ax^δ.*

Proof. The first assertion of this theorem follows from continuity of A and from the strong convergence of $\{x^\delta\}$ to $\bar{x}^*$ proved in Theorem 6.2.1. Let A be single-valued at the point $\bar{x}^*$. Since A is maximal monotone, then $\bar{x}^* \in int\ D(A)$. Consequently, A is locally bounded at $\bar{x}^*$. Therefore, there exists $g \in X^*$ such that $y^\delta \rightharpoonup g$. Now the equality $g = A\bar{x}^* = f$ is guaranteed by the fact that $gr A$ is demiclosed. ∎

We can establish the connection between sets $\mathcal{M}^\delta$ and Ω^δ. By the monotonicity of A, it has the inequality

$$\langle \zeta - f^\delta, w - v \rangle \leq \langle \xi - f^\delta, w - v \rangle \quad \forall \zeta \in Av, \quad \forall \xi \in Aw. \tag{6.2.11}$$

It is not difficult to see that $\mathcal{M}^\delta \subseteq \Omega^\delta$. Indeed, let $w_0 \in \mathcal{M}^\delta$, that is, there exists $\tilde{f} \in Aw_0$ such that $\|\tilde{f} - f^\delta\|_* \le \delta$ (see (6.2.2)). The inequality (6.2.11) implies

$$\langle \zeta - f^\delta, w_0 - v \rangle \le \delta C.$$

Thus, $w_0 \in \Omega^\delta$.

Note that the inverse inclusion $\Omega^\delta \subseteq \mathcal{M}^\delta$ does not necessarily hold. Let us give a corresponding example.

Example 6.2.3 Let $A : R^1 \to R^1$, $Ax = x - 3$, $f = 0$, $f^\delta = \delta$, $\delta \in (0, \delta^*]$, $G = [2, 4]$, $C = 2$. Then it is not difficult to make certain that $\mathcal{M}^\delta = [2, 4] \cap [3, 3 + 2\delta]$ and

$$\Omega^\delta = [2, 4] \cap [3 + \delta - 2\sqrt{2\delta}, 3 + \delta + 2\sqrt{2\delta}].$$

If $\delta = \dfrac{1}{2}$, then $M^\delta = [3, 4] \subset \Omega^\delta$.

Remark 6.2.4 An element x^δ can also be defined by means of the minimization problem

$$\|x^\delta - z^0\|^s = min \{\|x - z^0\|^s \mid x \in \Omega^\delta\}$$

with some fixed element $z^0 \in X$. Then in Theorem 6.2.1, $\bar{x}^*$ is the nearest element from N_G to z^0. In addition, if in (6.2.7), instead of $\|x\|^s$, uniformly convex functional $\omega(x)$ is minimized, then a solution x^δ is also unique and $\{x^\delta\}$ strongly converges to $\bar{x}^*$ as $\delta \to 0$. Moreover, the condition

$$\omega(\bar{x}^*) = min\{\omega(x^*) \mid x^* \in N_G\}$$

holds.

Generally speaking, convergence of the method (6.2.4), (6.2.7) does not imply solvability of the equation (6.1.1). In order to confirm this fact, we present the example of monotone operator $A : R^1 \to R^1$ such that the method (6.2.4), (6.2.7) converges on each bounded set G and, at the same time, $N = \emptyset$.

Example 6.2.5 Let $Ax = 1$ for all $x \in R^1$, $G = [0, a]$ with $a > 0$, $f = 0$, $f^\delta = \delta$ and $\delta > 0$ be sufficiently small. Then $\Omega^\delta = [0, a\delta(1 - \delta)^{-1}]$, $x^\delta = 0$ for all $\delta \in (0, \delta^*]$. Thus, $x^\delta \to 0$ as $\delta \to 0$, but the set $N = \emptyset$.

However, we are able to be sure that solvability of the variational inequality

$$\langle Ax - f, z - x \rangle \le 0 \quad \forall x \in G, \quad z \in G, \tag{6.2.12}$$

is equivalent to convergence of the residual method. In fact, if $x^\delta \rightharpoonup \bar{x} \in G$ as $\delta \to 0$, i.e., the method (6.2.4), (6.2.7) converges, then passing in (6.2.8) to the limit as $\delta \to 0$ we get (6.2.9). Therefore, $\bar{x}$ is a solution of the variational inequality (6.2.12).

Assume now that the variational inequality (6.2.12) is solvable and $w_0 \in G$ is its solution. Then

$$\langle \zeta - f, w_0 - x \rangle \le 0 \quad \forall x \in G, \quad \forall \zeta \in Ax, \tag{6.2.13}$$

and (6.2.5) holds. Thus, we have established that the set Ω^δ is nonempty. Further the convergence $x^\delta \to \bar{x}^*$ as $\delta \to 0$, where $\bar{x}^*$ is the minimal norm solution of (6.2.12), is proved as in Theorem 6.2.1.

2. Next we study the convergence of projection methods for the problem (6.2.4), (6.2.7).

Theorem 6.2.6 *Suppose that the conditions of Theorem 6.2.1 hold, A is an operator continuous on G and $\{X_n\}$, $n = 1, 2, ...$, is an ordered sequence of finite-dimensional subspaces of X. Let $Q_n : X \to X_n$ and $P_n : X_n \to X$ be linear operators, $|Q_n| \leq 1$, $f_n^\delta = Q_n^* f^\delta$, $G_n = G \cap X_n$, $A_n = Q_n^* A$,*

$$\lim_{n \to \infty} \|Q_n x - x\| = 0 \quad \forall x \in G, \tag{6.2.14}$$

and

$$\limsup_{n \to \infty} (\|P_n x_n\| - \|x_n\|) \leq 0 \quad \forall x_n \in \Omega_n^\delta, \tag{6.2.15}$$

where a constant C in (6.2.4) is such that diam $G_n \leq C$ for all $n > 0$, and

$$\Omega_n^\delta = \{w \in G_n \mid \langle A_n x - f_n^\delta, w - x \rangle \leq \delta C \quad \forall x \in G_n\}. \tag{6.2.16}$$

Let x_n^δ be defined as a solution of the following minimization problem:

$$\|x_n^\delta\|^s = min\{\|x\|^s \mid x \in \Omega_n^\delta\}. \tag{6.2.17}$$

Then $x_n^\delta \to x^\delta$ in X as $n \to \infty$.

Proof. First of all, observe that the problem (6.2.16), (6.2.17) is a finite-dimensional approximation of the residual method. The monotonicity condition of the operators $A_n : X_n \to X_n^*$ is simply verified. Since G_n is bounded for all $n \geq 0$, the inequality

$$\langle A_n y - f_n^\delta, y - x \rangle \leq 0 \quad \forall x \in G_n, \quad y \in G_n,$$

has a solution. It is not difficult now to see that the problem (6.2.16), (6.2.17) is uniquely solvable. Then, similarly to Theorem 6.2.1, the weak convergence $x_n^\delta \rightharpoonup \bar{x} \in G$ follows as $n \to \infty$. Since $x_n^\delta \in \Omega_n^\delta$, the inequality

$$\langle A_n x_n - f_n^\delta, x_n^\delta - x_n \rangle \leq \delta C, \quad x_n = Q_n x \in G_n$$

holds for all $x \in G$. Therefore,

$$\langle A x_n - f^\delta, x_n^\delta - x_n \rangle \leq \delta C.$$

Letting $n \to \infty$ we get

$$\langle A x - f^\delta, \bar{x} - x \rangle \leq \delta C \quad \forall x \in G, \quad \bar{x} \in G.$$

This means that $\bar{x} \in \Omega^\delta$. It follows from (6.2.14) and (6.2.15) that (6.2.16) and (6.2.17) approximate the problem (6.2.4), (6.2.7) with the result

$$\lim_{n \to \infty} \|x_n^\delta\| = \|x^\delta\|.$$

Taking into account the unique solvability of the problem (6.2.4), (6.2.7) and weak convergence of x_n^δ to $\bar{x}$, we deduce that $\bar{x} = x^\delta$. The proof is accomplished because X is E-space. ∎

3. We discuss the connection between the residual method and regularization method. Let x_* be a minimal norm element of G and $x_* \neq x^\delta$ for all $\delta \in (0, \delta^*]$. Consider a functional

$$\varphi^\delta(u) = \sup\{\langle \zeta - f^\delta, u - y\rangle - \delta C \mid y \in G, \ \zeta \in Ay\}.$$

It is obvious that the set Ω^δ in (6.2.4) can be defined as

$$\Omega^\delta = \{u \mid \varphi^\delta(u) \leq 0\}. \tag{6.2.18}$$

Study the properties of $\varphi^\delta(u)$. To this end, write down the inequalities

$$\langle \zeta - f^\delta, u - y\rangle \leq \langle \xi - f^\delta, u - y\rangle \leq C\|\xi - f^\delta\|_*$$

valid for all $u, y \in G$, $\xi \in Au$ and $\zeta \in Ay$. Moreover, $\varphi^\delta(u) \geq -\delta C$. Hence, the functional $\varphi^\delta(u)$ is proper and $G \subset int\ dom\ \varphi^\delta$. Then, by the inequality

$$\begin{aligned}\langle \zeta - f^\delta, tu_1 + (1-t)u_2 - y\rangle &= t\langle \zeta - f^\delta, u_1 - y\rangle + (1-t)\langle \zeta - f^\delta, u_2 - y\rangle \\ &\leq t\varphi^\delta(u_1) + (1-t)\varphi^\delta(u_2) \quad \forall \zeta \in Ay, \quad \forall u_1, u_2, y \in G,\end{aligned}$$

we prove that φ^δ is convex on G. Show that it is lower semi-continuous. Indeed, let $\lim_{n\to\infty} u_n = u$, where $u_n, u \in G$. Passing to the limit in the inequality

$$\langle \zeta - f^\delta, u_n - y\rangle - \delta C \leq \varphi^\delta(u_n) \quad \forall \zeta \in Ay, \quad \forall y \in G,$$

we obtain

$$\varphi^\delta(u) \leq \liminf_{n\to\infty} \varphi^\delta(u_n).$$

In addition, by Theorem 1.2.8, we can make sure that φ^δ has a subdifferential on G.

Due to Theorem 1.11.14, the residual method (6.2.4), (6.2.7) can be reduced to the variational inequality

$$\langle J^s x^\delta, x^\delta - x\rangle \leq 0 \quad \forall x \in \Omega^\delta, \quad x^\delta \in \Omega^\delta, \tag{6.2.19}$$

where J^s is a duality mapping with the gauge function $\mu(t) = t^{s-1}$. On the basis of Theorem 1.11.18, the problem (6.2.18), (6.2.19) is equivalent to the system

$$\langle J^s x^\delta + p\partial\varphi^\delta(x^\delta), x^\delta - y\rangle \leq 0 \quad \forall y \in G, \quad x^\delta \in G, \tag{6.2.20}$$

$$\varphi^\delta(x^\delta)(q - p) \leq 0 \quad \forall q \in R_+^1, \quad p \in R_+^1, \tag{6.2.21}$$

with a solution $\{p, x^\delta\}$ provided that the following Slater condition is satisfied: for every $p \in R_+^1$, there exists $\bar{x} \in \Omega^\delta$ such that

$$p\varphi^\delta(\bar{x}) < 0. \tag{6.2.22}$$

Let us establish (6.2.22) for our problem. To this end, construct a maximal monotone operator $\bar{A}$ with $D(\bar{A}) = G$. It is possible in view of Theorem 1.8.5. Recall that G is bounded. Therefore, by Corollary 1.7.6, there exists at least one solution $v^\delta \in G$ of the equation $\bar{A}v = f^\delta$. Hence, for $\xi^\delta = f^\delta \in \bar{A}v^\delta$, we have

$$p\langle \xi^\delta - f^\delta, v^\delta - x \rangle = 0 \quad \forall p > 0, \quad \forall x \in G.$$

From the monotonicity of $\bar{A}$, one gets

$$p\langle \xi - f^\delta, v^\delta - x \rangle \le p\langle \xi^\delta - f^\delta, v^\delta - x \rangle = 0 \quad \forall \xi \in \bar{A}x, \quad \forall p > 0, \quad \forall x \in G.$$

This means that (6.2.22) holds with $\bar{x} = v^\delta$. Since $x^\delta \ne x_*$ as $\delta \in (0, \delta^*]$, it results from (6.2.20) that $p \ne 0$. It is easily established by contradiction. Indeed, if $p = 0$ then (6.2.20) implies the inequality

$$\langle J^s x^\delta, x^\delta - y \rangle \le 0 \quad \forall y \in G, \quad x^\delta \in G,$$

that is, $\|x^\delta\| \le \|y\|$ for all $y \in G$. Hence, $x^\delta = x_*$ for $\delta \in (0, \delta^*]$, which contradicts our assumption that $x^\delta \ne x_*$. Thus, the following theorem is proved.

Theorem 6.2.7 *Suppose that the conditions of Theorem 6.2.1 are satisfied. Let x^δ be a solution of the problem (6.2.4), (6.2.7) and $x^\delta \ne x_*$ as $\delta \in (0, \delta^*]$, where x_* is a minimal norm element of G. Then there exists $\alpha > 0$ such that the pair $\{\alpha, x^\delta\}$ is a solution of the system of inequalities*

$$\langle \partial\varphi^\delta(x^\delta) + \alpha J^s x^\delta, x^\delta - x \rangle \le 0 \quad \forall x \in G, \quad x^\delta \in G, \tag{6.2.23}$$

$$\varphi^\delta(x^\delta)(q - \alpha^{-1}) \le 0 \quad \forall q \in R_+^1. \tag{6.2.24}$$

Conversely, the pair $\{\alpha, x^\delta\}$, which is a solution of the system (6.2.23), (6.2.24), determines a solution x^δ of the problem (6.2.4), (6.2.7).

We show now that, under the conditions of Theorem 6.2.7, the method (6.2.4), (6.2.7) is equivalent to regularization method (6.2.23) constructed for the variational inequality

$$\langle \partial\varphi^0(v), v - x \rangle \le 0 \quad \forall x \in G, \quad v \in G. \tag{6.2.25}$$

According to the definition of subdifferential $\partial\varphi^0$ at a point v, we have

$$sup\,\{\langle \zeta - f, x - y \rangle \mid y \in G, \zeta \in Ay\} \quad - \quad sup\,\{\langle \zeta - f, v - y \rangle \mid y \in G, \zeta \in Ay\}$$

$$\ge \quad \langle \partial\varphi^0(v), x - v \rangle \quad \forall x, v \in G. \tag{6.2.26}$$

It is obvious that

$$\varphi^0(x) = sup\,\{\langle \zeta - f, x - y \rangle \mid y \in G, \zeta \in Ay\} \ge 0 \quad \forall x \in G, \tag{6.2.27}$$

because $\langle \zeta - f, x - y \rangle = 0$ when $y = x$.

Let $\bar{x} \in G$ be a solution of the variational inequality

$$\langle Ay - f, y - x \rangle \leq 0 \quad \forall x \in G, \tag{6.2.28}$$

that is, there exists $\eta \in A\bar{x}$ such that

$$\langle \eta - f, \bar{x} - x \rangle \leq 0 \quad \forall x \in G. \tag{6.2.29}$$

Then, by Lemma 1.11.4, $\bar{x}$ is a solution of the inequality

$$\langle \eta - f, \bar{x} - x \rangle \leq 0 \quad \forall x \in G, \quad \forall \eta \in Ax.$$

This means that $\varphi^0(\bar{x}) \leq 0$. Taking into account (6.2.27), one gets that $\varphi^0(\bar{x}) = 0$. Then (6.2.26) for $v = \bar{x}$ gives

$$\varphi^0(x) \geq \langle \partial \varphi^0(\bar{x}), x - \bar{x} \rangle \quad \forall x \in G.$$

Hence, $g \in \partial \varphi^0(\bar{x})$ if $\varphi^0(x) \geq \langle g, x - \bar{x} \rangle$. However, by the definition of φ^0, it follows that

$$\varphi^0(x) \geq \langle \bar{\eta} - f, x - \bar{x} \rangle \quad \forall x \in G, \quad \forall \bar{\eta} \in A\bar{x}, \quad \bar{x} \in G,$$

which implies the inclusion $A\bar{x} - f \subset \partial \varphi^0(\bar{x})$. Then due to (6.2.29), we obtain that $\bar{x}$ is a solution of (6.2.25).

Let now $\bar{x}$ be a solution of (6.2.25), that is,

$$\varphi^0(\bar{x}) = min\{\varphi^0(x) \mid x \in G\}.$$

Since $\varphi^0(x) \geq 0$ for all $x \in G$ and since $\varphi^0(x)$ attains the null value at a solution $\bar{x}$ of (6.2.28), we have $\varphi^0(\bar{x}) = 0$. Thus, the inequalities (6.2.25) and (6.2.28) are equivalent. Therefore, solutions of the regularized inequality

$$\langle Ay + \alpha J^s y - f^\delta, y - x \rangle \leq 0 \quad \forall x \in G, \quad y \in G,$$

and regularized inequality (6.2.23) approximate one and the same solution of the equation (6.1.1). Using (6.2.24) and following the proof of Theorem 1.11.18 we come to the equality

$$\varphi^\delta(x^\delta) = 0, \tag{6.2.30}$$

which allows us to find α in (6.2.23) and define a connection between α and δ. The quantity $\varphi^\delta(x) = \varphi^\delta(x) + \delta C$ may be regarded as the quasi-residual of the inequality

$$\langle Ay - f^\delta, x_0 - y \rangle \leq 0 \quad \forall y \in G, \quad x_0 \in G,$$

on an element x^δ. Therefore, the equality (6.2.30) may be considered as the residual principle for the problem actually being solved

$$\langle Ay - f, \hat{x} - y \rangle \leq 0 \quad \forall y \in G, \quad \hat{x} \in G.$$

Hence, the method (6.2.4), (6.2.7) is equivalent to the regularization method with the operator $\partial \varphi^\delta$, where the regularization parameter α is defined by (6.2.30).

Consider again Example 6.2.3. If $1 + \delta - 2\sqrt{2\delta} > 0$ and $x^\delta \neq 2$, then an element $x^\delta = 3 + \delta - 2\sqrt{2\delta}$ is a solution of (6.2.23) with

$$\alpha = \frac{2\sqrt{2\delta}}{3 + \delta - 2\sqrt{2\delta}}.$$

Moreover,

$$max \ \{(x - 3 - \delta)(x^\delta - x) \mid x \in [2, 4]\} = 2\delta.$$

Remark 6.2.8 *Let in place of operator A, its maximal monotone approximations A^h be known such that for all $x \in D(A) = D(A^h)$ and $h \in (0, h^*]$,*

$$\mathcal{H}_{X^*}(Ax, A^h x) \leq g(\|x\|)h, \tag{6.2.31}$$

where $g(t)$ is a non-negative and non-decreasing function for $t \geq 0$. Replace Ω^δ by

$$\Omega^{\delta,h} = \{w \in G \mid \langle \zeta^h - f^\delta, w - v \rangle \leq C(\delta + g(C_1)h) \quad \forall x \in G, \quad \forall \zeta^h \in A^h v\},$$

where $C_1 > \|v\|$ for all $v \in G$. Then all the assertions of the present section can be obtained by the same arguments.

6.3   Penalty Method

In this section we study one more regularization method for solving variational inequalities, the so-called penalty method.

Let X be a reflexive strictly convex space with strictly convex dual space X^*, $A : X \rightarrow 2^{X^*}$ be a maximal monotone operator, $\Omega \subset int \, D(A)$ be a convex closed bounded set. Consider the variational inequality (7.1.1). Let it have a solution set $N \neq \emptyset$ and let $\bar{x}^* \in N$ be a minimal norm solution. Assume that operator A, element f and set Ω are given approximately, namely, their approximations A^h, f^δ and Ω_σ are known for all $(\delta, h, \sigma) \in \Re$, where $\Re = (0, \delta^*] \times (0, h^*] \times (0, \sigma^*]$ with some positive δ^*, h^*, σ^*, such that $f^\delta \in X^*$, $\|f - f^\delta\|_* \leq \delta$,

$$\mathcal{H}_X(\Omega, \Omega_\sigma) \leq \sigma \tag{6.3.1}$$

and

$$\mathcal{H}_{X^*}(Ax, A^h x) \leq hg(\|x\|) \quad \forall x \in \Omega \cup \Omega_\sigma. \tag{6.3.2}$$

In (6.3.1) and (6.3.2), $A^h : X \rightarrow 2^{X^*}$ are maximal monotone operators, $D(A^h) = D(A)$, $\Omega_\sigma \subset int \, D(A^h)$ are convex closed and bounded sets and $g(t)$ is a continuous, non-negative and non-decreasing function for all $t \geq 0$.

Given convex closed set Ω, the penalty operator $S : X \rightarrow X^*$ is defined by the formula

$$Sx = J(x - P_\Omega x),$$

where J is a normalized duality mapping in X, P_Ω is a metric projection operator onto Ω. As it has been proved in Lemma 1.5.18, S is a monotone demicontinuous and bounded mapping. We establish some of its additional properties. Rewrite (1.5.14) as

$$\langle Sx - Sy, P_\Omega x - P_\Omega y \rangle \geq 0 \quad \forall x \in X, \quad \forall y \in X.$$

Then the definition of P_Ω implies the estimate

$$\langle J(x - P_\Omega x), P_\Omega x - y \rangle \geq 0 \quad \forall y \in \Omega. \tag{6.3.3}$$

Lemma 6.3.1 *An element $\bar{x} \in \Omega$ is a metric projection of x onto Ω if and only if the following inequality is fulfilled:*

$$\|x - \bar{x}\|^2 \leq \langle J(x - \bar{x}), x - y \rangle \quad \forall y \in \Omega. \tag{6.3.4}$$

Proof. Let (6.3.4) hold. By the Cauchy–Schwarz inequality,

$$\|x - \bar{x}\| \leq \|x - \bar{x}\|^{-1} \langle J(x - \bar{x}), x - y \rangle \leq \|x - y\| \quad \forall y \in \Omega,$$

i.e., $\bar{x} = P_\Omega x$. Assume now that $\bar{x} = P_\Omega x$. Then (6.3.3) implies that for all $y \in \Omega$,

$$0 \leq \langle J(x - \bar{x}), \bar{x} - y \rangle = -\|x - \bar{x}\|^2 + \langle J(x - \bar{x}), x - y \rangle,$$

and (6.3.4) follows. ∎

Lemma 6.3.2 *Let X be a uniformly convex Banach space, X^* be strictly convex, Ω_1 and Ω_2 be convex closed sets in X, $\mathcal{H}_X(\Omega_1, \Omega_2) \leq \sigma$, $\delta_X(\epsilon)$ be a modulus of convexity of X. Then the following estimate holds:*

$$\|P_{\Omega_1} x - P_{\Omega_2} x\| \leq c_2 \delta_X^{-1}(2Lc_1\sigma), \tag{6.3.5}$$

where $1 < L < 1.7$, $c_1 = 2max\{\|x - \bar{x}_1\|, \|x - \bar{x}_2\|\}$, $\bar{x}_1 = P_{\Omega_1} x$, $\bar{x}_2 = P_{\Omega_2} x$, $c_2 = 2max\{1, \|x - \bar{x}_1\|, \|x - \bar{x}_2\|\}$.

Proof. Due to Theorem 1.6.4, we have

$$\langle J(x - \bar{x}_1) - J(x - \bar{x}_2), \bar{x}_2 - \bar{x}_1 \rangle \geq (2L)^{-1} \delta_X(c_2^{-1}\|\bar{x}_1 - \bar{x}_2\|). \tag{6.3.6}$$

Since $\mathcal{H}_X(\Omega_1, \Omega_2) \leq \sigma$, for any element $\bar{x}_2 \in \Omega_2$ there exists a $z_1 \in \Omega_1$ such that $\|\bar{x}_2 - z_1\| \leq \sigma$. Furthermore, (6.3.3) yields the inequality

$$\langle J(x - \bar{x}_1), z_1 - \bar{x}_1 \rangle \leq 0.$$

Then

$$\langle J(x - \bar{x}_1), \bar{x}_2 - \bar{x}_1 \rangle = \langle J(x - \bar{x}_1), \bar{x}_2 - z_1 \rangle + \langle J(x - \bar{x}_1), z_1 - \bar{x}_1 \rangle \leq \sigma \|x - \bar{x}_1\|.$$

By analogy, we assert that there exists $z_2 \in \Omega_2$ such that $\|\bar{x}_1 - z_2\| \leq \sigma$ and

$$\langle J(x - \bar{x}_2), \bar{x}_1 - \bar{x}_2 \rangle \leq \sigma \|x - \bar{x}_2\|.$$

Therefore,

$$\langle J(x - \bar{x}_1) - J(x - \bar{x}_2), \bar{x}_2 - \bar{x}_1 \rangle \leq \sigma(\|x - \bar{x}_1\| + \|x - \bar{x}_2\|) \leq \sigma c_1. \tag{6.3.7}$$

By (6.3.6) and (6.3.7), we deduce (6.3.5). ∎

Consider in X the equation

$$A^h x + \epsilon^{-1} S_\sigma x + \alpha J x = f^\delta, \quad \alpha > 0, \quad \epsilon > 0, \tag{6.3.8}$$

where $S_\sigma x = J(x - P_{\Omega_\sigma} x)$. Since, in view of Theorem 1.4.6, the operator S_σ is maximal monotone with $D(S_\sigma) = X$, we conclude that the operator $A^h + \epsilon^{-1} S_\sigma$ is so (see Theorem 1.8.3). A single-valued solvability of the equation (6.3.8) is guaranteed by Theorem 1.7.4. Thus, in place of the variational inequality (4.5.2) on Ω, it is proposed to solve operator equation (6.3.8) on $D(A)$. Therefore, (6.3.8) can be regarded as the regularized penalty method. Let $x_{\alpha,\epsilon}^\gamma$ with $\gamma(\delta, h, \sigma)$ be a (unique) solution of this equation.

Theorem 6.3.3 *Assume that X is a uniformly convex and uniformly smooth Banach space, X^* is a dual space, $\delta_X(\epsilon)$ and $\delta_{X^*}(\epsilon)$ are moduli of convexity of X and X^*, respectively. Denote $g_{X^*}(\epsilon) = \epsilon^{-1} \delta_{X^*}(\epsilon)$ and suppose that*

$$\lim_{\alpha \to 0} \frac{\delta + h + \epsilon}{\alpha} = 0, \quad \lim_{\alpha \to 0} \frac{g_{X^*}^{-1}(\delta_X^{-1}(\sigma))}{\alpha \epsilon} = 0. \tag{6.3.9}$$

Then $x_{\alpha,\epsilon}^\gamma \to \bar{x}^*$ as $\alpha \to 0$.

Proof. Since $x_{\alpha,\epsilon}^\gamma$ is a solution of (6.3.8), there exists an element $\zeta_{\alpha,\epsilon}^\gamma \in A^h x_{\alpha,\epsilon}^\gamma$ such that

$$\zeta_{\alpha,\epsilon}^\gamma + \epsilon^{-1} S_\sigma x_{\alpha,\epsilon}^\gamma + \alpha J x_{\alpha,\epsilon}^\gamma = f^\delta. \tag{6.3.10}$$

Let $\bar{x}_{\alpha,\epsilon}^\gamma = P_{\Omega_\sigma} x_{\alpha,\epsilon}^\gamma$. Then, by Lemma 6.3.1, we have

$$\|x_{\alpha,\epsilon}^\gamma - \bar{x}_{\alpha,\epsilon}^\gamma\|^2 \leq \langle S_\sigma x_{\alpha,\epsilon}^\gamma, x_{\alpha,\epsilon}^\gamma - u \rangle \quad \forall u \in \Omega_\sigma.$$

Taking into account (6.3.10), we obtain the inequality

$$\|x_{\alpha,\epsilon}^\gamma - \bar{x}_{\alpha,\epsilon}^\gamma\|^2 \leq \epsilon \langle f^\delta - \zeta_{\alpha,\epsilon}^\gamma - \alpha J x_{\alpha,\epsilon}^\gamma, x_{\alpha,\epsilon}^\gamma - u \rangle \quad \forall u \in \Omega_\sigma. \tag{6.3.11}$$

Let z_0 be some fixed point of Ω. Due to (6.3.1), for any $\sigma \in (0, \sigma^*]$, there exists $z_\sigma \in \Omega_\sigma$ such that $\|z_0 - z_\sigma\| \leq \sigma$, and then $z_\sigma \to z_0$ as $\sigma \to 0$. Putting in (6.3.11) $u = z_\sigma$, we get

$$\begin{aligned}\|x_{\alpha,\epsilon}^\gamma - \bar{x}_{\alpha,\epsilon}^\gamma\|^2 \quad &\leq \quad \epsilon \langle f^\delta - \xi_\sigma^h - \alpha J z_\sigma, x_{\alpha,\epsilon}^\gamma - z_\sigma \rangle \\[2mm] &\quad - \quad \epsilon \langle \zeta_{\alpha,\epsilon}^\gamma + \alpha J x_{\alpha,\epsilon}^\gamma - \xi_\sigma^h - \alpha J z_\sigma, x_{\alpha,\epsilon}^\gamma - z_\sigma \rangle \end{aligned} \tag{6.3.12}$$

for all $\zeta_{\alpha,\epsilon}^\gamma \in A^h x_{\alpha,\epsilon}^\gamma$ and for all $\xi_\sigma^h \in A^h z_\sigma$. The monotonicity property of the operator $A^h + \alpha J$ yields the relation

$$\langle \zeta_{\alpha,\epsilon}^\gamma + \alpha J x_{\alpha,\epsilon}^\gamma - \xi_\sigma^h - \alpha J z_\sigma, x_{\alpha,\epsilon}^\gamma - z_\sigma \rangle \geq 0.$$

By (6.3.2), for $\xi_\sigma^h \in A^h z_\sigma$, we find $\eta_\sigma \in A z_\sigma$ such that $\|\xi_\sigma^h - \eta_\sigma\| \le hg(\|z_\sigma\|)$. Then the following estimates are obtained:

$$\|\xi_\sigma^h\|_* \le \|\xi_\sigma^h - \eta_\sigma\|_* + \|\eta_\sigma\|_* \le hg(\|z_\sigma\|) + \|\eta_\sigma\|_*. \tag{6.3.13}$$

After this, (6.3.12) and (6.3.13) imply the relation

$$\|x_{\alpha.\epsilon}^\gamma - \bar{x}_{\alpha.\epsilon}^\gamma\|^2 \le \epsilon \Big(\|f\|_* + \delta + hg(\|z_\sigma\|) + \|\eta_\sigma\|_* + \alpha\|z_\sigma\| \Big)$$
$$\times \Big(\|x_{\alpha.\epsilon}^\gamma - \bar{x}_{\alpha.\epsilon}^\gamma\| + \|\bar{x}_{\alpha.\epsilon}^\gamma - z_\sigma\| \Big). \tag{6.3.14}$$

Since $\Omega \subset int\, D(A)$, we conclude that the operator A is locally bounded at every point of Ω, hence, at the chosen point z_0. Therefore, the sequence $\{\eta_\sigma\}$ is bounded as $\sigma \to 0$. By the properties of the function $g(t)$ and by the boundedness of the sets Ω_σ, there exist constants $c_1 > 0$ and $c_2 > 0$ such that (6.3.14) is evaluated as

$$\|x_{\alpha.\epsilon}^\gamma - \bar{x}_{\alpha.\epsilon}^\gamma\|^2 \le \epsilon \Big(c_1 \|x_{\alpha.\epsilon}^\gamma - \bar{x}_{\alpha.\epsilon}^\gamma\| + c_2 \Big).$$

From this, one deduces that a sequence $\{x_{\alpha.\epsilon}^\gamma\}$ is bounded as $\alpha \to 0$, say, $\|x_{\alpha.\epsilon}^\gamma\| \le r$, $r > 0$.

Introduce in X the auxiliary equation

$$Ax + \epsilon^{-1} Sx + \alpha Jx = f, \tag{6.3.15}$$

and denote its unique solution by $x_{\alpha.\epsilon}^0$. Making use of the above arguments, it is not difficult to verify that a sequence $\{x_{\alpha.\epsilon}^0\}$ is also bounded as $\alpha \to 0$ and $\|x_{\alpha.\epsilon}^0\| \le r$. This allows us to assert that $x_{\alpha.\epsilon}^0 \rightharpoonup \bar{x} \in X$. Furthermore, since $x_{\alpha.\epsilon}^0$ is a solution of (6.3.15), there exists $\eta_{\alpha.\epsilon}^0 \in A x_{\alpha.\epsilon}^0$ such that

$$\eta_{\alpha.\epsilon}^0 + \epsilon^{-1} S x_{\alpha.\epsilon}^0 + \alpha J x_{\alpha.\epsilon}^0 = f. \tag{6.3.16}$$

Then Lemma 6.3.1 with $y = z_0 \in \Omega$ implies

$$\langle \eta_{\alpha.\epsilon}^0 + \alpha J x_{\alpha.\epsilon}^0 - f, z_0 - x_{\alpha.\epsilon}^0 \rangle = \epsilon^{-1} \langle S x_{\alpha.\epsilon}^0, x_{\alpha.\epsilon}^0 - z_0 \rangle \ge 0. \tag{6.3.17}$$

In view of (6.3.17), it is quite easy to derive the estimate

$$\langle \eta_{\alpha.\epsilon}^0, x_{\alpha.\epsilon}^0 - z_0 \rangle \le (\|f\|_* + \alpha r) \|x_{\alpha.\epsilon}^0 - z_0\|. \tag{6.3.18}$$

Moreover, according to Lemma 1.5.14 for $x_0 \in \Omega \subset int\, D(A)$, there exists constant $r_0 > 0$ and a ball $B(z_0, r_0)$ such that for every $x_{\alpha.\epsilon}^0 \in D(A)$ and $\eta_{\alpha.\epsilon}^0 \in A x_{\alpha.\epsilon}^0$, the inequality

$$\langle \eta_{\alpha.\epsilon}^0, x_{\alpha.\epsilon}^0 - z_0 \rangle \ge r_0 \|\eta_{\alpha.\epsilon}^0\|_* - c_0(\|x_{\alpha.\epsilon}^0 - z_0\| + r_0) \tag{6.3.19}$$

holds, where

$$c_0 = sup\, \{\|\xi\|_* \mid \xi \in Ax, \ x \in B(z_0, r_0)\} < \infty.$$

Then, by (6.3.18) and (6.3.19), we have

$$\|\eta_{\alpha.\epsilon}^0\|_* \le r_0^{-1} (c_0 + \|f\|_* + \alpha r) \|x_{\alpha.\epsilon}^0 - z_0\| + c_0,$$

from which it results that the sequence $\{\eta^0_{\alpha.\epsilon}\}$ is bounded for all $\alpha > 0$. Now we find from (6.3.16) that

$$\|Sx^0_{\alpha.\epsilon}\|_* \leq \epsilon \|f - \eta^0_{\alpha.\epsilon} - \alpha Jx^0_{\alpha.\epsilon}\|_* \leq \epsilon(\|f\|_* + \|\eta^0_{\alpha.\epsilon}\|_* + \alpha\|x^0_{\alpha.\epsilon}\|).$$

Consequently, $Sx^0_{\alpha.\epsilon} \to \theta_{X^*}$ because $\epsilon \to 0$ as $\alpha \to 0$ (see (6.3.9)). Since S is demiclosed, the equality $S\bar{x} = \theta_{X^*}$ holds. This means that $\bar{x} \in \Omega$.

By (6.3.16), we calculate

$$\langle \eta^0_{\alpha.\epsilon} - \xi, x^0_{\alpha.\epsilon} - x \rangle \; + \; \langle \xi - f, x^0_{\alpha.\epsilon} - x \rangle + \alpha \langle Jx^0_{\alpha.\epsilon}, x^0_{\alpha.\epsilon} - x \rangle$$
$$+ \; \epsilon^{-1}\langle Sx^0_{\alpha.\epsilon} - Sx, x^0_{\alpha.\epsilon} - x \rangle = 0 \quad \forall x \in \Omega, \;\; \forall \xi \in Ax.$$

Then the monotonicity property of A and S leads to the inequality

$$\langle \xi - f, x^0_{\alpha.\epsilon} - x \rangle + \alpha \langle Jx^0_{\alpha.\epsilon}, x^0_{\alpha.\epsilon} - x \rangle \leq 0 \quad \forall x \in \Omega, \quad \forall \xi \in Ax. \tag{6.3.20}$$

Assuming that $\alpha \to 0$ in (6.3.20) and taking into account the weak convergence of $x^0_{\alpha.\epsilon}$ to $\bar{x} \in \Omega$, one gets

$$\langle \xi - f, x - \bar{x} \rangle \geq 0 \quad \forall x \in \Omega, \quad \forall \xi \in Ax. \tag{6.3.21}$$

The maximal monotonicity of A and Lemma 1.11.4 allow us to assert that there exists $\bar{\xi} \in A\bar{x}$ such that

$$\langle \bar{\xi} - f, x - \bar{x} \rangle \geq 0 \quad \forall x \in \Omega.$$

Hence, $\bar{x} \in N$.

Show that $x^0_{\alpha.\epsilon} \to \bar{x}^*$ as $\alpha \to 0$. Since $\bar{x}^0_{\alpha.\epsilon} = P_\Omega x^0_{\alpha.\epsilon} \in \Omega$, one has

$$\langle \xi - f, \bar{x}^0_{\alpha.\epsilon} - x \rangle \geq 0 \quad \forall x \in N, \quad \forall \xi \in Ax. \tag{6.3.22}$$

Rewrite (6.3.20) in the following form:

$$\langle \xi - f, x^0_{\alpha.\epsilon} - \bar{x}^0_{\alpha.\epsilon} \rangle + \langle \xi - f, \bar{x}^0_{\alpha.\epsilon} - x \rangle + \alpha \langle Jx^0_{\alpha.\epsilon}, x^0_{\alpha.\epsilon} - x \rangle \leq 0 \quad \forall \xi \in Ax. \tag{6.3.23}$$

Then, by (6.3.22) and by the Cauchy−Schwartz inequality, we obtain

$$\langle Jx^0_{\alpha.\epsilon}, x^0_{\alpha.\epsilon} - x \rangle \leq \alpha^{-1} \|\xi - f\|_* \|x^0_{\alpha.\epsilon} - \bar{x}^0_{\alpha.\epsilon}\| \quad \forall x \in N, \quad \forall \xi \in Ax.$$

Observe that (6.3.11) with $u = \bar{x}^0_{\alpha.\epsilon}$ gives for solution $x^0_{\alpha.\epsilon}$ of (6.3.15) the estimate

$$\|x^0_{\alpha.\epsilon} - \bar{x}^0_{\alpha.\epsilon}\| \leq c_3\epsilon,$$

where c_3 is a positive constant. That is to say, there exists $c_4 > 0$ such that

$$\langle Jx^0_{\alpha.\epsilon}, x^0_{\alpha.\epsilon} - x \rangle \leq c_4\epsilon\alpha^{-1} \quad \forall x \in N.$$

Due to the weak convergence of $x^0_{\alpha.\epsilon}$ to $\bar{x} \in N$, monotonicity of J and (6.3.9), one gets

$$\langle Jx, \bar{x} - x \rangle \leq 0 \quad \forall x \in N.$$

This means that $x_{\alpha,\epsilon}^0 \rightharpoonup \bar{x}^*$ as $\alpha \to 0$. Furthermore, similarly to Section 2.2,

$$x_{\alpha,\epsilon}^0 \to \bar{x}^* \quad \text{as} \quad \alpha \to 0. \tag{6.3.24}$$

Combining (6.3.10) and (6.3.16), we come to the following equality:

$$\langle \zeta_{\alpha,\epsilon}^\gamma - \eta_{\alpha,\epsilon}^0, x_{\alpha,\epsilon}^\gamma - x_{\alpha,\epsilon}^0 \rangle \quad + \quad \epsilon^{-1}\langle S_\sigma x_{\alpha,\epsilon}^\gamma - S x_{\alpha,\epsilon}^0, x_{\alpha,\epsilon}^\gamma - x_{\alpha,\epsilon}^0 \rangle$$

$$+ \quad \alpha\langle J x_{\alpha,\epsilon}^\gamma - J x_{\alpha,\epsilon}^0, x_{\alpha,\epsilon}^\gamma - x_{\alpha,\epsilon}^0 \rangle$$

$$= \quad \langle f^\delta - f, x_{\alpha,\epsilon}^\gamma - x_{\alpha,\epsilon}^0 \rangle. \tag{6.3.25}$$

Using Theorem 1.6.4, one can write

$$\langle J x_{\alpha,\epsilon}^\gamma - J x_{\alpha,\epsilon}^0, x_{\alpha,\epsilon}^\gamma - x_{\alpha,\epsilon}^0 \rangle \geq (2L)^{-1}\delta_X(c_5^{-1}\|x_{\alpha,\epsilon}^\gamma - x_{\alpha,\epsilon}^0\|), \tag{6.3.26}$$

where $1 < L < 1.7$ and $c_5 = 2max\{1, r\}$. Let $\zeta_{\alpha,\epsilon}^h \in A^h x_{\alpha,\epsilon}^0$ such that

$$\|\eta_{\alpha,\epsilon}^0 - \zeta_{\alpha,\epsilon}^h\|_* \leq hg(\|x_{\alpha,\epsilon}^0\|).$$

This element exists because of the condition (6.3.2). Since operators A^h are monotone and function $g(t)$ is non-decreasing, we obtain

$$\langle \zeta_{\alpha,\epsilon}^\gamma - \eta_{\alpha,\epsilon}^0, x_{\alpha,\epsilon}^\gamma - x_{\alpha,\epsilon}^0 \rangle \quad = \quad \langle \zeta_{\alpha,\epsilon}^\gamma - \zeta_{\alpha,\epsilon}^h, x_{\alpha,\epsilon}^\gamma - x_{\alpha,\epsilon}^0 \rangle + \langle \zeta_{\alpha,\epsilon}^h - \eta_{\alpha,\epsilon}^0, x_{\alpha,\epsilon}^\gamma - x_{\alpha,\epsilon}^0 \rangle$$

$$\geq \quad -hg(\|x_{\alpha,\epsilon}^0\|)\|x_{\alpha,\epsilon}^\gamma - x_{\alpha,\epsilon}^0\|$$

$$\geq \quad -hg(r)\|x_{\alpha,\epsilon}^\gamma - x_{\alpha,\epsilon}^0\|. \tag{6.3.27}$$

The monotonicity property of S_σ makes it possible to evaluate the second term in the left-hand side of (6.3.25):

$$\langle S_\sigma x_{\alpha,\epsilon}^\gamma - S x_{\alpha,\epsilon}^0, x_{\alpha,\epsilon}^\gamma - x_{\alpha,\epsilon}^0 \rangle \quad = \quad \langle S_\sigma x_{\alpha,\epsilon}^\gamma - S_\sigma x_{\alpha,\epsilon}^0, x_{\alpha,\epsilon}^\gamma - x_{\alpha,\epsilon}^0 \rangle$$

$$+ \quad \langle S_\sigma x_{\alpha,\epsilon}^0 - S x_{\alpha,\epsilon}^0, x_{\alpha,\epsilon}^\gamma - x_{\alpha,\epsilon}^0 \rangle$$

$$\geq \quad \langle S_\sigma x_{\alpha,\epsilon}^0 - S x_{\alpha,\epsilon}^0, x_{\alpha,\epsilon}^\gamma - x_{\alpha,\epsilon}^0 \rangle.$$

By Corollary 1.6.8,

$$\langle S_\sigma x_{\alpha,\epsilon}^0 \quad - \quad S x_{\alpha,\epsilon}^0, x_{\alpha,\epsilon}^\gamma - x_{\alpha,\epsilon}^0 \rangle = \langle J(x_{\alpha,\epsilon}^0 - P_{\Omega_\sigma} x_{\alpha,\epsilon}^0) - J(x_{\alpha,\epsilon}^0 - P_\Omega x_{\alpha,\epsilon}^0), x_{\alpha,\epsilon}^\gamma - x_{\alpha,\epsilon}^0 \rangle$$

$$\geq \quad - \|J(x_{\alpha,\epsilon}^0 - P_{\Omega_\sigma} x_{\alpha,\epsilon}^0) - J(x_{\alpha,\epsilon}^0 - P_\Omega x_{\alpha,\epsilon}^0)\|_*\|x_{\alpha,\epsilon}^\gamma - x_{\alpha,\epsilon}^0\|$$

$$\geq \quad - c_6 g_{X^*}^{-1}\left(2c_6 L\|P_{\Omega_\sigma} x_{\alpha,\epsilon}^0 - P_\Omega x_{\alpha,\epsilon}^0\|\right)\|x_{\alpha,\epsilon}^\gamma - x_{\alpha,\epsilon}^0\|, \tag{6.3.28}$$

where $c_6 = 2max\{1, c_3\epsilon^*, r + d + \sigma^*\}$, $d = max\{\|x\| \mid x \in \Omega\}$. Now Lemma 6.3.2 enables us to specify the obtained inequality. Indeed,

$$\langle S_\sigma x^0_{\alpha,\epsilon} - S x^0_{\alpha,\epsilon}, x^\gamma_{\alpha,\epsilon} - x^0_{\alpha,\epsilon}\rangle$$

$$\geq -c_6 g_{X^*}^{-1}\left(2Lc_6^2\delta_X^{-1}(2Lc_7\sigma)\right)\|x^\gamma_{\alpha,\epsilon} - x^0_{\alpha,\epsilon}\|, \qquad (6.3.29)$$

where $c_7 = 2max\{c_3\epsilon^*, r + d + \sigma^*\}$. It results from (6.3.25) - (6.3.29) that

$$(\delta + hg(r))\|x^\gamma_{\alpha,\epsilon} - x^0_{\alpha,\epsilon}\| + \epsilon^{-1}c_6 g_{X^*}^{-1}\left(2Lc_6^2\delta_X^{-1}(2Lc_7\sigma)\right)\|x^\gamma_{\alpha,\epsilon} - x^0_{\alpha,\epsilon}\|$$

$$\geq \alpha(2L)^{-1}\delta_X(c_5^{-1}\|x^\gamma_{\alpha,\epsilon} - x^0_{\alpha,\epsilon}\|). \qquad (6.3.30)$$

Finally, (6.3.30) yields the estimate

$$\|x^\gamma_{\alpha,\epsilon} - x^0_{\alpha,\epsilon}\| \leq c_5 g_X^{-1}\left[2Lc_5\left(\frac{\delta + hg(r)}{\alpha} + \frac{c_6}{\alpha\epsilon}g_{X^*}^{-1}(2Lc_6^2\delta_X^{-1}(2Lc_7\sigma))\right)\right].$$

In view of (6.3.9), this result together with (6.3.24) completes the proof. ∎

Remark 6.3.4 *If X is a Hilbert space, then the second condition in (6.3.9) is written as*

$$\lim_{\alpha \to 0} \frac{\sqrt{\sigma}}{\alpha\epsilon} = 0.$$

6.4   Proximal Point Method

In this section we study the proximal point algorithm for solving the equation (6.1.1).

1. Consider first the Hilbert space case. Let $A : H \to 2^H$ be a maximal monotone operator, equation (6.1.1) has a nonempty solution set N in the sense of inclusion. We assume that, in place of f, a sequence $\{f_n\}$, $n = 1, 2, ...$, is known such that

$$\|f - f_n\| \leq \delta_n. \qquad (6.4.1)$$

The proximal point algorithm is described as follows: Take some sequence of positive numbers $\{c_n\}$. For a constant $c_1 > 0$ and $x_0 \in D(A)$, we find the unique solution x_1 of the equation

$$c_1(Ax - f_1) + x = x_0$$

(see Theorem 1.7.4). Proceeding with this process further, we obtain in $D(A)$ a sequence $\{x_n\}$ which is defined recurrently, namely, x_n is a solution of the equation

$$c_n(Ax - f_n) + x = x_{n-1}, \quad c_n > 0, \quad n = 1, 2, ... \qquad (6.4.2)$$

Theorem 6.4.1 *Let $A : H \to 2^H$ be a maximal monotone operator, a solution set of (6.1.1) be nonempty,*

$$0 < c \leq c_n \leq C \quad and \quad \sum_{n=1}^{\infty} \delta_n < \infty. \qquad (6.4.3)$$

Then a sequence $\{x_n\}$ generated by (6.4.2) weakly converges to some element of N.

Proof. Let $x^* \in N$, $y_n \in Ax_n$ and

$$c_n(y_n - f_n) + x_n = x_{n-1}. \tag{6.4.4}$$

By (6.4.4), one gets

$$c_n(y_n - f, x_n - x^*) \quad + \quad c_n(f - f_n, x_n - x^*) + \|x_n - x^*\|^2$$
$$= \quad (x_{n-1} - x^*, x_n - x^*). \tag{6.4.5}$$

Due to the monotonicity of A, we have from (6.1.1), (6.4.3) and (6.4.5)

$$\|x_n - x^*\| \le \|x_{n-1} - x^*\| + c_n \delta_n \le \|x_{n-1} - x^*\| + C\delta_n. \tag{6.4.6}$$

This implies the inequality

$$\|x_n - x^*\| \le \|x_0 - x^*\| + C \sum_{i=1}^{n} \delta_i.$$

In view of convergence of the series $\sum_{n=1}^{\infty} \delta_n$, the sequence $\{x_n\}$ is bounded. Hence, there exists a subsequence $\{x_k\} \subset \{x_n\}$ which converges weakly to some $\bar{x} \in H$. We show that $\bar{x} \in N$. First of all, it is well known that if the sequence of positive numbers $\{a_n\}$ satisfies the recurrent inequality

$$a_n \le a_{n-1} + \beta_n, \quad \sum_{n=1}^{\infty} \beta_n < \infty,$$

then there exists $\lim_{n \to \infty} a_n = a \ge 0$. For this reason, (6.4.6) and the condition $\sum_{n=1}^{\infty} \delta_n < \infty$ imply existence of a limit for $\|x_n - x^*\|$ as $n \to 0$. We can write down that

$$\lim_{n \to \infty} \|x_n - x^*\| = \mu(x^*) < \infty. \tag{6.4.7}$$

Now (6.4.4) yields

$$\|x_n - x^*\|^2 + c_n^2 \|y_n - f_n\|^2 + 2c_n(y_n - f_n, x_n - x^*) = \|x_{n-1} - x^*\|^2. \tag{6.4.8}$$

Again, by the monotonicity of A, the inequality $(y_n - f, x_n - x^*) \ge 0$ holds. Transform (6.4.8) into the following relation:

$$\|x_n - x^*\|^2 + c_n^2 \|y_n - f_n\|^2 \le \|x_{n-1} - x^*\|^2 + 2C\delta_n M, \tag{6.4.9}$$

where a positive constant $M > 0$ satisfies the estimate $\|x_n - x^*\| \le M$. It follows from the convergence criterion of the series above that $\delta_n \to 0$ as $n \to \infty$. Hence, by virtue of (6.4.7) and (6.4.9),

$$\lim_{n \to \infty} c_n \|y_n - f_n\| = 0.$$

Since $c_n \ge c > 0$, we conclude that $y_n \to f$ as $n \to \infty$. Then Lemma 1.4.5 implies that $\bar{x} \in N$.

Show that $\bar{x}$ is uniquely defined. Suppose that there exist $\bar{x}_1 \neq \bar{x}$ such that $\bar{x}_1 \in N$, $x_m \rightharpoonup \bar{x}_1$, $\{x_m\} \subset \{x_n\}$,

$$\lim_{n \to \infty} \|x_n - \bar{x}_1\| = \mu(\bar{x}_1) < \infty$$

and

$$\lim_{n \to \infty} \|x_n - \bar{x}\| = \mu(\bar{x}) < \infty.$$

Write the obvious equality

$$\|x_n - \bar{x}_1\|^2 = \|x_n - \bar{x}\|^2 + 2(x_n - \bar{x}, \bar{x} - \bar{x}_1) + \|\bar{x}_1 - \bar{x}\|^2. \qquad (6.4.10)$$

Assuming $n = k$ and $n = m$ in (6.4.10) and passing to the limit as $k \to \infty$ and $m \to \infty$ we obtain, respectively,

$$\mu^2(\bar{x}_1) - \mu^2(\bar{x}) = \|\bar{x} - \bar{x}_1\|^2 > 0$$

and

$$\mu^2(\bar{x}_1) - \mu^2(\bar{x}) = -\|\bar{x} - \bar{x}_1\|^2 < 0,$$

because $\bar{x}$ and $\bar{x}_1$ are the weak limit points of $\{x_n\}$. Thus, we arrive at a contradiction. Consequently, $\bar{x}$ is unique. ∎

Theorem 6.4.2 *Let* $A : H \to 2^H$ *be a maximal monotone operator with* $D(A) = H$ *and conditions (6.4.3) hold. Then the weak convergence of the proximal point algorithm (6.4.2) is equivalent to solvability of the equation (6.1.1).*

Proof. By Theorem 6.4.1, strong (consequently, weak) convergence of the algorithm (6.4.2) follows from solvability of the equation (6.1.1). We show the contrary statement. Let the proximal algorithm (6.4.2) converge weakly, that is, $x_n \rightharpoonup \bar{x} \in H$. The weak convergence of $\{x_n\}$ implies its boundedness. Let $\|x_n\| \leq r$ for all $n > 0$. Construct a maximal monotone operator $\bar{A} = A + \partial I_{2r}$, where ∂I_{2r} is defined by (1.8.1). Note that $Ax = \bar{A}x$ for all $x \in B_0(\theta_X, 2r)$. Moreover, $D(\bar{A}) = B(\theta_X, 2r)$ is bounded in H. Therefore, by Corollary 1.7.6, the set $\bar{N} = \{x \mid f \in \bar{A}x\} \neq \emptyset$. Since a solution x_n of the equation (6.4.2) with maximal monotone operator belongs to $B_0(\theta_X, 2r)$ and it is uniquely defined, we conclude that a sequence $\{x_n\}$ coincides with the sequence constructed for the equation $\bar{A}x = f$ by the same proximal point algorithm (6.4.2). As in the proof of Theorem 6.4.1, it is established that $x_n \rightharpoonup \bar{x} \in \bar{N}$, where $\|\bar{x}\| < 2r$. In other words, $\bar{x} \in B_0(\theta_X, 2r)$. Hence $f \in A\bar{x}$. ∎

Theorem 6.4.3 *Let* $A : H \to 2^H$ *be a maximal monotone operator, a solution set* N *of (6.1.1) be nonempty and the conditions (6.4.3) hold. Assume that a sequence* $\{A^n\}$ *of maximal monotone operators* $A^n : H \to 2^H$ *is given in place of* A, $D(A) = D(A^n)$,

$$\mathcal{H}_H(Ax, A^n x) \leq g(\|x\|)h_n \quad \forall x \in D(A), \qquad (6.4.11)$$

where $g(t)$ *is a bounded positive function for* $t \geq 0$ *and* $\sum_{n=1}^{\infty} h_n < \infty$. *Then a solution sequence* $\{x_n\}$ *of the equations*

$$c_n(A^n x - f_n) + x = x_{n-1}, \quad n = 1, 2, \dots, \qquad (6.4.12)$$

converges weakly to some element of N.

Proof. Let $y_n \in A^n x_n$ be such that (6.4.4) is satisfied. If $x^* \in N$ then, by virtue of (6.4.11), we can choose $y_n^* \in A^n x^*$ which gives the estimate

$$\|y_n^* - f\| \leq g(\|x^*\|) h_n \tag{6.4.13}$$

because $f \in A x^*$. Rewrite (6.4.5) in the equivalent form

$$c_n(y_n - y_n^*, x_n - x^*) \quad + \quad c_n(y_n^* - f, x_n - x^*) + c_n(f - f_n, x_n - x^*)$$
$$+ \quad \|x_n - x^*\|^2 = (x_{n-1} - x^*, x_n - x^*). \tag{6.4.14}$$

Since A^n is monotone, the first term in the left-hand side of (6.4.14) is non-negative. Therefore, using (6.4.1), (6.4.3) and (6.4.13) we deduce from (6.4.14) that there exists $C_1 > 0$ such that

$$\|x_n - x^*\| \leq \|x_0 - x^*\| + C_1 \sum_{i=1}^{n} (\delta_i + h_i), \quad x^* \in N.$$

The rest of the proof follows the pattern of Theorem 6.4.1. ∎

Similarly to Theorem 6.4.2, it is possible to establish that the weak convergence of the proximal point algorithm (6.4.12) is equivalent to solvability of the equation (6.1.1). To this end, is is enough to construct the maximal monotone operator $\bar{A}^n + \partial I_{2r}$ and apply Theorem 6.4.3.

2. Consider further the case of Banach spaces. Let in (6.1.1) $A : X \to 2^{X^*}$ be a maximal monotone operator, $f \in X^*$, X be a reflexive strictly convex Banach space with strictly convex X^*. Introduce the condition

$$\mathcal{H}_{X^*}(Ax, A^n x) \leq g(\|x\|) h_n \quad \forall x \in D(A), \tag{6.4.15}$$

where $g(t)$ is a non-negative and increasing function for all $t \geq 0$. We study the proximal point algorithm defined by the following iterative scheme:

$$c_n(A^n x - f_n) + Jx = Jx_{n-1}, \quad c_n > 0 \quad n = 1, 2, \dots, \tag{6.4.16}$$

where $J : X \to X^*$ is the normalized duality mapping in X. Denote by x_n its (unique) solution when n is fixed (see Theorem 1.7.4).

Theorem 6.4.4 *Assume that $A : X \to 2^{X^*}$ and $A^n : X \to 2^{X^*}$ are maximal monotone operators, $D(A) = D(A^n)$, elements f, $f_n \in X^*$ for all $n > 0$,*

$$\|f - f_n\|_* \leq \delta_n \tag{6.4.17}$$

and (6.4.15) holds. If $c_n \to \infty$, $\delta_n \to 0$, $h_n \to 0$ as $n \to \infty$, and if a sequence $\{x_n\}$ defined recurrently by (6.4.16) is bounded, then equation (6.1.1) is solvable and every weak accumulation point of $\{x_n\}$ is a solution of (6.1.1).

Proof. Suppose that $y_n \in A^n x_n$ satisfies the equality

$$c_n(y_n - f_n) + Jx_n = Jx_{n-1}. \tag{6.4.18}$$

By (6.4.15), take an element $\bar{y}_n \in Ax_n$ such that

$$\|\bar{y}_n - y_n\|_* \leq g(\|x_n\|)h_n.$$

It is obvious that there exists a constant $C_2 > 0$ such that

$$\|\bar{y}_n - y_n\|_* \leq C_2 h_n. \tag{6.4.19}$$

Now (6.4.17) - (6.4.19) yield the relations

$$\|\bar{y}_n - f\|_* \leq \|\bar{y}_n - y_n\|_* + \|y_n - f_n\|_* + \|f_n - f\|_*$$

$$\leq (\|x_n\| + \|x_{n-1}\|)c_n^{-1} + \delta_n + \bar{C}h_n.$$

Due to the properties of $\{x_n\}$, $\{c_n\}$, $\{\delta_n\}$ and $\{h_n\}$, we conclude that $\bar{y}_n \to f$. Let $\{x_k\} \subset \{x_n\}$ and $x_k \rightharpoonup \bar{x} \in X$. Since A is demiclosed, we come to the required inclusion: $f \in A\bar{x}$. ∎

With some different assumptions, prove the following

Theorem 6.4.5 *Let $A : X \to 2^{X^*}$, $A^n : X \to 2^{X^*}$ be maximal monotone operators, $D(A) = D(A^n)$, (6.4.15) and (6.4.17) hold, and*

$$\sum_{n=1}^{\infty} c_n(\delta_n + h_n) < \infty. \tag{6.4.20}$$

Suppose that $c_n \to \infty$ as $n \to \infty$, a duality mapping J in X is weak-to-weak continuous, and a sequence $\{x_n\}$ defined by (6.4.16) is bounded. Then it weakly converges to some solution of (6.1.1).

Proof. It follows from the conditions of the theorem that $\delta_n \to 0$ and $h_n \to 0$ as $n \to \infty$. By Theorem 6.4.4, we have that a set $N = \{x \mid f \in Ax\} \neq \emptyset$. Let $x^* \in N$. Consider the functional (1.6.36) with $y = x^*$:

$$W(x, x^*) = 2^{-1}(\|x\|^2 - 2\langle Jx, x^* \rangle + \|x^*\|^2) \quad \forall x \in X.$$

In view of (1.6.41), there holds the inequality

$$W(x, x^*) - W(v, x^*) \leq \langle Jx - Jv, x - x^* \rangle \quad \forall x, v \in X. \tag{6.4.21}$$

If we put $x = x_n$ and $v = x_{n-1}$ then

$$W(x_n, x^*) - W(x_{n-1}, x^*) \leq \langle Jx_n - Jx_{n-1}, x_n - x^* \rangle.$$

By (6.4.15), we find $y_n^* \in A^n x^*$ such that for $f \in Ax^*$,

$$\|y_n^* - f\|_* \leq h_n g(\|x^*\|).$$

Taking into account (6.4.18), one gets

$$W(x_n, x^*) - W(x_{n-1}, x^*) \leq c_n \langle y_n - f_n, x^* - x_n \rangle$$

$$= c_n \langle y_n - y_n^*, x^* - x_n \rangle + c_n \langle y_n^* - f, x^* - x_n \rangle + c_n \langle f - f_n, x^* - x_n \rangle$$

$$\leq c_n (\delta_n + g(\|x^*\|) h_n) \|x_n - x^*\|,$$

because

$$c_n \langle y_n - y_n^*, x^* - x_n \rangle \leq 0.$$

Recall that $\{x_n\}$ is bounded. Therefore, there exists $M > 0$ such that

$$W(x_n, x^*) \leq W(x_{n-1}, x^*) + M c_n (\delta_n + h_n). \tag{6.4.22}$$

Then we have

$$W(x_n, x^*) \leq W(x_0, x^*) + M \sum_{k=1}^{n} c_k (\delta_k + h_k). \tag{6.4.23}$$

By (6.4.20), this means that the sequence $\{W(x_n, x^*)\}$ is also bounded. Moreover, (6.4.20) and (6.4.22) yield the limit relation

$$\lim_{n \to \infty} W(x_n, x^*) = \mu(x^*), \tag{6.4.24}$$

where $0 \leq \mu(x^*) < \infty$.

Theorem 6.4.4 asserts that the sequence $\{x_n\}$ has a weak accumulation point, that is, there exists a subsequence $\{x_k\} \subset \{x_n\}$ such that $x_k \rightharpoonup \bar{x} \in N$. We will show that there exists only one weak accumulation point. Suppose there is another point $\bar{x}_1$, $\bar{x}_1 \neq \bar{x}$, such that $x_m \rightharpoonup \bar{x}_1 \in N$, where $\{x_m\} \subset \{x_n\}$. By (6.4.24),

$$\lim_{n \to \infty} W(x_n, \bar{x}) = \mu(\bar{x})$$

and

$$\lim_{n \to \infty} W(x_n, \bar{x}_1) = \mu(\bar{x}_1).$$

It is easy to see that

$$\lim_{n \to \infty} \left(W(x_n, \bar{x}) - W(x_n, \bar{x}_1) \right) = \mu(\bar{x}) - \mu(\bar{x}_1) \tag{6.4.25}$$

$$= 2^{-1} (\|\bar{x}\|^2 - \|\bar{x}_1\|^2) + \lim_{n \to \infty} \langle Jx_n, \bar{x}_1 - \bar{x} \rangle.$$

It is clear that $\langle Jx_n, \bar{x}_1 - \bar{x} \rangle$ has a limit which we denote by l. Taking $n = k$ in the previous equality and using weak-to-weak continuity of J one gets that $l = \langle J\bar{x}, \bar{x}_1 - \bar{x} \rangle$. Repeating the same arguments with $n = m$, we derive that $l = \langle J\bar{x}_1, \bar{x}_1 - \bar{x} \rangle$. Consequently,

$$\langle J\bar{x}_1 - J\bar{x}, \bar{x}_1 - \bar{x} \rangle = 0.$$

Due to the strict monotonicity of J, we conclude that $\bar{x} = \bar{x}_1$. This fact establishes uniqueness of the weak accumulation point $\bar{x} \in N$ Therefore, the whole sequence $\{x_n\}$ weakly converges to $\bar{x}$. The proof is accomplished. ∎

Remark 6.4.6 *If N is the singleton, then the requirement of weak-to-weak continuity of J can be omitted.*

3. We study now the convergence of (6.4.2) for the equation (6.1.1) with a maximal accretive operator A.

Theorem 6.4.7 *Assume that X is a uniformly convex Banach space, $A : X \to 2^X$ is a maximal accretive operator, $A^n : X \to 2^X$ are m-accretive operators for all $n > 0$, duality mapping J is weak-to-weak continuous in X, $D(A) = D(A^n)$, $f \in X$, $f_n \in X$, such that (6.4.1) stays valid and*

$$\mathcal{H}_X(Ax, A^n x) \leq g(\|x\|)h_n \quad \forall x \in D(A), \tag{6.4.26}$$

where $g(t)$ is a non-negative and increasing function for all $t \geq 0$. Let $N = \{x^* \mid f \in Ax^*\} \neq \emptyset$, $c_n \to \infty$ and (6.4.20) hold. Then a sequence $\{x_n\}$ generated by (6.4.12) weakly converges to some $x^* \in N$.

Proof. Under the conditions of the theorem, the equation (6.4.2) is uniquely solvable. By analogy with (6.4.14), we have

$$c_n \langle J(x_n - x^*), y_n - y_n^* \rangle + c_n \langle J(x_n - x^*), y_n^* - f \rangle$$
$$+ c_n \langle J(x_n - x^*), f - f_n \rangle + \|x_n - x^*\|^2$$
$$= \langle J(x_n - x^*), x_{n-1} - x^* \rangle,$$

where $x^* \in N$ and $y_n^* \in A^n x^*$ satisfy (6.4.13). It is not difficult to deduce from this the estimate

$$\|x_n - x^*\| \leq \|x_0 - x^*\| + M_1 \sum_{k=1}^{n} c_k(\delta_k + h_k)$$

with some constant $M_1 > 0$. It implies the boundedness of $\{x_n\}$. Let $\{x_k\} \subset \{x_n\}$ and $x_k \rightharpoonup \bar{x} \in X$. If $y_n \in A^n x_n$ satisfy (6.4.4) then one can show that $y_n \to f$. Let $\xi^n \in Ax_n$ such that $\|y_n - \xi^n\| \leq g(\|x_n\|)h_n$. Then $\xi^n \to f$. It follows from Lemma 1.15.12 that $\bar{x}$ is a solution of equation (6.1.1).

In these circumstances, there exists a limit of $\|x_n - x^*\|$ for any fixed $x^* \in N$. Hence,

$$\lim_{n \to \infty} \|x_n - x^*\| = \mu(x^*),$$

where $0 \leq \mu(x^*) < \infty$. Let $\{x_m\} \subset \{x_n\}$ and $x_m \rightharpoonup \bar{x}_1 \neq \bar{x}$. It is obvious that

$$\|x_k - \bar{x}\|^2 \leq \|x_k - \bar{x}\|\|x_k - \bar{x}_1\| + \langle J(x_k - \bar{x}), \bar{x}_1 - \bar{x} \rangle.$$

Indeed,

$$
\begin{aligned}
\|x_k - \bar{x}\|^2 &= \langle J(x_k - \bar{x}), x_k - \bar{x} \rangle \\
&= \langle J(x_k - \bar{x}), x_n - \bar{x}_1 \rangle + \langle J(x_k - \bar{x}), \bar{x}_1 - \bar{x} \rangle \\
&\leq \|x_k - \bar{x}\|\|x_k - \bar{x}_1\| + \langle J(x_k - \bar{x}), \bar{x}_1 - \bar{x} \rangle.
\end{aligned}
$$

If $k \to \infty$ one gets in a limit

$$\mu^2(\bar{x}) \leq \mu(\bar{x})\mu(\bar{x}_1).$$

Analogously, the relation

$$\|x_m - \bar{x}_1\|^2 \leq \|x_m - \bar{x}_1\|\|x_m - \bar{x}\| + \langle J(x_m - \bar{x}_1), \bar{x} - \bar{x}_1 \rangle$$

implies

$$\mu^2(\bar{x}_1) \leq \mu(\bar{x}_1)\mu(\bar{x})$$

as $m \to \infty$. Consequently, $\mu(\bar{x}) = \mu(\bar{x}_1)$. By virtue of Lemma 1.5.7, we conclude that $\bar{x} = \bar{x}_1$. Hence, uniqueness of the weak limit point $\bar{x}$ is proved. ∎

4. To obtain the strongly convergent approximating sequence, construct regularized proximal point algorithms for monotone and accretive equations. Let H be a Hilbert space, $A : H \to H$ be single-valued and $A^n : H \to 2^H$ be maximal monotone operators, $D(A) = D(A^n)$, $n > 0$. Assume that the conditions (6.4.1) and (6.4.11) are fulfilled and

$$\lim_{n \to \infty} \alpha_n = 0, \quad \lim_{n \to \infty} \frac{\delta_n + h_n}{\alpha_n} = 0. \tag{6.4.27}$$

Then, due to Theorem 2.1.3, solution sequence $\{v_n\}$ of the equation

$$A^n v + \alpha_n v = f_n, \quad \alpha_n > 0, \quad n \geq 1, \tag{6.4.28}$$

is bounded and it converges as $n \to \infty$ to the minimal norm solution $\bar{x}^* \in N$. Given some element $x_0 \in H$, construct a sequence $\{x_n\}$, where x_{n+1} is calculated from the following equation:

$$c_n(A^n x + \alpha_n x - f_n) + x = x_n, \quad c_n > 0. \tag{6.4.29}$$

Rewrite the equation (6.4.29) in the equivalent form

$$\mu_n(A^n x - f_n) + x = \beta_n x_n, \tag{6.4.30}$$

where

$$\mu_n = \frac{c_n}{1 + c_n \alpha_n}$$

and

$$\beta_n = \frac{1}{1 + c_n \alpha_n}.$$

Then (6.4.28) can be represented as

$$\mu_n(A^n v - f_n) + v = \beta_n v. \tag{6.4.31}$$

Let $y^{i,j} \in A^i x_j$ and $w^{i,j} \in A^i v_j$. According to (6.4.30) and (6.4.31), there are elements $y^{n,n+1} \in A^n x_{n+1}$ and $w^{n,n} \in A^n v_n$ such that

$$\mu_n(y^{n,n+1} - f_n) + x_{n+1} = \beta_n x_n \tag{6.4.32}$$

and

$$\mu_n(w^{n,n} - f_n) + v_n = \beta_n v_n. \tag{6.4.33}$$

Subtracting (6.4.33) from (6.4.32) side by side and multiplying the obtained result by $x_{n+1} - v_n$, one gets

$$\mu_n(y^{n,n+1} - w^{n,n}, x_{n+1} - v_n) \quad + \quad (x_{n+1} - v_n, x_{n+1} - v_n)$$
$$= \quad \beta_n(x_n - v_n, x_{n+1} - v_n).$$

Since A^n are monotone, we have

$$\|x_{n+1} - v_n\|^2 \le \beta_n(x_n - v_n, x_{n+1} - v_n).$$

Now the Cauchy–Schwartz inequality yields

$$\|x_{n+1} - v_n\| \le \beta_n \|x_n - v_n\|. \tag{6.4.34}$$

It is clear from (6.4.28) that

$$w^{n+1,n+1} + \alpha_{n+1} v_{n+1} = f_{n+1}.$$

Then

$$(w^{n+1,n+1} - w^{n,n}, v_{n+1} - v_n) \quad + \quad (\alpha_{n+1} v_{n+1} - \alpha_n v_n, v_{n+1} - v_n)$$
$$= \quad (f_{n+1} - f_n, v_{n+1} - v_n) \tag{6.4.35}$$

or

$$(w^{n+1,n+1} - w^{n+1,n}, v_{n+1} \quad - \quad v_n) + (w^{n+1,n} - w^{n,n}, v_{n+1} - v_n)$$

$$+ \quad \alpha_{n+1} \|v_{n+1} - v_n\|^2 + (\alpha_{n+1} - \alpha_n)(v_n, v_{n+1} - v_n)$$

$$= \quad (f_{n+1} - f_n, v_{n+1} - v_n). \tag{6.4.36}$$

We know that there exists $d > 0$ such that $\|v_n\| \le d$ for all $n > 0$. Since A is single-valued, one has

$$\|w^{n+1,n} - w^{n,n}\| \quad \le \quad \|w^{n+1,n} - Av_n\| + \|Av_n - w^{n,n}\|$$

$$\le \quad g(\|v_n\|)(h_{n+1} + h_n) \le M(h_{n+1} + h_n), \tag{6.4.37}$$

where $M = \sup\{g(t) \mid 0 \le t \le d\}$. Further,

$$\|f_{n+1} - f_n\| \le \|f_{n+1} - f\| + \|f_n - f\| \le \delta_{n+1} + \delta_n. \tag{6.4.38}$$

By the monotonicity of A^{n+1}, we derive from (6.4.36) and (6.4.38)

$$\|v_{n+1} - v_n\| \le \frac{d\,|\alpha_{n+1} - \alpha_n|}{\alpha_{n+1}} + \frac{\delta_{n+1} + \delta_n}{\alpha_{n+1}} + M\frac{h_{n+1} + h_n}{\alpha_{n+1}} = \epsilon_n. \tag{6.4.39}$$

Furthermore, (6.4.34) and (6.4.39) allow us to obtain

$$\|x_{n+1} - v_{n+1}\| \leq \|x_{n+1} - v_n\| + \|v_n - v_{n+1}\| \leq \beta_n \|x_n - v_n\| + \epsilon_n.$$

Thus,

$$\|x_{n+1} - v_{n+1}\| \leq (1 - \gamma_n)\|x_n - v_n\| + \epsilon_n, \tag{6.4.40}$$

where

$$\gamma_n = \frac{c_n \alpha_n}{1 + c_n \alpha_n}, \quad c_n > 0, \quad \alpha_n > 0. \tag{6.4.41}$$

Applying Lemma 7.1.2 to (6.4.40) and taking into account the inequality

$$\|x_n - \bar{x}^*\| \leq \|x_n - v_n\| + \|v_n - \bar{x}^*\|,$$

we thus come to the following result:

Theorem 6.4.8 *Let the equation (6.1.1) be solvable in a Hilbert space H, $A : H \to H$ and $A^n : H \to 2^H$ be maximal monotone operators for all $n > 0$, $D(A) = D(A^n)$, $f \in H$, $f_n \in H$. Let the conditions (6.4.1), (6.4.11) and (6.4.27) be satisfied, where $\{\delta_n\}$ and $\{h_n\}$ are non-increasing sequences of positive numbers. Assume also that $c_n > 0$,*

$$\sum_{n=0}^{\infty} \gamma_n = \infty, \tag{6.4.42}$$

and

$$\lim_{n \to \infty} \frac{|\alpha_{n+1} - \alpha_n| + \delta_n + h_n}{\alpha_{n+1} \gamma_n} = 0, \tag{6.4.43}$$

where γ_n is defined by (6.4.41). Then a sequence $\{x_n\}$ generated by the iterative process (6.4.29) converges strongly to the minimal norm solution $\bar{x}^*$ of the equation (6.1.1).

5. We proceed to the regularized proximal algorithm in a Banach space X. Assume that $A : X \to X^*$ and $A^n : X \to 2^{X^*}$ are maximal monotone, $f \in X^*$, $f_n \in X^*$ such that (6.4.15) and (6.4.17) are still valid. Denote by v_n, $n = 1, 2, ...$, solutions of the equation

$$A^n v + \alpha_n Jv = f_n, \quad \alpha_n > 0, \tag{6.4.44}$$

where $J : X \to X^*$ is the normalized duality mapping. Consider the regularized proximal point algorithm defining an approximate sequence $\{x_n\}$ in the following form:

$$c_{n-1}(A^{n-1}x + \alpha_{n-1}Jx - f_{n-1}) + Jx = Jx_{n-1}, \quad c_{n-1} > 0, \quad n = 1, 2, ... \tag{6.4.45}$$

It is clear that there exist $y^{n,n+1} \in A^n x_{n+1}$ and $w^{n,n} \in A^n v_n$ such that

$$c_n(y^{n,n+1} + \alpha_n Jx_{n+1} - f_n) + Jx_{n+1} = Jx_n \tag{6.4.46}$$

and

$$w^{n,n} + \alpha_n Jv_n = f_n. \tag{6.4.47}$$

In the sequel, we assume that X is a uniformly convex Banach space and $\delta_X(\epsilon)$ is its modulus of convexity. Recall that function $\delta_X(\epsilon)$ is continuous and increasing on the interval $[0, 2]$ and $\delta_X(0) = 0$. At the same time, the function $g_X(\epsilon)$ defined by (1.1.13) is continuous and non-decreasing on $[0, 2]$ and $g_X(0) = 0$. However, we assume for simplicity that $g_X(\epsilon)$ is the increasing function (see also Remarks 1.6.9 and 6.5.3).

As before, suppose that solution set N of (6.1.1) is not empty, and let $\bar{x}^*$ be its minimal norm solution. According to Theorem 2.2.1, under the conditions (6.4.15), (6.4.17) and (6.4.27), the sequence $\{v_n\}$ is bounded, say, $\|v_n\| \leq d$ for all $n > 0$ and $v_n \to \bar{x}^*$ as $n \to \infty$. Along with (6.4.47), the equality

$$w^{n+1,n+1} + \alpha_{n+1} J v_{n+1} = f_{n+1}, \tag{6.4.48}$$

holds, where $w^{n+1,n+1} \in A^{n+1} v_{n+1}$. Subtracting the equality (6.4.47) from (6.4.48) and calculating the values of the obtained functionals on the element $v_{n+1} - v_n$, we have

$$\begin{aligned} \langle w^{n+1,n+1} - w^{n,n}, v_{n+1} - v_n \rangle &+ \alpha_{n+1} \langle J v_{n+1} - J v_n, v_{n+1} - v_n \rangle \\ &+ (\alpha_{n+1} - \alpha_n) \langle J v_n, v_{n+1} - v_n \rangle \\ &= \langle f_{n+1} - f_n, v_{n+1} - v_n \rangle. \end{aligned} \tag{6.4.49}$$

By the property (1.6.19), evaluate the second term in (6.4.49) as

$$\alpha_{n+1} \langle J v_{n+1} - J v_n, v_{n+1} - v_n \rangle \geq \alpha_{n+1} (2L)^{-1} \delta_X(c_0^{-1} \|v_{n+1} - v_n\|), \tag{6.4.50}$$

where $1 < L < 1.7$ and $c_0 = 2max\{1, d\}$. The monotonicity of A^{n+1} yields

$$\begin{aligned} \langle w^{n+1,n+1} - w^{n,n}, v_{n+1} - v_n \rangle &= \langle w^{n+1,n+1} - w^{n+1,n}, v_{n+1} - v_n \rangle \\ &+ \langle w^{n+1,n} - w^{n,n}, v_{n+1} - v_n \rangle \\ &\geq \langle w^{n+1,n} - w^{n,n}, v_{n+1} - v_n \rangle. \end{aligned} \tag{6.4.51}$$

Return again to (6.4.37) and (6.4.38) assuming that $\{\delta_n\}$ and $\{h_n\}$ are non-increasing sequences of positive numbers. We have

$$\|w^{n+1,n} - w^{n,n}\|_* \leq 2M h_n$$

and

$$\|f_{n+1} - f_n\|_* \leq 2\delta_n.$$

Then the estimates (6.4.50) and (6.4.51) allow us to deduce from (6.4.49) the inequality

$$\alpha_{n+1} (2L)^{-1} \delta_X(c_0^{-1} \|v_{n+1} - v_n\|)$$

$$\leq \Big(2(\delta_n + M h_n) + |\alpha_{n+1} - \alpha_n| \Big) \|v_{n+1} - v_n\|.$$

Now some simple algebra leads to the estimate

$$\|v_{n+1} - v_n\| \leq c_0 \epsilon_n, \tag{6.4.52}$$

where

$$\epsilon_n = g_X^{-1} \Big(L_1 \frac{\delta_n + h_n + |\alpha_{n+1} - \alpha_n|}{\alpha_{n+1}} \Big), \quad L_1 = 2L c_0 max\{2, 2M, d\}. \tag{6.4.53}$$

Theorem 6.4.9 *Let X be a uniformly convex Banach space, $A : X \to X^*$ and $A^n : X \to 2^{X^*}$ be maximal monotone operators for all $n > 0$, $D(A) = D(A^n)$, $f \in X^*$, $f_n \in X^*$. Assume that a solution set N of the equation (6.1.1) is nonempty, the conditions (6.4.15), (6.4.17), (6.4.27) and (6.4.42) are satisfied, where $\{\delta_n\}$ and $\{h_n\}$ are non-increasing sequences of positive numbers. Let γ_n and ϵ_n are defined by (6.4.41) and (6.4.53), respectively, and*

$$\lim_{n \to \infty} \frac{\epsilon_n}{\gamma_n} = 0. \tag{6.4.54}$$

If a sequence $\{x_n\}$ generated by (6.4.45) is bounded, then $x_n \to \bar{x}^*$, where $\bar{x}^* \in N$ is the minimal norm solution of (6.1.1).

Proof. Using the property (1.6.43) of the functional $W(x,z)$ we have

$$W(x_{n+1}, v_{n+1}) \quad \leq \quad W(x_{n+1}, v_n) + \langle Jv_{n+1} - Jx_{n+1}, v_{n+1} - v_n \rangle$$

$$\leq \quad W(x_{n+1}, v_n) + \|Jv_{n+1} - Jx_{n+1}\|_* \|v_{n+1} - v_n\|. \tag{6.4.55}$$

Recall that $\|v_n\| \leq d$ and assume that $\|x_n\| \leq d_1$ for all $n > 0$. Then (6.4.52) implies

$$W(x_{n+1}, v_{n+1}) \leq W(x_{n+1}, v_n) + c_0(d_1 + d)\epsilon_n. \tag{6.4.56}$$

Further, the property (1.6.41) of the functional $W(x,z)$ gives

$$W(x_{n+1}, v_n) \leq W(x_n, v_n) + \langle Jx_{n+1} - Jx_n, x_{n+1} - v_n \rangle. \tag{6.4.57}$$

Evaluate the last term in the right-hand side of (6.4.57). By virtue of (6.4.46) and (6.4.47), one gets the following result:

$$Jx_{n+1} - Jx_n = -c_n(y^{n,n+1} + \alpha_n Jx_{n+1} - f_n)$$

$$= -c_n(y^{n,n+1} - w^{n,n} + \alpha_n Jx_{n+1} - \alpha_n Jv_n).$$

Applying now the monotonicity of A^n we have

$$\langle Jx_{n+1} - Jx_n, x_{n+1} - v_n \rangle \quad = \quad -c_n \langle y^{n,n+1} - w^{n,n}, x_{n+1} - v_n \rangle$$

$$- \quad c_n \alpha_n \langle Jx_{n+1} - Jv_n, x_{n+1} - v_n \rangle$$

$$\leq \quad -c_n \alpha_n \langle Jx_{n+1} - Jv_n, x_{n+1} - v_n \rangle.$$

Taking into account (1.6.44), it is easy to see that

$$\langle Jx_{n+1} - Jx_n, x_{n+1} - v_n \rangle \leq -c_n \alpha_n W(x_{n+1}, v_n).$$

Then (6.4.57) can be rewritten as

$$W(x_{n+1}, v_n) \leq W(x_n, v_n) - c_n \alpha_n W(x_{n+1}, v_n).$$

Thus,
$$W(x_{n+1}, v_n) \leq (1 - \gamma_n)W(x_n, v_n).$$
Substitute this estimate for (6.4.56). Then
$$W(x_{n+1}, v_{n+1}) \leq W(x_n, v_n) - \gamma_n W(x_n, v_n) + c_0(d + d_1)\epsilon_n.$$
By the conditions (6.4.42) and (6.4.54) and by Lemma 7.1.2, we obtain that
$$\lim_{n \to \infty} W(x_n, v_n) = 0.$$

Finally, the left inequality of (1.6.48) and the properties of $\delta_X(\epsilon)$ allow us to conclude that $\|x_n - v_n\| \to 0$ as $n \to \infty$. Then the assertion to be proved follows from the inequality
$$\|x_n - \bar{x}^*\| \leq \|x_n - v_n\| + \|v_n - \bar{x}^*\|. \quad \blacksquare$$

In conclusion, we present a result like Theorem 6.4.8 for the equation (6.1.1) with accretive operator A.

Theorem 6.4.10 *Let in a Banach space* X, *the equation (6.1.1) be solvable, that is,* $N = \{x^* \mid f \in Ax^*\} \neq \emptyset$, $A : X \to X$ *be a maximal accretive operator,* $f \in X$, *duality mapping* $J : X \to X^*$ *be continuous and weak-to-weak continuous. Suppose that* $A^n : X \to 2^X$ *are m-accretive operators,* $D(A^n) = D(A)$, $f_n \in X$, *and the conditions (6.4.1), (6.4.26), (6.4.27), (6.4.42) and (6.4.43) hold and* γ_n *is defined by (6.4.41). Then a sequence* $\{x_n\}$ *generated by (6.4.29) converges strongly in* X *to the unique solution* $\bar{x}^* \in N$ *satisfying the inequality (2.7.7).*

6.5   Iterative Regularization Method

In order to obtain approximations to the minimal norm solution $\bar{x}^*$ of the equation (6.1.1) by the operator regularization method, we needed to solve a sequence of regularized equations (2.2.4) with corresponding parameters $\alpha_n \to 0$. The iterative regularization method which we study in this section does not solve regularized equations, while it gives a new approximation to $\bar{x}^*$ on every iteration step of the algorithm. It is sufficient for this to calculate values of the given operator on the current iteration.

1. Assume that X is a real uniformly convex and uniformly smooth Banach space. We solve the equation (6.1.1) with maximal monotone bounded operator $A : X \to 2^{X^*}$ having χ-growth order, that is,
$$\|\xi\|_* \leq \chi(\|x\|) \quad \forall \xi \in Ax, \quad \forall x \in X, \tag{6.5.1}$$
where $\chi(t)$ is a continuous non-decreasing function for $t \geq 0$. As usual, suppose that the equation (6.1.1) with $f \in X^*$ has a nonempty solution set N, and that, in place of f and A, approximations $f^{\delta_n} \in X^*$ and $A^{h_n} : X \to 2^{X^*}$ are given such that A^{h_n} are maximal monotone operators, $D(A) = D(A^{h_n}) = X$,
$$\|f^{\delta_n} - f\|_* \leq \delta_n \tag{6.5.2}$$

and

$$\mathcal{H}_{X^*}(A^{h_n}x, Ax) \leq \phi(\|x\|)h_n \quad \forall x \in X, \tag{6.5.3}$$

where a function $\phi(t)$ is continuous non-negative and non-decreasing for all $t \geq 0$. Thus, in reality, instead of (6.1.1), the perturbed equation

$$A^{h_n}x = f^{\delta_n} \tag{6.5.4}$$

is solved. In general, it not necessarily has a solution. In the case when A and A^{h_n} are hemicontinuous, the inequality (6.5.3) is replaced by

$$\|A^{h_n}x - Ax\|_* \leq \phi(\|x\|)h_n \quad \forall x \in X. \tag{6.5.5}$$

Study the following iterative regularization algorithm:

$$Jx^{n+1} = Jx^n - \epsilon_m(\xi_n^{h_m} + \alpha_m Jx^n - f^{\delta_m}), \quad n = 0, 1, 2, \dots , \tag{6.5.6}$$

where $\xi_n^{h_m} \in A^{h_m}x^n$, $m = n + n_0$, n_0 is defined below by (6.5.11) and positive parameters α_m and ϵ_m and non-negative parameters δ_m and h_m satisfy the inequalities

$$\alpha_m \leq \bar{\alpha}, \ \epsilon_m \leq \bar{\epsilon}, \ \delta_m \leq \bar{\delta}, \ h_m \leq \bar{h}. \tag{6.5.7}$$

As before, we denote by $\delta_X(t)$ the modulus of convexity of X and by $\rho_X(\tau)$ its modulus of smoothness. Since the functions $\delta_X(t)$ and $\rho_X(\tau)$ are increasing on the interval $[0,2]$ and $[0,\infty)$, respectively, and since $\delta_X(0) = \rho_X(0) = 0$, the inverse functions $\delta_X^{-1}(\cdot)$ and $\rho_X^{-1}(\cdot)$ are also increasing and $\delta_X^{-1}(0) = \rho_X^{-1}(0) = 0$. All these function are continuous. We introduce $g_X(t)$ by the formula (1.1.13) and assume that it is increasing for all $t \in [0, 2]$ (see Remarks 1.6.9 and 6.5.3).

Consider the unperturbed regularized equation

$$Az + \alpha Jz = f, \quad \alpha > 0. \tag{6.5.8}$$

As it was proved in Section 2.2, its solutions z_α exist for all $\alpha > 0$, $\|z_\alpha\| \leq \|\bar{x}^*\| \leq K_0$ and $z_\alpha \to \bar{x}^*$ as $\alpha \to 0$, where $\bar{x}^*$ is the minimal norm element of N. Denote $z_m = z_{\alpha_m}$ and assume that $\alpha_m \to 0$ as $m \to \infty$. It is obvious that

$$\|x^n - \bar{x}^*\| \leq \|z_m - \bar{x}^*\| + \|x^n - z_m\|. \tag{6.5.9}$$

Therefore, to establish strong convergence of the method (6.5.6) we need only to prove that

$$\lim_{n \to \infty} \|x^n - z_m\| = 0.$$

Let R_0 be any non-negative number. Assume

$$K_1 = \sqrt{2R_0} + K_0, \quad K_2 = K_1 + \bar{\epsilon}c_1, \quad c = 2c_0 LK_0,$$

$$c_0 = 2max\{1, K_0\}, \quad c_1 = \chi(K_1) + \bar{h}\phi(K_1) + \bar{\alpha}K_1 + \bar{\delta} + \|f\|_*,$$

$$c_2 = 8max\{L, K_2\}, \quad c_3 = K_0 + K_2, \quad c_4 = 2max\{1, K_1\}.$$

Here $1 < L < 1.7$ is the Figiel constant. Construct the following sequence:

$$\gamma_k = \epsilon_k(K_0 + K_1)\Big(h_k\phi(K_1) + \delta_k\Big) + 8c_1^2\epsilon_k^2 + c_2\rho_{X^*}(c_1\epsilon_k)$$

$$+ \quad c_0c_3g_X^{-1}\Big(\frac{c|\alpha_k - \alpha_{k+1}|}{\alpha_k}\Big), \quad k = 0, 1, 2, \dots . \tag{6.5.10}$$

Choose n_0 according to the condition

$$n_0 = min\Big\{k \mid c_4\delta_X^{-1}\Big(\frac{2L\gamma_k}{\epsilon_k\alpha_k}\Big) \le \triangle^{-1}(R_0)\Big\}, \tag{6.5.11}$$

where $\triangle(\tau) = 8\tau^2 + c_2\rho_X(\tau)$. It is clear that the functions $\triangle(t)$ and its inverse function $\triangle^{-1}(s)$ are positive for all $t, s \in [0, \infty)$, increasing and $\triangle(0) = \triangle^{-1}(0) = 0$. Therefore, n_0 is well defined if

$$\Big\{\frac{\gamma_k}{\epsilon_k\alpha_k}\Big\} \to 0 \quad \text{as} \quad k \to \infty. \tag{6.5.12}$$

Introduce the functional defined by (1.6.36):

$$W(x, z) = 2^{-1}(\|x\|^2 - 2\langle Jx, z\rangle + \|z\|^2) \quad \forall x, z \in X, \tag{6.5.13}$$

where J is the normalized duality mapping. Let z_{n_0} be a solution of the equation (6.5.8) with $\alpha = \alpha_{n_0}$ and let an initial point x^0 in (6.5.6) satisfy the inequality $W(x^0, z_{n_0}) \le R_0$. Such x^0 exists, because $W(z_{n_0}, z_{n_0}) = 0$. For example, we can put $x^0 = z_{n_0}$.

We premise the main result of this section on the following lemma.

Lemma 6.5.1 *Let $\bar{x}^*$ be a minimal norm solution of the equation (2.1.1) and let z_n and z_{n+1} be solutions of the equation (6.5.8) with $\alpha = \alpha_n$ and $\alpha = \alpha_{n+1}$, respectively. Then the following inequality holds:*

$$\|z_n - z_{n+1}\| \le c_0g_X^{-1}\Big(\frac{c|\alpha_n - \alpha_{n+1}|}{\alpha_n}\Big). \tag{6.5.14}$$

Proof. Since z_n and z_{n+1} are solutions of (6.5.8), there exist $\zeta_n \in Az_n$ and $\zeta_{n+1} \in Az_{n+1}$ such that the equalities

$$\zeta_n + \alpha_nJz_n = f \tag{6.5.15}$$

and

$$\zeta_{n+1} + \alpha_{n+1}Jz_{n+1} = f \tag{6.5.16}$$

are satisfied. Evaluate from below the expression

$$D = \langle \zeta_n + \alpha_nJz_n - \zeta_{n+1} - \alpha_nJz_{n+1}, z_n - z_{n+1}\rangle.$$

Applying the estimate (1.6.19) and taking into account the fact that A is monotone, one gets

$$D = \langle \zeta_n - \zeta_{n+1}, z_n - z_{n+1} \rangle + \alpha_n \langle Jz_n - Jz_{n+1}, z_n - z_{n+1} \rangle$$

$$\geq \alpha_n \langle Jz_n - Jz_{n+1}, z_n - z_{n+1} \rangle$$

$$\geq \alpha_n (2L)^{-1} \delta_X (c_0^{-1} \| z_n - z_{n+1} \|). \tag{6.5.17}$$

Next we evaluate D from above. Since z_n and z_{n+1} satisfy (6.5.15) and (6.5.16),

$$D = \langle \zeta_n + \alpha_n Jz_n - f, z_n - z_{n+1} \rangle - \langle \zeta_{n+1} + \alpha_n Jz_{n+1} - f, z_n - z_{n+1} \rangle$$

$$= -\langle \zeta_{n+1} + \alpha_n Jz_{n+1} - f, z_n - z_{n+1} \rangle$$

$$= \langle \zeta_{n+1} + \alpha_{n+1} Jz_{n+1} - f, z_n - z_{n+1} \rangle - \langle \zeta_{n+1} + \alpha_n Jz_{n+1} - f, z_n - z_{n+1} \rangle$$

$$= (\alpha_{n+1} - \alpha_n) \langle Jz_{n+1}, z_n - z_{n+1} \rangle$$

$$\leq |\alpha_{n+1} - \alpha_n| \| z_{n+1} \| \| z_n - z_{n+1} \|$$

$$\leq K_0 |\alpha_{n+1} - \alpha_n| \| z_n - z_{n+1} \|. \tag{6.5.18}$$

By (6.5.17) and (6.5.18), the estimate (6.5.14) follows. ∎

In a Hilbert space $\delta_X(\epsilon) \geq 8^{-1}\epsilon^2$, $g_X(\epsilon) \geq 8^{-1}\epsilon$. Therefore, (6.5.14) accepts the form

$$\| z_n - z_{n+1} \| \leq \frac{8K_0 |\alpha_n - \alpha_{n+1}|}{\alpha_n}.$$

However, if we prove Lemma 6.5.1 directly in a Hilbert space, then one gets

$$\| z_n - z_{n+1} \| \leq \frac{K_0 |\alpha_n - \alpha_{n+1}|}{\alpha_n}. \tag{6.5.19}$$

On the basis of (1.6.43), we write down the inequality

$$W(x^{n+1}, z_{m+1}) - W(x^{n+1}, z_m) \leq \langle Jz_{m+1} - Jx^{n+1}, z_{m+1} - z_m \rangle$$

$$\leq \| Jz_{m+1} - Jx^{n+1} \|_* \| z_{m+1} - z_m \|. \tag{6.5.20}$$

Furthermore, (1.6.41) yields

$$W(x^{n+1}, z_m) - W(x^n, z_m) \leq \langle Jx^{n+1} - Jx^n, x^{n+1} - z_m \rangle$$

$$= \langle Jx^{n+1} - Jx^n, x^n - z_m \rangle + \langle Jx^{n+1} - Jx^n, x^{n+1} - x^n \rangle. \tag{6.5.21}$$

Combining (6.5.20) and (6.5.21) we have

$$W(x^{n+1}, z_{m+1}) - W(x^n, z_m) \leq \langle Jx^{n+1} - Jx^n, x^{n+1} - x^n \rangle$$

$$+ \|Jx^{n+1} - Jz_{m+1}\|_* \|z_{m+1} - z_m\| + \langle Jx^{n+1} - Jx^n, x^n - z_m \rangle. \qquad (6.5.22)$$

Estimate every summand forming the right-hand side of (6.5.22).

1) By (6.5.1) and (6.5.3), there exists $\xi^n \in Ax^n$ such that

$$\|\xi_n^{h_m}\|_* \leq \|\xi_n^{h_m} - \xi^n\|_* + \|\xi^n\|_* \leq h_m \phi(\|x^n\|) + \chi(\|x^n\|),$$

and

$$\|f^{\delta_m}\|_* \leq \|f\|_* + \|f^{\delta_m} - f\|_* \leq \|f\|_* + \delta_m.$$

Returning to the regularization method (6.5.6), we calculate

$$\|Jx^{n+1} - Jx^n\|_* = \epsilon_m \|\xi_n^{h_m} + \alpha_m Jx^n - f^{\delta_m}\|_* \leq \epsilon_m c_1(n),$$

where

$$c_1(n) = \chi(\|x^n\|) + h_m \phi(\|x^n\|) + \alpha_m \|x^n\| + \|f\|_* + \delta_m. \qquad (6.5.23)$$

Now we apply (1.6.17) and get the following result:

$$\langle Jx^{n+1} - Jx^n, x^{n+1} - x^n \rangle \leq 8\|Jx^{n+1} - Jx^n\|_*^2 + c_2(n)\rho_{X^*}(\|Jx^{n+1} - Jx^n\|_*),$$

where

$$c_2(n) = 8max\{L, \|x^{n+1}\|, \|x^n\|\}. \qquad (6.5.24)$$

Then

$$\langle Jx^{n+1} - Jx^n, x^{n+1} - x^n \rangle \leq 8\epsilon_m^2 c_1^2(n) + c_2(n)\rho_{X^*}(\epsilon_m c_1(n)).$$

2) It is obvious that

$$\|Jx^{n+1} - Jz_{m+1}\|_* \leq \|x^{n+1}\| + \|z_{m+1}\| = c_3(n). \qquad (6.5.25)$$

By Lemma 6.5.1,

$$\|z_{m+1} - z_m\| \leq c_0 g_X^{-1}\Big(\frac{c|\alpha_m - \alpha_{m+1}|}{\alpha_m}\Big).$$

Consequently,

$$\|Jx^{n+1} - Jz_{m+1}\|_* \|z_{m+1} - z_m\| \leq c_0 c_3(n) g_X^{-1}\Big(\frac{c|\alpha_m - \alpha_{m+1}|}{\alpha_m}\Big).$$

3) Evaluate the last term in (6.5.22):

$$\langle Jx^{n+1} - Jx^n, x^n - z_m \rangle = -\epsilon_m \langle \xi_n^{h_m} + \alpha_m Jx^n - f^{\delta_m}, x^n - z_m \rangle$$

$$= -\epsilon_m \langle \xi_n^{h_m} - \xi^n, x^n - z_m \rangle - \epsilon_m \langle f - f^{\delta_m}, x^n - z_m \rangle$$

$$- \quad \epsilon_m \langle \xi^n - \zeta_m, x^n - z_m \rangle$$

$$- \quad \epsilon_m \langle \zeta_m + \alpha_m J z_m - f, x^n - z_m \rangle$$

$$- \quad \epsilon_m \alpha_m \langle J x^n - J z_m, x^n - z_m \rangle,$$

where $\xi^n \in A x^n$, $\zeta_m \in A z_m$ and $\zeta_m + \alpha_m J z_m = f$. Recalling that the operator A is monotone, we deduce

$$\langle J x^{n+1} - J x^n, x^n - z_m \rangle \quad \leq \quad \epsilon_m \Big(h_m \phi(\|x^n\|) + \delta_m \Big) \|x^n - z_m\|$$

$$- \quad \epsilon_m \alpha_m (2L)^{-1} \delta_X \Big(\frac{\|x^n - z_m\|}{c_4(n)} \Big),$$

where

$$c_4(n) = 2max\{1, \ \|x^n\|, \|z_m\|\}. \tag{6.5.26}$$

Rewrite (6.5.22) using the estimates obtained in 1) - 3). We have

$$W(x^{n+1}, z_{m+1}) \quad \leq \quad W(x^n, z_m) - \epsilon_m \alpha_m (2L)^{-1} \delta_X \Big(\frac{\|x^n - z_m\|}{c_4(n)} \Big)$$

$$+ \quad \epsilon_m \Big(h_m \phi(\|x^n\|) + \delta_m \Big) \|x^n - z_m\| + 8\epsilon_m^2 c_1^2(n)$$

$$+ \quad c_2(n) \rho_{X^*}(\epsilon_m c_1(n)) + c_0 c_3(n) g_X^{-1} \Big(\frac{c|\alpha_m - \alpha_{m+1}|}{\alpha_m} \Big). \tag{6.5.27}$$

By hypotheses, $W(x^0, z_{n_0}) \leq R_0$. Consider arbitrary $n \geq 0$ such that $W(x^n, z_m) \leq R_0$. Then (1.6.37) implies

$$\|x^n\| \leq \|z_m\| + \sqrt{2R_0} \leq K_0 + \sqrt{2R_0} = K_1.$$

In this case, the following estimates hold:

$$c_1(n) \leq \chi(K_1) + \bar{h}\phi(K_1) + \bar{\alpha}K_1 + \|f\|_* + \bar{\delta} = c_1,$$

$$\|x^{n+1}\| \leq \|x^n\| + \|J x^{n+1} - J x^n\|_* \leq K_1 + \epsilon_m c_1(n) \leq K_1 + \bar{\epsilon} c_1 = K_2, \tag{6.5.28}$$

$$c_2(n) \leq 8max\{L, \ K_2\} = c_2,$$

$$c_3(n) \leq K_2 + K_0 = c_3,$$

$$c_4(n) \leq 2max\{1, \ K_1\} = c_4.$$

After that, the inequality (6.5.27) is obtained in the final form:

$$W(x^{n+1}, z_{m+1}) \leq W(x^n, z_m) - \epsilon_m \alpha_m (2L)^{-1} \delta_X (c_4^{-1} \|x^n - z_m\|) + \gamma_m, \tag{6.5.29}$$

where γ_m coincides with (6.5.10) if $k = m$. It results from (6.5.29) that

$$W(x^{n+1}, z_{m+1}) \leq W(x^n, z_m) + \bar{\gamma},$$

with

$$\bar{\gamma} = \bar{\epsilon}(\bar{h}\phi(K_1) + \bar{\delta})(K_0 + K_1) + 8c_1^2\bar{\epsilon}^2 + c_2\rho_{X^*}(c_1\bar{\epsilon}) + c_0c_3g_X^{-1}(cd),$$

where we assume that

$$\frac{|\alpha_n - \alpha_{n+1}|}{\alpha_n} \leq d \tag{6.5.30}$$

with some constant $d > 0$. Thus, if $W(x^n, z_{n+n_0}) \leq R_0$ for all $n \geq 0$ then $\|x_n\| \leq K_1$. Otherwise, if $W(x^n, z_m) \leq R_0$ for $0 \leq n \leq n^* < \infty$, where n^* is an integer, then

$$R_0 < W(x^{n^*+1}, z_{n^*+1+n_0}) \leq R_0 + \bar{\gamma} = R_1. \tag{6.5.31}$$

We show that inequality $W(x^n, z_m) \leq R_1$ holds for all $n \geq n^* + 1$. To this end, consider the following alternative: either

$$(H_1): \quad (2L)^{-1}\delta_X\left(\frac{\|x^{n^*+1} - z_{n^*+1+n_0}\|}{c_4}\right) > \frac{\gamma_{n^*+1+n_0}}{\alpha_{n^*+1+n_0}\epsilon_{n^*+1+n_0}}$$

or

$$(H_2): \quad (2L)^{-1}\delta_X\left(\frac{\|x^{n^*+1} - z_{n^*+1+n_0}\|}{c_4}\right) \leq \frac{\gamma_{n^*+1+n_0}}{\alpha_{n^*+1+n_0}\epsilon_{n^*+1+n_0}}.$$

If (H_1) is true then we deduce from (6.5.29) and (6.5.31) that

$$W(x^{n^*+2}, z_{n^*+2+n_0}) < W(x^{n^*+1}, z_{n^*+1+n_0}) \leq R_1. \tag{6.5.32}$$

At the same time, the hypothesis (H_2) can not be held. Indeed, assuming the contrary, we obtain

$$\|x^{n^*+1} - z_{n^*+1+n_0}\| \leq c_4\delta_X^{-1}\left(\frac{2L\gamma_{n^*+1+n_0}}{\alpha_{n^*+1+n_0}\epsilon_{n^*+1+n_0}}\right).$$

If the sequence in (6.5.12) decreases, we come to the inequality

$$\|x^{n^*+1} - z_{n^*+1+n_0}\| \leq \triangle^{-1}(R_0) \tag{6.5.33}$$

by reason of (6.5.11).

We estimate $W(x^n, z_m)$ through $\|x^n - z_m\|$. Due to the inequality (1.6.48), one gets

$$W(x^n, z_m) \leq 8\|x^n - z_m\|^2 + c_6(n)\rho_X(\|x^n - z_m\|),$$

where

$$c_6(n) = 8max\{L, \|x^n\|, \|z_m\|\}.$$

It is clear from (6.5.28) that

$$c_6(n) \leq 8max\{L, K_2\} = c_2.$$

and
$$W(x^n, z_m) \le \triangle(\|x^n - z_m\|).$$
Then
$$W(x^{n^*+1}, z_{n^*+1+n_0}) \le \triangle(\|x^{n^*+1} - z_{n^*+1+n_0}\|).$$
Taking into account (6.5.33) we deduce
$$W(x^{n^*+1}, z_{n^*+1+n_0}) \le R_0, \tag{6.5.34}$$
which contradicts the assumption (6.5.31). Consequently, by induction, the estimate $W(x^n, z_m) \le R_1$ is satisfied for all $n \ge 0$ and then
$$\|x^n\| \le \|z_m\| + \sqrt{2R_1} \le K_0 + \sqrt{2R_1} = K_3, \tag{6.5.35}$$
that is, the sequence $\{x^n\}$ is bounded.

We introduce the new denotation:
$$\Psi(W(x^n, z_m)) = (2L)^{-1}\delta_X\left(c_4^{-1}\triangle^{-1}(W(x^n, z_m))\right).$$

It is easy to verify that the function $\Psi(t)$ is positive for all $t > 0$, continuous, increasing and $\Psi(0) = 0$. If we repeat now deduction of (6.5.29) replacing K_1 by K_3 everywhere above, then we shall obtain the following recursive inequality:
$$W(x^{n+1}, z_{m+1}) \le W(x^n, z_m) - \epsilon_m \alpha_m \Psi(W(Jx^n, z_m)) + \gamma_m,$$
where γ_m is defined by (6.5.10). It can be written as
$$\lambda_{n+1} \le \lambda_n - \rho_n \Psi(\lambda_n) + \kappa_n, \tag{6.5.36}$$
where
$$\lambda_n = W(x^n, z_{n+n_0}), \quad \rho_n = \epsilon_{n+n_0}\alpha_{n+n_0}, \quad \kappa_n = \gamma_{n+n_0}.$$
By the properties of the functional $W(x, z)$, $\lambda_n \ge 0$ for all $n \ge 0$. Consequently, we can apply Lemma 7.1.3 which gives the sufficient conditions to assert that $\lim_{n\to\infty} W(x^n, z_m) = 0$. The left inequality of (1.6.48) implies
$$\|x^n - z_m\| \le 2c_4\delta_X^{-1}(LW(x^n, z_m)).$$
Then
$$\lim_{n\to\infty} \|x^n - z_m\| = 0.$$
By (6.5.9), this enables us to state the following result:

Theorem 6.5.2 *Suppose that X is a uniformly convex Banach space and*
1) A solution set N of the equation (6.1.1) is not empty and $\bar{x}^*$ is its minimal norm solution;
2) $A : X \to 2^{X^*}$ is a maximal monotone bounded operator with χ-growth (6.5.1);
3) Instead of the equation $Ax = f$, in fact, the perturbed equations (6.5.4) with maximal monotone operators A^{h_n} are solved, the estimates (6.5.2) and (6.5.3) hold and $D(A) =$

$D(A^{h_n}) = X$ *for all* $n \geq 0$;

4) *An initial approximation* x^0 *in the iterative regularization method (6.5.6) satisfies the inequality* $W(x^0, z_{n_0}) \leq R_0$, *where* $W(x, z)$ *is defined by (6.5.13) and* R_0 *is an arbitrary non-negative number;*

5) z_{n_0} *is a solution of the operator equation (6.5.8) with* $\alpha = \alpha_{n_0}$, *where* n_0 *obeys the rule (6.5.11);*

6) *Positive parameters* α_n *and* ϵ_n *and non-negative parameters* δ_n *and* h_n *approach zero as* $n \to \infty$, *such that (6.5.7) and (6.5.30) are valid. Besides, let*

7) $$\sum_{n=1}^{\infty} \alpha_n \epsilon_n = \infty;$$

8) $$\lim_{n \to \infty} \frac{\delta_n + h_n + \epsilon_n}{\alpha_n} = 0;$$

9) $$\lim_{n \to \infty} \frac{\rho_{X^*}(\epsilon_n)}{\epsilon_n \alpha_n} = 0;$$

10) $$\lim_{n \to \infty} \frac{g_X^{-1}\left(\alpha_n^{-1}|\alpha_n - \alpha_{n+1}|\right)}{\alpha_n \epsilon_n} = 0.$$

Then the sequence $\{x^n\}$ *generated by (6.5.6) converges strongly to* $\bar{x}^*$ *as* $n \to \infty$.

Remark 6.5.3 *If* $g_X(t)$ *does not increase strictly for all* $t \in [0,2]$ *but there exists a non-negative increasing continuous function* $\tilde{g}_X(t)$ *such that* $g_X(t) \geq \tilde{g}_X(t)$, *then Theorem 6.5.2 remains still valid if in its conditions, proof and conclusions* $g(\cdot)$ *and* $g^{-1}(\cdot)$ *are replaced by* $\tilde{g}(\cdot)$ *and* $\tilde{g}^{-1}(\cdot)$, *respectively. As it was already mentioned in Section 1.6, in most cases of uniformly convex Banach spaces the modulus of convexity* $\delta_X(t) \geq d_1 t^{\gamma}$, $\gamma \geq 2$, $d_1 > 0$. *Consequently,* $\tilde{g}_X(t) = d_1 t^{\gamma-1}$ *and* $\tilde{g}_X^{-1}(\xi) = (d_1^{-1}\xi)^{\kappa}$, *where* $\kappa = (\gamma - 1)^{-1}$.

Recall that if $\delta_X(t) \geq d_1 t^{\gamma}, \gamma \geq 2, d_1 > 0$, then the modulus of smoothness $\rho_{X^*}(\tau) \leq d_2 \tau^{\frac{\gamma}{\gamma-1}}$, $d_2 > 0$. In this situation, the requirements 8) - 10) are simplified, namely:

$$\lim_{n \to \infty} \frac{\delta_n + h_n + \epsilon_n^{\kappa}}{\alpha_n} = 0$$

and

$$\lim_{n \to \infty} \frac{|\alpha_n - \alpha_{n+1}|^{\kappa}}{\alpha_n^{\gamma \kappa} \epsilon_n} = 0.$$

Let us give some examples.

1) $X = L^p$, $p \leq 2$, $X^* = L^q$, $p^{-1} + q^{-1} = 1$, $q \geq 2$. Here $\delta_X(t) \geq d_1 t^2$, that is, $\gamma = 2$, $\tilde{g}_X^{-1}(\xi) = d_1^{-1}\xi$, $\rho_{X^*}(\tau) \leq d_2 \tau^2$, d_1, d_2 are positive constants. Hence, the convergence of (6.5.6) to $\bar{x}^*$ is guaranteed if

$$\sum_{n=1}^{\infty} \alpha_n \epsilon_n = \infty, \quad \lim_{n \to \infty} \alpha_n = 0,$$

$$\lim_{n \to \infty} \frac{\delta_n + h_n + \epsilon_n}{\alpha_n} = 0$$

and

$$\lim_{n \to \infty} \frac{|\alpha_n - \alpha_{n+1}|}{\alpha_n^2 \epsilon_n} = 0,$$

that includes the case of a Hilbert space.

2) $X = L^p$, $p > 2$, $X^* = L^q$, $p^{-1} + q^{-1} = 1$, $q \le 2$. In this case $\delta_X(t) \ge d_3 t^p$, $\gamma = p$, $\tilde{g}_X^{-1}(\xi) = d_3^{1/(1-p)} \xi^{1/(p-1)}$, $\rho_{X^*}(\tau) \le d_4 \tau^q$, d_3, d_4 are positive constants. Then the convergence conditions are the following:

$$\sum_{n=1}^{\infty} \alpha_n \epsilon_n = \infty, \qquad \lim_{n \to \infty} \alpha_n = 0,$$

$$\lim_{n \to \infty} \frac{\delta_n + h_n + \epsilon_n^{1/(p-1)}}{\alpha_n} = 0$$

and

$$\lim_{n \to \infty} \frac{|\alpha_n - \alpha_{n+1}|^{1/(p-1)}}{\alpha_n^{p/(p-1)} \epsilon_n} = 0.$$

2. Results like Theorem 6.5.2 can also be obtained for the equation (2.1.1) with an accretive operator $A : X \to X$. For simplicity of the proof, we assume that the sequence $\{x^n\}$ generated by the iterative algorithm

$$x^{n+1} = x^n - \epsilon_n (A^{h_n} x^n + \alpha_n x^n - f^{\delta_n}), \quad n = 0, 1, 2, \dots, \tag{6.5.37}$$

is bounded. Suppose that equation (2.1.1) is solvable, N is its solution set and the (unique) solution $\bar{x}^* \in N$ satisfies the inequality (2.7.7). Let A be a demicontinuous operator, $D(A) = X$, perturbed operators $A^{h_n} : X \to X$ and right-hand sides $f^{\delta_n} \in X$ satisfy the previous conditions (6.5.2) and (6.5.5) (with the norm of X), $D(A^{h_n}) = X$ and (6.5.7) holds for all $m = n \ge 0$.

Introduce the intermediate equation

$$Az + \alpha z = f, \quad \alpha > 0. \tag{6.5.38}$$

It was proved in Section 2.7 that if X possesses an approximation and duality mapping $J : X \to X^*$ is continuous and weak-to-weak continuous, then solutions z_α of (6.5.38) converge to $\bar{x}^*$ as $\alpha \to 0$ and $\|z_\alpha\| \le 2\|\bar{x}^*\|$. By analogy with Lemma 6.5.1, it is possible to show that for solutions $z_n = z_{\alpha_n}$ and $z_{n+1} = z_{\alpha_{n+1}}$ the following estimate holds:

$$\|z_{n+1} - z_n\| \le 2\|\bar{x}^*\| \frac{|\alpha_{n+1} - \alpha_n|}{\alpha_n}. \tag{6.5.39}$$

Convergence analysis of the method (6.5.37) is done by a scheme like the monotone case, but now, in place of (6.5.13), the functional $V(x, z) = 2^{-1}\|x - z\|^2$ is studied.

Indeed, assuming that $\|x^n\| \leq R < \infty$ one gets

$$\|x^{n+1} - z_{n+1}\|^2 \leq \|x^{n+1} - z_n\|^2 + 2\left\langle J(z_{n+1} - x^{n+1}), z_{n+1} - z_n\right\rangle$$

$$\leq \|x^{n+1} - z_n\|^2 + 2\|z_{n+1} - x^{n+1}\|\|z_{n+1} - z_n\|. \qquad (6.5.40)$$

Similarly to (6.5.21), we derive

$$\|x^{n+1} - z_n\|^2 \leq \|x^n - z_n\|^2 + 2\left\langle J(x^{n+1} - z_n), x^{n+1} - x^n\right\rangle$$

$$= \|x^n - z_n\|^2 + 2\left\langle J(x^n - z_n), x^{n+1} - x^n\right\rangle$$

$$+ 2\left\langle J(x^{n+1} - z_n) - J(x^n - z_n), x^{n+1} - x^n\right\rangle. \qquad (6.5.41)$$

Consequently, (6.5.40) and (6.5.41) imply

$$\|x^{n+1} - z_{n+1}\|^2 \leq \|x^n - z_n\|^2 + 2\|z_{n+1} - x^{n+1}\|\|z_{n+1} - z_n\|$$

$$+ 2\left\langle J(x^n - z_n), x^{n+1} - x^n\right\rangle$$

$$+ 2\left\langle J(x^{n+1} - z_n) - J(x^n - z_n), x^{n+1} - x^n\right\rangle. \qquad (6.5.42)$$

By (6.5.37), we calculate

$$\left\langle J(x^n - z_n), x^{n+1} - x^n\right\rangle = -\epsilon_n\left\langle J(x^n - z_n), A^{h_n}x^n + \alpha_n x^n - f^{\delta_n}\right\rangle$$

$$= -\epsilon_n\alpha_n\|x^n - z_n\|^2 - \epsilon_n\left\langle J(x^n - z_n), A^{h_n}x^n + \alpha_n z_n - f^{\delta_n}\right\rangle.$$

In their turn, (6.5.2) and (6.5.5) yield the inequality

$$\left\langle J(x^n - z_n), A^{h_n}x^n + \alpha_n z_n - f^{\delta_n}\right\rangle \geq -\left(\delta_n + h_n\phi(\|x^n\|)\right)\|x^n - z_n\|$$

$$+ \left\langle J(x^n - z_n), Ax^n - Az_n\right\rangle$$

$$+ \left\langle J(x^n - z_n), Az_n + \alpha_n z_n - f\right\rangle.$$

By virtue of the equality $Az_n + \alpha_n z_n = f$ and by the accretiveness of A, we have

$$\left\langle J(x^n - z_n), A^{h_n}x^n + \alpha_n z_n - f^{\delta_n}\right\rangle \geq -\left(\delta_n + h_n\phi(\|x^n\|)\right)\|x^n - z_n\|.$$

Thus,

$$\left\langle J(x^n - z_n), x^{n+1} - x^n\right\rangle \leq -\epsilon_n\alpha_n\|x^n - z_n\|^2 + \epsilon_n\left(\delta_n + h_n\phi(\|x^n\|)\right)\|x^n - z_n\|.$$

Let the sequences $\{x^n\}$ and $\{z_n\}$ be bounded by constants R and K_0, respectively. Then we conclude that

$$\|z_n - x^n\| \le K_0 + R = d_5, \quad n = 0, 1, 2, \dots .$$

There exist constants $d_6 > 0$ and $d_7 > 0$ such that

$$\langle J(x^{n+1} - z_n) - J(x^n - z_n), x^{n+1} - x^n \rangle \le 8\|x^{n+1} - x^n\|^2 + d_6 \rho_X(\|x^{n+1} - x^n\|)$$

and

$$\|x^{n+1} - x^n\| \le (\|Ax^n\| + \bar{h}\phi(R) + \bar{\alpha}R + \|f\| + \bar{\delta})\epsilon_n \le d_7 \epsilon_n,$$

provided that the operator A is bounded. Finally, we deduce from (6.5.42) the following inequality:

$$
\begin{aligned}
\|x^{n+1} - z_{n+1}\|^2 \ \le \ & \|x^n - z_n\|^2 - 2\epsilon_n \alpha_n \|x^n - z_n\|^2 \\[2mm]
&+ \ 2\epsilon_n d_5 \Big(\delta_n + h_n \phi(R)\Big) + 16 d_7^2 \epsilon_n^2 \\[2mm]
&+ \ 2 d_6 \rho_X(d_7 \epsilon_n) + 4 d_5 K_0 \frac{|\alpha_{n+1} - \alpha_n|}{\alpha_n}.
\end{aligned}
\tag{6.5.43}
$$

Finally, we use Lemma 7.1.3 again to obtain the following theorem.

Theorem 6.5.4 *A bounded sequence $\{x^n\}$ generated by the algorithm of iterative regularization (6.5.37) in a uniformly smooth Banach space X, which possesses an approximation, converges strongly to the unique solution $\bar{x}^*$ defined by the inequality (2.7.7) if the equation $Ax = f$ is solvable and*
1) A and A^{h_n} are bounded demicontinuous accretive operators, $D(A) = D(A^{h_n}) = X$, conditions (6.5.2) (6.5.5) and (6.5.7) are satisfied;
2) J is the continuous and weak-to-weak continuous mapping;

$$3) \quad \lim_{n \to \infty} \alpha_n = 0;$$

$$4) \quad \sum_{n=1}^{\infty} \alpha_n \epsilon_n = \infty;$$

$$5) \quad \lim_{n \to \infty} \frac{\delta_n + h_n + \epsilon_n}{\alpha_n} = 0;$$

$$6) \quad \lim_{n \to \infty} \frac{\rho_X(\epsilon_n)}{\epsilon_n \alpha_n} = 0;$$

$$7) \quad \lim_{n \to \infty} \frac{|\alpha_n - \alpha_{n+1}|}{\alpha_n^2 \epsilon_n} = 0.$$

Remark 6.5.5 *If it is not known a priori that the sequence $\{x^n\}$ is bounded, then it is necessary to use the proof scheme of Theorem 6.5.2.*

6.6   Iterative-Projection Regularization Method

Suppose that X is uniformly convex Banach space, $A : X \to 2^{X^*}$ is a maximal monotone bounded operator with domain $D(A)$ and χ-growth (6.5.1). In place of the equation (6.1.1), we study in this section the variational inequality problem: To find $y \in \Omega$ such that

$$\langle Ay - f, x - y \rangle \geq 0 \quad \forall x \in \Omega, \tag{6.6.1}$$

where $\Omega \subset int\ D(A)$ is a convex closed set and $f \in X^*$. As in the previous section, we suppose that (6.1.1) has a nonempty solution set N and that A, f and Ω are given with perturbations which we denote, respectively, by A^{h_n}, f^{δ_n} and Ω_n, $n = 0, 1, 2, ...$, such that $D(A^{h_n}) = D(A)$, the inequalities (6.5.3) and (6.5.2) hold as $x \in D(A)$, $\Omega_n \subset int\ D(A)$ are convex closed sets and

$$\mathcal{H}_X(\Omega_n, \Omega) \leq \sigma_n. \tag{6.6.2}$$

We assume that there exist a convex function $\tilde{\delta}_X(t)$ such that $\delta_X(t) \geq \tilde{\delta}_X(t)$ for all $0 \leq t \leq 2$ and increasing function $g_X(t) = t^{-1}\delta_X(t)$.

1. We construct the iterative regularization method in the form:

$$Jx^{n+1} = Jx^n - \epsilon_m(\bar{\eta}_n^{h_m} + \alpha_m J\bar{x}^n - f^{\delta_m} + q_m), \quad n = 0, 1, 2, ... , \tag{6.6.3}$$

where $\bar{\eta}_n^{h_m} \in A^{h_m}\bar{x}^n$, $m = n + n_0$,

$$q_m = (\alpha_m + \alpha_m\|\bar{x}^n\| + \|\bar{\eta}_n^{h_m} - f^{\delta_m}\|_*)\frac{J(x^n - \bar{x}^n)}{\|x^n - \bar{x}^n\|}, \tag{6.6.4}$$

and $\bar{x}^n = P_{\Omega_m}x^n$ is the metric projection of $x^n \in X$ onto Ω_m, $n_0 \geq 0$ is defined by (6.6.7) below.

Suppose that step parameters ϵ_n, regularization parameters α_n and perturbation parameters δ_n, h_n and σ_n, describing, respectively, (6.5.2), (6.5.3) and (6.6.2), are positive for all $n \geq 0$ and vanish as $n \to \infty$, $\alpha_{n+1} \leq \alpha_n$, $\epsilon_{n+1} \leq \epsilon_n$, and (6.5.7) holds together with $\sigma_m \leq \bar{\sigma}$.

We introduce the intermediate variational inequality problem: To find $z \in \Omega$ such that

$$\langle Az + \alpha Jz - f, x - z \rangle \geq 0 \quad \forall x \in \Omega, \quad z \in \Omega, \quad \alpha > 0. \tag{6.6.5}$$

In Section 4.1 we have shown that its solution z_α exists and is unique for each $\alpha > 0$, the sequence $\{z_\alpha\}$ is bounded for all $\alpha > 0$, say $\|z_\alpha\| \leq K_0$, and $\lim_{\alpha \to 0} z_\alpha = \bar{x}^*$, where $\bar{x}^* \in N$ is a minimal norm solution.

Let R_0 be any non-negative number. We introduce the following constants:

$$K_1 = \sqrt{2R_0} + K_0, \quad K_2 = K_1 + \bar{\epsilon}c_1, \quad K_3 = K_0 + 2K_1,$$

$$c_0 = 2max\{1, K_0\}, \quad c_1 = \chi(K_3) + \bar{h}\phi(K_3) + \bar{\alpha}K_3 + \bar{\delta} + \|f\|_*,$$

$$c_2 = 8max\{L, K_2\}, \quad 1 < L < 1.7, \quad c_3 = 2K_2,$$

$$c_4 = 2Lmax\{1, K_0, 4L(\chi(K_0) + \bar{\alpha}K_0)\}, \quad c_5 = 2max\{1, K_3\},$$

$$c_6 = 2c_0 L K_0, \quad C = min\left\{(2L)^{-1}, \ c_5\left[g_X\left(c_5^{-1}(K_1 + K_3)\right)\right]^{-1}\right\}.$$

Construct the numerical sequence γ_k, $k = 0, 1, 2, \dots$ as follows:

$$\gamma_k = \epsilon_k\left(h_k\phi(K_3) + \delta_k\right)(K_0 + K_3) + 8c_1^2\epsilon_k^2 + c_2\rho_{X^*}(c_1\epsilon_n)$$

$$+ \quad c_0 c_3 c_4 g_X^{-1}\left[c_6\frac{\alpha_k - \alpha_{k+1}}{\alpha_k} + c\left(\sqrt{\frac{\sigma_k}{\alpha_k}} + \sqrt{\frac{\sigma_{k+1}}{\alpha_{k+1}}}\right)\right]. \tag{6.6.6}$$

Choose n_0 according to the condition

$$n_0 = min\{k \mid 2c_4\tilde{\delta}_X^{-1}\left(\frac{\gamma_k}{2C\alpha_k\epsilon_k}\right) \leq \triangle^{-1}(R_0)\}, \tag{6.6.7}$$

where $\triangle(\tau) = 8\tau^2 + c_2\rho_{X^*}(\tau)$. Let z_{n_0} be a solution of the variational inequality (6.6.5) with $\alpha = \alpha_{n_0}$ and let the initial point x^0 in (6.6.3) satisfy the inequality $W(x^0, z_{n_0}) \leq R_0$, where $W(x, z)$ is defined by (6.5.13).

We need a more general statement than Lemma 6.5.1.

Lemma 6.6.1 *Suppose that*
1) X is a uniformly convex Banach space;
2) sequences $\{z_{\alpha_1}\}$ and $\{z_{\alpha_2}\}$ of solutions of variational inequalities

$$\langle T_1 z + \alpha_1 J z, x - z\rangle \geq 0 \quad \forall x \in \Omega_1, \quad z \in \Omega_1, \tag{6.6.8}$$

and

$$\langle T_2 z + \alpha_2 J z, x - z\rangle \geq 0 \quad \forall x \in \Omega_2, \quad z \in \Omega_2, \tag{6.6.9}$$

are bounded for all $\alpha_1 > 0$ and for all $\alpha_2 > 0$, respectively, that is, there exists $K_0 > 0$ such that $\|z_{\alpha_1}\| \leq K_0$, $\|z_{\alpha_2}\| \leq K_0$;
3) an operator $T_1 : X \to 2^{X^*}$ is maximal monotone on $D(T_1) \subseteq X$ and bounded on the sequences $\{z_{\alpha_1}\}$ and $\{z_{\alpha_2}\}$, i.e., there exists a constant $M > 0$ such that $\|\zeta_1\|_* \leq M$ for all $\zeta_1 \in T_1 z_{\alpha_1}$ and $\|\xi_1\|_* \leq M$ for all $\xi_1 \in T_1 z_{\alpha_2}$;
4) an operator $T_2 : X \to 2^{X^*}$ is maximal monotone on $D(T_2) \subseteq X$, $D(T_1) = D(T_2) = Q$ and $\mathcal{H}_{X^*}(T_1 z, T_2 z) \leq \omega$ for all $z \in \Omega_2$;
5) $\Omega_i \subset int \ Q$ $(i = 1, 2)$ are convex closed sets such that $H_X(\Omega_1, \Omega_2) \leq \sigma$.
Then the following estimate holds:

$$\|z_{\alpha_1} - z_{\alpha_2}\| \leq c_0 g_X^{-1}\left(c_6\frac{|\alpha_1 - \alpha_2|}{\alpha_1} + c_7\frac{\omega}{\alpha_1} + c_8\sqrt{\frac{\sigma}{\alpha_1}}\right), \tag{6.6.10}$$

where $c_6 = 2c_0 L K_0$, $c_7 = 2c_0 L$, $c_8 = max\{1, 2Lc_9\}$ and $c_9 = 2M + (\alpha_1 + \alpha_2)K_0 + \omega$.

Proof. Since z_{α_1} is a solution of (6.6.8) and z_{α_2} is a solution of (6.6.9), there exist $\zeta_1 \in T_1 z_{\alpha_1}$ and $\zeta_2 \in T_2 z_{\alpha_2}$ such that

$$\langle \zeta_1 + \alpha_1 J z_{\alpha_1}, x - z_{\alpha_1}\rangle \geq 0 \quad \forall x \in \Omega_1 \tag{6.6.11}$$

and
$$\langle \zeta_2 + \alpha_2 J z_{\alpha_2}, x - z_{\alpha_2} \rangle \geq 0 \quad \forall x \in \Omega_2. \tag{6.6.12}$$

We compose the following expression:
$$D = \langle \zeta_1 + \alpha_1 J z_{\alpha_1} - \zeta_2 - \alpha_2 J z_{\alpha_2}, z_{\alpha_1} - z_{\alpha_2} \rangle.$$

Regarding the condition 4), there exists $\xi_1 \in T_1 z_{\alpha_2}$ such that $\|\zeta_2 - \xi_1\|_* \leq \omega$. Then, due to the monotonicity of T_1 and properties of duality mapping J, we deduce

$$
\begin{aligned}
D &= \langle \zeta_1 - \xi_1 + \alpha_1(J z_{\alpha_1} - J z_{\alpha_2}) + \xi_1 - \zeta_2 + (\alpha_1 - \alpha_2) J z_{\alpha_2}, z_{\alpha_1} - z_{\alpha_2} \rangle \\[2mm]
&\geq \alpha_1 (2L)^{-1} \delta_X(c_0^{-1} \|z_{\alpha_1} - z_{\alpha_2}\|) - \|\zeta_2 - \xi_1\|_* \|z_{\alpha_1} - z_{\alpha_2}\| \\[2mm]
&\quad - |\alpha_1 - \alpha_2| \|z_{\alpha_2}\| \|z_{\alpha_1} - z_{\alpha_2}\| \\[2mm]
&\geq -\|z_{\alpha_1} - z_{\alpha_2}\|(\omega + K_0|\alpha_1 - \alpha_2|) + \alpha_1 (2L)^{-1} \delta_X \left(c_0^{-1} \|z_{\alpha_1} - z_{\alpha_2}\| \right). \tag{6.6.13}
\end{aligned}
$$

On the other hand, since $\mathcal{H}_X(\Omega_1, \Omega_2) \leq \sigma$, we assert that for $z_{\alpha_2} \in \Omega_2$ there exists $z_1 \in \Omega_1$ such that $\|z_{\alpha_2} - z_1\| \leq \sigma$ and
$$\langle \zeta_1 + \alpha_1 J z_{\alpha_1}, z_{\alpha_1} - z_1 \rangle \leq 0$$

because of (6.6.11). Therefore,
$$
\begin{aligned}
\langle \zeta_1 + \alpha_1 J z_{\alpha_1}, z_{\alpha_1} - z_{\alpha_2} \rangle &= \langle \zeta_1 + \alpha_1 J z_{\alpha_1}, z_{\alpha_1} - z_1 \rangle + \langle \zeta_1 + \alpha_1 J z_{\alpha_1}, z_1 - z_{\alpha_2} \rangle \\[2mm]
&\leq (\|\zeta_1\|_* + \alpha_1 \|z_{\alpha_1}\|)\sigma \leq (M + \alpha_1 K_0)\sigma.
\end{aligned}
$$

Analogously, by (6.6.12), the following estimate is obtained:
$$\langle \zeta_2 + \alpha_2 J z_{\alpha_2}, z_{\alpha_2} - z_{\alpha_1} \rangle \leq (\|\zeta_2\|_* + \alpha_2 K_0)\sigma.$$

It is obvious that
$$\|\zeta_2\|_* \leq \|\xi_1\|_* + \|\zeta_2 - \xi_1\|_* \leq M + \omega.$$

From this one gets that
$$D \leq \left(2M + (\alpha_1 + \alpha_2) K_0 + \omega \right) \sigma = c_9 \sigma.$$

The last inequality and (6.6.13) yield
$$c_9 \sigma + \|z_{\alpha_1} - z_{\alpha_2}\|(\omega + K_0|\alpha_1 - \alpha_2|) \geq \alpha_1 (2L)^{-1} \delta_X \left(c_0^{-1} \|z_{\alpha_1} - z_{\alpha_2}\| \right). \tag{6.6.14}$$

Consider two possible cases:
$$(i) \ \|z_{\alpha_1} - z_{\alpha_2}\| \leq c_0 g_X^{-1} \left(c_6 \frac{|\alpha_1 - \alpha_2|}{\alpha_1} + c_7 \frac{\omega}{\alpha_1} + \sqrt{\frac{\sigma}{\alpha_1}} \right), \tag{6.6.15}$$

$$(ii) \; \|z_{\alpha_1} - z_{\alpha_2}\| \; > \; c_0 g_X^{-1} \left(c_6 \frac{|\alpha_1 - \alpha_2|}{\alpha_1} + c_7 \frac{\omega}{\alpha_1} + \sqrt{\frac{\sigma}{\alpha_1}} \right).$$

Since $\delta_X(t) \leq \delta_H(t) \leq t^2$, where H is a Hilbert space, case (ii) implies

$$\frac{\|z_{\alpha_1} - z_{\alpha_2}\|}{c_0} \geq g_H \left(\frac{\|z_{\alpha_1} - z_{\alpha_2}\|}{c_0} \right) \geq g_X \left(\frac{\|z_{\alpha_1} - z_{\alpha_2}\|}{c_0} \right) > \sqrt{\frac{\sigma}{\alpha_1}}.$$

Taking into account (6.6.14) we obtain

$$\frac{c_0 \delta_X(c_0^{-1} \|z_{\alpha_1} - z_{\alpha_2}\|)}{\|z_{\alpha_1} - z_{\alpha_2}\|} \; \leq \; c_6 \frac{|\alpha_1 - \alpha_2|}{\alpha_1} + c_7 \frac{\omega}{\alpha_1} + \frac{2L c_0 c_9}{\|z_{\alpha_1} - z_{\alpha_2}\|} \frac{\sigma}{\alpha_1}$$

$$\leq \; c_6 \frac{|\alpha_1 - \alpha_2|}{\alpha_1} + c_7 \frac{\omega}{\alpha_1} + 2L c_9 \sqrt{\frac{\sigma}{\alpha_1}}.$$

It follows from this relation that

$$\|z_{\alpha_1} - z_{\alpha_2}\| \leq c_0 g_X^{-1} \left(c_6 \frac{|\alpha_1 - \alpha_2|}{\alpha_1} + c_7 \frac{\omega}{\alpha_1} + 2L c_9 \sqrt{\frac{\sigma}{\alpha_1}} \right).$$

Comparing this estimate with (6.6.15) we see that the conclusion of the lemma is true. ∎

Return now to the method (6.6.3), (6.6.4) and denote solutions of the variational inequalities

$$\langle Az + \alpha_m Jz - f, x - z \rangle \geq 0 \quad \forall x \in \Omega_m, \; z \in \Omega_m, \tag{6.6.16}$$

and

$$\langle Az + \alpha_{m+1} Jz - f, x - z \rangle \geq 0 \quad \forall x \in \Omega_{m+1}, \; z \in \Omega_{m+1}, \tag{6.6.17}$$

respectively, by z_m and z_{m+1}. Suppose that X is a uniformly convex and uniformly smooth Banach space. Similarly to (6.5.22), one gets

$$W(x^{n+1}, z_{m+1}) - W(x^n, z_m) \; \leq \; \langle Jx^{n+1} - Jx^n, x^{n+1} - x^n \rangle$$

$$+ \; \|Jx^{n+1} - Jz_{m+1}\|_* \|z_{m+1} - z_m\|$$

$$+ \; \langle Jx^{n+1} - Jx^n, \bar{x}^n - z_m \rangle$$

$$+ \; \langle Jx^{n+1} - Jx^n, x^n - \bar{x}^n \rangle. \tag{6.6.18}$$

Evaluate each of the four terms in the right-hand side of (6.6.18).
A. By (6.6.3) and (6.6.4), we have

$$\|Jx^{n+1} - Jx^n\|_* = \epsilon_m \|\bar{\eta}_n^{h_m} - f^{\delta_m} + \alpha_m J\bar{x}^n + q^m\|_*$$

$$\leq \; 2\epsilon_m(\|\bar{\eta}_n\|_* + \|\bar{\eta}_n^{h_m} - \bar{\eta}_n\| + \|f^{\delta_m}\|_* + \alpha_m \|\bar{x}^n\| + \alpha_m) \leq c_1(n)\epsilon_m,$$

where $\bar{\eta}_n \in A\bar{x}^n$ such that $\|\bar{\eta}_n - \bar{\eta}_n^{h_m}\|_* \le h_m \phi(\|\bar{x}^n\|)$ and

$$c_1(n) = 2(\chi(\|\bar{x}^n\|) + h_m\phi(\|\bar{x}^n\|) + \alpha_m\|\bar{x}^n\| + \|f\|_* + \alpha_m + \delta_m).$$

Due to (1.6.17),

$$\langle Jx^{n+1} - Jx^n, x^{n+1} - x^n \rangle \le 8\|Jx^{n+1} - Jx^n\|_*^2 + c_2(n)\rho_{X^*}(\|Jx^{n+1} - Jx^n\|_*),$$

where $\rho_{X^*}(\tau)$ is the modulus of smoothness of X^* and

$$c_2(n) = 8max\{L, \|x^{n+1}\|, \|x^n\|\}.$$

Therefore,

$$\langle Jx^{n+1} - Jx^n, x^{n+1} - x^n \rangle \le 8c_1^2(n)\epsilon_m^2 + c_2(n)\rho_{X^*}(c_1(n)\epsilon_m).$$

B.   In its turn, by Lemma 6.6.1,

$$\|Jx^{n+1} - Jz_{m+1}\|_* \|z_{m+1} - z_m\|$$

$$\le c_0 c_3(n) g_X^{-1} \left[c_6 \frac{\alpha_m - \alpha_{m+1}}{\alpha_m} + c_4 \left(\sqrt{\frac{\sigma_m}{\alpha_m}} + \sqrt{\frac{\sigma_{m+1}}{\alpha_{m+1}}} \right) \right],$$

where

$$c_3(n) = 2max\{\|x^{n+1}\|, \|z_{m+1}\|\}.$$

We assumed here that the sequence $\{\alpha_n\}$ is non-increasing, also used the estimate (6.6.10) with $\omega = 0$ and the following relations:

$$\mathcal{H}_X(\Omega_n, \Omega_{n+1}) \le \sigma_n + \sigma_{n+1}$$

and

$$\sqrt{\frac{\sigma_{n+1} + \sigma_n}{\alpha_n}} \le \sqrt{\frac{\sigma_n}{\alpha_n}} + \sqrt{\frac{\sigma_{n+1}}{\alpha_{n+1}}}.$$

C.   It is possible to verify that

$$\langle Jx^{n+1} - Jx^n, \bar{x}^n - z_m \rangle = -\epsilon_m \langle \bar{\eta}_n^{h_m} - f^{\delta_m} + \alpha_m J\bar{x}^n + q_m, \bar{x}^n - z_m \rangle$$

$$= -\epsilon_m \langle \bar{\eta}_n^{h_m} - \bar{\eta}_n, \bar{x}^n - z_m \rangle - \epsilon_m \alpha_m \langle J\bar{x}^n - Jz_m, \bar{x}^n - z_m \rangle$$

$$- \epsilon_m \langle \bar{\eta}_n - \xi_m, \bar{x}^n - z_m \rangle - \epsilon_m \langle \xi_m + \alpha_m Jz_m - f, \bar{x}^n - z_m \rangle$$

$$- \epsilon_m \langle f - f^{\delta_m}, \bar{x}^n - z_m \rangle - \epsilon_m \langle q_m, \bar{x}^n - z_m \rangle,$$

where $\xi_m \in Az_m$. Recall that

$$\langle \xi_m + \alpha_m Jz_m - f, z_m - z \rangle \le 0 \quad \forall z \in \Omega_m. \tag{6.6.19}$$

Take into consideration the following inequalities:

$$\langle \bar{\eta}_n - \xi_m, \bar{x}^n - z_m \rangle \geq 0,$$

$$\langle \xi_m + \alpha_m J z_m - f, \bar{x}^n - z_m \rangle \geq 0,$$

$$\langle J \bar{x}^n - J z_m, \bar{x}^n - z_m \rangle \geq (2L)^{-1} \delta_X \left(\frac{\| \bar{x}^n - z_m \|}{c_5(n)} \right),$$

where

$$c_5(n) = 2 max\{1, \| \bar{x}^n \|, \| z_m \|\},$$

and

$$\langle q_m, \bar{x}^n - z_m \rangle \geq 0,$$

valid due to the monotonicity of operator A, (6.6.19), (1.6.28), and (1.5.12), respectively. Then we come to the estimate

$$\langle J x^{n+1} - J x^n, \bar{x}^n - z_m \rangle \leq \epsilon_m \left(h_m \phi(\| \bar{x}^n \|) \| + \delta_m \right) \| \bar{x}^n - z_m \|$$

$$- \epsilon_m \alpha_m (2L)^{-1} \delta_X \left(\frac{\| \bar{x}^n - z_m \|}{c_5(n)} \right).$$

D. Finally, the last term in (6.6.18) is evaluated as

$$\langle J x^{n+1} - J x^n, x^n - \bar{x}^n \rangle = -\alpha_m \epsilon_m \langle J \bar{x}^n, x^n - \bar{x}^n \rangle$$

$$- \epsilon_m \langle q_m, x^n - \bar{x}^n \rangle - \epsilon_m \langle \bar{\eta}_n^{h_m} - f^{\delta_m}, x^n - \bar{x}^n \rangle$$

$$\leq \epsilon_m \alpha_m \| \bar{x}^n \| \| \bar{x}^n - x^n \| - \epsilon_m \alpha_m \| \bar{x}^n - x^n \|$$

$$- \epsilon_m \alpha_m \| \bar{x}^n \| \| x^n - \bar{x}^n \| - \epsilon_m \| \bar{\eta}_n^{h_m} - f^{\delta_m} \|_* \| \bar{x}^n - x^n \|$$

$$+ \epsilon_m \| \bar{\eta}_n^{h_m} - f^{\delta_m} \|_* \| x^n - \bar{x}^n \|$$

$$= -\epsilon_m \alpha_m \| \bar{x}^n - x^n \|.$$

Combination of A - D gives the following result:

$$W(x^{n+1}, z_{m+1}) - W(x^n, z_m)$$

$$\leq -\epsilon_m \alpha_m \left[(2L)^{-1} \delta_X \left(\frac{\| \bar{x}^n - z_m \|}{c_5(n)} \right) + \| x^n - \bar{x}^n \| \right] + \Upsilon_m, \qquad (6.6.20)$$

where

$$
\begin{aligned}
\Upsilon_m & = \epsilon_m h_m \phi(\|\bar{x}^n\|)\|\bar{x}^n - z_m\| + \epsilon_m \delta_m \|\bar{x}^n - z_m\| + 8\epsilon_m^2 c_1^2(n) + c_2(n)\rho_{X^*}(c_1(n)\epsilon_m) \\
& \quad + c_0 c_3(n) g_X^{-1} \left[c_6 \frac{\alpha_m - \alpha_{m+1}}{\alpha_m} + c_4 \left(\sqrt{\frac{\sigma_m}{\alpha_m}} + \sqrt{\frac{\sigma_{m+1}}{\alpha_{m+1}}} \right) \right].
\end{aligned} \tag{6.6.21}
$$

Since $W(x^0, z_{n_0}) \leq R_0$, it is possible to consider $n \geq 0$ such that $W(x^n, z_{n+n_0}) \leq R_0$. For these n we have $\|x^n\| \leq K_1$. Observe that $\bar{x}^n \in \Omega_m$ and $z_m \in \Omega_m$. This enables us to derive the estimate

$$
\|\bar{x}^n\| \leq \|\bar{x}^n - x^n\| + \|x^n\| \leq \|z_m - x^n\| + \|x^n\| \leq 2\|x^n\| + \|z_m\| \leq 2K_1 + K_0 = K_3.
$$

Therefore, if (6.5.7) is satisfied then

$$
c_1(n) \leq 2(\chi(K_3) + \bar{h}\phi(K_3) + \bar{\alpha}K_3 + \bar{\alpha} + \bar{\delta} + \|f\|_*) = c_1.
$$

Moreover, similarly to (6.5.28),

$$
\|x^{n+1}\| \leq K_1 + \epsilon_m c_1(n) \leq K_1 + \bar{\epsilon}c_1 = K_2.
$$

Now it can be verified that $c_s(n) \leq c_s$, $s = 2, 3, 5$. Thus, it follows from (6.6.20) and (6.6.21)) that

$$
W(x^{n+1}, z_{m+1}) \leq W(x^n, z_m) - \epsilon_m \alpha_m \left((2L)^{-1}\delta_X(c_5^{-1}\|\bar{x}^n - z_m\|) + \|x^n - \bar{x}^n\| \right) + \gamma_m,
$$

where γ_m is calculated by the formula (6.6.6) with $k = m$.

It is obvious that

$$
\frac{\|x^n - \bar{x}^n\|}{c_5} = \frac{\delta_X(c_5^{-1}\|x^n - \bar{x}^n\|)}{g_X(c_5^{-1}\|x^n - \bar{x}^n\|)}.
$$

At the same time, one gets

$$
g_X(c_5^{-1}\|x^n - \bar{x}^n\|) \leq g_X(c_5^{-1}(K_1 + K_3)) = c_{10}
$$

because $\|x^n - \bar{x}^n\| \leq K_1 + K_3$. Hence,

$$
\|x^n - \bar{x}^n\| \geq c_5 c_{10}^{-1}\delta_X(c_5^{-1}\|x^n - \bar{x}^n\|) \geq c_5 c_{10}^{-1}\tilde{\delta}_X(c_5^{-1}\|x^n - \bar{x}^n\|).
$$

The last inequality implies

$$
\begin{aligned}
(2L)^{-1}\delta_X(c_5^{-1}\|\bar{x}^n - z_m\|) & \quad + \quad \|x^n - \bar{x}^n\| \geq (2L)^{-1}\tilde{\delta}_X(c_4^{-1}\|\bar{x}^n - z_m\|) \\
& \quad + \quad c_5 c_{10}^{-1}\tilde{\delta}_X(c_4^{-1}\|x^n - \bar{x}^n\|) \\
& \geq \quad C\left(\tilde{\delta}_X(c_4^{-1}\|\bar{x}^n - z_m\|) + \tilde{\delta}_X(c_4^{-1}\|x^n - \bar{x}^n\|) \right).
\end{aligned}
$$

Since the function $\tilde{\delta}_X(t)$ is convex and increasing for all $0 \leq t \leq 2$, we can write

$$(2L)^{-1}\delta_X\left(\frac{\|\bar{x}^n - z_m\|}{c_5}\right) \quad + \quad \|x^n - \bar{x}^n\| \geq 2C\tilde{\delta}_X\left(\frac{\|\bar{x}^n - z_m\| + \|x^n - \bar{x}^n\|}{2c_5}\right)$$

$$\geq \quad 2C\tilde{\delta}_X\left(\frac{\|x^n - z_m\|}{2c_5}\right).$$

Thus, we come to the following numerical inequality:

$$W(x^{n+1}, z_{m+1}) \leq W(x^n, z_m) - 2C\epsilon_m\alpha_m\tilde{\delta}_X\left(\frac{\|x^n - z_m\|}{2c_5}\right) + \gamma_m.$$

Assume that for all $m \geq n_0$,

$$\frac{\alpha_m - \alpha_{m+1}}{\alpha_m} \leq d_1, \quad \frac{\sigma_m}{\alpha_m} \leq d_2. \tag{6.6.22}$$

Then

$$W(x^{n+1}, z_{m+1}) \leq R_0 + \bar{\gamma},$$

where

$$\bar{\gamma} = \bar{\epsilon}(\bar{h}\phi(K_3) + \bar{\delta})(K_0 + K_3) + 8c_1^2\bar{\epsilon}^2 + c_2\rho_{X^*}(c_1\bar{\epsilon}) + c_0c_3g_X^{-1}(c_6d_1 + 2c_4d_2).$$

By analogy with the proof of (6.5.35), one can show that the sequence $\{x^n\}$ is bounded by a constant which does not depend on n. Further the proof can be done by the same scheme as in Theorem 6.5.2. Thus, the following statement holds:

Theorem 6.6.2 *Suppose that*
1) X *is a uniformly convex and uniformly smooth Banach space with the modulus of convexity* $\delta_X(\epsilon)$ *and there exists a convex increasing function* $\tilde{\delta}_X(\epsilon)$ *such that* $\delta_X(\epsilon) \geq \tilde{\delta}_X(\epsilon)$ *for all* $0 \leq t \leq 2$;
2) $\bar{x}^*$ *is the minimal norm solution of variational inequality* (6.6.1);
3) $A : X \to 2^{X^*}$ *is a maximal monotone bounded operator with* χ-*growth* (6.5.1), $f \in X^*$;
4) A *and* f *are given with perturbations as* $A^{h_n} : X \to 2^{X^*}$ *and* $f^{\delta_n} \in X^*$ *satisfying the conditions* (6.5.2) *and* (6.5.3), *where* $D(A^{h_n}) = D(A)$;
5) Ω *is also known approximately with the estimate* $\mathcal{H}_X(\Omega, \Omega_n) \leq \sigma_n$, *where* Ω *and* Ω_n *are convex closed sets in* X, $\Omega \subset \text{int } D(A)$ *and* $\Omega_n \subset \text{int } D(A)$;
6) *in the method of iterative regularization* (6.6.3), (6.6.4), *the initial approximation* x^0 *satisfies the inequality* $W(x^0, z_{n_0}) \leq R_0$, *where* z_{n_0} *is a solution of the variational inequality* (6.6.16) *with* $m = n_0$ *and* n_0 *obeys the rule* (6.6.7);
7) *parameters* α_n, ϵ_n, δ_n, h_n, σ_n *are such that* (6.5.7) *is valid with* $\sigma_n \leq \bar{\sigma}$ *and*

$$8) \quad \lim_{n \to \infty} \alpha_n = 0;$$

$$9) \quad \sum_{n=1}^{\infty} \alpha_n\epsilon_n = \infty;$$

$$10) \quad \lim_{n \to \infty} \frac{\rho_{X^*}(\epsilon_n)}{\epsilon_n \alpha_n} = 0;$$

$$11) \quad \lim_{n \to \infty} \frac{h_n + \delta_n + \sigma_n + \epsilon_n}{\alpha_n} = 0;$$

$$12) \quad \lim_{n \to \infty} \frac{g_X^{-1}\left(\alpha_n^{-1}|\alpha_n - \alpha_{n+1}|\right)}{\alpha_n \epsilon_n} = 0.$$

Moreover, $\{\alpha_n\}$ does not increase and (6.6.22) holds. Then the sequence $\{x^n\}$ generated by (6.6.3) and (6.6.4) strongly converges to $\bar{x}^*$ as $n \to \infty$.

Remark 6.6.3 *The requirement $\delta_X(\epsilon) \geq \tilde{\delta}_X(\epsilon)$ in 1) is not too restrictive. For instance, in the spaces l^p, L^p and W_m^p, $1 < p < \infty$, the moduli of convexity $\delta_X(\epsilon) \geq c_1 \epsilon^2$ if $1 < p \leq 2$ and $\delta_X(\epsilon) \geq c_2 \epsilon^p$ if $2 \leq p < \infty$ with some constants c_1, $c_2 > 0$, that is, $\tilde{\delta}_X(\epsilon) = c_1 \epsilon^2$ and $\tilde{\delta}_X(\epsilon) = c_2 \epsilon^p$, respectively.*

Next we omit the assumption 2) of the previous theorem and prove the existence of a solution to variational inequality (6.6.1) provided that the regularization process (6.6.3), (6.6.4) converges.

Theorem 6.6.4 *Assume that in a uniformly convex and uniformly smooth Banach space X, the iterative sequence $\{x^n\}$ generated by (6.6.3), (6.6.4), where non-negative parameters h_n, δ_n, σ_n and positive parameters α_n, ϵ_n approach zero as $n \to \infty$ and $\sum_{n=0}^{\infty} \alpha_n \epsilon_n = \infty$, strongly converges to an element $x^* \in \Omega$. Then x^* is a solution of the variational inequality (6.6.1).*

Proof. Without loss of generality, we suppose that $h_n = \delta_n = 0$, the operator A is one-to-one and $n_0 = 0$. Let x^* be not a solution of (6.6.1). Then there exists $x \in \Omega$ and $c > 0$ such that

$$\langle Ax - f, x - x^* \rangle = -c. \tag{6.6.23}$$

Represent (6.6.23) in the equivalent form

$$-c = \langle Ax - f + \alpha_n Jx, x - x^* \rangle - \alpha_n \langle Jx, x - x^* \rangle.$$

Since $\alpha_n \to 0$, there exists a number $k_1 > 0$ such that for every $n \geq k_1$

$$\alpha_n \langle Jx, x - x^* \rangle \leq \frac{c}{2}.$$

In this case, for all $n \geq k_1$ we have

$$\langle Ax + \alpha_n Jx - f, x - x^* \rangle \leq -\frac{c}{2}. \tag{6.6.24}$$

Represent now (6.6.24) in the equivalent form

$$\langle Ax + \alpha_n Jx - f, x - \bar{x}^n \rangle + \langle Ax + \alpha_n Jx - f, \bar{x}^n - x^* \rangle \leq -\frac{c}{2}, \tag{6.6.25}$$

where $\bar{x}^n = P_{\Omega_n} x^n$. By the conditions, $x^n \to x^*$. Therefore, on account of the inclusion $x^* \in \Omega$, the limit relation $\bar{x}^n \to x^*$ holds too (see Lemma 6.3.1). Then there exists an integer $k_2 > 0$ such that for all $n \geq k_2$,

$$|\langle Ax + \alpha_n Jx - f, \bar{x}^n - x^* \rangle| \leq \|Ax + \alpha_n Jx - f\|_* \|\bar{x}^n - x^*\| \leq \frac{c}{4}$$

because $\|Ax + \alpha_n Jx - f\|_*$ is bounded. Hence, by (6.6.25), for all $n \geq max\{k_1, k_2\}$,

$$\langle Ax + \alpha_n Jx - f, x - \bar{x}^n \rangle \leq -\frac{c}{4}.$$

Finally, using the monotonicity of A and the property (1.5.3) of J, we obtain the inequality

$$-\frac{c}{4} \geq \langle A\bar{x}^n + \alpha_n J\bar{x}^n - f, x - \bar{x}^n \rangle + \alpha_n(\|x\| - \|\bar{x}^n\|)^2.$$

Then there exists an integer $k_3 > 0$ such that for all $n \geq k_3$,

$$-\frac{c}{4} \geq \langle A\bar{x}^n + \alpha_n J\bar{x}^n - f, x - x^* \rangle + \alpha_n(\|x\| - \|x^*\|)^2 + \beta_n,$$

where

$$\beta_n = \langle A\bar{x}^n + \alpha_n J\bar{x}^n - f, x^* - \bar{x}^n \rangle - 2\alpha_n(\|x\| - \|x^*\|)(\|\bar{x}^n\| - \|x^*\|)$$

$$+ \alpha_n(\|x^*\| - \|\bar{x}^n\|)^2 \leq \frac{c}{8}.$$

Hence, for all $n \geq max\{k_1, k_2, k_3\}$,

$$-\frac{c}{8} \geq \langle A\bar{x}^n + \alpha_n J\bar{x}^n - f, x - x^* \rangle + \alpha_n(\|x\| - \|x^*\|)^2,$$

or

$$-\frac{c}{8} \geq \langle A\bar{x}^n + \alpha_n J\bar{x}^n - f + q_n, x - x^* \rangle + \alpha_n(\|x\| - \|x^*\|)^2 - \langle q_n, x - x^* \rangle.$$

Estimate the last term. By virtue of the fact that $\mathcal{H}_X(\Omega, \Omega_n) \leq \sigma_n$, there exists $v_n \in \Omega_n$ such that $\|v_n - x\| \leq \sigma_n$, where x satisfies (6.6.23). Let

$$c_n = \alpha_n + \alpha_n \|\bar{x}^n\| + \|A\bar{x}^n - f\|_*.$$

Then

$$\langle q_n, x - x^* \rangle = c_n \left\langle \frac{J(x^n - \bar{x}^n)}{\|\bar{x}^n - x^n\|}, x - x^* \right\rangle = c_n \left\langle \frac{J(x^n - \bar{x}^n)}{\|\bar{x}^n - x^n\|}, \bar{x}^n - x^* \right\rangle$$

$$+ c_n \left\langle \frac{J(x^n - \bar{x}^n)}{\|\bar{x}^n - x^n\|}, v_n - \bar{x}^n \right\rangle + c_n \left\langle \frac{J(x^n - \bar{x}^n)}{\|\bar{x}^n - x^n\|}, x - v_n \right\rangle.$$

In view of Lemma 1.5.17,
$$\langle J(x^n - \bar{x}^n), v_n - \bar{x}^n \rangle \leq 0.$$

Therefore,
$$\langle q_n, x - x^* \rangle \leq c_n \Big(\|\bar{x}^n - x^*\| + \|x - v_n\| \Big) \leq c_n \Big(\|\bar{x}^n - x^*\| + \sigma_n \Big),$$

and there exists $k_4 > 0$ such that for all $n \geq k_4$,
$$\langle q_n, x - x^* \rangle \leq \frac{c}{16}.$$

Consequently, if $n \geq max\{k_1, \ k_2, \ k_3, \ k_4\} = k_5$ then
$$0 \geq -\frac{c}{16} \geq \langle A\bar{x}^n + \alpha_n J\bar{x}^n + q_n - f, x - x^* \rangle + \alpha_n (\|x\| - \|x^*\|)^2.$$

By (6.6.3), for $n \geq k_5$,
$$\langle Jx^{n+1} - Jx^n, x - x^* \rangle \geq \alpha_n \epsilon_n (\|x\| - \|x^*\|)^2. \tag{6.6.26}$$

Due to the strong convergence of $\{x^n\}$ to x^* and continuity of J, the limit relation $Jx^n \to Jx^*$ holds. Hence, the series standing in the left-hand side of the inequality
$$\sum_{n=0}^{\infty} \langle Jx^{n+1} - Jx^n, x - x^* \rangle \leq \|x - x^*\| \sum_{n=0}^{\infty} \|Jx^{n+1} - Jx^n\|_*$$

converges. Then (6.6.26) implies that
$$\sum_{n=0}^{\infty} \alpha_n \epsilon_n \ < \ \infty,$$

which contradicts the assumption of the theorem. Hence, x^* is a solution of the variational inequality (6.6.1). ∎

3. Constructing the sequence $\{x^n\}$, we make use of the projection operation onto a set Ω. If Ω is given in the form of functional inequalities, then the effective method of solving variational inequalities is the method of indefinite Lagrange multipliers. Consider the general situation when $\Omega = \Omega_1 \cap \Omega_2$, where it is assumed that the projection operation onto the first set Ω_1 is done quite easily, and the second set Ω_2 is given by the following constraint system:
$$\Omega_2 = \{x \in X \mid \varphi_i(x) \leq 0, \ i = 1, 2, ..., l\},$$

where $\varphi_i(x)$ are continuous convex functionals on X. It is possible that $\Omega_1 = X$. Then we define the iterative process of finding a solution $\bar{x}^*$ of the variational inequality (6.6.1) as
$$Jx^{n+1} = Jx^n - \epsilon_n (\Phi(x^n) + \alpha_n Jx^n) \tag{6.6.27}$$

and
$$\lambda^{n+1} = \{\lambda_i^n + \epsilon_n (\varphi_i(x) - \alpha_n \lambda_i^n)\}_+,$$

where

$$\Phi(x) = Ax - f + \sum_{i=1}^{l} \lambda_i^n \varphi_i'(x)$$

and $\{a\}_+ = max\{0, a\}$. If $\Omega_1 \neq X$ then we replace (6.6.27) by the expression

$$Jx^{n+1} = Jx^n - \epsilon_n \Big(\Phi(\bar{x}^n) + \alpha_n J\bar{x}^n + q_n \Big),$$

where $\bar{x}^n = P_{\Omega_1} x^n$ and

$$q_n = (\alpha_n + \alpha_n \|\bar{x}^n\| + \|\Phi(\bar{x}^n)\|_*) \frac{J(x^n - \bar{x}^n)}{\|x^n - \bar{x}^n\|}.$$

The convergence proof is similar to the proof of Theorem 6.5.2 including the situation when an operator A, element f and sets Ω_1 and Ω_2 are given with perturbations.

Observe that the iterative process (6.6.27) with $\alpha_n = 0$ is not stable, in general, even if A is strongly monotone.

6.7   Continuous Regularization Method

Continuous regularization processes for solving ill-posed equation (6.1.1) deal with ordinary differential equations in which the role of "regularization" parameter is performed by a certain positive function $\alpha(t)$ with $t \geq t_0$. The stated methods are reduced to the Cauchy problem for a differential equation of some order. The order of the differential equation is called the order of the continuous method. In the present section we study the first order continuous regularization methods. We emphasize that, in the sequel, existence of solutions to any differential equation of this section is assumed.

We investigate the continuous regularization methods separately for equations with monotone and accretive operators in Hilbert and Banach spaces.

1. Let the equation (6.1.1) be given in a Hilbert space H, $A : H \to H$ be a monotone continuous operator, $D(A) = H$ and $f \in H$. Let (6.1.1) have a nonempty solution set N and $\bar{x}^*$ be its minimal norm solution. Under these conditions, the set N is convex and closed and there exists a unique element $\bar{x}^* \in N$. Assume that monotone continuous operator $A(t) : H \to H$ and right-hand side $f(t) \in H$ are perturbations of A and f, respectively, such that

$$\|A(t)x - Ax\| \leq h(t)g(\|x\|) \quad \forall x \in H, \quad \forall t \geq t_0, \tag{6.7.1}$$

and

$$\|f(t) - f\| \leq \delta(t) \quad \forall t \geq t_0. \tag{6.7.2}$$

We suppose that $\alpha(t) \to 0$, $\delta(t) \to 0$, $h(t) \to 0$ as $t \to \infty$ and $g(s)$ is a continuous, non-negative and non-decreasing function for all $s \geq 0$.

Consider the differential equation

$$\frac{dy(t)}{dt} + A(t)y(t) + \alpha(t)y(t) = f(t), \quad t \geq t_0, \quad y(t_0) = y_0. \tag{6.7.3}$$

Our aim is to prove the strong convergence of $y(t)$ to $\bar{x}^*$ as $t \to \infty$. The research scheme is the same as in the iterative regularization algorithms. We introduce the intermediate equation

$$Aw(t) + \alpha(t)w(t) = f. \tag{6.7.4}$$

It is known that $\|w(t)\| \leq \|\bar{x}^*\|$ and

$$\lim_{t \to \infty} w(t) = \bar{x}^*. \tag{6.7.5}$$

Obviously,

$$\|y(t) - \bar{x}^*\| \leq \|y(t) - w(t)\| + \|w(t) - \bar{x}^*\|, \tag{6.7.6}$$

so that we need only to prove that

$$\|y(t) - w(t)\| \to 0 \quad \text{as} \quad t \to \infty$$

and then

$$\lim_{t \to \infty} \|y(t) - \bar{x}^*\| = 0.$$

Theorem 6.7.1 *Suppose that*
(i) all the conditions of the present subsection are fulfilled;
(ii) a solution $y(t)$ of the differential equation (6.7.3) exists on the interval $[t_0, +\infty)$;
(iii) a solution $w(t)$ of the operator equation (6.7.4) (which necessarily exists) is differentiable on the interval $[t_0, \infty)$;
(iv) $\alpha(t)$ is a positive continuous and differentiable function satisfying the following limit relations:

$$\lim_{t \to \infty} \alpha(t) = 0, \quad \lim_{t \to \infty} \frac{|\alpha'(t)|}{\alpha^2(t)} = 0, \quad \lim_{t \to \infty} \frac{h(t) + \delta(t)}{\alpha(t)} = 0, \quad \int_{t_0}^{\infty} \alpha(\tau)d\tau = \infty. \tag{6.7.7}$$

Then $\lim_{t \to \infty} y(t) = \bar{x}^*$.

Proof. Denote

$$V(y(t), w(t)) = 2^{-1}\|y(t) - w(t)\|^2.$$

It is easy to see that

$$\frac{\partial V(y(t), w(t))}{\partial y} = y(t) - w(t)$$

and

$$\frac{\partial V(y(t), w(t))}{\partial w} = w(t) - y(t).$$

Then

$$
\begin{aligned}
\frac{dV(y(t), w(t))}{dt} &= \left(\frac{\partial V(y(t), w(t))}{\partial y}, \frac{dy(t)}{dt} \right) + \left(\frac{\partial V(y(t), w(t))}{\partial w}, \frac{dw(t)}{dt} \right) \\
&= -(A(t)y(t) - A(t)w(t), y(t) - w(t)) - \alpha(t)\|y(t) - w(t)\|^2 \\
&\quad - (Aw(t) + \alpha(t)w(t) - f, y(t) - w(t)) + (f(t) - f, y(t) - w(t)) \\
&\quad + (Aw(t) - A(t)w(t), y(t) - w(t)) + (w(t) - y(t), w'(t)).
\end{aligned}
$$

By the monotonicity of $A(t)$, we have

$$(A(t)y(t) - A(t)w(t), y(t) - w(t)) \geq 0.$$

Since $w(t)$ satisfies (6.7.4), one gets

$$(Aw(t) + \alpha(t)w(t) - f, y(t) - w(t)) = 0.$$

Using (6.7.1) and (6.7.2) we deduce

$$(Aw(t) - A(t)w(t), y(t) - w(t)) \leq h(t)g(\|\bar{x}^*\|)\|y(t) - w(t)\|$$

and

$$(f(t) - f, y(t) - w(t)) \leq \delta(t)\|y(t) - w(t)\|.$$

Hence,

$$\frac{dV(y(t), w(t))}{dt} \leq -\alpha(t)\|y(t) - w(t)\|^2 + \Big(\delta(t) + h(t)g(\|\bar{x}^*\|) + \|w'(t)\|\Big)\|y(t) - w(t)\|. \quad (6.7.8)$$

Evaluate further $\|w'(t)\|$ from above. By (6.5.19),

$$\|w(t_1) - w(t_2)\| \leq \frac{\|\bar{x}^*\| \, |\alpha(t_1) - \alpha(t_2)|}{\alpha(t_1)}.$$

This yields the inequality

$$\left\|\frac{w(t_1) - w(t_2)}{t_1 - t_2}\right\| \leq \frac{\|\bar{x}^*\|}{\alpha(t_1)}\left|\frac{\alpha(t_1) - \alpha(t_2)}{t_1 - t_2}\right|.$$

Owing to the differentiability of $w(t)$ and $\alpha(t)$, we obtain

$$\left\|\frac{dw(t)}{dt}\right\| \leq \|\bar{x}^*\|\frac{|\alpha'(t)|}{\alpha(t)}. \quad (6.7.9)$$

Then it follows from (6.7.8) that

$$\frac{d\|y(t) - w(t)\|}{dt} \leq -\alpha(t)\|y(t) - w(t)\| + \delta(t) + h(t)g(\|\bar{x}^*\|)) + \|\bar{x}^*\|\frac{|\alpha'(t)|}{\alpha(t)}.$$

Denoting here $\lambda(t) = \|y(t) - w(t)\|$ we come to the differential inequality

$$\frac{d\lambda(t)}{dt} \leq -\alpha(t)\lambda(t) + \gamma(t),$$

where

$$\gamma(t) = \delta(t) + h(t)g(\|\bar{x}^*\|) + \|\bar{x}^*\|\frac{|\alpha'(t)|}{\alpha(t)}.$$

In order to show that $\lambda(t) \to 0$, it is necessary to be sure that the conditions of Lemma 7.2.2 hold. Indeed, we see that $\alpha(t)$ and $\gamma(t)$ satisfy the lemma. Next, taking into account condition (iv) of the theorem, we obtain

$$\lim_{t \to \infty} \frac{\gamma(t)}{\alpha(t)} = 0. \quad (6.7.10)$$

Thus, $\|y(t) - w(t)\| \to 0$ as $t \to \infty$. By (6.7.6), the theorem is proved. ∎

Corollary 6.7.2 *Under the conditions of the previous theorem, if the function $\alpha(t)$ is decreasing, then the last relation in (iv) can be omitted.*

Proof. Show that the limit equality

$$\lim_{t \to \infty} \frac{|\alpha'(t)|}{\alpha^2(t)} = 0 \tag{6.7.11}$$

implies

$$\int_{t_0}^{\infty} \alpha(\tau) d\tau = +\infty. \tag{6.7.12}$$

Indeed, since $\alpha(t)$ is decreasing, we conclude from (6.7.11) that there exists a constant $c > 0$ such that

$$-\frac{\alpha'(t)}{\alpha^2(t)} \leq c.$$

Then

$$\frac{1}{\alpha(t)} - \frac{1}{\alpha(t_0)} \leq c(t - t_0),$$

that leads to the estimate

$$\alpha(t) \geq \frac{\alpha(t_0)}{1 + c\alpha(t_0)(t - t_0)}.$$

Thus, (6.7.12) results. ∎

The following functions $\alpha(t)$ and $\gamma(t)$ satisfy (6.7.10) - (6.7.12):
1) $\alpha(t) = c_1 t^{-r}$, $\gamma(t) = c_2 e^{-st}$, $0 < r < 1$, $s > 0$, $c_1 > 0$, $c_2 > 0$;
2) $\alpha(t) = c_1 t^{-r}$, $\gamma(t) = c_2 t^{-s}$, $0 < r < 1$, $s > r$, $c_1 > 0$, $c_2 > 0$;
3) $\alpha(t) = c_1 t^{-r}$, $\gamma(t) \equiv 0$, $0 < r < 1$, $c_1 > 0$.

2. Next we assume that $\alpha(t)$ is a convex function. In this case we do not need to require differentiability of a solution $w(t)$ to the equation (6.7.4).

So, let $A : H \to H$ be a monotone continuous operator, $\alpha(t)$ be a positive differentiable convex and decreasing function for all $t \geq t_0$ such that $\lim_{t \to \infty} \alpha(t) = 0$ and (6.7.11) holds. Let $A(t) : H \to H$, $t \geq t_0$, be a family of monotone continuous operators with $D(A(t)) = H$ satisfying, as earlier, the conditions (6.7.1) and (6.7.2). Consider the Cauchy problem (6.7.3) again and suppose for simplicity that it has a unique solution $y(t)$ defined for all $t \geq t_0$.

We rewrite the regularized operator equation (6.7.4) in the form:

$$Aw(\tau) + \alpha(\tau)w(\tau) = f. \tag{6.7.13}$$

For every fixed $\tau \geq t_0$, we also construct the auxiliary Cauchy problem

$$\frac{dz(t, \tau)}{dt} + Az(t, \tau) + \alpha(\tau)z(t, \tau) = f, \quad z(t_0, \tau) = y_0. \tag{6.7.14}$$

Denote

$$r(t, \tau) = 2^{-1}\|z(t, \tau) - w(\tau)\|^2.$$

It is clear that

$$\frac{dr(t,\tau)}{dt} = \left(\frac{dz(t,\tau)}{dt}, z(t,\tau) - w(\tau)\right).$$

The equations (6.7.13) and (6.7.14) involve the scalar equality

$$\left(\frac{dz(t,\tau)}{dt}, z(t,\tau) - w(\tau)\right) \quad + \quad (Az(t,\tau) - Aw(\tau), z(t,\tau) - w(\tau))$$

$$+ \quad \alpha(\tau)\|z(t,\tau) - w(\tau)\|^2 = 0. \qquad (6.7.15)$$

Since A is a monotone operator, the following differential inequality is obtained from (6.7.15):

$$\frac{dr(t,\tau)}{dt} \le -2\alpha(\tau)r(t,\tau)$$

with

$$r(t_0,\tau) = 2^{-1}\|y_0 - w(\tau)\|^2 = r_0(\tau).$$

Now we have from Lemma 7.2.2 the estimate

$$r(t,\tau) \le r_0(\tau)exp\Big(-2\alpha(\tau)(t - t_0)\Big). \qquad (6.7.16)$$

By the hypotheses, $w(\tau)$ is bounded by $\|\bar{x}^*\|$, therefore, there exists a constant $c > 0$ such that $r_0(\tau) \le c$ for any $\tau \ge t_0$. Then, in view of (6.7.16), we conclude that the trajectories $z(t,\tau)$ are bounded for all $t \ge t_0$ and all $\tau \ge t_0$. Consequently, there exists $c_1 > 0$ such that $\|z(t,\tau)\| \le c_1$ for all $t \ge t_0$ and for all $\tau \ge t_0$. Using now (6.7.16) with $t = \tau$, one gets

$$r(\tau,\tau) \le c\,exp\Big(-2\alpha(\tau)(\tau - t_0)\Big), \qquad (6.7.17)$$

where the argument of the exponential function becomes indefinite as $\tau \to +\infty$. Taking into account (6.7.7) and applying L'Hopital's rule, we obtain from (6.7.17) the limit relation

$$r(\tau,\tau) \to 0 \quad \text{as} \quad \tau \to \infty. \qquad (6.7.18)$$

Along with (6.7.3), consider the following problem with the exact data A and f :

$$\frac{du(t)}{dt} + Au(t) + \alpha(t)u(t) = f, \quad u(t_0) = y_0. \qquad (6.7.19)$$

Define the functions

$$v(t,\tau) = \|u(t) - z(t,\tau)\|$$

and

$$\bar{v}(t,\tau) = 2^{-1}v^2(t,\tau).$$

Then

$$\frac{d\bar{v}(t,\tau)}{dt} = \left(\frac{du(t)}{dt} - \frac{dz(t,\tau)}{dt}, u(t) - z(t,\tau)\right) = v(t,\tau)\frac{dv(t,\tau)}{dt}.$$

Further, (6.7.14) and (6.7.19) yield

$$\frac{d\bar{v}(t,\tau)}{dt} \quad + \quad (Au(t) - Az(t,\tau), u(t) - z(t,\tau)) + \alpha(t)v^2(t,\tau)$$
$$+ \quad (\alpha(t) - \alpha(\tau))(z(t,\tau), u(t) - z(t,\tau)) = 0. \tag{6.7.20}$$

By the monotonicity of A and boundedness of $z(t,\tau)$, (6.7.20) is reduced to the inequality

$$\frac{d\bar{v}(t,\tau)}{dt} \leq -\alpha(t)v^2(t,\tau) + c_1|\alpha(t) - \alpha(\tau)|v(t,\tau), \tag{6.7.21}$$

which implies a similar differential inequality for the function $v(t,\tau)$, namely,

$$\frac{dv(t,\tau)}{dt} \leq -\alpha(t)v(t,\tau) + c_1|\alpha(t) - \alpha(\tau)| \quad \forall \tau \geq t_0 \tag{6.7.22}$$

with the initial condition $v(t_0, \tau) = 0$.

The following expression is obtained from convexity of $\alpha(t)$:

$$|\alpha(t) - \alpha(\tau)| \leq |\alpha'(t)|(\tau - t), \quad \tau > t,$$

and (6.7.22) can be rewritten as

$$\frac{dv(t,\tau)}{dt} \leq -\alpha(t)v(t,\tau) + c_1|\alpha'(t)|(\tau - t).$$

It is not difficult to see that Lemma 7.2.2 gives now the estimate

$$v(\tau, \tau) \leq \frac{c_1 \int_{t_0}^{\tau} |\alpha'(s)|(\tau - s) exp\left(\int_{t_0}^{s} \alpha(\theta)d\theta\right) ds}{exp\left(\int_{t_0}^{\tau} \alpha(\theta)d\theta\right)}. \tag{6.7.23}$$

If there exists a constant $C > 0$ such that

$$\int_{t_0}^{\tau} |\alpha'(s)|(\tau - s) exp\left(\int_{t_0}^{s} \alpha(\theta)d\theta\right) ds \leq C,$$

then the condition (6.7.12) guarantees that

$$v(\tau, \tau) \to 0 \quad \text{as} \quad \tau \to \infty. \tag{6.7.24}$$

Otherwise, (6.7.24) can be satisfied if we twice apply L'Hopital's rule to the right-hand side of the inequality (6.7.23) and use (6.7.11). Observe that boundedness of the trajectory $u(t)$ for all $t \geq t_0$ also follows from (6.7.24).

Let

$$p(t) = \|y(t) - u(t)\| \quad \text{and} \quad \bar{p}(t) = 2^{-1}p^2(t).$$

By (6.7.3) and (6.7.19), we obtain

$$\frac{d\bar{p}(t)}{dt} \quad + \quad (A(t)y(t) - A(t)u(t), y(t) - u(t))$$
$$+ \quad (A(t)u(t) - Au(t), y(t) - u(t)) + \alpha(t)p^2(t)$$
$$= \quad (f(t) - f, y(t) - u(t)).$$

Since $A(t)$ is a monotone operator, one has

$$\frac{d\bar{p}(t)}{dt} \leq (A(t)u(t) - Au(t), y(t) - u(t)) + \alpha(t)p^2(t) + (f - f(t), y(t) - u(t)).$$

Now (6.7.1) and (6.7.2) imply

$$\frac{d\bar{p}(t)}{dt} \leq -2\alpha(t)\bar{p}(t) + \Big(\delta(t) + h(t)g(\|u(t)\|)\Big)p(t).$$

The trajectory $u(t)$ is bounded, therefore, there exists a constant $c_2 > 0$ such that the previous inequality is rewritten as follows:

$$\frac{d\bar{p}(t)}{dt} \leq -2\alpha(t)\bar{p}(t) + c_2\Big(\delta(t) + h(t)\Big)p(t) \tag{6.7.25}$$

or

$$\frac{dp(t)}{dt} \leq -\alpha(t)p(t) + c_2\Big(\delta(t) + h(t)\Big). \tag{6.7.26}$$

Assume that

$$\lim_{t \to \infty} \frac{\delta(t) + h(t)}{\alpha(t)} = 0. \tag{6.7.27}$$

Exploiting Lemma 7.2.2 again, it is easy to verify that

$$\lim_{t \to \infty} p(\tau) = 0. \tag{6.7.28}$$

Now the relations (6.7.5), (6.7.18), (6.7.24), (6.7.28) and the obvious inequality

$$\|y(\tau) - \bar{x}^*\| \leq \|y(\tau) - u(\tau)\| + \|u(\tau) - z(\tau, \tau)\|$$

$$+ \|z(\tau, \tau) - w(\tau)\| + \|w(\tau) - \bar{x}^*\|$$

allow us to formulate the following theorem:

Theorem 6.7.3 *Let A and $A(t)$, $t \geq t_0$, be monotone continuous operators defined on a Hilbert space H, $f \in H$ and $f(t) \in H$ for all $t \geq t_0$. Let (6.7.1), (6.7.2) and (6.7.27) be satisfied, the operator equation (6.1.1) have a nonempty solution set N and $\bar{x}^*$ be its minimal norm solution. Suppose that $\alpha(t)$ is a positive convex differentiable and decreasing to zero function with the property (6.7.11) and that the Cauchy problems (6.7.3), (6.7.14) and (6.7.19) have unique solutions on the interval $[t_0, \infty)$. Then the trajectory $y(t)$ of the equation (6.7.3) converges strongly to $\bar{x}^*$ as $t \to \infty$.*

3. Assume that the non-monotone perturbations $A(t) : H \to H$ of the operator A satisfy the condition

$$(A(t)x_1 - A(t)x_2, x_1 - x_2) \geq -h_1(t)\Psi(x_1, x_2) \quad \forall x_1,\, x_2 \in H, \tag{6.7.29}$$

where a continuous function $h_1(t) \geq 0$, $h_1(t) \to 0$ as $t \to \infty$, the function $\Psi(x_1, x_2)$ is non-negative continuous and bounded, i.e., it carries bounded sets to bounded sets. Then if the rest of the conditions of Theorem 6.7.1 hold and if the trajectory $y(t)$ is bounded, we obtain, in place of (6.7.25), the following inequality:

$$\frac{d\bar{p}(t)}{dt} \leq -2\alpha(t)\bar{p}(t) + c_2\Big(\delta(t) + h(t)\Big)p(t) + c_3 h_1(t) \tag{6.7.30}$$

with some constant $c_3 > 0$. Since $2ab \leq a^2 + b^2$, we have

$$
\begin{aligned}
c_2\Big(\delta(t) + h(t)\Big)p(t) &= c_2 p(t)\sqrt{\delta(t) + h(t)}\sqrt{\delta(t) + h(t)} \\
&\leq 2^{-1}c_2^2\Big(\delta(t) + h(t)\Big) + \Big(\delta(t) + h(t)\Big)\bar{p}(t).
\end{aligned}
$$

Then it follows from (6.7.30) that

$$\frac{d\bar{p}(t)}{dt} \leq -\alpha(t)\Big(2 - \frac{\delta(t) + h(t)}{\alpha(t)}\Big)\bar{p}(t) + c_4\Big(\delta(t) + h(t) + h_1(t)\Big),$$

where $c_4 = max\ \{2^{-1}c_2^2, c_3\}$. By (6.7.27), there exists $c_5 > 0$ such that

$$\frac{d\bar{p}(t)}{dt} \leq -c_5\alpha(t)\bar{p}(t) + c_4\left(\delta(t) + h(t) + h_1(t)\right).$$

Consequently, Theorem 6.7.3 is still valid for non-monotone perturbations $A(t)$ provided that (6.7.29) holds with the additional requirement

$$\lim_{t\to\infty} \frac{h_1(t)}{\alpha(t)} = 0.$$

4. Assume that the operator $A : H \to H$ satisfies the Lipschitz condition and there exists $r_0 > 0$ such that

$$(Ax - f, x) \geq 0 \quad \text{as} \quad \|x\| \geq r_0. \tag{6.7.31}$$

Then the equation (6.7.19) has a unique solution $u(t)$ on an interval $[t_0, \bar{t})$, $\bar{t} \leq +\infty$. We show that in this case $\bar{t} = +\infty$. Set the contrary assumption: $\bar{t} < +\infty$. First of all, we prove the inclusion $u(t) \in B_0(\theta_H, r_0)$. Let $u(t) \notin B_0(\theta_H, r_0)$ for $t_1 \leq t \leq t_2 < \bar{t}$ and $\|u(t_1)\| = r_0$. It is clear that $u(t) \in H \setminus B_0(\theta_H, r_0)$. Calculate the scalar product of (6.7.19) and $u(t)$. We have

$$\frac{1}{2}\frac{d\|u(t)\|^2}{dt} + (Au(t) - f, u(t)) + \alpha(t)\|u(t)\|^2 = 0.$$

In view of (6.7.31),

$$\frac{d\|u(t)\|}{dt} \leq -\alpha(t)\|u(t)\| \quad \forall t \in [t_1, t_2].$$

Now Lemma 7.2.2 yields the estimate

$$\|u(t)\| \leq \|u(t_1)\| exp\left(-\int_{t_1}^{t} \alpha(\tau)d\tau\right) < \|u(t_1)\|,$$

which contradicts the claim that $u(t) \notin B_0(\theta_H, r_0)$ as $t \in [t_1, t_2]$.

Since t_2 is arbitrary and $t_2 > t_1$, we conclude that $u(t) \in B_0(\theta_H, r_0)$ on the semi-interval $[t_1, \bar{t})$. By (6.7.19), this fact implies the boundedness of $\|u'(t)\|$ on $[t_0, \bar{t})$. Then for all $t, t' \in [t_0, \bar{t})$ one has

$$\|u(t) - u(t')\| \leq L|t - t'|, \quad L = max\{\|u'(t)\| \mid t \in [t_0, \bar{t})\}.$$

Therefore, there exists $\tilde{u}$ such that

$$\lim_{t \to \bar{t}} u(t) = \tilde{u} \in H.$$

Now we can again apply the existence theorem to (6.7.19) with the initial condition $u(\bar{t}) = \tilde{u}$ in order to be sure that $u(t)$ is defined for $t \geq \bar{t}$. The obtained contradiction proves the first claim: $\bar{t} = +\infty$. In addition, these arguments imply a boundedness of $u(t)$ when $t \geq t_0$.

Similar results can be established for the equation (6.7.14). It is possible also to prove the existence theorem of a unique bounded solution of the differential equation (6.7.3) if we make the corresponding assumptions for the perturbed operator $A(t)$.

Next we will study continuous regularization in Banach spaces.

5. Let X be a reflexive Banach space, X and X^* be strictly convex Banach spaces, X possesses an approximation, duality mapping $J : X \to X^*$ be continuous and weak-to-weak continuous. Consider the equation (6.1.1) with continuous accretive operator $A : X \to X$ and regularized differential equation (6.7.3) with continuous accretive operator $A(t) : X \to X$, with this $D(A) = D(A(t)) = X$. As in Subsection 1 of this section, we assume that solution $w(t)$ of the intermediate equation (6.7.4) is differentiable on the interval $[t_0, \infty)$.

Since (6.5.39) holds in the accretive case, an estimate like (6.7.9) is deduced in the following form:

$$\left\| \frac{dw(t)}{dt} \right\| \leq 2\|\bar{x}^*\| \frac{|\alpha'(t)|}{\alpha(t)}.$$

Let $r(t) = y(t) - w(t)$. Since

$$\frac{d\|\omega(t)\|^2}{dt} = 2\Big\langle J\omega(t), \frac{d\omega(t)}{dt} \Big\rangle,$$

we use (6.7.3) and (6.7.4) and come to the inequality

$$\frac{1}{2}\frac{d\|r(t)\|^2}{dt} \quad + \quad \langle J(y(t) - w(t)), A(t)y(t) - A(t)w(t)\rangle + \alpha(t)\|r(t)\|^2$$

$$\leq \ \Big(\delta(t) + h(t)g(2\|\bar{x}^*\|) + \Big\| \frac{dw(t)}{dt} \Big\|\Big)\|r(t)\|.$$

Therefore,

$$\frac{d\|r(t)\|}{dt} \leq -\alpha(t)\|r(t)\| + \delta(t) + h(t)g(2\|\bar{x}^*\|) + 2\|\bar{x}^*\|\frac{|\alpha'(t)|}{\alpha(t)}$$

because $A(t)$ is accretive. Then from Lemma 7.2.1 the following assertion arises:

Theorem 6.7.4 *Under the assumptions of this subsection, if the equation (6.1.1) has a nonempty solution set N and the conditions (6.7.1), (6.7.2) and (6.7.7) hold, then a solution $y(t)$ of the Cauchy problem (6.7.3) converges strongly to $\bar{x}^* \in N$ as $t \to \infty$, where $\bar{x}^*$ is a unique element satisfying (2.7.7).*

6. Suppose now that X is a uniformly convex and uniformly smooth Banach space, $A : X \to X^*$ is a monotone continuous operator, $D(A) = X$. Let $x(t)$ and $y(t)$ be functions defined on $[t_0, +\infty)$ with values in X. Introduce the Lyapunov functional

$$W(x(t), y(t)) = 2^{-1}(\|x(t)\|^2 - 2\langle Jx(t), y(t)\rangle + \|y(t)\|^2). \qquad (6.7.32)$$

Lemma 6.7.5 *Let a function $x(t)$ be continuous, $y(t)$ be differentiable and $Jx(t)$ be Gâteaux differentiable on $[t_0, +\infty)$. Then the functional (6.7.32) is differentiable and the equality*

$$\frac{dW(x(t), y(t))}{dt} = \left\langle \frac{dJx(t)}{dt}, x(t) - y(t)\right\rangle + \left\langle Jy(t) - Jx(t), \frac{dy(t)}{dt}\right\rangle$$

holds.

Proof. Since spaces X and X^* are uniformly smooth and $\|x(t)\| = \|Jx(t)\|_*$, we have

$$\frac{\partial W(x(t), y(t))}{\partial Jx} = x(t) - y(t) \qquad (6.7.33)$$

and

$$\frac{\partial W(x(t), y(t))}{\partial y} = Jy(t) - Jx(t). \qquad (6.7.34)$$

Convexity of $W(x, y)$ with respect to Jx and y (see Section 1.6) implies the following inequalities for $t_0 < s < t$:

$$W(x(t), y(t)) \geq W(x(s), y(t)) + \left\langle Jx(t) - Jx(s), \frac{\partial W(x(s), y(t))}{\partial Jx}\right\rangle$$

and

$$W(x(s), y(t)) \geq W(x(s), y(s)) + \left\langle \frac{\partial W(x(s), y(s))}{\partial y}, y(t) - y(s)\right\rangle.$$

According to (6.7.33) and (6.7.34), one gets

$$W(x(t), y(t)) \geq W(x(s), y(s)) + \langle Jx(t) - Jx(s), x(s) - y(t)\rangle$$

$$+ \langle Jy(s) - Jx(s), y(t) - y(s)\rangle$$

and

$$W(x(s), y(s)) \geq W(x(t), y(t)) + \langle Jx(s) - Jx(t), x(t) - y(s)\rangle$$

$$+ \langle Jy(t) - Jx(t), y(s) - y(t)\rangle,$$

from which it results that

$$\left\langle \frac{Jx(t) - Jx(s)}{t - s}, x(t) - y(s) \right\rangle \quad + \quad \left\langle Jy(t) - Jx(t), \frac{y(t) - y(s)}{t - s} \right\rangle$$

$$\geq \quad \frac{W(x(t), y(t)) - W(x(s), y(s))}{t - s}$$

$$\geq \quad \left\langle \frac{Jx(t) - Jx(s)}{t - s}, x(s) - y(t) \right\rangle$$

$$+ \quad \left\langle Jy(s) - Jx(s), \frac{y(t) - y(s)}{t - s} \right\rangle. \qquad (6.7.35)$$

Duality mapping J is continuous in a uniformly smooth space X. Therefore, the conditions of the lemma allow us to pass in (6.7.35) to the limit when $s \to t$. We obtain

$$\left\langle \frac{dJx(t)}{dt}, x(t) - y(t) \right\rangle \quad + \quad \left\langle Jy(t) - Jx(t), \frac{dy(t)}{dt} \right\rangle \geq \frac{dW(x(t), y(t))}{dt}$$

$$\geq \quad \left\langle \frac{dJx(t)}{dt}, x(t) - y(t) \right\rangle + \left\langle Jy(t) - Jx(t), \frac{dy(t)}{dt} \right\rangle.$$

The lemma is proved. ∎

Suppose that the equation (6.1.1) is given with perturbed date $f(t)$ and $A(t)$. Moreover, $D(A(t)) = D(A) = X$ for $t \geq t_0$ and, as before, $A(t)$ is a family of monotone and continuous operators such that

$$\|Ax - A(t)x\|_* \leq g(\|x\|)h(t) \quad \forall x \in X, \qquad (6.7.36)$$

and

$$\|f - f(t)\|_* \leq \delta(t), \qquad (6.7.37)$$

where $h(t)$, $\delta(t)$ and $g(s)$ have the same properties as in (6.7.1) and (6.7.2).

We study the Cauchy problem for the following differential equation:

$$\frac{dJy(t)}{dt} + A(t)y(t) + \alpha(t)Jy(t) = f(t), \quad t \geq t_0, \quad y(t_0) = y_0 \in X. \qquad (6.7.38)$$

We introduce the intermediate equation

$$Aw(t) + \alpha(t)Jw(t) = f. \qquad (6.7.39)$$

Theorem 6.7.6 *Suppose that the following conditions are satisfied:*
1) the equation (6.1.1) has a solution set N and $\bar{x}^* \in N$ is its minimal norm solution;
2) the solution $y(t)$ of the Cauchy problem (6.7.38) exists and is bounded for all $t \geq t_0$;
3) the inequalities (6.7.37) and (6.7.36) hold;
4) the properties (6.7.7) are fulfilled;
5) either $\delta_X(\epsilon) \geq C_1\epsilon^2$, $C_1 > 0$, and the solution $w(t)$ of the equation (6.7.39) is differentiable on the interval $[t_0, \infty)$ or

6) the equation (6.7.39) is differentiable by t in the strong sense, and there exists strictly increasing and continuous for all $\xi \geq 0$ function $\psi(\xi)$ such that $\psi(0) = 0$,

$$\left\langle \frac{dJw(t)}{dt}, \frac{dw}{dt} \right\rangle \geq \left\| \frac{dw}{dt} \right\| \psi\left(\left\| \frac{dw}{dt} \right\| \right)$$

and

$$\psi\left(\frac{|\alpha'(t)|}{\alpha(t)} \right) / \alpha(t) \to 0.$$

Then $y(t) \to \bar{x}^*$ as $t \to \infty$.

Proof. By the hypotheses, there exists a constant $R_1 > 0$ such that $\|y(t)\| \leq R_1$. In its turn, it is known from Section 2.2 that $w(t) \to \bar{x}^*$ as $t \to \infty$ and $\|w(t)\| \leq \|\bar{x}^*\|$. Rewrite (6.7.33) and (6.7.34) as

$$\frac{\partial W(y(t), w(t))}{\partial Jy} = y(t) - w(t)$$

and

$$\frac{\partial W(y(t), w(t))}{\partial w} = Jw(t) - Jy(t).$$

Calculate the following derivative:

$$\frac{dW(y(t), w(t))}{dt} = \left\langle \frac{dJy(t)}{dt}, \frac{\partial W(y(t), w(t))}{\partial Jy(t)} \right\rangle + \left\langle \frac{\partial W(y(t), w(t))}{\partial w}, \frac{dw(t)}{dt} \right\rangle$$

$$= -\langle A(t)y(t) - A(t)w(t), y(t) - w(t) \rangle$$

$$- \alpha(t)\langle Jy(t) - Jw(t), y(t) - w(t) \rangle$$

$$- \langle Aw(t) + \alpha(t)Jw(t) - f, y(t) - w(t) \rangle + \langle f(t) - f, y(t) - w(t) \rangle$$

$$+ \langle Aw(t) - A(t)w(t), y(t) - w(t) \rangle + \left\langle Jw(t) - Jy(t), \frac{dw(t)}{dt} \right\rangle.$$

By (1.6.19), we have

$$\langle Jx - Jy, x - y \rangle \geq (2L)^{-1}\delta_X(c_6^{-1}\|x - y\|), \tag{6.7.40}$$

where $1 < L < 1.7$, $c_6 = 2max\{1, R_1, \|\bar{x}^*\|\}$. Then (6.7.38), (6.7.39), (6.7.40), Lemma 6.7.5 and monotonicity of $A(t)$ yield

$$\frac{dW(y(t), w(t))}{dt} \leq -\alpha(t)(2L)^{-1}\delta_X(c_6^{-1}\|y(t) - w(t)\|)$$

$$+ \left(\delta(t) + h(t)g(\|\bar{x}^*\|) \right)\|y(t) - w(t)\| + \|Jw(t) - Jy(t)\|_* \left\| \frac{dw(t)}{dt} \right\|.$$

Estimate $\|w'(t)\|$ from above. With the help of Lemma 6.5.1, we can write

$$\|w(t_1) - w(t_2)\| \leq c_0 g_X^{-1}\left(\frac{C|\alpha(t_1) - \alpha(t_2)|}{\alpha(t_1)}\right), \quad C = 2Lc_0\|\bar{x}^*\|, \tag{6.7.41}$$

where $c_0 = 2max\{1, \|\bar{x}^*\|\}$. If 5) holds then $g_X^{-1}(\zeta) \leq C_1^{-1}\zeta$. Hence, for the differentiable function $w(t)$, there exists a constant $R_2 > 0$ such that

$$\left\|\frac{dw}{dt}\right\| \leq R_2\frac{|\alpha'(t)|}{\alpha(t)}. \tag{6.7.42}$$

The rest of the proof follows the pattern of Theorem 6.7.1.

Let now $\Delta(\tau) = 8\tau^2 + c_7\rho_X(\tau)$, where $c_7 = 8max\{L, R_1, \|\bar{x}^*\|\}$, and $\Delta^{-1}(\cdot)$ is its inverse function. By inequality (1.6.45), we then conclude that

$$\Delta^{-1}\Big(W(x(t), y(t))\Big) \leq \|x(t) - y(t)\|. \tag{6.7.43}$$

Hence, (6.7.42) and (6.7.43) lead to the final inequality

$$\begin{aligned}
\frac{dW(x(t), y(t))}{dt} \leq & \; -\alpha(t)(2L)^{-1}\delta_X\Big(c_6^{-1}\Delta^{-1}\big(W(x(t), y(t))\big)\Big) \\
& + \Big(\delta(t) + h(t)g(\|\bar{x}^*\|) + R_2\frac{|\alpha'(t)|}{\alpha(t)}\Big)(R_1 + \|\bar{x}^*\|)
\end{aligned}$$

or

$$\begin{aligned}
\frac{dW(x(t), y(t))}{dt} \leq & \; -\alpha(t)(2L)^{-1}\varphi\Big(W(x(t), y(t))\Big) \\
& + \Big(\delta(t) + h(t)g(\|\bar{x}^*\|) + R_2\frac{|\alpha'(t)|}{\alpha(t)}\Big)(R_1 + \|\bar{x}^*\|),
\end{aligned}$$

where $\varphi(\xi) = c_6^{-1}\delta_X\Big(\Delta^{-1}(\xi)\Big)$ is a positive continuous non-decreasing function for all $\xi \geq 0$ and $\varphi(0) = 0$. Then the conditions of the theorem allow us to conclude on the basis of Lemma 7.2.1 that

$$\lim_{t \to \infty} W(x(t), y(t)) = 0. \tag{6.7.44}$$

Observe, that we are not able to estimate $\|w'(t)\|$ from (6.7.41) when $\delta_X(\epsilon) < C_1\epsilon^2$. Let 6) hold. Differentiating (6.7.39) by t, we obtain

$$A'(w)\frac{dw(t)}{dt} + \alpha(t)\frac{dJw(t)}{dt} + \alpha'(t)Jw(t) = 0, \tag{6.7.45}$$

where A' is a Fréchet derivative of A. Then, by (6.7.45), the equality

$$\Big\langle A'(w)\frac{dw(t)}{dt}, z\Big\rangle + \alpha'(t)\langle Jw(t), z\rangle + \alpha(t)\Big\langle\frac{dJw(t)}{dt}, z\Big\rangle = 0$$

appears for every $z \in X$. Assuming $z = w'(t)$ in this equality and making use of Definition 1.3.8 we have

$$\alpha(t) \left\langle \frac{dJw(t)}{dt}, \frac{dw(t)}{dt} \right\rangle \leq |\alpha'(t)| \|w(t)\| \left\| \frac{dw(t)}{dt} \right\|,$$

that is,

$$\alpha(t)\psi \left(\left\| \frac{dw(t)}{dt} \right\| \right) \leq |\alpha'(t)| \|\bar{x}^*\|.$$

Consequently,

$$\left\| \frac{dw(t)}{dt} \right\| \leq \psi^{-1} \left(\|\bar{x}^*\| \frac{|\alpha'(t)|}{\alpha(t)} \right),$$

and (6.7.44) arises from Lemma 7.2.1 again. Finally, (1.6.48) implies the inequality

$$\|x - y\| \leq 2c_6 \delta_X^{-1} \left(LW(Jx, y) \right).$$

Thus, $\|y(t) - w(t)\| \to 0$ as $t \to \infty$. The proof is accomplished by (6.7.6). ∎

Remark 6.7.7 *Along with (6.7.3) and (6.7.38), we are able to study differential equations*

$$\frac{dy(t)}{dt} + A(t)y(t) + \alpha(t)(y(t) - u^0) = f(t), \quad t \geq t_0, , \quad y(t_0) = y_0$$

and

$$\frac{dJy(t)}{dt} + A(t)y(t) + \alpha(t)J(y(t) - u^0) = f(t), \quad t \geq t_0, \quad y(t_0) = y_0,$$

were u^0 is some fixed point in H and X, respectively. By simple additional algebra, one can establish convergence of $y(t)$ to unique solution $\bar{x} \in N$, such that

$$\|\bar{x} - u^0\| = min\{\|x - u^0\| \mid x \in N\}.$$

6.8   Newton–Kantorovich Regularization Method

The convergence of the Newton–Kantorovich classical approximations for the nonlinear equation (6.1.1) has been studied by many authors (see, for instance, [104]), mainly, in the case when an operator A is invertible. Some results deal with the investigations of influence of the monotonicity of A on the behavior of the Newton–Kantorovich method. For example, in [221], convergence of this method was established under the assumption that A is strongly monotone and potential. For operator A being arbitrarily monotone, the question of convergence and numerical realization of the Newton–Kantorovich algorithm was open a long time. Note that in this situation only regularizing processes constructed on the basis of discrete and continuous Newton–Kantorovich schemes enable us to prove strong convergence to a solution of the equation (6.1.1). In the present section we study iterative and continuous Newton–Kantorovich regularization methods.

1. Let X be a reflexive Banach space, X and X^* be strictly convex Banach spaces, $A : X \to X^*$ be a monotone twice Fréchet differentiable (hence, wittingly continuous) on

X operator, N be a nonempty solution set of (6.1.1), $\bar{x}^*$ be a solution of (6.1.1) with a minimal norm. Let x_n^α be a unique solution of the equation

$$Ax_n^\alpha + \alpha_n J^s x_n^\alpha = f, \tag{6.8.1}$$

where $\alpha_n > 0$, $n = 0, 1, 2, ...$, $\alpha_n \to 0$ as $n \to \infty$, $J^s : X \to X^*$ be duality mapping with the gauge function $\mu(t) = t^{s-1}$, $s \geq 2$. Then it is known (see Section 2.2) that $x_n^\alpha \to \bar{x}^*$ as $n \to \infty$. Assume that an operator J^s possesses the property: there exists $c > 0$ such that

$$\langle J^s x - J^s y, x - y \rangle \geq c\|x - y\|^s \quad \forall x, y \in X. \tag{6.8.2}$$

The Newton–Kantorovich method for equation (6.1.1) takes the following form:

$$Az_n + A'(z_n)(z_{n+1} - z_n) = f, \quad n = 0, 1, 2, ..., \tag{6.8.3}$$

where $A'(z_n)$ is a non-negative operator by reason of Definition 1.3.8. Consequently, the equation (6.8.3) is linear with respect to z_{n+1} and it belongs to the class of ill-posed problems. Including into (6.8.3) the regularizing operator connected with duality mapping J^s, we form the following equation:

$$Ax_n + A'(x_n)(x_{n+1} - x_n) + \alpha_n J^s x_{n+1} = f. \tag{6.8.4}$$

The latter equation may be considered as some generalization of the Newton–Kantorovich method for (6.8.1), however, in contrast to the classical Newton–Kantorovich method the linearization process of $J^s x$ in (6.8.4) has not been realized. This is accounted for by the fact that as $s \neq 2$ the operator $(J^s)'(x_n)$ does not have properties necessary for well-posedness of the obtained equation. We analyze the behavior of a sequence $\{x_n\}$ as $n \to \infty$. Assume that

$$\|A''(x)\| \leq \varphi(\|x\|) \quad \forall x \in X, \tag{6.8.5}$$

where $\varphi(t)$ is a non-negative and non-decreasing function for all $t \geq 0$. Using the Taylor formula (1.1.16) we have from (6.8.1),

$$\langle Ax_n, x_{n+1} - x_n^\alpha \rangle \quad + \quad \langle A'(x_n)(x_n^\alpha - x_n), x_{n+1} - x_n^\alpha \rangle$$

$$+ \quad \frac{1}{2}\langle A''(\xi_n)(x_n^\alpha - x_n)^2, x_{n+1} - x_n^\alpha \rangle$$

$$+ \quad \alpha_n \langle J^s x_n^\alpha, x_{n+1} - x_n^\alpha \rangle = \langle f, x_{n+1} - x_n^\alpha \rangle, \tag{6.8.6}$$

where $\xi_n = x_n^\alpha + \Theta(x_n - x_n^\alpha)$, $\Theta = \Theta(x_{n+1} - x_n^\alpha)$, $0 < \Theta < 1$ (see (1.1.16)). Calculating the values of the functionals, that are in both parts of equation (6.8.4), on the element $x_{n+1} - x_n^\alpha$ and subtracting the equality (6.8.6) from the obtained expression, one gets

$$\langle A'(x_n)(x_{n+1} - x_n^\alpha), x_{n+1} - x_n^\alpha \rangle \quad + \quad \alpha_n \langle J^s x_{n+1} - J^s x_n^\alpha, x_{n+1} - x_n^\alpha \rangle$$

$$- \quad \frac{1}{2}\langle A''(\xi_n)(x_n^\alpha - x_n)^2, x_{n+1} - x_n^\alpha \rangle = 0.$$

Taking into account non-negativity of the first term and the conditions (6.8.2) and (6.8.5) we come to the estimate

$$\|x_{n+1} - x_n^\alpha\| \leq \left(\frac{\varphi(r_n)}{2c\alpha_n}\right)^\tau \lambda_n^{2\tau}, \tag{6.8.7}$$

where $\lambda_n = \|x_n - x_n^\alpha\|$, $\tau = \dfrac{1}{s-1}$, $r_n \geq max\{\|x_n^\alpha\|, \lambda_n\}$. Write down (6.8.1) for $n = n+1$:

$$Ax_{n+1}^\alpha + \alpha_{n+1}J^s x_{n+1}^\alpha = f. \tag{6.8.8}$$

From (6.8.1) and (6.8.8) follows the equality

$$\langle Ax_n^\alpha - Ax_{n+1}^\alpha, x_n^\alpha - x_{n+1}^\alpha \rangle \quad + \quad \alpha_n \langle J^s x_n^\alpha - J^s x_{n+1}^\alpha, x_n^\alpha - x_{n+1}^\alpha \rangle$$

$$+ \quad (\alpha_n - \alpha_{n+1})\langle J^s x_{n+1}^\alpha, x_n^\alpha - x_{n+1}^\alpha \rangle = 0.$$

The monotonicity of A and (6.8.2) yield now the estimate

$$\|x_{n+1}^\alpha - x_n^\alpha\| \leq \left(\frac{|\alpha_n - \alpha_{n+1}|}{c\alpha_n}\right)^\tau \|x_{n+1}^\alpha\|. \tag{6.8.9}$$

Let $\|\bar{x}^*\| \leq d$. Then, by (2.2.9), $\|x_n^\alpha\| \leq \|\bar{x}^*\| \leq d$ for all $n > 0$ and $r_n \geq max\{d, \|x_n\|\}$. It is not difficult to verify that (6.8.7) and (6.8.9) imply

$$\lambda_{n+1} = \|x_{n+1} - x_{n+1}^\alpha\| \leq \|x_{n+1} - x_n^\alpha\| + \|x_n^\alpha - x_{n+1}^\alpha\|$$

$$\leq e_n^\tau \lambda_n^{2\tau} + \frac{d}{c^\tau}\left(\frac{|\alpha_n - \alpha_{n+1}|}{\alpha_n}\right)^\tau, \tag{6.8.10}$$

where $e_n = \varphi(r_n)(2c\alpha_n)^{-1}$. Further we assume that $2 \leq s < 3$ and

a) $\{\alpha_n\}$ is a monotone decreasing sequence, moreover, there exists $\sigma > 0$ such that the inequality $\alpha_{n+1} \geq \sigma\alpha_n$ holds as $n = 0, 1, \ldots$;

b) $\left(\dfrac{c_1^{\tau_1}\lambda_0}{\eta\alpha_0^{\tau_1}}\right)^\tau \leq q < 1$, $\tau_1 = \dfrac{1}{3-s}$, $\eta = \sigma^\kappa$, $\kappa = \dfrac{s-1}{(3-s)^2}$, $c_1 = \dfrac{\varphi(d+\gamma)}{2c}$,

where $\gamma > 0$ is found from the estimate

$$\eta\left(\frac{\alpha_0}{c_1}\right)^{\tau_1} \leq \gamma; \tag{6.8.11}$$

c) $\left(\dfrac{|\alpha_n - \alpha_{n+1}|}{\alpha_n^{2\tau_1}}\right)^\tau \leq \dfrac{q^{s-1} - q^2}{c_2}$, $c_2 = \dfrac{dc_1^{\tau_1}}{c^\tau \sigma^{\tau_1+\kappa}}$.

Note that from the condition a) we have $\sigma < 1$, and, hence, $\eta = \sigma^\kappa < 1$. By (6.8.11) and by the property of $\{\alpha_n\}$, it follows that

$$\eta c_1^{-\tau_1}\alpha_n^{\tau_1} \leq \gamma \quad \forall n > 0. \tag{6.8.12}$$

Therefore, the condition b) and (6.8.11) imply $\lambda_0 \leq \gamma$. Hence, $r_0 = d + \gamma$. Show that the inequality

$$\frac{c_1^{\tau_1} \lambda_n}{\eta \alpha_n^{\tau_1}} \leq q^{s-1} < 1 \qquad (6.8.13)$$

results from the assumptions a) - c) with $2 \leq s < 3$. Indeed, if $n = 0$ then (6.8.13) is true because of b). Since $\tau + \tau_1 = 2\tau\tau_1$, we deduce from (6.8.7) with $n = 0$ and from b) that

$$\frac{c_1^{\tau_1} \|x_1 - x_0^{\alpha}\|}{\eta \alpha_0^{\tau_1}} \leq \left(\frac{c_1}{\alpha_0}\right)^{2\tau\tau_1} \left(\frac{\lambda_0}{\eta}\right)^{2\tau} \leq q^2 < 1,$$

that is, $\|x_1 - x_0^{\alpha}\| < \gamma$ and $r_1 = d + \gamma$. Let then inequality (6.8.13) hold with $n = k$. Establish its validity for $n = k + 1$. On the basis of (6.8.12) and (6.8.13), it can be easily verified that $\lambda_k \leq \gamma$, that is, $r_k = d + \gamma$. Therefore (6.8.10) gives

$$\lambda_{k+1} \leq \frac{c_1^{\tau}}{\alpha_k^{\tau}} \lambda_k^{2\tau} + c_2 \left(\frac{\alpha_k - \alpha_{k+1}}{\alpha_k}\right)^{\tau}.$$

Then, by making use of a) - c) and (6.8.13) with $2 \leq s < 3$, one gets

$$\frac{c_1^{\tau_1} \lambda_{k+1}}{\eta \alpha_{k+1}^{\tau_1}} \leq \left(\frac{c_1^{\tau_1} \lambda_k}{\eta \alpha_k^{\tau_1}}\right)^{2\tau} + c_2 \left(\frac{\alpha_k - \alpha_{k+1}}{\alpha_k^{2\tau_1}}\right)^{\tau} \leq q^2 + q^{s-1} - q^2 = q^{s-1} < 1,$$

because $\tau + \tau_1 = 2\tau\tau_1$. Since $\{\alpha_n\} \to 0$ as $n \to \infty$, we have from (6.8.13) that $\lambda_n = \|x_n - x_n^{\alpha}\| \to 0$. Finally, the the strong convergence $\{x_n\}$ to $\bar{x}^*$ follows from the inequality

$$\|x_n - \bar{x}^*\| \leq \|x_n - x_n^{\alpha}\| + \|x_n^{\alpha} - \bar{x}^*\|.$$

Thus, we have proved the following theorem:

Theorem 6.8.1 *Let equation (6.1.1) have a nonempty solution set N, $\bar{x}^*$ be its solution with the minimal norm, $\|\bar{x}^*\| \leq d$. A be a twice differentiable monotone operator, $\alpha_n \to 0$ as $n \to \infty$ and the conditions (6.8.2), (6.8.5) and a) - c) be satisfied as $2 \leq s < 3$. Then a sequence $\{x_n\}$ generated by iterative process (6.8.4) strongly converges as $n \to \infty$ to $\bar{x}^* \in N$.*

Remark 6.8.2 *The class of Banach spaces, in which the property (6.8.2) takes place, is nonempty. Indeed, in the Lebesgue spaces with $1 < p \leq 2$ and $p > 2$ one can assume $s = 2$ and for $s = p$, respectively (see Section 1.6).*

Conditions a) - c) of Theorem 6.8.1 impose requirements on the choice of the initial approximation x_0 in (6.8.4) and on the choice of the sequence $\{\alpha_n\}$. Show that it is possible to achieve the fulfillment of these conditions. Let

$$\alpha_n = \frac{1}{(m+n)^{\beta}}, \qquad 0 < \beta < \frac{1}{2\tau_1 - 1}, \qquad (6.8.14)$$

where m is a fixed positive number. Let us choose $\sigma < 1$, $q < 1$ and γ. Choose $m > 0$ such that

$$\alpha_0 = \left(\frac{\gamma}{\eta}\right)^{1/\tau_1} c_1$$

(see the condition (6.8.11)). If b) is satisfied then we do not change γ. Otherwise, we choose γ such that

$$\frac{\lambda_0}{\gamma} \leq q^{1/\tau},$$

and then b) holds. It is not difficult to be verified that

$$\frac{\alpha_n - \alpha_{n+1}}{\alpha_n^{2\tau_1}} \to 0 \quad as \quad n \to \infty,$$

if β is defined by (6.8.14). We are able to take m large enough, that c) holds for all $n \geq 0$. Then for every initial approximation x_0 the conditions of Theorem 6.8.1 are satisfied.

2. Suppose that, in place of f and A, the sequences of perturbed data $\{f_n\}$ and $\{A_n\}$ are known such that for all $n \geq 0$ elements $f_n \in X^*$ and operators $A_n : X \to X^*$ have the same properties as A, and then

$$\|f - f_n\|_* \leq \delta_n$$

and

$$\|A_n x - Ax\|_* \leq g(\|x\|)h_n \quad \forall x \in X,$$

where $g(s)$ is a non-negative continuous function for all $s \geq 0$.

Define approximations x_n by the equation

$$A_n x_n + (A_n)'(x_n)(x_{n+1} - x_n) + \alpha_n J^s x_{n+1} = f_n, \tag{6.8.15}$$

and the intermediate regularized equation as follows:

$$A_n x_n^\alpha + \alpha_n J^s x_n^\alpha = f_n.$$

Under these conditions, the validity of Theorem 6.8.1 can be established if there holds the additional relation

$$\lim_{n \to \infty} \frac{\delta_n + h_n}{\alpha_n} = 0.$$

3. If X is a Hilbert space H, then $s = 2$, $\tau = \tau_1 = \kappa = c = 1$, $\eta = \sigma$, J^s is the identity operator I, and the Newton$-$Kantorovich regularization method (6.8.4) has the form:

$$Ax_n + A'(x_n)(x_{n+1} - x_n) + \alpha_n x_{n+1} = f. \tag{6.8.16}$$

Besides, another sort of the Newton$-$Kantorovich regularization methods can be constructed on the basis of the classical Newton$-$Kantorovich scheme applied to the following regularization problem:

$$Ax + \alpha_n Sx = f,$$

where $S : H \to H$ is a twice differentiable operator such that

$$(Sx - Sy, x - y) \geq \|x - y\|^2.$$

In this case, approximations x_n are found from the equation

$$Ax_n + \alpha_n Sx_n + (A'(x_n) + \alpha_n S'(x_n))(x_{n+1} - x_n) = f. \tag{6.8.17}$$

If S is the identity operator I, then $S' = I$ and (6.8.17) coincides with (6.8.16). Let

$$\|S''(x)\| \leq \varphi_1(\|x\|),$$

the function $\varphi_1(t)$ be of the same class as $\varphi(t)$. Then sufficient conditions for the convergence of the iterative process (6.8.17) are given by Theorem 6.8.1 with $s = 2$.

4. Let $A : H \to H$ be a monotone Fréchet differentiable operator, the equation (6.1.1) have a nonempty solution set N, $\bar{x}^* \in N$ be the minimal norm solution. We present the Newton−Kantorovich iterative regularization method (6.8.16) with inexact right-hand side, namely,

$$Ax_n + A'(x_n)(x_{n+1} - x_n) + \alpha_{n+1}x_{n+1} = f_n.$$

It is obvious that its continuous analogue can be written as a differential equation

$$Ax(t) + A'(x(t))x'_t(t) + \alpha(t)x'(t) + \big(\alpha'(t) + \alpha(t)\big)x(t) = f(t), \quad t \geq t_0 \geq 0, \qquad (6.8.18)$$

with the initial condition

$$x(t_0) = x_0, \quad x_0 \in H. \qquad (6.8.19)$$

We assume that a function $\alpha(t)$ is positive continuously differentiable for all $t \geq t_0$, continuous function $f(t)$ is a certain approximation of the right-hand side f in (6.1.1) such that (6.7.2) holds. In addition, a function $\delta(t)$ is continuous and non-negative and

$$\int_{t_0}^{\infty} \delta(\tau)d\tau \neq 0. \qquad (6.8.20)$$

It is easy to see that (6.8.18) can be rewritten in the equivalent form as

$$\big(Ax(t) + \alpha(t)x(t)\big)'_t + Ax(t) + \alpha(t)x(t) = f(t).$$

Denote $v(t) = Ax(t) + \alpha(t)x(t)$. Then we obtain the following Cauchy problem:

$$\frac{dv(t)}{dt} + v(t) = f(t), \quad v(t_0) = Ax_0 + \alpha_0 x_0, \quad \alpha_0 = \alpha(t_0), \qquad (6.8.21)$$

which has a unique solution $v(t)$ on any finite interval $[t_0, T]$, $T > t_0$. In this situation, the solution of problem (6.8.18), (6.8.19) can be uniquely defined by the formula

$$x(t) = \big(A + \alpha(t)I\big)^{-1}v(t) \quad \forall t \in [t_0, T].$$

Solving the linear equation (6.8.21) we have

$$Ax(t) + \alpha(t)x(t) = exp(-t)\Big(\int_{t_0}^{t} f(\tau)exp(\tau)d\tau + (Ax_0 + \alpha_0 x_0)exp(t_0)\Big). \qquad (6.8.22)$$

Further, for each $t \geq t_0$ we consider in H the operator equation

$$Ax_\alpha(t) + \alpha(t)x_\alpha(t) = f(t). \qquad (6.8.23)$$

It is known (see Section 2.1) that if

$$\lim_{t \to \infty} \alpha(t) = 0, \quad \lim_{t \to \infty} \frac{\delta(t)}{\alpha(t)} = 0, \tag{6.8.24}$$

then $\lim_{t \to \infty} \|x_\alpha(t) - \bar{x}^*\| = 0$, where $\bar{x}^*$ is the solution of equation (6.1.1) with minimal norm. Let

$$\beta_0 = \|Ax_0 + \alpha_0 x_0\|, \quad \sigma(t) = \int_{t_0}^t \delta(\tau) exp(\tau) d\tau.$$

From (6.8.22) and (6.8.23), it appears that

$$(Ax(t) - Ax_\alpha(t), x(t) - x_\alpha(t)) + \alpha(t)\|x(t) - x_\alpha(t)\|^2$$

$$= \Big(exp(-t)\Big[\int_{t_0}^t f(\tau) exp(\tau) d\tau + (Ax_0 + \alpha_0 x_0) exp(t_0)\Big] - f(t), x(t) - x_\alpha(t) \Big).$$

The monotonicity property of A implies

$$\|x(t) - x_\alpha(t)\| \le \gamma_1(t) + \gamma_2(t),$$

where

$$\gamma_1(t) = \beta_0 exp(t_0 - t)\alpha^{-1}(t)$$

and

$$\gamma_2(t) = \|exp(-t)\int_{t_0}^t f(\tau) exp(\tau) d\tau - f(t)\|\alpha^{-1}(t).$$

Suppose that

$$\lim_{t \to \infty} \frac{exp(-t)\sigma(t)}{\alpha(t)} = 0. \tag{6.8.25}$$

The following should be noted. Since $\sigma(t) > 0$ for sufficiently large t, it follows from (6.8.25) that

$$\lim_{t \to \infty} \frac{exp(-t)}{\alpha(t)} = 0. \tag{6.8.26}$$

Thus, $\gamma_1(t)$ is infinitely small as $t \to \infty$. Moreover,

$$\gamma_2(t) = \|exp(-t)\int_{t_0}^t (f(\tau) - f) exp(\tau) d\tau - exp(t_0 - t)f + f - f(t)\|\alpha^{-1}(t)$$

$$\le \Big(exp(-t)\sigma(t) + exp(t_0 - t)\|f\| + \delta(t)\Big)\alpha^{-1}(t).$$

It shows that $\gamma_2(t) \to 0$ as $t \to \infty$ provided that the conditions (6.8.24) - (6.8.26) are satisfied. This proves the following theorem:

Theorem 6.8.3 *Suppose $A : H \to H$ is a Fréchet differentiable monotone operator, equation (6.1.1) has a nonempty solution set in H, for $t \ge t_0$ the function $\alpha(t)$ is positive and continuously differentiable, $\delta(t)$ is non-negative and continuous, $f(t) : [t_0, \infty) \to H$ is continuous, and the conditions (6.7.2), (6.8.20), (6.8.24) and (6.8.25) are satisfied. Then $x(t) \to \bar{x}^*$ as $t \to \infty$, where $x(t)$ is the unique solution of the Cauchy problem (6.8.18), (6.8.19).*

Note that the class of functions satisfying (6.8.24) and (6.8.25) is not empty. For example, we could have $\alpha(t) = (t + \lambda)^{-\gamma}$, $\delta(t) = exp(-ct)$, $\gamma > 0$, $c > 0$, $\lambda > 0$ for $t_0 = 0$ and $\lambda \geq 0$ for $t_0 > 0$.

Let the operator A in (6.1.1) be given with an error, the approximations

$$A(t, x) : [t_0,\ \infty) \times H \to H$$

being monotone with respect to the second argument be differentiable in the strong sense with respect to each argument, and

$$\|A(t, x) - Ax\| \leq g(\|x\|)h(t) \quad \forall t \geq t_0, \quad \forall x \in H,$$

here $h(t)$ and $g(t)$ are non-negative continuous functions. In this situation, we replace the equation (6.8.18) by the following:

$$A'_t(t, x(t)) + A'_x(t, x(t))x'_t(t) + (\alpha'(t) + \alpha(t))x(t) + \alpha(t)x'_t(t) + A(t, x(t)) = f(t)$$

or

$$\Big(A(t, x(t)) + \alpha(t)x(t)\Big)'_t + A(t, x(t)) + \alpha(t))x(t) = f(t). \tag{6.8.27}$$

The solution of Cauchy problem (6.8.27), (6.8.19) converges to the solution $\bar{x}^*$ of equation (6.1.1) as $t \to \infty$, if to the conditions of Theorem 6.8.3 the following relation is added:

$$\lim_{t \to \infty} \frac{h(t)}{\alpha(t)} = 0.$$

In conclusion, we provide discrete and continuous schemes for the regularized Gauss–Newton method which were studied in [41] and [3] for the operator equation $Ax = 0$ in a Hilbert space.

1. Discrete scheme:

$$x_{n+1} = x_n - \Big(A'^*(x_n)A'(x_n) + \alpha_n I\Big)^{-1}\Big(A'^*(x_n)A(x_n) + \alpha_n(x_n - z^0)\Big),$$

where $A'(x_n) = A'_x(x_n)$, $A'^*(x_n) = A'^*_x(x_n)$ and z^0 is some element of H.

2. Continuous scheme:

$$\frac{dx}{dt} = -\Big(A'^*(x(t))A'(x(t)) + \alpha(t)I\Big)^{-1}\Big(A'^*(x(t))A(x(t)) + \alpha(t)(x(t) - z^0)\Big).$$

Bibliographical Notes and Remarks

The classical quasi-solution method and residual method for potential equations are due to Ivanov and Liskovets [97, 98, 99, 130]. The convergence conditions of the residual method for a monotone potential and weakly closed operator A with a non-convex, generally speaking, set $\mathcal{M}^\delta$ have been obtained in [183]. However, the requirement of the sequential weak closedness of a nonlinear mapping is quite strong. Therefore, in a number of works, the operator conditions have been essentially weakened. So, for a potential operator A the residual method has been studied rather fully in [130]. Modifications of these methods for monotone problems have been presented in [202]. Certainly, problem (6.1.1) can be solved by the operator regularization method described in Chapters 2 and 3 as well, but interest in the residual method is connected with extension of tools for numerical solutions of ill-posed problems, taking into account availability of different initial information about the problem. The connection of the quasi-solution and residual methods with the regularization methods was also established in [195].

Strong convergence and stability of the penalty method were proved in [15]. The proximal point method was studied in [181, 184, 204, 212, 227]. Another approach is developed in [105]. The results of Section 6.5 concerning iterative regularization of monotone and accretive equations in Banach spaces first appeared in [10]. The case of Hilbert spaces was earlier considered by Bakushinskii [40] and Bruck [62]. The proof of Lemma 6.6.1 and Theorems 6.6.2, 6.6.4 can be found in [11, 21]. The other algorithms of the iterative regularization are studied in [20, 23]. The special iterations for ill-posed problems are also described in [93]. The continuous regularization method of the first order in Hilbert and Banach spaces was investigated in [14, 31, 36]. The linear case was earlier studied in [4]. The high order methods are considered, for instance, in [59, 60, 124]. The convergence of the regularized Newton−Kantorovich in the iterative form was proved in [194] and in the differential form in [200]. Many results of Section 6.8 can be transferred to variational inequalities.

Chapter 7

APPENDIX

7.1 Recurrent Numerical Inequalities

Lemma 7.1.1 *Let $\{\lambda_n\}$ and $\{\gamma_n\}$ be sequences of non-negative real numbers, $\{\alpha_n\}$ be a sequence of positive real numbers such that*

$$\lambda_{n+1} \leq \lambda_n - \alpha_n \lambda_n + \gamma_n \quad \forall\, n \geq 0,$$

$$\frac{\gamma_n}{\alpha_n} \leq c_1 \quad \text{and} \quad \alpha_n \leq \alpha. \tag{7.1.1}$$

Then $\lambda_n \leq max\{\lambda_0, K_*\}$, where $K_* = (1 + \alpha)c_1$.

Proof. Similarly to the proof of Lemma 7.1.3, consider the following alternative for all $n \geq 0$: either

$$H_1 : \lambda_n \leq \frac{\gamma_n}{\alpha_n}$$

or

$$H_1 : \lambda_n > \frac{\gamma_n}{\alpha_n}.$$

The hypothesis H_1 gives the estimate $\lambda_n \leq c_1$. In turn, the hypothesis H_2 implies $\lambda_{n+1} < \lambda_n$. At intermediate indexes we have

$$\lambda_{n+1} \leq \lambda_n + \gamma_n \leq c_1 + \alpha_n c_1 \leq (1 + \alpha)c_1.$$

From this the claim follows. ∎

Lemma 7.1.2 *Let $\{\lambda_n\}$ and $\{\gamma_n\}$ be sequences of non-negative real numbers, $\{\alpha_n\}$ be a sequence of positive real numbers satisfying the inequality*

$$\lambda_{n+1} \leq \lambda_n - \alpha_n \lambda_n + \gamma_n \quad \forall n \geq 0, \quad \alpha_n \leq 1, \tag{7.1.2}$$

where $\alpha_n \to 0$ as $n \to \infty$,

$$\sum_{n=0}^{\infty} \alpha_n = \infty, \tag{7.1.3}$$

$$\lim_{n \to \infty} \frac{\gamma_n}{\alpha_n} = 0. \tag{7.1.4}$$

Then

$$\lim_{n \to \infty} \lambda_n = 0. \tag{7.1.5}$$

Proof. From (7.1.2) it is obvious that

$$\lambda_{n+1} \leq \prod_{i=0}^{n}(1 - \alpha_i)\lambda_0 + \sum_{i=0}^{n} \gamma_i \prod_{k=i+1}^{n}(1 - \alpha_k). \tag{7.1.6}$$

Since

$$\prod_{i=0}^{n}(1 - \alpha_i) \leq exp\left(-\sum_{i=0}^{n} \alpha_i\right),$$

then by virtue of the condition (7.1.3), the first term in the right-hand side of (7.1.6) tends to zero as $n \to \infty$. Further,

$$\sum_{i=0}^{n} \gamma_i \prod_{k=i+1}^{n}(1 - \alpha_k) \quad \leq \quad \sum_{i=0}^{n} \gamma_i exp\left(-\sum_{k=i+1}^{n} \alpha_k\right)$$

$$= \quad \sum_{i=0}^{n} \gamma_i \frac{exp\left(\sum_{k=0}^{i} \alpha_k\right)}{exp\left(\sum_{k=0}^{n} \alpha_k\right)}. \tag{7.1.7}$$

Applying the Stolz theorem (see [82]) we obtain

$$\lim_{n \to \infty} \sum_{i=0}^{n} \gamma_i \frac{exp\left(\sum_{k=0}^{i} \alpha_k\right)}{exp\left(\sum_{k=0}^{n} \alpha_k\right)} \quad = \quad \lim_{n \to \infty} \frac{\gamma_{n+1} exp\left(\sum_{k=0}^{n+1} \alpha_k\right)}{exp\left(\sum_{k=0}^{n+1} \alpha_k\right) - exp\left(\sum_{k=0}^{n} \alpha_k\right)}$$

$$= \quad \lim_{n \to \infty} \frac{\gamma_{n+1}}{1 - exp\left(-\alpha_{n+1}\right)}$$

$$= \quad \lim_{n \to \infty} \frac{\gamma_{n+1}}{\alpha_{n+1}}.$$

Then (7.1.5) follows from (7.1.4), (7.1.6) and (7.1.7). ∎

We prove more a general statement.

Lemma 7.1.3 *If sequences of non-negative real numbers $\{\lambda_n\}$ and $\{\gamma_n\}$ and a bounded sequence $\{\rho_n\}$ of positive numbers satisfy the inequality*

$$\lambda_{n+1} \leq \lambda_n - \rho_n \Psi(\lambda_n) + \gamma_n, \quad n \geq 0, \tag{7.1.8}$$

where $\Psi(t)$ is a continuous increasing function, $\Psi(0) = 0$,

$$\sum_{n=0}^{\infty} \rho_n = \infty, \quad \lim_{n \to \infty} \frac{\gamma_n}{\rho_n} = 0,$$

then

$$\lim_{n \to \infty} \lambda_n = 0.$$

Proof. Consider the following alternative for all $n \geq 0$: either

$$H_1 : \Psi(\lambda_n) \leq \left(\sum_{i=0}^{n} \rho_i \right)^{-1} + \frac{\gamma_n}{\rho_n}$$

or

$$H_2 : \Psi(\lambda_n) > \left(\sum_{i=0}^{n} \rho_i \right)^{-1} + \frac{\gamma_n}{\rho_n}.$$

Introduce sets I_1 and I_2 as the totalities of numbers $n \geq 0$ such that the hypotheses H_1 and H_2 hold for all $n \in I_1$ and $n \in I_2$, respectively. It is clear that a union of these sets is a set of all positive integers. Show that I_1 is infinite. Indeed, assuming the contrary, it is not difficult to be sure that there exists $N_0 \geq 0$ such that for all $n \geq N_0$,

$$\lambda_n \leq \lambda_{n-1} - \frac{\rho_{n-1}}{\sum_{i=0}^{n-1} \rho_i}$$

or

$$\lambda_n \leq \lambda_{N_0} - \sum_{j=N_0}^{n-1} \frac{\rho_j}{A_j},$$

where

$$A_j = \sum_{i=0}^{j} \rho_i.$$

By the Abel–Dini test [82],

$$\sum_{j=0}^{\infty} \frac{\rho_j}{A_j} = \infty.$$

Therefore, beginning with some n all λ_n becomes negative which should not be because every $\lambda_n \geq 0$.

Assume $I_1 = \{n_1, ..., n_l, ...\}$ and consider two following cases:

1) $n_1 = 0$. It is obvious that on an arbitrary interval $\tilde{I}_l = [n_l, n_{l+1}]$, where $n_{l+1} > n_l + 1$,

$$\lambda_{n_l+1} \leq \lambda_{n_l} + \gamma_{n_l} \leq \Psi^{-1} \left(\frac{1}{A_{n_l}} + \frac{\gamma_{n_l}}{\rho_{n_l}} \right) + \gamma_{n_l}.$$

At the same time, for all $n_l < n < n_{l+1}$ we have

$$\lambda_n \leq \lambda_{n_l+1} - \sum_{i=n_l+1}^{n-1} \frac{\rho_i}{A_i} < \lambda_{n_l+1}.$$

By virtue of unboundedness of the set I_1, it results that $\lim_{n \to \infty} \lambda_n = 0$.

2) $n_1 > 1$. It is clear that for $n > n_1$ it is possible to use the previous reasoning, while $\{1, 2, ..., n_1 - 1\} \subset I_2$. Hence,

$$\lambda_{n+1} \leq \lambda_0 - \sum_{j=0}^{n} \frac{\rho_j}{A_j} \leq \lambda_1, \quad 1 \leq n \leq n_1 - 1,$$

and the equality $\lim_{n \to \infty} \lambda_n = 0$ holds again. ∎

7.2   Differential Inequality

Next we study the asymptotic behavior of solutions to the ordinary differential inequalities.

Lemma 7.2.1 *Let a non-negative function $\lambda(t)$ satisfy the differential inequality*

$$\frac{d\lambda(t)}{dt} \leq -\alpha(t)\psi(\lambda(t)) + \gamma(t), \quad \lambda(t_0) = \lambda_0, \quad t \geq t_0, \tag{7.2.1}$$

where a function $\alpha(t)$ is continuous positive for $t \geq t_0$, $\gamma(t)$ is continuous non-negative, $\psi(\lambda)$ is positive continuous and non-decreasing for $\lambda > 0$, $\psi(0) = 0$. Moreover, let

$$\int_{t_0}^{\infty} \alpha(\tau)d\tau = \infty \tag{7.2.2}$$

and

$$\lim_{t \to \infty} \frac{\gamma(t)}{\alpha(t)} = 0. \tag{7.2.3}$$

Then $\lambda(t) \to 0$ as $t \to \infty$.

Proof. Consider the following alternative:

$$H_1 : \psi(\lambda(t)) < q(t);$$

$$H_2 : \psi(\lambda(t)) \geq q(t),$$

where

$$q(t) = \Big(\int_{t_0}^{t} \alpha(\tau)d\tau \Big)^{-1} + \frac{\gamma(t)}{\alpha(t)}.$$

Define the sets

$$T_1^i = \{t_0 \leq t \in (t_i, \bar{t}_i) \subseteq R_+^1 \mid H_1 \text{ is true}\}, \quad T_1 = \cup_i T_1^i, \tag{7.2.4}$$

$$T_2^j = \{t_0 \leq t \in [t_j, \bar{t}_j] \subseteq R_+^1 \mid H_2 \text{ is true}\}, \quad T_2 = \cup_j T_2^j. \tag{7.2.5}$$

It is easy to see that $T = T_1 \cup T_2 = [t_0, \infty)$. Prove that T_1 is an unbounded set. For that assume the contrary. Then there exists $t = \tau_1$ such that for all $t \geq \tau_1$ the hypothesis H_2 holds, and (7.2.1) yields the inequality

$$\frac{d\lambda(t)}{dt} \leq -\frac{\alpha(t)}{\int_{t_0}^{t} \alpha(\tau)d\tau}, \quad t \geq \tau_1. \tag{7.2.6}$$

Hence,

$$\lambda(t) \leq \lambda(\tau_1) - \int_{\tau_1}^{t} \frac{\alpha(\tau)}{S(\tau)}d\tau, \tag{7.2.7}$$

where

$$S(\tau) = \int_{t_0}^{\tau} \alpha(s)ds > 0.$$

It is obvious that

$$\int_{\tau_1}^{t} \frac{\alpha(t)}{S(t)} dt = \ln S(t) - \ln S(\tau_1) \to \infty \quad as \quad t \to \infty.$$

Then we deduce from the inequality (7.2.7) that there exists a point $t = \tau_2$, for which $\lambda(\tau_2) < 0$. This contradicts the conditions of the lemma. By (7.2.2) and (7.2.3), the positive function $\psi(\lambda(t)) \to 0$ as $t \to \infty$ and $t \in T_1$. Now the convergence of $\lambda(t)$ to zero as $t \in T_1$ and $t \to \infty$ is guaranteed due to the properties of $\psi(t)$. At the same time the function $\lambda(t)$ decreases on sets T_2^j because of (7.2.6). Thus, the lemma is proved. ∎

Lemma 7.2.2 *Let a non-negative function $\lambda(t)$ satisfy the differential inequality*

$$\frac{d\lambda(t)}{dt} \le -\alpha(t)\lambda(t) + \gamma(t), \quad t \ge t_0, \tag{7.2.8}$$

where $\alpha(t)$ is positive and $\gamma(t)$ are non-negative continuous functions on the interval $[t_0, \infty)$. Then the inequality

$$\lambda(t) \le \lambda(t_0) exp\left(-\int_{t_0}^{t} \alpha(s)ds\right) + \int_{t_0}^{t} \gamma(\theta) exp\left(-\int_{\theta}^{t} \alpha(s)ds\right) d\theta \tag{7.2.9}$$

holds. If (7.2.2) and (7.2.3) are satisfied, then $\lambda(t) \to 0$ as $t \to \infty$.

Proof. Multiplying both parts of (7.2.8) by $z(t) = exp\left(\int_{t_0}^{t} \alpha(s)ds\right)$ we obtain

$$\frac{d}{dt}\left(\lambda(t)z(t)\right) \le \gamma(t)z(t).$$

Then

$$\lambda(t)z(t) \le \lambda(t_0) + \int_{t_0}^{t} \gamma(\tau)z(\tau)d\tau,$$

that is equivalent to (7.2.9). The first term in the right-hand side of (7.2.9) tends to zero by the equality (7.2.2). We find the limit of the second term as $t \to \infty$. Denote the anti-derivative of $\alpha(t)$ by $\bar{\alpha}(t)$. If the integral

$$\int_{t_0}^{\infty} \gamma(\theta)e^{\bar{\alpha}(\theta)} d\theta \tag{7.2.10}$$

is divergent then applying L'Hopital's rule and (7.2.3), one gets

$$\lim_{t\to\infty} \int_{t_0}^{t} \gamma(\theta)e^{-\int_{t_0}^{t} \alpha(s)ds} d\theta = \lim_{t\to\infty} \frac{\int_{t_0}^{t} \gamma(\theta)e^{\bar{\alpha}(\theta)} d\theta}{e^{\bar{\alpha}(t)}}$$

$$= \lim_{t\to\infty} \frac{\gamma(t)}{\alpha(t)} = 0. \tag{7.2.11}$$

If the integral (7.2.10) is convergent, then (7.2.11) holds in view of the equality (7.2.2) again. The lemma is proved. ∎

Bibliographical Notes and Remarks

Lemmas 7.1.2, 7.1.3 and 7.1.1 were proved in [9, 29]. Lemmas 7.2.1 and 7.2.2 can be found in [4, 11]. The considerable part of the book [229] deals with the linear recurrent numerical inequalities and their applications.

BIBLIOGRAPHY

[1] A.A. Abramov and A.N. Gaipova, The existence of solutions of certain equations that contain monotone discontinuous transformations, *Zh. Vychisl. Mat. i Mat. Fiz.*, 12 (1972), 525-528.

[2] S. Adli, D. Goeleven, and M. Thera, Recession mappings and noncoercive variational inequalities, *Nonl. Anal., Theory, Meth. and Appl.*, 26 (1996), 1573-1603.

[3] R.G. Airapetyan, A.G. Ramm, and A.B. Smirnova, Continuous methods for solving nonlinear ill-posed problems, *Operator Theory and Applications*, Amer. Math. Soc., Fields Institute Communications, Providence, RI, 2000, 111-138.

[4] Ya.I. Alber, A continuous regularization of linear operators equations in Hilbert spaces, *Mat. Zametki*, 9 (1968), 42-54.

[5] Ya.I. Alber, The solution of nonlinear equations with monotone operators in Banach spaces, *Siberian Math. J.*, 16 (1975), 1-8.

[6] Ya.I. Alber, The solution by the regularization method of operator equations of the first kind with accretive operators, *Differential Equations*, 11 (1975), 1665-1670.

[7] Ya.I. Alber, The solution of nonlinear equations with monotone operators on sets of Banach space, *Differential Equations*, 13 (1977), 1300-1303.

[8] Ya.I. Alber, The monotonicity method and the approximate computation of the value of the nonlinear unbounded operator, *Siberian Math. J.*, 19 (1978), 179-183.

[9] Ya.I. Alber, The solution of equations and variational inequlities with maximal monotone operators, *Soviet Math. Dokl.*, 20 (1979), 871-876.

[10] Ya.I. Alber, Itertive regularization in Banach spases, *Soviet Math. (Iz. VUZ)*, **30** (1986), 1-8.

[11] Ya.I. Alber, *"Methods for Solving Nonlinear Operator Equations and Variational Inequalities in Banach Spaces"*, D.Sc. Thesis, Gorky, 1986.

[12] Ya.I. Alber, The regularization method for variational inequalities with nonsmooth unbounded operators in Banach space, *Appl. Math. Lett.*, 6 (1993), 63-68.

[13] Ya.I. Alber, Generalized projection operators in Banach spaces: properties and applications, *Funct. Differential Equations, Proceedings of the Israel Seminar*, 1 (1994), 1-21.

[14] Ya.I. Alber, A new approach to investigation of evolution differential equations in Banach space, *Nonl. Anal., Theory, Meth. and Appl.*, 23 (1994), 1115-1134.

[15] Ya.I. Alber, On the penalty method for variational inequalities with nonsmooth unbounded operators in Banach space, *Numer. Funct. Anal. and Optim.*, 16 (1995), 1111-1125.

[16] Ya.I. Alber, Metric and generalized projection operators in Banach spaces: properties and applications, *Theory and Applications of Nonlinear Operators of Accretive and Monotone Type (A. Kartsatos, Ed.)*, 15-50, Marcel Dekker, inc., 1996, 15-50.

[17] Ya.I. Alber, D. Butnariu, and G. Kassay, Convergence and stability of a regularization method for maximal monotone inclusions and its applications to convex optimization, *Variational Analysis and Applications (F. Giannessi and A. Maugeri, Eds.)*, 1-44. Kluwer Acad. Publ., Dordrecht, 2004.

[18] Ya.I. Alber, D. Butnariu, and I. Ryazantseva, Regularization methods for ill-posed inclusions and variational inequalities with domain perturbations. *J. Nonlinear and Convex Analysis*, 2 (2001), 53-79.

[19] Ya.I. Alber, D. Butnariu, and I. Ryazantseva, Regularization of monotone variational inequalities with Mosco approximations of the constraint sets. *Set-Valued Analysis*, 13 (2005), 265-290.

[20] Ya.I. Alber, D. Butnariu, and I. Ryazantseva, Regularization and resolution of monotone variational inequalities with operators given by hypomonotone approximations, *J. Nonlinear and Convex Analysis*, 6 (2005), 23-53.

[21] Ya.I. Alber, A. Kartsatos, and E. Litsyn, Iterative solution of unstable variational inequalities on approximately given sets, *Abstr. Appl. Anal.*, 1 (1996), 45-64.

[22] Ya.I. Alber and O.A. Liskovets, The principle of the smoothing functional for solution of equations of the first kind with monotone operators, *Differential Equations*, 20 (1984), 603-608.

[23] Ya.I. Alber and M. Nashed, Iterative-projection regularization of unstable variational inequalities, *Analysis*, 24 (2004), 19-39.

[24] Ya.I. Alber and A.I. Notik, Iterative processes in Orlicz spaces, *Methods of Optimization and Operation Research*, 1984, 114-123.

[25] Ya.I. Alber and A.I. Notik, Geometric properties of Banach spaces and approximate methods for solving nonlinear operator equations, *Soviet Math. Dokl.*, 29 (1984), 611-615.

[26] Ya.I. Alber and A.I. Notik, Parallelogram inequalities in Banach spaces and some properties of the duality mapping, *Ukrainian Math. J.*, 40 (1988), 650-652.

[27] Ya.I. Alber and A.I. Notik, Perturbed unstable variational inequalities with unbounded operator on approximately given sets, *Set-Valued Anal.*, 1 (1993), 393-402.

[28] Ya.I. Alber and A.I. Notik, On some estimates for projection operator in a Banach space, *Commun. Appl. Nonl. Anal.*, 2 (1995), 47-56.

[29] Ya.I. Alber and S. Reich, An iterative method for solving a class of nonlinear operator equations in Banach spaces, *Panamer. Math. J.*, 4 (1994), 39-54.

[30] Ya.I. Alber, S. Reich, and I. Ryazantseva. Nonlinear problems with $d-$accretive operators, *Preprint*, 2003.

[31] Ya.I. Alber and I.P. Ryazantseva, Minimization of convex functionals, *Proceeding of the VI Conference on Extremal Problems and Their Applications*, Tallin, 1973.

[32] Ya.I. Alber and I.P. Ryazantseva, Regularization of nonlinear equations with monotone operators. *USSR Comput. Math. and Math. Phys.*, 15 (1975), 1-7.

[33] Ya.I. Alber and I.P. Ryazantseva, The principle of the residual in nonlinear problems with monotone discontinuous mappings as a regularizing algorithm. *Soviet Math. Dokl.*, 19 (1978), 437-440.

[34] Ya.I. Alber and I.P. Ryazantseva, The solution of nonlinear problems with monotone discontinuous mappings. *Differential Equations*, 15 (1979), 228-237.

[35] Ya.I. Alber and I.P. Ryazantseva, Variational inequalities with discontinuous monotone operators. *Soviet Math. Dokl.*, 25 (1982), 206-210.

[36] Ya.I. Alber and I.P. Ryazantseva, On regularizired evolution equations with operators of monotone type. *Funct. Differential Equations*, 7 (2000), 177-187.

[37] Yu.T. Antohin, Ill-posed problems in Hilbert space and stable methods of their solution, *Differential Equations*, 3 (1967), 1135-1156.

[38] E. Asplund, Positivity of duality mappings, *Bull. Amer. Math. Soc.*, 73 (1967), 200-203.

[39] A.B. Bakushinskii, Regularization algorithms for linear equations with unbounded operators, *Soviet Math. Dokl.*, 9 (1968), 1298-1300.

[40] A.B. Bakushinskii, Methods for solution of monotone variational inequalities that are based on the principle of itertive regularization, *USSR Comput. Math. and Math. Phys.*, 16 (1976), 1350-1362.

[41] A.B. Bakushinskii, The problem of the convergence of the iterative regularized Gauss-Newton method, *Comput. Math. and Math. Phys.*, 32 (1992), 1353-1359.

[42] A.B. Bakushinskii and A.G. Goncharskii, *Ill-Posed Problems, Numerical Methods and Applications*, Moskow University Publishers, 1989.

[43] A.B. Bakushinskii and A.G. Goncharskii, *Ill-Posed Problems: Theory and Applications*, Kluwer Acad. Publ., Dordrecht, 1994.

[44] L. Bers, F. John, and M. Schechter, *Partial Differential Equations*, Interscience, London, 1964.

[45] Y. Binyamini and J. Lindenstrauss, *Geometric Nonlinear Functional Analysis*, Amer. Math. Soc., Providence, RI, 2000.

[46] J.F. Bonnans and A. Shapiro, *Perturbation Analysis of Optimization Problems*, Springer Verlag, New York, 2000.

[47] H. Brézis, Équations et inéquations non-linéaires dans les espaces véctoriels en dualité, *Ann. Institut Fourier Grenoble* 18 (1968), 115-176.

[48] H. Brézis, *Opérateurs maximaux monotones*, North-Holland, Amsterdam, 1973.

[49] H. Brézis, *Opérateurs maximaux monotones et semi-groupes de contractions dans les espaces de Hilbert*, Math. Studies, b. 5, 1973.

[50] H. Brézis and F.E. Browder, Some new results about Hammerstein equations, *Bull. Amer. Math. Soc.*, 80 (1974), 567-572.

[51] H. Brézis, M.G. Crandall, and A. Pazy, Perturbations of nonlinear maximal monotone sets in Banach spaces, *Communic. Pure Appl. Math.*, 23 (1970), 123-144.

[52] F.E. Browder, Nonlinear elliptic boundary value problems. I, *Bull. Amer. Math. Soc.*, 69 (1963), 862-874.

[53] F.E. Browder, Nonlinear elliptic boundary value problems. II, *Trans. Amer. Math. Soc.*, 117 (1965), 530-550.

[54] F.E. Browder, Existence and approximation of solutions of nonlinear variational inequations, *Proc. Nat. Acad. Sci. USA*, 56 (1966), 1080-1086.

[55] F.E. Browder, Nonlinear maximal monotone operators in Banach spaces, *Math. Ann.*, 175 (1968), 89-113.

[56] F.E. Browder, The fixed point theory of multivalued mappings in topological vector spaces, *Math. Ann.*, 177 (1968), 283-301.

[57] F.E. Browder, *Nonlinear Operators and Nonlinear Equations of Evolution in Banach Spaces*, Providence, 1976.

[58] F.E. Browder and P. Hess, Nonlinear mappings of monotone type in Banach spaces, *J. Funct. Anal.*, 11 (1972), 251-294.

[59] F.E. Browder and B.An. Ton, Nonlinear functional equations in Banach spaces and elliptic superregularization, *Math. Z.*, 105 (1968), 177-195.

[60] F.E. Browder and B.An. Ton, Convergence of approximants by regularization for solutions of nonlinear functional equations in Banach spaces, *Math. Z.*, 106 (1968), 1-16.

[61] R.E. Bruck, Jr., Nonexpansive projections on subsets of Banach spaces, *Pacific. J. of Math.*, 47 (1973), 341-356.

[62] R.E. Bruck, Jr., A strongly convergent iterative solution of $0 \in U(x)$ for a maximal monotone operator U in Hilbert space, *J. Math. Anal. Appl.*, 48 (1974), 114-126.

[63] W.L. Bynim, Weak parallelogram lows for Banach spaces, *Can. Math. Bull.*, 19 (1976), 269-275.

[64] J. Céa, *Optimisation. Théorie et Algorithmes*, Dunod, Paris, 1971.

[65] A. Cernes, Ensembles maximaux accretive et m-accretifs, *Isr. J. Math.*, 19 (1974), 335-348.

[66] C.E. Chidume and H. Zegeye, Iterative approximation of solutions of nonlinear equations of Hammerstein type, *Abstr. Appl. Anal.*, 2003 (2003), 353-365.

[67] A. Corduneanu, Some remarks on the sum of two m-accretive mappings, *Rev. Roum. Math. Pures et Appl.*, 20 (1975), 411-414.

[68] S. Cruceanu, Regularization pour les problems a operateurs monotones et la methode de Galerkine, *Comment. Math. Univ. Carolinae*, 12 (1971), 1-13.

[69] M.M. Day, *Normed Linear Spaces*, Springer - Verlag, New York, 1973.

[70] H. Debrunner and P. Flor, Ein Erweiterungssatz fur monotone Mengen, *Arch. Math.*, 15 (1964), 445-447.

[71] J. Diestel, *The Geometry of Banach Spaces*, Lecture Notes Math., Vol. 485. Springer Verlag, New York - Berlin, 1975.

[72] X.P. Ding and E. Tarafgar, Monotone generalized variational inequalities and generalized complementary problems, *J. Optim. Theory and Appl.*, 88 (1996), 107-122.

[73] P. Doktor and M. Kucera, Perturbations of variational inequalities and rate convergence of solution, *Czech. Math. J.*, 30 (1980), 426-437.

[74] A.L. Dontchev and T. Zolezzi, *Well-Posed Optimization Problems*, Springer Verlag, Berlin, 1993.

[75] N. Dunford and J.T. Schwartz, *Linear Operators. Pt. 2: Spectral Theory: Self-adjoint Operators in Hilbert Space*, Interscience Publishers, New York, London, 1963.

[76] G. Duvaut and J.-L. Lions, *Inequalities in Mechanics and Physics*, Springer, Berlin, 1972.

[77] R.E. Edwards, *Functional Analysis: Theory and Applications*, New York, Holt, Rinehart and Winston, 1965.

[78] Yu.V. Egorov, Some problems in the theory of optimal control, *Dokl. Acad. Nauk SSSR*, 145 (1962), 720-723.

[79] I. Ekeland and R. Temam, *Convex Analysis and Variational Problems*, Studies in Mathematics and Its Applications, Amsterdam - New York, (1) 1976.

[80] H.W. Engl, K. Kunisch, and A. Neubauer, Convergence rates for Tikhonov regularization of non-linear ill-posed problems, *Inverse Problems*, 5 (1989), 523-540.

[81] T. Figiel, On the moduli of convexity and smoothness, *Studia Mathematica*, 56 (1976), 121-155.

[82] G.M. Fichtenholz, *The Fundamentals of Mathematical Analysis*, Vol. I,II, Oxford, Pergamon, 1965.

[83] X. Gaewskli, K. Greger, and K. Zacharias, *Nichtlineare Operatorgleihungen and Operatordifferentialgleihungen*, Mathematishe Monographien, Band 38, Academie-Verlag, Berlin, 1974.

[84] I.M. Gel'fand and S.V. Fomin, *Calculus of Variations*, Englewood Cliffs, N.J., Prentice-Hall, 1963.

[85] R.G. Glowinski, J.-L. Lions, and R. Tremolieres, *Analyse Numerique des Inequations Variationnelles*, Vol. 1, Dunod, Paris, 1976.

[86] D. Goeleven, On a class of hemivariational inequalities involving hemicontinuous monotone operators, *Numer. Funct. Anal. and Optim.*, 17 (1996), 77-92.

[87] K. Goebel and W.A. Kirk, *Topics in Metric Fixed Point Theory*, Cambridge studies in advanced mathematics, Vol. 28, Cambridge University Press, 1990.

[88] K. Goebel and S. Reich, *Uniform Convexity, Hyperbolic Geometry and Nonexpansive Mappings*, Marcel Dekker, New York and Basel, 1984.

[89] J.-P. Gossez, Operateurs monotones non lineares dans les espaces de Banach non reflexifs, *J. Math. Anal. and Appl.*, 34 (1971), 371-395.

[90] E.I. Grigolyuk and V.M. Tolkachev, *Contact Problems of Plates and Shells*, Mashinostroenie, Moscow, 1980.

[91] J. Hadamard, *Le probléme de Cauchy et les équations aux dérivées partielles hyperboliques*, Paris, Hermann, 1932.

[92] B. Halpern, Fixed points of nonexpanding maps, *Bull. Amer. Math. Soc.*, 73 (1967), 957-961.

[93] M. Hanke, A. Neubauer, and O. Scherzer, A convergence analysis of the Landweber iteration for nonlinear ill-posed problems, *Numer. Math.*, 72 (1995), 21-37.

[94] O. Hanner, On the uniform convexity of L^p and l^p, *Ark. Math.*, 3 (1956), 239-244.

[95] E. Hille and R.S. Phillips, *Functional Analysis and Semi-groups*, Providence, R.I., Amer. Math. Soc., 1957.

[96] P.R. Holmes, *Geometric Functional Analysis*, Springer Verlag, New York, 1975.

[97] V.K. Ivanov, On linear problem with are not well-posed, *Dokl. Acad. Nauk SSSR*, 145 (1962), 270-272.

[98] V.K. Ivanov, On ill-posed problems, *Mat. Sb. (N. S.)* , 61 (1963), 211-223.

[99] V.K. Ivanov, V.V. Vasin, and V.P. Tanana, *Theory of Ill-posed Linear Problems and its Applications*, Nauka, Moskow, 1978.

[100] Chin-Rong Jou and Jen-Chih Yao, Extension of generalized multy-valued variational inequalities, *Appl. Math. Lett.*, 6 (1993), 21-25.

[101] L.M. Kachanov, *Foundations of the Theory of Plasticity*, North-Holland Publ. Co., Amsterdam, 1971.

[102] R.I. Kachurovskii, On monotone operators and convex functionals, *Uspekhi Mat. Nauk*, 15 (1960), 213-215.

[103] R.I. Kachurovskii, Nonlinear monotone operators in Banach spaces, *Uspekhi Mat. Nauk*, 23 (1968), 121-168.

[104] L.V. Kantorovich and G.P. Akilov, *Functional Analysis in normed spaces*, Pergamon Press, New York - London, 1964.

[105] A. Kaplan and R. Tichatschke, *Stable Methods for Ill-posed Variational Problems: Prox-Regularization of Elliptic Variational Inequalities and Semi-infinite Problems*, Akademie Verlag, Berlin, 1994.

[106] T. Kato, Demicontinuity, hemicontinuity and monotonicity, *Bull. Amer. Math. Soc.*, 70 (1964), 548-550.

[107] T. Kato, *Perturbation Theory for Linear Operators*, Springer, Berlin, 1966.

[108] T. Kato, Nonlinear semigroups and evalution equation, *J. Math. Soc. Japan*, 19 (1967), 508-520.

[109] T. Kato, Accretive operators and nonlinear evolution equations in Banach spaces, *Proc. Symp. Pure. Math.*, 13 (1970), 133-161.

[110] N. Kennmochi, Accretive mappings in Banach spaces, *Hirishima Math. J.*, 2 (1972), 163-177.

[111] D. Kinderlehrer and G. Stampacchia, *An Introduction to Variational Inequalities and Their Applications*, Academic Press, New York - London - Toronto, 1980.

[112] V. Klee, Convex bodies and periodic homeomorphisms in Hilbert space, *Trans. Amer. Math. Soc.*, 74 (1953), 10-43.

[113] R. Kluge, *Nichtlinear Variationsungleihungen und Extremalaufgaben. Theory and Naherungsverfahren*, Verl. der Wiss., Berlin, 1979.

[114] M.Yu. Kokurin, On the use of regularization for correcting monotone variational inequalities that are given approximately,*Izv. VUZov. Matematika*, 2 (1992), 49-56.

[115] M.Yu. Kokurin, A method for the operator regularization of equations of the first kind that minimize the residual, *Izv. VUZov. Matematika*, 12 (1993), 59-69.

[116] M.Yu. Kokurin, On the regularization of problems of the optimal control of solutions of some ill-posed variational inequalities of monotone type, *Siberian Math. J.*, 38 (1997), 84-91.

[117] A.N. Kolmogorov and S.V. Fomin, *Elements of the Theory of Functions and Functional Analysis, Vol. II*, Academic Press, New York, 1961.

[118] Ya.V. Konstantinova and O. A. Liskovets, Regularization of equation with arbitrarily perturbed accretive operators, *Dokl. Acad. Nauk BSSR*, 23 (1983), 680-683.

[119] M.A. Krasnosel'skii and Ya.B. Rutickii, *Convex Functions and Orlicz Spaces*, Groningen, the Netherlands, Noordhoff, 1961.

[120] A.S. Kravchyk, *Variational and Quasi-variational Inequalities in Mechanics*, MGAPI, Moscow, 1997.

[121] K. Kunisch and W. Ring, Regularization of nonlinear ill-posed problems with closed operators, *Numer. Funct. Anal. and Optimiz.*, 14 (1993), 389-404.

[122] A.G. Kurosh, *Course of General Algebra*, Moscow, 1955.

[123] A.V. Lapin, An investigation of some nonlinear problems of filtration theory, *Zh. Vychisl. Mat. i Mat. Fiz.*, 19 (1979), 689-700.

[124] R. Lattes and J.-L. Lions, *Méthode de Quasi-réversibilité et Applications*, Dunod, Paris, 1967.

[125] M.M. Lavrent'ev, *Some Ill-posed Problems of Mathemitical Physics*, Nauka, Novosibirsk, 1962.

[126] A.S. Leonov, Optimality with respect to the order of accuracy of the generalized principle of the residual and of some other algorithms for the solution of nonlinear ill-posed problems with approximate data, *Siberian Math. J.*, 29 (1988), 940-947.

[127] J. Lindenstrauss and L. Tzafriri, *Classical Banach Spaces II*, Springer Verlag, Berlin-Heidelberg-New York, 1979.

[128] J.-L. Lions, *Quelques methodes de resolution des problems aux limites non linéaires*, Dunod, Paris, 1969.

[129] O.A. Liskovets, The connection of the principle of the residual with the regularization method, *Vesci. AN BSSR, Ser. Fiz.-Mat. Navuk*, 3 (1972), 30-34.

[130] O.A. Liskovets, *Variational methods for the solution of unstable problems*, Nauka i Tekhnika, Minsk, 1981.

[131] O.A. Liskovets, *Theory and methods of solving ill-posed problems, Mathematical analysis*, 20 (1982), 116-178, Itogi Nauki i Tekhniki, VINITI, Moskow.

[132] O.A. Liskovets, Solution of equations of the first kind with monotone operator under nonmonotone perturbations, *Dokl. AN BSSR*, 27 (1983), 101-104.

[133] O.A. Liskovets, Regularization of problems with discontinuous monotone, arbitrarily perturbed operators, *Soviet Math. Dokl.*, 28 (1983), 324-327.

[134] O.A. Liskovets, Finite-dimensional projection regularization of ill-posed problems with monotone operators. I , Monotone approximating operators, Institut Mat. Akad. Nauk BSSR, Preprint No. 15 (172), 1984.

[135] O.A. Liskovets, Finite-dimensional projection regularization of ill-posed problems with monotone operators. II , Arbitrary approximating operators, Institut Mat. Akad. Nauk BSSR, Preprint No. 20 (205), 1984.

[136] O.A. Liskovets, Finite-dimensional discrete regularization for integral Hammerstein equation of the first kind with monotone operators, In: "Differential Equations", Minsk, 1985, 1-33.

[137] O.A. Liskovets, Regularization of variational inequalities with pseudomonotone operators an approximately defined domains, *Differential Equations*, 25 (1989), 1970-1977.

[138] O.A. Liskovets, Regularization of ill-posed variational inequalities on approximately given sets, In: "Differential Equations", Minsk, 1991, 1-53.

[139] O.A. Liskovets, Regularization of ill-posed mixed variational inequalities, *Soviet Math. Dokl.*, 43 (1991), 384-387.

[140] F. Liu and M.Z. Nashed, Regularization of nonlinear ill-posed variational inequalitie s and convergence rates, *Set-Valued Anal.*, 6 (1998), 113-344.

[141] L. A. Liusternik and V. I. Sobolev, Elements of Functional Analysis, Frederick Ungar Publishing Company, New York, 1961.

[142] A.D. Ljashko, I.B. Badriev, and M.M. Karchevskii, The variational method for equations with monotone discontinuous operators, *Izv. VUZov. Matematika,* 11 (1978), 63-69.

[143] A.D. Ljashko and M.M. Karchevskii, Difference methods for solving nonlinear problems of filtration theory, *Izv. VUZov. Matematika,* 7 (1983), 28-45.

[144] V.P. Maslov, The existence of a solution of an ill-posed problem is equivalent to the convergence of a regularization process, *Uspehi Mat. Nauk,* 23 (1968), 183-184.

[145] K. Maurin, Methods of Hilbert Spaces, P.W.N., Warszawa, 1972.

[146] S. Mazur, Über konvexe Mengen in linearen normierten Räumen, *Studia Mathematica,* 4 (1933), 70-84.

[147] S.G. Mikhlin, *The Numerical Performance of Variational Methods,* Wolters-Noordhoff Publishing Groningen, The Netherlands, 1971.

[148] G.J. Minty, Monotone (nonlinear) operators in Hilbert space, *Duke Math. J.,* 29 (1962), 341-346.

[149] G.J. Minty, On a "monotonicity" method for the solution of nonlinear equations in Banach spaces, *Proc. Nat. Acad. Sci.,* 50 (1963), 1038-1041.

[150] G.J. Minty, On the monotonicity of the gradient of a convex function, *Pacific J. Math.,* 14 (1964),243-247.

[151] V.A. Morozov, Pseudosolutions, *Zh. Vychisl. Mat. i Mat. Fiz.,* 9 (1969), 1387-1391.

[152] V.A. Morozov, Linear and nonlinear ill-posed problems, Mathematical analysis. Vol 11, 112-178, Acad. Nauk SSSR, VINITI, 1973.

[153] V.A. Morozov, *Regularization Methods of Unstable Problems,* Izd. Moskov. Univ., Moskow, 1987.

[154] U. Mosco, Convergence of convex sets and of solutions of variational inequations, *Advances in Math.,* 3 (1969), 510-585.

[155] M.Z. Nashed and F. Liu, On nonlinear ill-posed problems II: Monotone operator equations and monotone variational inequalities, *Theory and Applications of Nonlinear Operators of Accretive and Monotone Type (A. Kartsatos, Ed.)* Marcel Dekker, inc., 1996, 223-240.

[156] A. Neubauer, Tikhonov regularization for nonlinear ill-posed problems: optimal convergence rates and finite-dimensional appraximation, *Inverse Problems,* 5 (1989), 541-557.

[157] Nguen Byong, Approximate solutions of an equation Hammerstein type in Banach spaces, *USSR Comput. Math. and Math. Phys.*, 25 (1985), 1256-1260.

[158] Nguen Byong, Solutions of the Hammerstein equation in Banach spaces, *Ukrainian Math. J.*, 37 (1985), 159-162.

[159] L. Nirenberg, *Topics in Nonlinear Functional Analysis*, Courant Institute of Mathematical Sciences, New York, 1974.

[160] A.I. Notik, Properties of a duality mapping with a scale function, *Soviet Math.*, 29 (1985), 96-98.

[161] Z. Opial, Weak convergence of the sequence of successive approximations for nonexpansive mappings, *Bull. Amer. Math. Sos.*, 73 (1967), 591-597.

[162] D. Pascali and S. Sburlan, *Nonlinear Operators of Monotone Type*, R.S.R., Bucuresti,1978.

[163] V.N. Pavlenko, Existence theorems for elliptic variational inequalities with quasipotential operators, *Differential Equations*, 24 (1988), 913-916.

[164] V.N. Pavlenko, On the solvability of variational inequalities with discontinuous semi-monotone operators, *Ukrainian Math. J.*, 45 (1993), 475-480.

[165] A.I. Perov and Yu.V. Trubnikov, *Differential Equations with Monotone Nonlinearities*, Nauka i Tekhnika, Minsk, 1986.

[166] R.R. Phelps, *Convex Functions, Monotone operators and Differentiability*, Springer Verlag, 2nd Edition, Berlin, 1993.

[167] B.L. Phillips, A technique for the numerical solution of certain equations of the first kind, *J. ACM*, 9 (1962), 84-97.

[168] G. Pisier, Martingales with values in uniformly convex Banach spaces, *Isr. J. Math.*, 20 (1975),326-350.

[169] M. Reeken, General theorem on bifurcation and its applications to the Hartree equations of the Helium atom, *J. Math. Phys.*, 112 (1902), 2502-2512.

[170] S. Reich, Approximating zeros of accretive operators, *Proc. Amer. Math. Soc.*, 51 (1972), 381-384.

[171] S. Reich, Extension problems for accretive sets in Banach spaces, *J. Funct. Anal.*, 26 (1977), 378-395.

[172] S. Reich, Approximating fixed points of nonexpansive mappings, *Panamer. Math. J.*, 4 (1994), 23-28.

[173] F. Riesz, Uber lineare Funktionalgleichungen, *Acta Math.*, 41 (1918), 71-98.

[174] F. Riesz and B. Sz.-Nagy, *Functional Analysis,* Frederick Ungar Pablishing Co., New York, 1955.

[175] R.T. Rockafellar, Characterization of the subdufferentials of convex functions, *Pacific. J. Math.,* 17 (1966), 497-510.

[176] R.T. Rockafellar, Convexity properties of nonlinear maximal monotone operators, *Bull. Amer. Math. Soc.,* 75 (1969), 74-77.

[177] R.T. Rockafellar, Local boundedness of nonlinear monotone operators, *Michigan Math. J.,* 16 (1969), 397-407.

[178] R.T. Rockafellar, On the maximal monotonicity of subdifferential mappings, *Pacific. J. Math.,* 33 (1970), 209-216.

[179] R.T. Rockafellar, On the maximality of sums of nonlinear monotone operators, *Trans. Amer. Math. Soc.,* 149 (1970), 75-88.

[180] R.T. Rockafellar, Monotone operators and augmented Lagrangina methods in nonlinear programming, *Nonl. Programming,* 3 (1972), 1-25.

[181] R.T. Rockafellar, Monotone operators and the proximal point algorithm, *SIAM J. Contr. and Optim.,* 14 (1976), 877-898.

[182] W. Rudin, *Functional analysis,* McGraw-Hill, New York, 1973.

[183] I.P. Ryazantseva, The solution of nonlinear equations with discontinuous monotone operators, *Siberian Math. J.,* 20 (1979), 144-147.

[184] I.P. Ryazantseva, Regularization of equations with accretive operators by the method of successive approximations, *Siberian Math. J.,* 21 (1980), 223-226.

[185] I.P. Ryazantseva, On equations with semimonotone discontinuous mappings, *Mat. Zametki,* 30 (1981), 143-152.

[186] I.P. Ryazantseva, Value computation of semimonotone unbounded operator, *Siberian Math. J.,* 1981, VINITI, N3277-81, 1-11.

[187] I.P. Ryazantseva, The choice of the regularization parameter for nonlinear equations with an approximately specified monotone operator, *Soviet Math.,* 29 (1982), 65-70.

[188] I.P. Ryazantseva, The principle of the residual for nonlinear problems with monotone operators, *Differential Equations,* 19 (1983), 1079-1080.

[189] I.P. Ryazantseva, Solution of variational inequalities with monotone operators by the regularization method, *USSR Comput. Math. and Math. Phys.,* 23 (1983), 479-483.

[190] I.P. Ryazantseva, Variational inequalities with monotone operators on the approximately given sets, *USSR Comput. Math. and Math. Phys.,* 24 (1984), 932-936.

[191] I.P. Ryazantseva, Nonlinear operator equations with accretive mappings. *Soviet Math.*, 29 (1985), 52-57.

[192] I.P. Ryazantseva, The choice of the regularization parameter in the solution of nonlinear problems with monotone operators. *Izv. VUZov. Matematika*, 4 (1985), 55-57.

[193] I.P. Ryazantseva, The quasioptimal choice of the regularization parameter in the solution of nonlinear equations with monotone operators, *USSR Comput. Math. and Math. Phys.*, 26 (1986), 1731-1735.

[194] I.P. Ryazantseva, Iterative methods of the Newton - Kantorovich type for solving nonlinear ill-posed problems with monotone operators, *Differential Equations*, 23 (1987), 2012-2014.

[195] I.P. Ryazantseva, Residual method in nonlinear monotone problems, *Izv. VUZov. Matematika*, 1987, VINITI, N7550-B87, 1-14.

[196] I.P. Ryazantseva, The stable method to determine the construction of pseudosolutions of nonlinear equations with monotone operators, *Differential Equations*, 25 (1989), 1457-1459.

[197] I.P. Ryazantseva, The algorithm of the solution of nonlinear monotone equations with an unknown estimate of the initial data error, *USSR Comput. Math. and Math. Phys.*, 29 (1989), 1572-1576.

[198] I.P. Ryazantseva, The minimal residual principle in nonlinear monotone problems, *USSR Comput. Math. and Math. Phys.*, 31 (1991), 99-103.

[199] I.P. Ryazantseva, The Tikhonov method in nonlinear monotone problems, *USSR Comput. Math. and Math. Phys.*, 32 (1992), 1189-1190.

[200] I.P. Ryazantseva, Some continuous regularization methods for monotone equations, *USSR Comput. Math. and Math. Phys.*, 34 (1994), 1-7.

[201] I.P. Ryazantseva, The operator method of regularization of nonlinear monotone ill-posed problems, *Ill-Posed Problems in Natural Sciences. Proc. Intern. Conf.*, Moscow, 1992, 149-154.

[202] I.P. Ryazantseva, *Stable methods of the solutions of nonlinear monotone ill-posed problems*, D.Sc. Thesis, Nizhnii Novgorod, 1996.

[203] I.P. Ryazantseva, Solvability of variational inequalities with unbounded semimonotone operators, *Izv. VUZov. Matematika*, 7 (1999), 49-53.

[204] I.P. Ryazantseva, Regularized proximal algorithm for nonlinear equations of monotone type in a Banach space, *USSR Comput. Math. and Math. Phys.*, 42 (2002), 1247-1255.

[205] H. Schaefer, *Topological Vector Spaces*. The MacMillan Company, New York, London, 1966.

[206] L. Schwartz, *Courx d'analyse*, Hermann, Paris, 1967.

[207] S. Serb, Some estimates for the modulus of smoothness and convexity of a Banach space, *IMathematica(Cluj)*, No. 1 (1992), 61-70.

[208] S.S. Sim and M.G. Kim, Existence and convergence of regular solution of ill-posed nonlinear operator equation, *Cyxak Math.*, 4 (1987), 20-25.

[209] V.P. Šmulian, Sur la structure de la sphere unitaire dans l'espace de Banach, *Math. Sbornik*, 9 (1941), 545-561.

[210] S.L. Sobolev, *Applications of Functional Analysis in Mathematical Physics*, Providence, R.I., AMS, 1963.

[211] S.L. Sobolev and V.L. Vaskevich, *The Theory of Cubature Formulas*, Kluwer Academic Publishers, Dordrecht - Boston - London, 1997.

[212] J.E. Spingar, Submonotone mappings and the proximal point algorithm, *Numer. Funct. Anal. and Optim.*, 4 (1981-1982), 123-150.

[213] I.V. Sragin, Conditions for the measurability of superpositions, *Soviet Math. Dokl.*, 12 (1971), 465-470.

[214] W. Takahashi and G.-E. Kim, Strong convergence of approximants to fixed points of nonexpansive nonself-mappings in Banach spaces, *Nonlinear Anal.*, 32 (1998), 447-454.

[215] V.P. Tanana, V.A. Korshunov and A.A. Shtarkman, Principle of minimal residuals for solving ill-posed problems, *Studing on Functional Analysis*, USU, 1978, 99-104.

[216] A.N. Tikhonov, On the stability of inverse problems, *Acad. Sci. URSS*, 39 (1943), 176-179.

[217] A.N. Tikhonov and V.Ya. Arsenin, *Solutions of Ill-posed Problems*, Wiley, New York, 1977.

[218] A.N. Tikhonov, A.S. Leonov, and A.G. Yagola, *Nonlinear Ill-posed Problems*, Nauka, Moscow, 1995.

[219] M.M. Vainberg, On the convergence of the process of steepest descent for nonlinear equations, *Siberian Math. J.*, 2 (1961), 201-220.

[220] M.M. Vainberg, *Variational Methods for the Study of Nonlinear Operators*, Holden-Day, San Francisco, 1964.

[221] M.M. Vainberg, *Variational Methods and Method of Monotone Operators*, Wiley, New York, 1973.

[222] G. Vainikko, Error estimates of the successive approximation method for ill-posed problems, *Automat. Renome Control.*, 41 (1980), p.1, 356-363.

[223] G. Vainikko, Error bounds in regularization methods for normally solvable problems, *Zh. Vychisl. Mat. i Mat. Fiz.*, 12 (1972), 481-483.

[224] F.P. Vasil'ev, *Numerical Methods for Solving Extremal Problems*, Nauka, Moskow, 1980.

[225] F.P. Vasil'ev, *Methods for Solving Extremal Problems*, Nauka, Moskow, 1981.

[226] F.P. Vasil'ev, An estimate for the rate of convergence of A.N.Tikhonov's regularization method for nonstable minimization problems, *Soviet Math. Dokl.*, 37 (1988), 452-455.

[227] F.P. Vasil'ev and O. Obradovich, A regularized proximal method for minimization problems, *USSR Comput. Math. and Math. Phys.*, 33 (1993), 157-164.

[228] V.V. Vladimirov, Yu.E. Nesterov, and Yu.N. Chekanov, Uniformly convex functionals, *Vestnik Moskov. Univ. Ser. 15*, 3 (1978), 12-23.

[229] M.T. Wasan, *Stochastic Approximation*. Cambridge at the University Press, 1969.

[230] J. Weyer, Liklische monotone lins nichtlinearen Operators and Symmetric sowie Selbstungierkeit seiner Linearisierung, Koln, Diplomarbeit, 1974.

[231] J. Weyer, Maximal monotonicity of operators with sufficiently large domain and application on the Hartree problem, *Manuscripta Math.*, 38 (1982), 163-174.

[232] F. Wille, Galerkins Lozungsnaherungen bei monotone Abbildungen, *Math. J.*, 127 (1972), 10-16.

[233] Xu Zong-Ben and G.F. Roach, Characteristic inequalities of uniformly convex and uniformly smooth Banach spaces, *J. Math. Anal. Appl.*, 157 (1991), 189-210.

[234] Chi-Lin Yen, The range of m-dissipative sets, *Bull. Amer. Math. Soc.*, 78 (1972), 197-199.

[235] K. Yosida, *Functional Analysis*, Springer, Berlin, 1971.

[236] V.V. Yurgelas, Methods of approximate solving equations with monotone operators, PhD Thesis, Voronezh State University, 1983.

[237] E. Zeidler, *Nonlinear Functional Analysis and Applications IIB: Nonlinear Monotone Operators*, Springer Verlag, New York, 1990.

[238] E. Zeidler, *Applied Functional Analysis. Main Principles and Their Applications*, Springer Verlag, New York, 1995.

INDEX

Printed in the United States
57308LVS00001B/43-48